Handbook of Occupational Safety and Health

Human Factors and Ergonomics

Series Editor

Gavriel Salvendy

Professor Emeritus

School of Industrial Engineering

Purdue University

Chair Professor & Head

Dept. of Industrial Engineering

Tsinghua Univ., P.R. China

Published Titles

Conceptual Foundations of Human Factors Measurement, *D. Meister*

Content Preparation Guidelines for the Web and Information Appliances: Cross-Cultural Comparisons, *H. Liao, Y. Guo, A. Savoy, and G. Salvendy*

Designing for Accessibility: A Business Guide to Countering Design Exclusion, *S. Keates*

Handbook of Cognitive Task Design, *E. Hollnagel*

The Handbook of Data Mining, *N. Ye*

Handbook of Digital Human Modeling: Research for Applied Ergonomics and Human Factors Engineering, *V. G. Duffy*

Handbook of Human Factors and Ergonomics in Health Care and Patient Safety, *P. Carayon*

Handbook of Human Factors in Web Design, *R. Proctor and K. Vu*

Handbook of Occupational Safety and Health, *D. Koradecka*

Handbook of Standards and Guidelines in Ergonomics and Human Factors, *W. Karwowski*

Handbook of Virtual Environments: Design, Implementation, and Applications, *K. Stanney*

Handbook of Warnings, *M. Wogalter*

Human-Computer Interaction: Designing for Diverse Users and Domains, *A. Sears and J. A. Jacko*

Human-Computer Interaction: Design Issues, Solutions, and Applications, *A. Sears and J. A. Jacko*

Human-Computer Interaction: Development Process, *A. Sears and J. A. Jacko*

The Human-Computer Interaction Handbook: Fundamentals, Evolving Technologies, and Emerging Applications, Second Edition, *A. Sears and J. A. Jacko*

Human Factors in System Design, Development, and Testing, *D. Meister and T. Enderwick*

Introduction to Human Factors and Ergonomics for Engineers, *M. R. Lehto and J. R. Buck*

Macroergonomics: Theory, Methods and Applications, *H. Hendrick and B. Kleiner*

Smart Clothing: Technology and Applications, *Gilsoo Cho*

Theories and Practice in Interaction Design, *S. Bagnara and G. Crampton-Smith*

The Universal Access Handbook, *C. Stephanidis*

Usability and Internationalization of Information Technology, *N. Aykin*

User Interfaces for All: Concepts, Methods, and Tools, *C. Stephanidis*

Forthcoming Titles

Computer-Aided Anthropometry for Research and Design, *K. M. Robinette*

Foundations of Human-Computer and Human-Machine Systems, *G. Johannsen*

Handbook of Human Factors in Web Design, Second Edition, *K. Vu and R. Proctor*

Human Performance Modeling: Design for Applications in Human Factors and Ergonomics, *D. L. Fisher, R. Schweickert, and C. G. Drury*

Practical Speech User Interface Design, *James R. Lewis*

Handbook of Occupational Safety and Health

Edited by

Danuta Koradecka

CRC Press
Taylor & Francis Group
Boca Raton London New York

CRC Press is an imprint of the
Taylor & Francis Group, an **informa** business

CRC Press
Taylor & Francis Group
6000 Broken Sound Parkway NW, Suite 300
Boca Raton, FL 33487-2742

Printed in the United States of America on acid-free paper
10 9 8 7 6 5 4 3 2 1

International Standard Book Number: 978-1-4398-0684-5 (Hardback)

Library of Congress Cataloging-in-Publication Data

Handbook of occupational safety and health / editor, Danuta Koradecka.
p. cm. -- (Human factors and ergonomics)
Includes bibliographical references and index.
ISBN 978-1-4398-0684-5 (alk. paper)
1. Industrial hygiene. 2. Industrial safety. I. Koradecka, D. (Danuta) II. Title. III. Series.

RC967.H262 2009
613.6'2--dc22 2009037860

Visit the Taylor & Francis Web site at
http://www.taylorandfrancis.com

and the CRC Press Web site at
http://www.crcpress.com

Contents

PART IV The Effects of Hazards on Work Processes

PART V Basic Directions for Shaping Occupational Safety and Ergonomics

Preface

Occupational safety and health have considerable value for the employee and employer alike. As work processes become more flexible, this branch of knowledge becomes more important for society as a whole. This knowledge is both fascinating and complex, encompassing achievements in the technical, biological and social sciences fields, which have experienced rapid growth during the past decade. Practical use of this body of knowledge—due to globalization of production and deregulation of labour markets—should be similar among individual countries.

Poland, a member state of the European Community since 2004, has harmonised its regulations and practice with the EU's required standards for occupational safety and health. This process covered the entirety of working conditions with a goal of preventing occupational accidents and diseases and satisfying ergonomic requirements.

Modern companies must create working conditions that are not only safe and maintain health and life but are also optimal for the needs and psychosocial capacities of workers. Hence, this manual also will be interesting for readers outside Poland.

Ultimately, it is the human being, with his or her limited psychophysical capacities, who should be of the utmost importance.

Professor Danuta Koradecka, PhD, D.Med.Sc.

Director of the Central Institute for Labour Protection–National Research Institute

Introduction

Occupational Safety and Health: From the Past, through the Present, and into the Future

Danuta Koradecka

Post-twentieth century society is convinced of the unique position of our civilisation, and we are proud of the scientific and technical progress in shaping work processes. At the same time, we are amazed at the discoveries of work processes solutions and products from many centuries or even millennia ago, such as the ergonomic handles of axes or stone tools, the aqueducts in Rome or Istanbul that we still admire, and the way the mighty pyramids were built. People remain somewhat in the background of these achievements, although building the magnificent structures of Egypt, China or Persia took the lives of tens of thousands. Skeletons from those times reveal pathological changes associated with the work people did, for example, simple tasks in the Neolithic period (about 3000 years ago), when human societies shifted from hunter–gatherer to farming civilisations. During a 1972–1973 archaeological excavation in Aber-Hureyra (today's northern Syria), Andrew M.T. Moore found the remains of 162 people from two settlements. An analysis of the women's bones showed work-related changes (Molleson 1994). Many hours of daily monotonous work, for example, grinding grain using a saddle quern-stone, were performed in a kneeling and flexed posture. This led to significant degenerative changes in the lumbar spine (as a result of flexion while the body is bent forward), knee joints (caused by the pressure of the ground) and big toes (as a result of hyperextension while kneeling). Carrying loads on their heads led to changes in the first cervical vertebra—the addition of lateral processes of the vertebrae to stabilise the position of the neck. Changes caused by daily forced postures over a long period of time gradually resulted in degenerative changes in other organs; these can be considered work-related pathological changes.

These changes intensified with a decrease in the egalitarianism of communities—people began to specialise in specific tasks in order to increase the quantity of goods produced and the associated income. People who performed the same type of work all the time reached a high level of excellence in that work; however, the price was often high, with the work resulting in deterioration of health or even death (Chapanis 1951). These hazards did not disappear with industrialisation; their types simply changed. Excessive dynamic physical workload was replaced with static workload, excess of signal stimuli (Paluszkiewicz 1975), noise and chemical hazards and, later, radiation. Automation, introduced thanks to technical progress, has resulted in monotonous work tasks and mental processes, which are dangerous for the musculoskeletal system (Rahimi and Karwowski 1992).

In the past decade, we have been experiencing another revolution in workstations and work processes (Ozok and Salvendy 1996). Computerisation, while increasing the possibilities for controlling and carrying out work processes, has made work even more monotonous and has increased the eye strain and static workload associated with a forced sitting posture (Strasser 2007). Computerisation has also increased the overload of some muscle groups (Christensen 1960; Dul and Hildebrandt 1987; Grandjean and Hunting 1977; Kidd and Karwowski 1994). Occupational risk is also associated with biological factors. Biotechnologies are yet another challenge for humankind. All of these hazards and cases of strain are inherently accompanied by stress, which is universal among workers who are striving to be the best in order to maintain their position at work or even just to keep their jobs (European Agency for Safety and Health at Work 2002). When stress is too great, workers may become passive and escape into alcohol or the world of 'wonder' pills. Thus, substantial technical progress has not solved the problems of occupational safety and health, but has only shifted the core of the problems from chemical and physical hazards to psychophysical and biological ones. Labour protection—like art in the Renaissance—must now focus on people with limited psychophysical abilities in the workplace.

Workers' abilities are limited due to the requirements of homeostasis, that is, the need to maintain a constant internal environment of parameters such as the internal temperature or pH of the blood. These parameters must be at a constant level in order for biochemical and enzymatic processes, which are necessary for health and life (Figure 1), to occur. In the living environment, and especially in the work environment, humans are exposed to extreme levels of factors such as temperature (from –20°C to + 70°C) and noise (up to 140 dB). In the course of phylogenetic development, our bodies have developed mechanisms to prevent an imbalance in the internal environment by physiological processes such as increasing heart rate, breath rate and sweating and changing the placenta of the peripheral blood vessels (Astrand and Rodahl 1977). These mechanisms, however, have a limited ability to compensate for harmful factors in the work environment (Koradecka 1982). Moreover, long-term involvement of these mechanisms results in a substantial increase in the physical work capacity (Brouha 1962; Lehmann 1962), which, in turn, leads to chronic fatigue.

These processes influence the development of *occupational diseases* (Ramazzini 2009), defined as diseases associated with exposure to harmful work conditions. *Paraoccupational diseases* are those associated indirectly with work conditions (the so-called civilisation diseases such as hypertension, obesity, diabetes) and are often rooted indirectly in unsuitable work and living conditions. We often assume that work conditions may constitute a 'trigger mechanism', which increases the onset of diseases to which the human body has a genetic predisposition and which would not have developed under different conditions (e.g., carcinogenic diseases).

We tend to perceive the conditions of work and life of humans from a broader perspective because of these factors. This is consistent with the definition provided in the Constitution of the World Health Organization, which states that 'health is the state of complete physical, mental and social well-being and not merely the absence of disease or infirmity' (Stellman 1998). In our efforts to meet the requirements resulting from such a perception of health in the work environment, ergonomics brings us closer to the objective (Stanton 2005). *Ergonomics* is defined as an adaptation of

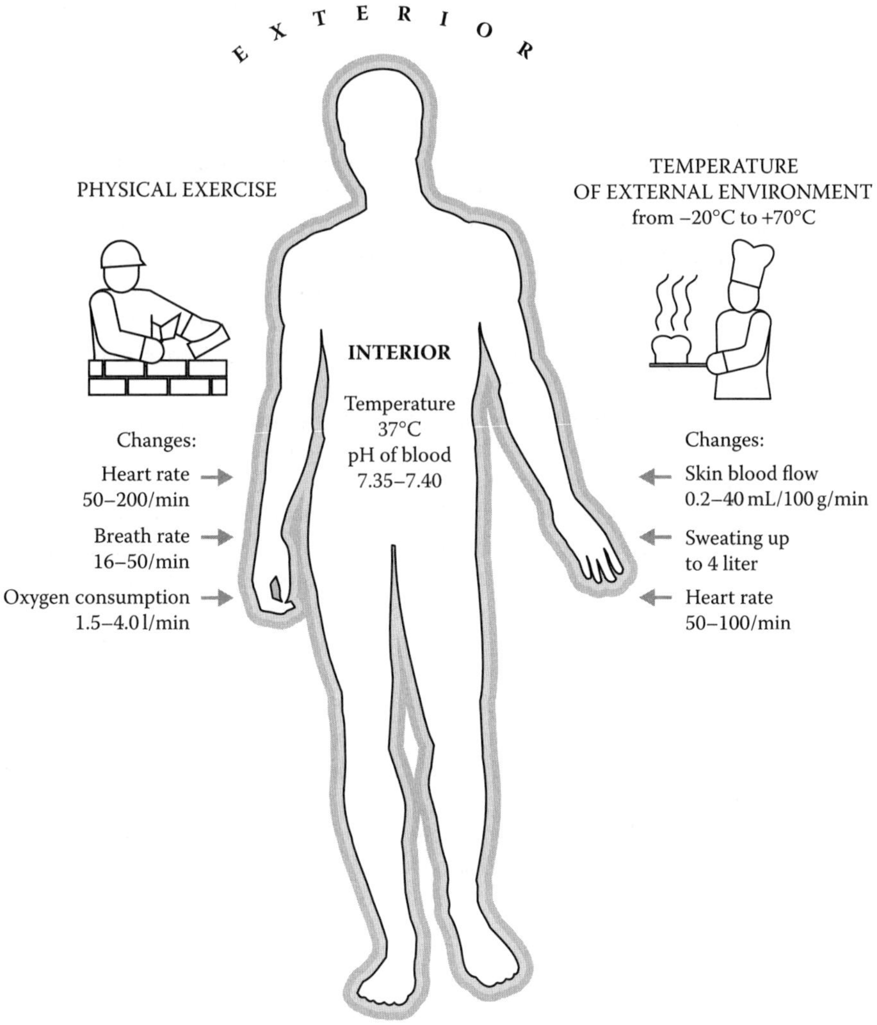

FIGURE 1 Response of an organism to maintain homeostasis.

workstations, work processes and the work environment to the psychophysical abilities of the human body.

The concept of ergonomics has its own history and methodological bases (Franus 1991). The term 'ergonomics' was first used by Wojciech Jastrzębowski in 1857 in his treatise *An Outline of Ergonomics, or the Science of Work, Based upon the Truths Drawn from the Science of Nature*. Here we find many thoughts consistent with modern knowledge on modifying work conditions (Jastrzębowski 2004). I would like to show the timeless nature of the problems of shaping work conditions by commenting on a quotation from the treatise. Jastrzębowski writes: '... for it is well-known that our vital *forces* grow weak and impoverished as much by the *lack* of their exercise as by their *abuse*; and they are maintained in their proper condition, growing and increasing by their proper and moderated exercise, which we call work.' This quotation illustrates the very popular—and often disregarded—principle of planning work processes in a way that reduces excessive effort and monotonous work tasks.

Wojciech Jastrzębowski defines ergonomics as follows: 'By the term Ergonomics, derived from Greek word *ergon* (εργον)—*work*, and *nomos* (νομος)—*principle* or *law*, we mean the Science of Work, that is the use of Man's forces or faculties with which he has been endowed by his Maker.' He also praises training and education: 'The second chief advantage which we draw from work is that through it we acquire the skill to perform work itself more and more easily and with an ever-growing satisfaction, accuracy, and liking for it. In other words that we can are able to undertake work at the expense of a lesser and lesser amount of toil and drudgery, but to the ever-increasing gain of ourselves and the common good.'

Jastrzębowski also mentions the need to develop one's personality through work, which is emphasised so often these days: '*Perfection* on the other hand, the advantage now under discussion, is always seen as one of our inner properties, a thing strictly connected with us and a direct consequence of *Ability* (…) Apart from their absolute value, by which our being is endowed with a similar value, these Perfections also have a relative value, which concerns the objectives of our active, improving and productive life.' Thus, we have made a full circle in the causes and effects of actions, going back to the term 'perfection called health', which does not differ substantially from the World Health Organization's definition of health. At present, ergonomics aims to optimise the adaptation of workstations, processes and the work environment to the psychophysical abilities of humans, not only to protect human life and health, but also to provide humans with an opportunity to maximally develop their personality (Kim 2001).

Questions often arise about the relationship between ergonomics and occupational safety. The simplest answer is that occupational safety protects the workers' life, whereas ergonomics protects the workers' health (Karwowski 2006).

Another term very close to the concept of *occupational safety and ergonomics* is the concept of *occupational safety and health*, used often in legislation. Ensuring occupational health means shaping work conditions and the work environment in a manner that ensures health protection (Alli 2001). This includes a full range of physical, biological and chemical factors. Shaping the psychophysical climate at work is important, in addition to being able to participate in planning tasks and available support—everything that makes up the beautiful, traditional concept of *well-being*.

To sum up the analysis of these definitions, the somewhat artificial division between occupational safety, ergonomics, and occupational health is not very significant from the perspective of a practitioner. In fact, the logical sequence of tasks undertaken to protect a worker's health and life in modern complex work processes is more important (Koradecka 1997).

First, the highest admissible concentrations and intensities of harmful agents (chemical, physical and dusts) in the work environment must be established to protect workers' health and that of the next generations. In individual member states of the European Union (EU), admissible values have been established for an average of 500 harmful chemical substances (in Poland, the list now contains 523 items).

At present, there are only 104 substances on the list, agreed upon by all of the member states of the EU. This is mainly due to lack of knowledge about aspects of the harmfulness of chemical agents (e.g., their carcinogenic and mutagenic nature) and the associated difficulties in occupational risk assessment (Koradecka and Bugajska 1999). Apart from the health aspect, the economic aspect, associated with the cost of decreasing concentrations to the recommended values, should also be considered. Of course, placing an agent on the list along with the value of the highest admissible concentration or intensity is not enough, since this only constitutes information on the threat. To prevent hazards, information serving as a basis for establishing of the highest admissible concentration values should be obtained from expert documentation.

The next stage in creating an environmental safety and protection system is the development of standardised methods for determining harmful agents. These methods are necessary to control pollution of the environment by chemical and physical agents and dusts and to undertake preventive actions (Koradecka et al. 2006). Then—at the design stage—the values of harmful agent emissions resulting from the application of specific technologies must be determined (Benczek and Kurpiewska 1996). After exhausting the possibilities for protecting health and life through proper, modern design of products, workstations, and work processes, compliance with the basic safety requirements must be supervised (Salvendy and Karwowski 1994).

EU directives on testing and certification of products with regard to their compliance with safety and health and environmental protection requirements have enforced mandatory CE marking of products, which confirms their conformity with European standards, since 1995. A declaration from the manufacturer is sufficient for simple products; however, prior to marking personal and collective protective equipment and particularly dangerous machinery, listed in Appendix IV of directive 89/392/EEC, compliance with complex procedures is required throughout the several stages of creation of the product, such as design and approval of the prototype. Products must comply with these rules in order to be exported to the EU and to be approved for marketing in all member states.

After verifying whether testing laboratories and bodies certifying products and quality systems meet the requirements listed in European standards, they can be accredited. Poland has used the European certification system since January 1994; it was officially introduced by an act of 3 April 1994. The implementation of this system at research and testing laboratories in our country is invaluable. The EU recognises the test results and facilitates the export of Polish products into the European Economic Zone. Thus, compliance with the requirements of occupational safety and ergonomics is of economic significance.

Economic stimuli are equally important for stimulating healthy work conditions. In pre–World War II Poland, economic stimuli took the form of differentiated insurance premium rates. These are currently used in many developed countries (e.g., Germany and France). The European Foundation for the Improvement of Living and Working Conditions has also prepared a list of modern economic stimuli to motivate companies to comply with the requirements of occupational safety and health (Bailey et al. 1995; Rzepecki 2007). Poland has differentiated insurance premiums depending on occupational risk since 2003 (DzU no. 199, item 1673, with amendments).

Compliance with the requirements of occupational safety and ergonomics is thus no longer perceived as a humanitarian gesture of good will; it has become an economic category, indicating the further development of the science and practices associated with these issues.

In our research we have compared national statistics data on exposure to harmful physical and chemical factors in a population of 9225 persons. The data was obtained from two sources: the results of a survey on subjective assessment of the working environment, covering 1001 persons, and from measurements taken in the working environment, covering 823 persons selected out of this population (Figure 2).

Note that there are considerable differences between an objectively measured amount of exposure and the subjective perception of this exposure by workers, the latter being considerably worse. Against this background, national statistics based on employers' reports turned out to be significantly underestimated.

In light of the fact that subjective exposure assessment is dependent on the individual features of an employee, the psychosocial conditions of the work tasks performed, and the workers' perception of health hazards, the need to carry out risk assessment in the work environment by means of both objective and subjective methods is fully justified.

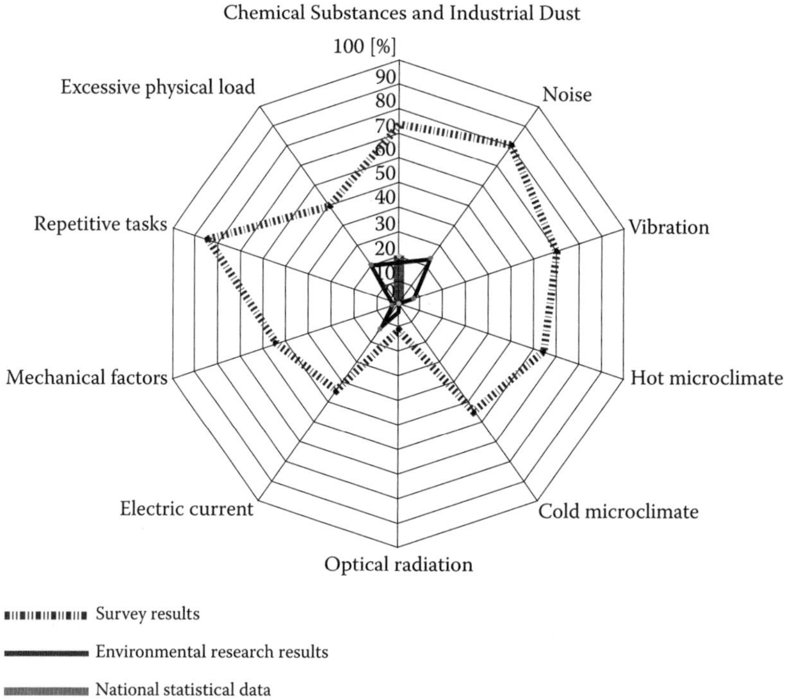

FIGURE 2 Comparison of survey results and environmental research results, according to CIOP-PIB research and national statistical data, of the number of workers employed in conditions of exposure to environmental factors. (From Koradecka, D. 2010. *Int J Occup Safety Ergonomics (JOSE)* 16(1):3–14. With permission.)

QUO VADIS?

It is difficult to make reliable forecasts for the future; we can only identify opportunities for development. In occupational safety and health, globalisation brings not only new technical risks, but also new problems associated with different models of employment. Conditions for the protection of life and health are also increasingly different. Associated risks, however, are due not only to changing work conditions but also to improper risk management (Karwowski 2003). Thus, we can forecast the following:

- Maintenance of the downward trend from the manufacturing sector towards the sector of services
- A high level of variability of entities on the labour market (particularly small- and medium-sized enterprises)
- An increase in the level of part-time employment and remote employment
- An increase in workers' age
- An increase in the number of women employed

At the same time, societies are undergoing many lifestyle changes and feeling consequences such as an increase of obese people, people addicted to alcohol or other substances, people suffering from sleep disorders or depression. These general social changes—in opposition with transformations in the world of work—will result in tensions and hazards not only to life and health, but also to socioeconomic development (Koradecka 1997). For instance, in the United Kingdom, according to the health and safety executive data, there will be 13 million new workers in the workplace by 2015; inexperienced workers are 40% times more likely to have accidents than experienced workers. Also in the United Kingdom, by 2010 most of the present small- and medium-sized enterprises will cease to exist, and 4.5 million new ones will replace them. Small organisations have a higher number of accidents; the risk decreases if the organisation has operated for a long period of time.

The number of older workers will also increase; they are absent from work less often, but their absences are longer. Obesity will increase, which is conducive to increased absenteeism due to musculoskeletal disorders and heart disease.

New technologies are another significant challenge for occupational safety and health, including

- Nanotechnologies
- Biotechnologies
- Spatial computing
- Alternative sources of energy

Work processes will need to become more effective. This will probably lead to the following:

- More frequent monitoring of the workplace
- Equipping workers with microchip ID cards

- A need to increase interactions between humans and independently working robots

Under such supervision, workers' levels of stress and depression will grow. Workers will react to these enhanced requirements by using risky medications more frequently to increase their efficiency, as these can improve memory and eliminate fatigue for up to 36 hours.

On the other hand, technological progress will be substantial, for example, in the following fields:

- Hydrogen infrastructure in households and in transport
- Robotisation, including in offices
- New-generation nuclear reactors
- Wind power

Trust in technologies will increase, and a new generation of workers will be highly independent. This may lead to changes in the perceptions of issues related to occupational safety and health.

REFERENCES

Alli, B. O. 2001. *Fundamental Principles of Occupational Health and Safety*. Geneva: ILO.

Astrand, P. O., and K. Rodahl. 1977. *Textbook of Work Physiology*. London: McGraw-Hill.

Bailey, S., K. Jorgensen, Ch. Koch, W. Kriiger, and H. Litske. 1995. *An Innovative Economic Incentive Model for Improvement of the Working Environment in Europe*. Dublin: European Foundation for the Improvement of Living and Working Conditions.

Benczek, K. M., and J. Kurpiewska. 1996. Emission of toxic substances during polystyrene processing by an injection molding machine. *Chemia i Inżynieria Ekologiczna* 3:239–246.

Brouha, L. 1962. *Physiology in Industry*. Warsaw: WNT.

Chapanis, A. 1951. Theory and methods for analyzing errors in man-machine systems. *Ann New York Acad Sci* 51:1179–1203.

Christensen, E. H. 1960. Muscular work and fatigue. In *Muscle as a Tissue*. eds. K. Rodahl, and S. M. Horwath, 176–189. New York: McGraw-Hill.

Dul, J., and V. H. Hildebrandt. 1987. Ergonomic guidelines for the prevention of low backpain at the work place. *Ergonomics* 30:419–429.

European Agency for Safety and Health at Work. 2002. *How to Tackle Psychosocial Issues and Reduce Work-Related Stress*. 25–85. Luxembourg: Office for Official Publication of the EC.

Franus, E. 1991. *Connective Networks in Ergonomics. General Methodological Considerations*. Amsterdam: Elsevier.

Grandjean, E., and W. Hunting. 1977. Ergonomics of posture. Review of various problems of standing and sitting posture. *Appl Ergonomics* 8:135–140.

Jastrzębowski, W. 2004. *An Outline of Ergonomics, or the Science of Work based upon the Truths Drawn from the Science of Nature—1857*. Warsaw: CIOP-PIB.

Karwowski, W. 2003. *Design and Management of Computer-Integrated Manufacturing Systems with the Use of Ergonomics*. Warsaw: CIOP-PIB.

Karwowski, W. 2006. *International Encyclopedia of Ergonomics and Human Factor*. 2nd ed. Vol. 1. Boca Raton, FL: Taylor & Francis.

Kidd, P. T., and W. Karwowski. 1994. *Advances in Agile Manufacturing. Integrating Technology, Organization and People*. Amsterdam: IOS Press.

Kim, J.V. 2001. Cognitive engineering research at Riso from 1962–1979. In *Advances in Human Performance and Cognitive Engineering Research*, Vol. 1. ed. E. Salas, 1–57. Amsterdam: Elsevier Science Ltd.

Koradecka, D. 1982. *Peripheral Blood Circulation Under the Influence of Occupational Exposure to Hand Transmitted Vibration*. Series: Schriftenreihe Arbeitsschutz No. 32. Dortmund: Bundesanstalt fur Arbeitsschutz und Unfallforschung.

Koradecka, D. 1997. Ergonomics and safety in societies in transfer. *Ergonomics* 40(10):1130–1147.

Koradecka, D. 2010. A comparative study of objective and subjective assessment of occupational risk. *Int J Occup Safety Ergonomics (JOSE)* 16(1):3–14.

Koradecka, D., and J. Bugajska. 1999. Physiological instrumentation. In *The Occupational Ergonomics Handbook*. eds. W. Karwowski and W. S. Marras. 525–547. Boca Raton, FL: CRC Press.

Koradecka, D., et al. 2006. Chemical, dust, biological, and electromagnetic radiation hazards. In *Handbook of Human Factors and Ergonomics*, 3rd ed. ed. G. Salvendy. 945–964. Hoboken, NJ: John Wiley & Sons.

Lehmann, G. 1962. *Occupational Psychology in Practice*. Stuttgart: Thieme.

Molleson, T. 1994. To what do bones from Abu Hureyra testify? *Świat Nauki* 10(38):68–73.

Ozok, A. F. and G. Salvendy, eds. 1996. Advances in applied ergonomics. *Proceedings of the 1st International Conference on Applied Ergonomics (ICAE'96)*, Istanbul, Turkey. Istanbul-West Lafayette: USA Publishing.

Paluszkiewicz, L. 1975. *Ergonomic Properties of Signaling and Control Devices*. Warsaw: Instytut Wydawniczy CRZZ.

Rahimi, M., and W. Karwowski, eds. 1992. *Human–Robot Interaction*. London: Taylor & Francis.

Ramazzini, B. 2009. *Works*, eds. F. Carnevale, et al., Verona: Cierre edizioni.

Rzepecki, J. 2007. Economic aspects of the development of working conditions. *Occupational Safety* 12:2–5.

Salvendy, G., and W. Karwowski, eds. 1994. *Design of Work and Development of Personnel in Advanced Manufacturing*. New York: John Wiley & Sons.

Stanton, N. A. 2005. Human factors and ergonomics methods. In *Handbook of Human Factors and Ergonomics Methods*. eds. N. Stanton, A. Hedge, K. Brookhuis, E. Salas, and H. Hendrick, 1.1–1.9. Boca Raton, FL: CRC Press.

Stellman, J. M., ed. 1998. *Encyclopedia of Occupational Health and Safety*. 4th ed. Vol. 1. 15.2–15.8. Geneva: ILO.

Strasser, H., ed. 2007. Assessment of the ergonomic quality of hand-held tools and computer input devices. *Series: Ergonomics, Human Factors and Safety*, Vol. 1, 57–74. Amsterdam: IOS Press.

About the Editor

Professor Danuta Koradecka, PhD, D.Med.Sc. and Director of the Central Institute for Labour Protection–National Research Institute (CIOP-PIB), is a specialist in occupational health. Her research interests include the human health effects of hand-transmitted vibration; ergonomics research on the human body's response to the combined effects of vibration, noise, low temperature and static load; assessment of static and dynamic physical load; development of hygienic standards and development and implementation of ergonomic solutions to improve work conditions in accordance with International Labour Organisation (ILO) conventions and European Union (EU) directives. She is the author of over 200 scientific publications.

Professor Koradecka is active in numerous national and international organizations working to protect human health in the work environment. She has been a World Health Organisation (WHO) and ILO expert for many years. Professor Koradecka has chaired Poland's Interdepartmental Commission for Maximum Admissible Concentrations and Intensities for Agents Harmful to Health in the Working Environment for over 20 years. Since 2003 she has represented Poland's government in the EU's Advisory Committee on Safety and Health at Work.

Professor Koradecka founded the *International Journal of Occupational Safety and Ergonomics* (*JOSE*) and has been its editor-in-chief since its inception. She also serves on the editorial boards of several international journals.

Contributors

Danuta Augustyńska, PhD
Central Institute for Labour Protection
National Research Institute
Warsaw, Poland

Grażyna Bartkowiak, PhD
Department of Personal Protective Equipment
Central Institute for Labour Protection
National Research Institute
Lodz, Poland

Krzysztof Baszczyński, PhD
Department of Personal Protective Equipment
Central Institute for Labour Protection
National Research Institute
Lodz, Poland

Anna Bogdan, PhD
Central Institute for Labour Protection
National Research Institute
Warsaw, Poland

Radosław Bojanowski, MA
Central Institute for Labour Protection
National Research Institute
Warsaw, Poland

Joanna Bugajska, PhD
Central Institute for Labour Protection
National Research Institute
Warsaw, Poland

Grzegorz Dudka, MSc
Central Institute for Labour Protection
National Research Institute
Warsaw, Poland

Jacek Dutkiewicz, PhD
Institute of Agricultural Medicine
Lublin, Poland

Władysław Dybczyński, PhD
Faculty of Electrical Engineering
Bialystok Technical University
Bialystok, Poland

Marek Dźwiarek, PhD
Central Institute for Labour Protection
National Research Institute
Warsaw, Poland

Zbigniew Engel, PhD, DSc
Central Institute for Labour Protection
National Research Institute
Warsaw, Poland

Józef Gierasimiuk, MSc
Central Institute for Labour Protection
National Research Institute
Warsaw, Poland

Zygmunt J. Grabarczyk, PhD
Central Institute for Labour Protection
National Research Institute
Warsaw, Poland

Krzysztof Gryz, PhD
Central Institute for Labour Protection
National Research Institute
Warsaw, Poland

Elżbieta Jankowska, PhD
Central Institute for Labour Protection
National Research Institute
Warsaw, Poland

Wiesława Kamińska, PhD
Department of Personal Protective Equipment
Central Institute for Labour Protection
National Research Institute
Lodz, Poland

Jolanta Karpowicz, PhD
Central Institute for Labour Protection
National Research Institute
Warsaw, Poland

Maria Konarska, PhD, DSc
Central Institute for Labour Protection
National Research Institute
Warsaw, Poland

Danuta Koradecka, PhD, D.Med.Sc.
Central Institute for Labour Protection
National Research Institute
Warsaw, Poland

Piotr Kowalski, PhD
Central Institute for Labour Protection
National Research Institute
Warsaw, Poland

Wiesław G. Kowalski, MD, PhD
Military Institute of Aviation Medicine
Warsaw, Poland

Barbara Krzyśków, PhD
Central Institute for Labour Protection
National Research Institute
Warsaw, Poland

Stefan M. Kwiatkowski, PhD
The Maria Grzegorzewska Academy of Special Education
Warsaw, Poland

Katarzyna Majchrzycka, PhD
Department of Personal Protective Equipment
Central Institute for Labour Protection
National Research Institute
Lodz, Poland

Kazimierz Marek, PhD, D.Med.Sc.
Institute of Occupational Medicine and Environmental Health (Retired)
Sosnowiec, Poland

Jerzy S. Michalik, PhD
Central Institute for Labour Protection
National Research Institute
Warsaw, Poland

Leszek Morzyński, PhD
Central Institute for Labour Protection
National Research Institute
Warsaw, Poland

Krystyna Myrcha, PhD
Central Institute for Labour Protection
National Research Institute
Warsaw, Poland

Grzegorz Owczarek, PhD
Department of Personal Protective Equipment
Central Institute for Labour Protection
National Research Institute
Lodz, Poland

Krzysztof A. Pachocki, PhD
National Institute of Public Health
Institute of Hygiene
Warsaw, Poland

Zofia Pawłowska, PhD
Central Institute for Labour Protection
National Research Institute
Warsaw, Poland

Piotr Pietrowski, PhD
Department of Personal Protective Equipment
Central Institute for Labour Protection
National Research Institute
Lodz, Poland

Daniel Podgórski, PhD, DSc
Central Institute for Labour Protection
National Research Institute
Warsaw, Poland

Małgorzata Pośniak, PhD
Central Institute for Labour Protection
National Research Institute
Warsaw, Poland

Danuta Roman-Liu, PhD, DSc
Central Institute for Labour Protection
National Research Institute
Warsaw, Poland

Jolanta Skowroń, PhD
Central Institute for Labour Protection
National Research Institute
Warsaw, Poland

Agnieszka Stefko, MSc
Department of Personal Protective Equipment
Central Institute for Labour Protection
National Research Institute
Lodz, Poland

Ryszard Studenski, PhD, DSc
University of Silesia (Retired)
Katowice, Poland

Iwona Sudoł-Szopińska, PhD, D.Med.Sc.
Central Institute for Labour Protection
National Research Institute
Warsaw, Poland

Krystyna Świder, MA
Central Institute for Labour Protection
National Research Institute
Warsaw, Poland

Maria Widerszal-Bazyl, PhD, DSc
Central Institute for Labour Protection
National Research Institute
Warsaw, Poland

Agnieszka Wolska, PhD
Central Institute for Labour Protection
National Research Institute
Warsaw, Poland

Jan Żera, PhD, DSc
Central Institute for Labour Protection
National Research Institute
Warsaw, Poland

Krystyna Zużewicz, PhD
Central Institute for Labour Protection
National Research Institute
Warsaw, Poland

Part I

Legal Labour Protection

1 Legal Labour Protection

Barbara Krzyśków

CONTENTS

1.1 INTRODUCTION

Labour protection is an ambiguous term. In theory, the definition of labour protection used in common language is different from that of labour protection used in the language of law.

In general usage, labour protection is the entirety of the rules of law, research, and both organisational and technical resources, and is for the protection of workers' rights, and their lives and health, from dangerous and harmful factors in a work environment. Labour protection also aims to create optimal work conditions in terms of ergonomics, physiology, and the psychology of work.

There are differences in the meaning of labour protection in the science of law; they concern the subject as well as the aim and scope of the protection offered. Labour protection can be either the protection of work as such, regarded by T. Kotarbiński as 'all activities aiming at overcoming difficulties to meet one's crucial needs', or it can be understood not as the protection of the working world or its interests, but as the protection of human beings and the labour force (the workers). In this case, labour protection means the protection of a worker from the negative phenomena that occur during the work process. Therefore, legal labour protection is the system of legal means that is used to provide both safety and health protection to workers.

The notion of labour protection is associated with the development of labour law as a branch of law. In the beginning of the nineteenth century, lack of regulations concerning labour protection led to a lower physical efficiency of workers, which bothered the public. In France and England, this was a reason for the state to interfere in the sphere of private life, which then included working relations. In

Prussia, the reason for state interference was the army's indication that it was dissatisfied with the physical development of its conscripts. State interference consisted of the creation of norms that reduced working time, at first for children in England in 1802, then for women, and finally for other employees. The set of legal norms concerning labour protection was variously called 'protective working law', 'factory legislation', and 'legislation on labour protection'. These phrases were used to separate from the law the norms that laid obligations and prohibitions upon employers. Sanctions were imposed on the employers if these laws were not obeyed. Over the course of time, especially in the German theory of law, a distinction was introduced between protection from risks that occurred during the work process and the regulation of working time and working relations. This led to the differentiation of labour protection, regarded as a subject of public law, from legislation on working relations, regarded as a subject of private law. The purpose of introducing these norms was not necessarily the protection of the health and life of workers, but rather a reasonable exploitation of the workforce by eliminating the conditions that lead to its premature exhaustion (Jaśkiewicz et al. 1970; Sanetra 1994).

After the Second World War, labour protection was defined as a set of activities aiming to eliminate the factors responsible for reducing the abilities of the workforce. It also aimed to provide optimal conditions for exploitation, appropriate recovery and development of the workforce. The subject was not protection of the health or life of the worker, but rather protection of his or her working ability. On the other hand, some people believed the appropriate subject of protection should be the worker as a person—his or her life, health and working ability, which should not be separated from the worker. Such a formulation of the subject of labour protection can be found in Article 207, paragraph 2 of the Polish Labour Code (LC), with an amendment passed on 2 February 1996. Pursuant to Article 207, an employer must protect the health and the life of employees (Świątkowski 2002; Wagner 2004).

Polish law for the protection of workers' health and life has a long tradition, beginning before the Second World War—and at first indirectly—with the foundation of the Labour Inspectorate. A temporary decree was issued on the establishment and activities of the Labour Inspectorate on 13 January 1919. In 1927, the decree was replaced by the Regulation on the Labour Inspectorate. Regulations that imposed obligations on the employer to provide for the protection of the life and health of workers then came into force on 16 March 1928, with Article 7 of the Regulation of the President of the Republic of Poland on an employment contract for blue-collar workers, and Article 461 of the Code of Obligations. The Regulation of the President of the Republic of Poland of 16 March 1929 refers directly to the occupational safety and health (OSH) of workers.

Safety and health laws were then repeatedly revised and amended over the course of time. The principle of providing safety and health protection for workers was reinforced by granting this principle the power of a constitutional provision in 1952. In the Polish Constitution, adopted in 1997, this principle acquired the status of a separate social law, indicating the importance and significance of labour protection recognised by the legislature. Article 66 of the Polish Constitution states that 'everyone shall have the right to safe and hygienic conditions of work'. The persons this law refers to are specified in the second sentence of this article, according to which 'the methods of implementing this right and the obligations of employers shall be specified by statute'. The constitutional right to safe and hygienic conditions of work was therefore fulfilled by the LC (Nycz 2000).

1.2 OCCUPATIONAL SAFETY AND HEALTH IN THE LABOUR CODE

Incorporating OSH provisions into the LC may suggest that these regulations concern only employers and employees because Article 1 of the LC restricts its provisions to them. The scope of persons protected in terms of OSH is much wider than that suggested by the provision quoted above and was extended under the provision of Article 304 of the LC. Pursuant to this article, the employer shall ensure safe and hygienic conditions of work as listed in Article 207, Paragraph 2 of the LC (concerning the basic duties of an employer) to natural persons who work under arrangements other than employment. However, this rule applies only when the persons are working in the employing establishment or at a location indicated by the employer. The duty of the employer to ensure safe and hygienic conditions for activities carried out within their establishment also applies to students and trainees who are not employees. This means that if students and trainees undergo training within the area of the employing establishment, the employer shall apply the necessary provisions to guarantee their protection. This amendment of the LC, adopted by the act of 13 April 2007 on the National Labour Inspectorate, is crucial in terms of broadening the scope of protected entities. Pursuant to the amended wording of Article 304, paragraph 1 of the LC, the employer shall ensure safe and hygienic conditions not only to persons who work under arrangements other than the employment relationship, but also to persons who carry out business activities in the establishment or at a location indicated by the employer. This duty also refers to entrepreneurs who are not employers. The consequence of establishing this duty is that natural persons who work under arrangements other than the employment relationship in an employing establishment or at a location indicated by the employer are obligated to comply with the duties imposed by an employer or entrepreneur on employees, according to Article 211 of the LC.

Part X of the LC is the basic provision that regulates issues concerning OSH. However, these are not the only provisions of the LC that refer to OSH issues. In Chapter 2 of Part I of the LC, entitled *Basic Principles of the Labour Law*, Article 15 states that an employer shall provide employees with safe and healthy conditions of work. The location of the provision mentioned above unambiguously suggests that it ranks as a basic principle of the labour law (Jończyk 1992). This duty is strict but not absolute, and deviations from this rule are permitted in cases when risks to life and health result from the type of work. This concerns persons who take part in rescue operations or other activities, such as military service, police, fire service, and so on, which expose the person to high risk (Wyka 2003). Article 145 of the LC is an example. Pursuant to this article, the working time may be shortened to less than the standards set out in the LC for employees working under conditions that are particularly arduous or harmful to their health, as well as in the case of monotonous work or work conducted at a predetermined work rate.

Parts VIII and IX of the LC, regarding the rights of employees in connection with parenthood and the employment of adolescents, are also crucial to OSH.

In Part VIII, immediately after the provisions that guarantee continuation of the employment contract with female employees during pregnancy or employees who benefit from maternity and parental leaves, there are many provisions for the protection of the health of pregnant or nursing employees. Among other things, there are provisions stating that a pregnant employee shall not be required to work overtime, at night or in an interrupted work-process system. However, the most important regulation in terms of protecting the health and life of women and their children is that pregnant women must not be employed to perform works that are particularly arduous or hazardous to health. The Council of Ministers specified the list of such works in the Regulations of 10 September 1996. The employer employing a pregnant or nursing employee performing the works listed in this regulation, which are forbidden for such an employee regardless of the degree of exposure, shall transfer the employee to other work or, if this is not possible, grant her a leave of absence for as long as necessary. A female employee shall preserve her right to her current remuneration during the leave of absence. In other cases, the employer shall adapt the conditions of work to the requirements specified in the regulations by the Council of Ministers, or reduce the working time so that the hazard to the health or safety of the employee is eliminated. If an adaptation of the conditions of work on the current work post or the reduction of working time is not possible or advisable, the employer shall transfer the employee to other work or grant her leave of absence for as long as necessary. The above-mentioned provisions shall accordingly apply to the employer in the event where counterindications to performing the current work stem from a medical certificate.

The legal protection of employment of adolescents is formulated slightly differently. In this case, the legislature put emphasis not only on health protection, but also on continuing education and correct psychophysical development of an adolescent. Hence, in the Regulation of the Council of Ministers of 24 August 2004, concerning the list of works prohibited to young persons, works that can negatively influence the psychological development of a young person, including working as a dogcatcher or serving alcohol, are included.

Article 283 of the LC should be also taken into consideration while discussing the subject of OSH. This article defines a catalogue of offences against the workers' rights to life and health protection. A crucial rule is that if the persons responsible for the conditions in an establishment or the management of employees fails to observe the provisions or rules of OSH, he or she shall be subject to a fine, regardless of whether negative consequences have been suffered in terms of the safety and health of the workers (paragraph 1 of Article 283, LC). This provision defines an employer or a person who manages employees as responsible. For obvious reasons, the employer takes on the responsibility as a private person. In cases where the employer is an organisational unit, the responsibility lies with the managing person of this unit who carries out the OSH legal activities (paragraph 1, Article 283, LC; Radecki 2001). The responsibility is on the person who manages the employees only when the person actually performs the managing function and regards the person only in the scope of this management (Iwulski and Sanetra 1999). Paragraph 2 of

Article 283 of the LC also contains a catalogue of offences concerning each person who is obliged to perform the duties defined in this regulation, such as the OSH service member, the physician who takes care of employees, and the social labour inspector. The catalogue includes the following offences:

- Failing to notify the relevant labour inspector and the relevant state sanitary inspector of the place, type, and scope of his or her activities; or of a change of the place, type, and scope of his or her activities; or of a change in technology that may lead to increased hazards to the health of employees.
- Failing to ensure that a construction or reconstruction of a building intended for working premises, or a part thereof, is based on designs that meet the requirements of OSH regulations, approved by authorised experts.
- Equipping work posts with machinery and other technical devices that do not meet the requirements of conformity assessment.
- Providing employees with personal protective equipment that does not meet the requirements of conformity assessment.
- Using materials and technological processes when the degree of harmfulness to the health of employees has not been identified, and failing to apply appropriate means of prevention; using chemical substances and preparations that are not clearly marked in a way that makes their identification possible; using dangerous substances and dangerous chemical preparations without material safety data sheets (MSDS) of those substances or without careful packaging; and lack of protection of workers from their harmful influences, fire, or explosion.
- Failing to notify the relevant labour inspector, prosecutor or other appropriate body of a fatal, grave, or collective accident at work, or of any other accident that had the above effects, if it is deemed an accident at work; failing to notify of a case or suspected case of occupational diseases; failing to disclose an accident at work or occupational disease or providing untrue information, evidence, or documents concerning such accidents and diseases.
- Failing to comply with an enforceable order of a body of the State Labour Inspectorate within the prescribed time limit.
- Obstructing the activities of a body of the State Labour Inspectorate, in particular, preventing a visit to an employing establishment or failing to provide information necessary to fulfilling its tasks.
- Granting permission to perform work or other gainful activities by a child under 16 years of age without a permit issued by a competent labour inspector.

Part X of the LC contains provisions that determine the duties of parties to the employment relationship: the employer, employee, and persons who manage employees. However, out of necessity, the provisions of this article are extended to other subjects such as OSH service members, OSH commission members and physicians who take care of the employees. Their duties and authority in the sphere of OSH are determined either directly by the provisions of the LC or by the executory provisions.

The basic duties of an employer are defined in Article 207 of the LC. The same article states that an employer is the person responsible for the OSH conditions at the employing establishment. The liability is statutory and complete. The liability of an employer concerns his or her employees and is also included in the criminal law provisions.

The employer shall protect the health and life of employees by assuring safe and hygienic work conditions while making proper use of scientific and technical advancements. This is a crucial provision in terms of OSH, which particularly indicates the following:

- The purpose of the binding OSH law is the protection of life and health of employees.
- The duties of an employer are derived not only from the OSH provisions but also from the OSH rules.
- The duties mentioned above are not static but require the use of scientific and technical advancement.

When an employer breaches his or her duty to assure safe and hygienic conditions of work, he or she is liable to the authorities responsible for the supervision of the conditions of work, the National Labour Inspectorate and the Sanitary Inspectorate, which can confirm breaches of OSH duties. These two bodies are regulated by separate legal acts determining the scope of their duties and powers. The National Labour Inspectorate is the most important body founded to control and supervise work conditions. The act of 13 April 2007 by the National Labour Inspectorate broadens the authority of this body. Apart from control and supervision of work conditions, it is also responsible for the inspection of the legality of employment (including the employment of foreigners), among other things. This act increases the number of entities that can be inspected to include employers and entrepreneurs for whom work is performed under the following conditions: persons performing work under arrangements other than the employment relationship or persons carrying out business activities in the employing establishment or at a location indicated by the employer or entrepreneur. The organisation's power to supervise and control extends to a competent body of the National Labour Inspectorate, which is authorised to give orders, address improvement notices and, in punishment proceedings, impose a fine up to 5000 Polish Zloties (PLN) for the offence defined in Article 283 of the LC.*

Disputes relating to civil and criminal liability over damages resulting from a failure to provide safety and health at work shall be resolved by a court.

1.2.1 Duties of an Employer

The duties of an employer defined in Article 207 of the LC are recognised as basic duties. The difference between basic and other duties of an employer described in

* In the event a case is brought before a magistrate's court, the amount of the fine imposed in the legal proceedings can increase to 30,000 PLN.

Part X of the LC is not content-related but rather distinguishes the basic duties that are general and those that concern the rights of an employer in the area of management, work organisation, and the administration of financial and economic resources (Wagner 2004). Basic duties include ensuring the observance of OSH regulations in an employing establishment, issuing orders to cure failures in that respect, and controlling the implementation of those orders. The employer shall also ensure compliance with orders, reports, decisions, and instructions issued by the authorities responsible for the supervision of the conditions of work and ensure compliance with the recommendations of the social labour inspector. The employer's basic duties also include appointing a coordinator to supervise the OSH of all employees, even if they are employed by different employers but perform work at the same time and in the same place. The employer who commences business activity shall also notify, in writing, the relevant labour inspector and the relevant state sanitary inspector of the place, type, and scope of that activity.

The remaining duties are more detailed and are of different types. These concern the following:

- Buildings and working premises
- Machinery and other technical devices
- Work factors and processes posing special hazards to health or life
- Preventive health protection
- Accidents at work and occupational diseases
- OSH training
- Personal protective equipment, work clothing and footwear
- OSH services
- Consultations on OSH, and the formation of an OSH committee

The complexity of the employer's OSH duties makes them difficult to classify. Depending on the chosen criteria and classification methods, the distinction between the direct and indirect duties or organisational and property-related duties is defined in legal literature (Wyka 2003, 226). The division of duties into direct and indirect prevention duties seems more appropriate (Wyka 2003, 227). The basis for the distinction of these duties is the possibility of assertion of claims, in the event that an employer is in breach of the duties imposed on him. In the first case, failure to observe the rules can be processed using procedures of collective dispute and in the second case, the same can be processed as an individual claim. Direct prevention duties include duties that concern employees as a group, such as the construction of buildings, the organisation and equipping of workplaces, and usage of materials and technological processes, as well as duties regarding the employee as a person, such as providing preventive examinations, OSH training, and personal protective equipment.

Indirect duties include educational, bureaucratic and organisational duties. This includes the obligation of an employer to know the rules of work safety, including OSH regulations, to the extent necessary to perform his duties (Article 207, paragraph 3, LC). Bureaucratic duties include maintaining various registers, documentations, and lists. Among other things, an employer shall maintain a register of tests and measurements of agents that are harmful to employee's health, a register of accidents at work, a register of cases of actual and suspected occupational diseases, a register of all types of work involving contact with carcinogenic substances (and a register of all employees performing such types of work), a register of completed OSH training programmes, documentation of occupational-risk assessment, lists of the types of work forbidden to adolescent workers, and a list of light works that are allowed. Organisational duties involve mainly the basic duty of an employer to organise work in a manner ensuring safe and hygienic work conditions, to create OSH services, to appoint an OSH committee, and to hold consultations. These duties cannot be executed as individual or group claims. They should be executed only as part of the procedure of state supervision of work conditions.

1.2.1.1 Duties of an Employer in Relation to Direct Prevention

A building with working premises must meet OSH requirements. Any reconstruction of that building must improve OSH conditions. The employer is allowed to construct or reconstruct a building intended to be working premises only when it is based on designs that meet OSH requirements and is approved by authorised experts. These requirements are specified mainly in the general provisions of OSH (Regulation of the Minister of Labour, and the Social Policy of 26 September 1997), and also in the specific provisions regarding branches of work or types of work activities. The provisions precisely regulate the requirements concerning the location and size of the working premises, lighting, heating, and ventilation.

Machinery and other technical devices that are used while performing the work or which are part of the equipment of employees or work stands shall be designed and built such that they assure safe and hygienic conditions of work. In particular, they shall protect the employees against injuries, the influence of dangerous chemical substances, electric shock, excessive noise, harmful shocks, the effects of vibrations and radiation, and any harmful or dangerous influence of other agents in the work environment. The construction of machines shall meet standards of ergonomics. Depending on the date when the work stand was equipped with the machinery or other technical devices, the machinery shall fulfil either the minimum requirements of OSH defined in the Regulations of the Minister of Economy of 30 October 2002 (for work stands equipped with machinery and other technical devices before 1 January 2003) or, if they are subject to a conformity assessment, the requirements defined in the Regulations of the Minister of Economy of 20 December 2005 (for the remaining machinery and technical devices).

The employer shall not use any materials or technological processes without determining their degree of harmfulness to the health of employees and adopting relevant means of prevention. The means of prevention defined in the LC include obligations to use dangerous substances and chemical preparations according to the instructions, protect against their harmful influence, and protect against fire or explosion. The use of dangerous substances and chemical preparations shall be permitted, provided the means of ensuring employees' health and life protection are applied. The dangerous substances or preparations used by the employer shall be classified, registered and have MSDSs. Additional obligations are imposed on employers who employ workers who are exposed to the influence of carcinogenic or mutagenic substances, agents, or technological processes, or to biological agents or ionising radiation. In such cases, the basic duty of an employer is to replace the substances, biological agents, or technological processes with those that are less harmful to health, or to use other available means to limit the degree of exposure by making proper use of scientific and technical advancements.

The employer shall provide employees with initial, periodical, and follow-up health assessment. An initial health assessment is required for newly recruited persons, adolescent workers, and other employees who are transferred to work stands that are either subject to the influence of agents harmful to health or that have arduous conditions.* In the event of an inability to work for more than 30 days due to illness, the employee shall be subject to further follow-up health assessment to specify his or her ability to perform work at the current work stand. The health assessment shall be carried out by occupational health physicians, with reference to the terms defined in the executory provisions of the LC, at the cost of the employer, and, as much as possible, during working hours. The employer shall provide employees with preventive health care during the entire period of employment. In the event that employees work in conditions that involve exposure to the influence of carcinogenic substances and agents or dust, thereby causing fibrosis, the employer shall also ensure periodical health assessments for these employees after they stop working in contact with such substances. The employer shall ensure a periodical health assessment after termination of the employment relationship, if the former employee requests such examinations.

The employee must not be allowed to perform work if he or she lacks the required qualifications or skills or adequate knowledge of the provisions and rules of OSH. The employer shall provide OSH training at his or her own cost and during working hours. The OSH training can be divided into initial training for the employee before allowing him or her to work, or periodical training sessions depending on the work performed and the substances to which the employee is exposed. The periods can vary from 1 year for employees employed at work stands with extremely high safety and health hazards to 6 years for office workers (Ordinance of the Minister of Economy and Labour of 27 July 2004).

The employer shall provide the employee with free personal equipment for protection against the influence of agents present in the work environment that are dangerous and harmful to their health (see attachment no. 2 of the Ordinance of the Minister of Labour and Social Policy of 26 September 1997). This personal protective equipment must meet the requirements of conformity assessment. The employer shall provide the employee with work clothing and footwear if the employee's own clothing could be destroyed or soiled considerably, because of technological and sanitary requirements or as demanded by OSH regulations. Personal protective

---

* The minimum scope of the initial and periodical health assessments, as well as the maximum period of validity for periodical health assessments, was determined by hints regarding the manner of carrying out the health assessments of the employees. These hints are described in attachment no. 1 to the Regulations of the Minister of Health and Social Welfare of 30 May 1996, which deals with the medical examinations of workers, the scope of the preventive health care and physician's notices issued in cases prescribed by the LC.

equipment should be used only so long as it has the required protective and functional properties. Any damage to personal protective equipment that diminishes its protective properties requires repair or exchange of the equipment by the employer. The regulations are different for work clothing and footwear. Both should be replaced by the employer after the anticipated service life and footwear used only at certain work stands. The employer may specify the work stands at which the employees, with their consent, will be allowed to use their own work clothing and footwear. However, this does not apply to those work stands that are used to perform works that are directly connected with the operation of machinery and other technical devices or tasks that cause strong soiling or contamination of work clothing and footwear with chemical or radioactive agents or with biologically infective agents. The personal protective equipment, work clothing, and footwear used by the employee and provided by the employer are the property of the employer.

1.2.1.2 Duties of an Employer in Relation to Indirect Prevention

The employer must be acquainted with the provisions of labour protection, including OSH provisions, within the scope necessary to perform his or her duties. For this purpose, the employer shall undergo OSH training, which will be repeated periodically. The employer must be informed about work conditions in his enterprise; the employer must conduct tests and measurements of agents in the work environment that are harmful to employees' health, assess occupational risk connected with the work performed, register and maintain the results of such tests and measurements, and make them available to employees. In the case of high occupational risk, the employer shall perform and apply the necessary means of prevention, thus mitigating such risks. The information acquired by the employer makes it possible to meet his obligation to notify the relevant labour inspector and the relevant state sanitary inspector of the place, type, and scope of that activity within 30 days of the date of commencement of the activity. The above duty shall apply to the employer in the event of changes in the place, type and scope of the activity carried out.

The employer also has obligations regarding registration and recording, resulting from the provisions of Part X of the LC and other provisions of the LC. They include obligations to prepare the following lists: types of work forbidden to adolescent workers and female employees; types of light work permitted for adolescents employed for purposes other than vocational training; work stands where an employee must use personal protective equipment, work clothing, and footwear; and work stands where an employee can use his own clothing and footwear after receiving the allowance paid by the employer instead of clothing provided by the employer. In addition, on the basis of the provisions of Part X of the LC, the employer is obliged to keep a register of the following:

- The types of work involving contact with carcinogenic or mutagenic substances, preparations, agents, or technological processes, and a register of all employees performing such types of work
- The types of work exposing employees to the influence of harmful biological agents and a register of the employees performing such types of work

- Tests and measurements of agents in the work environment that are harmful to health
- Accidents at work
- Cases of occupational diseases

The employer shall also maintain documentation regarding preventive examinations of employees, OSH training, and OSH instructions, which form the basis for occupational risk assessment (Dołęgowski and Janczała 1998; Rączkowski 1999).

Organisational duties include the basic duty of an employer to organise a workplace in a manner that provides safe and hygienic work conditions, as well as the obligation to create OSH services, hold consultations, and appoint an OSH commission.

OSH services has advisory and controlling capacities in matters of OSH. The range of duties and qualifications for OSH service employees is determined in the Ordinance of the Council of Ministers of 2 September 1997. An employer who employs up to 100 employees can entrust the duties of OSH services to an employee performing another work. In the absence of competent employees, the employer may entrust the duties of OSH services to third-party specialists. The employer himself may perform the tasks normally reserved for these services if he employs less than 10 employees or, in exceptional cases, if he employs fewer than 20 employees* and conducts business activities with low occupational risks. However, a competent labour inspector, in consultation with the relevant state sanitary inspector, may order the formation of OSH services even in the case of a lower number of employees if this is justified by the occupational risks encountered. The number of employees in OSH services is determined by the employer, who takes into account the employment rate and the hazards and nuisances that are to be found in the enterprise. The employees employed in OSH services shall have a certain level of education and period of service in OSH, specified in the executory provisions.

In 2004, provisions concerning the duty of an employer to consult the employees or their representatives on any measures related to OSH resulted when Polish law was adjusted to conform to European Union legislation, particularly the provisions of the framework directive 89/391/EEC concerning the introduction of measures to encourage improvements in the safety and health of workers. This directive concerns the participation of employees or their representatives in creating the conditions of the workplace through consultations, among other things. According to Polish law, the employer must consult employees or their representatives on any measures related to OSH, particularly concerning the following steps:

- Changes in the workplace organisation or fitting out of work stands; introduction of new technological processes and chemical substances or preparations if they may pose a hazard to employees' health or life
- Evaluation of the occupational risks associated with performing specific types of work, and informing the employees of such risks

* In cases where the business activity of an employer is classified among the activities that have not received a risk rating larger than the third category under the provisions of social insurance in relation to accidents at work or occurrence of occupational diseases.

- Formation of OSH services or entrusting these duties to other persons
- Designation of employees responsible for first aid
- Assignment of personal protective equipment, work clothing and footwear
- Training employees on all aspects of OSH

OSH consultations should be held during working hours. The employees or their representatives retain the right to remuneration for the time they are out of work due to their participation in OSH consultations. Employees or their representatives nominated to lead consultations with an employer in the area of OSH can freely express their opinions and cannot suffer any negative consequences deriving from their representative function, as guaranteed by law. Employees or their representatives may also submit requests to the employer to eliminate or limit the occupational risks in which they are involved.

When an employer employs more than 250 workers, an OSH commission shall be appointed to hold OSH consultations. The OSH commission is an advisory, opinion-giving body and shall be composed of an equal number of the employer's representatives, including OSH service members, the physicians responsible for the preventive health care of employees, and employees' representatives, including a social labour inspector. The duty of the OSH commission, apart from holding consultations, is to review the conditions of the workplace, periodically evaluate OSH status, provide opinions on the measures adopted by the employer to prevent accidents at work and occupational diseases, formulate recommendations for the improvement of workplace conditions, and cooperate with the employer in the fulfilment of his or her OSH responsibilities.

The employer has also specific obligations in cases of accidents at work or suspicion of occurrence of occupational diseases. In the event of an accident at work, the employer shall take the steps necessary to eliminate the hazard, provide first aid to the injured, and establish the circumstances and causes of the accident in accordance with the executory provisions of the LC. The employer shall nominate an accident investigation team composed of the employer's and employees' representatives. The accident investigation team will establish the circumstances and causes of the accident on the basis of evidence they collect and then write a report, which is to be approved by the employer. The duty of the employer is to take into consideration the conclusions from the report and adopt the recommended measures in order to prevent similar accidents.

The employer shall notify the relevant body of the State Sanitary Inspectorate if he suspects a case of occupational disease. When the occupational disease* has been diagnosed, the employer shall notify the relevant body of the State Sanitary Inspectorate and the relevant labour inspector. Furthermore, the employer shall establish the cause of occupational disease and the nature and degree of susceptibility to that disease, immediately remove agents causing the occupational disease, apply other necessary means of prevention, and ensure compliance with medical recommendations.

* Occupational disease is diagnosed by the relevant state sanitary inspector according to the executory provisions of the LC.

1.2.2 Duties of an Employee and a Person Who Manages Employees

Section 2 of Part X of the LC regulates the rights and basic duties of an employee and a person who manages employees. The duties of an employee here are the basic duties according to Article 211 of the LC. Serious violation of these duties (fulfilling the condition of guilt) can be the basis to terminate the contract of employment without notice (Article 52, paragraph 1, point 1, LC). According to these provisions, the basic duties of an employee are as follows:

- To know the OSH regulations, attend training programmes and briefings concerning the same, and take required examinations.
- To perform his or her work in accordance with OSH provisions and rules and comply with directions and instructions provided by superiors in such a context.
- To ensure the proper condition of machinery, equipment, tools, and devices, and the order and neatness of the workplace.
- To apply the provided collective protective equipment, personal protective equipment, work clothing, and footwear according to their intended purposes.
- To undergo an initial health assessment with periodical follow-ups and other recommended health assessments and comply with medical instructions.
- To immediately notify the superior of any accident or hazard to human life or health observed in the employing establishment and warn coworkers and other persons present in the area of imminent danger.
- To cooperate with the employer and superiors in the discharge of duties relating to OSH.

If an employee fails to observe OSH provisions, the employer may administer a penalty of admonition, a penalty of reprimand, or a penalty of a fine.

Apart from the duties of an employer and an employee, the duties of a person who manages employees (foreman, master, etc.) are also defined in the LC. A person who manages employees shall undertake the following responsibilities:

- Organise work stands in compliance with the OSH regulations.
- Ensure the efficiency and proper use of personal protective equipment.
- Organise, prepare, and carry out work with due consideration for the protection of employees against accidents at work, occupational diseases, and other diseases related to the conditions of the work environment.
- Ensure the safe and hygienic conditions of the work premises and technical facilities, as well as the efficiency and proper use of the collective protective equipment.
- Ensure employee compliance with OSH provisions and rules.
- Ensure compliance with the recommendations of the physician responsible for the medical care of employees.

1.3 SOURCES OF LAW THAT INCLUDE OCCUPATIONAL SAFETY AND HEALTH REQUIREMENTS, OTHER THAN THE LABOUR CODE

Apart from the provisions of the LC and the executory provisions of the LC, there are other rules concerning workplace safety and health that regulate obligations in administrative relations. These can be found in the Act on Chemical Substances and Preparations, the Construction Law, the Atomic Law, and the Geological and Mining Law, among other laws.

Collective labour agreements and workplace regulations should also be enumerated as sources of labour law, that is, OSH law. In the event of collective labour agreements, the requirements concerning OSH included therein shall not be less favourable than the provisions included in the generally binding law. The case of workplace regulations is somewhat different. Workplace regulations shall specify the organisation and order of the work process and the associated rights and duties of both employers and employees. According to the provisions of the LC, workplace regulations must be introduced by any employer who employs more than 20 persons. According to the LC, OSH issues are crucial among all the matters that should be defined in workplace regulations. The following matters are part of OSH: providing employees with personal protective and hygienic equipment; listing types of work forbidden to adolescent workers and female employees; listing types of work and work posts allowed to adolescent workers to complete vocational training; listing types of light works allowed to adolescent workers employed for a purpose other than vocational training; knowledge of duties relating to OSH and fire protection, including the method of informing the employees of the occupational risks involved in the work performed.

1.4 INTERNATIONAL LAW

OSH as a discipline of law is subject to frequent changes due to scientific progress in the field of technology, medicine, and work organisation. The International Labour Law is of major importance for OSH law. Poland is a member of several international organisations that issue regulations with great significance in Polish OSH requirements. The most important of these are the International Labour Organisation (ILO), the Council of Europe, and the European Union.

The ILO was founded in 1919. Since its inception, the ILO has been dealing intensively with the creation of work conditions, especially by performing normative activities. According to the Constitution of the ILO, the organisation creates norms of law in the form of international conventions and recommendations (Article 19, point 1 of the Constitution). These conventions are resolutions of the General Conference of the ILO, which contain norms regulating issues in the areas of working relations, social securities, and social policy. The conventions are also regulations of International Law, which come into force in any given country after their ratification by the member country. Recommendations are resolutions of the General Conference of the ILO, passed in cases that are not yet regulated by any type of conventions or that constitute a supplement or development of the general norms included in conventions. Recommendations do not have legal force, while the conventions do, but are nevertheless crucial components of law adopted by the ILO (Florek and Seweryński 1988).

The ILO conducts intensive legislative activity in the area of OSH: 187 conventions, including 42 directly concerned with OSH, have been enacted from the foundation of the ILO in 1919 through 2006.

According to the Constitution of the Republic of Poland (Article 87, point 1), the conventions ratified by Poland are a source of binding law. The ILO conventions are ratified after consent is obtained per a separate act of law (Article 89, point 1 of the constitution). This makes the ratified convention, enumerated in the *Journal of the Laws of the Republic of Poland*, part of the national legal order, and it can be then enforced directly. In the event that the ratified convention contradicts national legal acts, the convention is superior to the national legal acts. The convention is ratified only after the national law has been transposed to the requirements included in the convention.

The basic ILO convention concerning OSH and the work environment is convention no. 155, adopted in 1981. It obliges the signatory countries to formulate, implement, and periodically review their national policy on OSH in the work environment. The representative organisations of employers and employees should be consulted regarding the contents of the policy before putting the policy into practice. Convention no. 155 set the criteria to be met in order to guarantee the creation of a proper national OSH policy. The most important criterion is the determination of the functions and responsibilities of public authorities, employers, and workers, as well as their participation in the formulation of the policy.

Equally important for OSH is ILO convention no. 187, adopted in 2006 on the promotional framework for OSH. The member states that ratified convention no. 187 are obliged to formulate a national policy that promotes OSH and to lay out, develop, and implement a national system for OSH. Each country shall also formulate, implement, monitor, evaluate, and periodically review a national OSH programme. Poland has ratified 7 out of 15 ILO conventions that directly concerned OSH issues.

The Council of Europe is an international organisation that was founded in 1949 to protect human rights, pluralist democracy, and the rule of law. The council aims to find common solutions to the challenges facing European society and to consolidate democratic stability in Europe by backing political, legislative, and constitutional reforms. The Council of Europe has 47 member countries. Poland has been a member of the council since 1991. The legislative activity of the council consists of issuing international legal acts (treaties and conventions; 200 enacted through 2006). These acts are in force in the member states after ratification. From an OSH point of view, the most important act is the European Social Charter, adopted in 1961 and revised in 1996. The charter, ratified by Poland, consists of two parts. The first part includes declarations from the signatory countries concerning the aims of their social policy and consists of 19 points. The second part transposes the aims listed in the first part into international legal obligations and contains 19 articles (72 sections). OSH protection issues are regulated in points 2 and 3 of the first part and in Articles 2 and 3 of the second part of the charter (Henczel and Maciejewska 1997).

The ratification of the charter does not mean that it can be put into practice, unlike the ILO conventions. The provisions of the charter shall be implemented using Polish legislation in order to guarantee their enforcement. Poland is bound by the charter provisions included in the *Journal of Laws* of 1999 (no. 8, point 67). Article 3, among the charter regulations enumerated in the *Journal of Laws*, concerns the right to safe and hygienic workplace conditions.

Poland has been member of the European Union, an international organisation, since 1 May 2004. Despite the fact that European Union legislation is binding only to member states, Poland began to adapt its national law to thc standards of the European Union long before entering this institution by signing the 'Agreement Establishing an Association between the Republic of Poland and the European Communities and their Member States', also called the 'Europe Agreement', on 16 December 1991. One of the conditions of the Association Agreement that must be fulfilled in order to become a member state is adapting existing and future laws to European Union legislation. This obligation has had the greatest influence on Polish legislation in the last 10 years. Changes in legislation led to considerable conformity of Polish OSH legislation with the European Union law. The process of adapting to European Union law was not finished when Poland became a member state. The European Union constantly adopts new OSH norms that must become national law.

According to the Treaty of the European Union, the legal acts of the European Union are regulations, directives, decisions, recommendations, and opinions.

Regulations have general applications and concern all member states of the European Union. They are binding in their entirety and are directly applicable to all member states after they are published in the Official Journal of the European Union or on the date indicated in the regulation.

Directives are binding to each member state to which they are addressed with reference to the result to be achieved, but shall leave the choice of the form and methods to the corresponding national authorities.*

Decisions are binding in their entirety upon those to whom they are addressed. These legal acts are directed to particular institutions, enterprises, or member states.

Recommendations and opinions are not legal acts because they have no binding force, but they do have political implications. Only the institutions of the European Union are authorised to take a position on particular issues.

A directive is the basic legal act of the European Union regarding OSH. According to the treaty that established the European Union, a lack of transposition of the directive within the deadline determined by the European Union (usually included in the same directive), or erroneous transposition, can have directly binding legal consequences in the member state, despite the fact that the directive is a legal act that is only applicable after it has been transposed into the national law (Blanpain and Matey 1993). The most important OSH directive adopted by the European Union is the framework directive 89/391/EEC of 12 June 1989, regarding the introduction of measures to encourage improvements in the safety and health of workers at work, as well as 19 individual directives based on Article 16 of the framework directive. These directives constitute the core OSH legislation for the protection of workers but are not the only directives adopted by the European Union concerning this topic

* The principles of transposition of the directives into the legislation of the member states are determined in the rich case law history of the European Court of Justice.

(Florek 1993). The framework directive emphasises reducing the likelihood of health and/or life risks for employees and not on reducing the occupational accidents and diseases, as was the case previously. Hence, the prevention, reduction, and elimination of occupational risks have great significance in the directive concerning risks that cannot be avoided in a workplace. The general principles for prevention listed in the directive are as follows:

- Preventing occupational risks
- Informing workers of such risks
- Training workers to facilitate the reduction or elimination of the risks
- Providing appropriate means and organisation of work

Detailed directives regulate OSH principles depending on the workplace, methods of performing the work, and factors related to the work environment. Poland fulfils its obligation to transpose the OSH requirements included in the directives into national law within the time specified by the directives.

The very high standards defined by the directives and the large number of directives result in the European Union emphasising necessary limits to legislation in this area and focusing on the complete and proper implementation of already-adopted directives.

REFERENCES

Act of 13 April 2007 on National Labour Inspectorate. DzU no. 89, item 589.

Act of 26 June 1974—The Labour Code. DzU 1998 no. 21, item 94; as amended.

Blanpain, R., and M. Matey. 1993. *European Labour Law from the Polish Perspective.* Warsaw: Scholar Agency.

Council Directive 89/391/EEC of 12 June 1989 on the introduction of measures to encourage improvements in the safety and health of workers at work. JO L 183/1.

Dołęgowski, B., and S. Janczała. 1998. *A Practical Guide for OSH Services.* Gdansk: ODDK.

Florek, L. 1993. *European Communities Laws on Employment.* Warsaw: F. Ebert Foundation.

Florek, L., and M. Seweryński. 1988. *International Labour Law.* Warsaw: Publication Institute of the Trade Unions.

Henczel, R., and J. Maciejewska. 1997. *The Council of Europe's Basic Documents on Social Policy.* Warsaw: Scientific Scholar.

Iwulski, J., and W. Sanetra. 1999. *The Labour Code: A Commentary.* Warsaw: Librata.

Jáskiewicz, W., C. Jackowiak, and W. Piotrowski. 1970. *Labour Law: An Outline of a Lecture.* Warsaw: Scientific Publishers PWN.

Jończyk, J. 1992. *Labour Law.* Warsaw: Scientific Publishers PWN.

Nycz, T. 2000. *Constitutional Guarantees of Safe and Healthy Working Conditions.* Tarnobrzeg: Tarbonus.

Ordinance of the Council of Ministers of 10 September 1996 on the list of types of work particularly onerous or harmful for the health of women. DzU no. 114, item 545; as amended.

Ordinance of the Council of Ministers of 2 September 1997 on safety and health services. DzU no. 109, item 704; as amended.

Ordinance of the Council of Ministers of 24 August 2004 concerning the list of works prohibited to young persons and the terms of employment in some of the jobs listed. DzU no. 200, item 2047; as amended.

Ordinance of the Minister of Economy and Labour of 27 July 2004 on *training* activities in the field of occupational safety and health. DzU no. 180, item 1860.
Ordinance of the Minister of Economy of 20 December 2005 on basic requirements concerning machinery and safety components. DzU no. 259, item 2170.
Ordinance of the Minister of Economy of 30 October 2002 on minimum requirements concerning occupational safety and health on work stands equipped with machinery. DzU no. 191, item 1596; as amended.
Ordinance of the Minister of Labour and Social Policy of 26 Septemer 1997 on general provisions of occupational safety and health. DzU 2003 no. 169, item 1650, as amended.
Ordinance of the Minister of Labour and Social Policy of 26 September 1997 on general provisions of occupational safety and health]of the Minister of Health and Social Welfare of 30 May 1996 on medical examination of workers, scope of preventive health care and physician's notices issued in cases prescribed by the Labour Code. DzU no. 69, item 332; as amended.
Rączkowski, B. 1999. *OSH in Practice.* Gdansk: Consulting and HR Training Center.
Radecki, W. 2001. *Offenses Against the Rights of People in Gainful Employment.* Warsaw: C. H. Beck.
Sanetra, W. 1994. *Labour Law.* Bialystok: Temida.
Świątkowski, A. M. 2002. *A Commentary on the Labour Code.* Krakow: Scientific Paper Authors and Publishers Society Universitas.
Świątkowski, A. M. 2003. *Occupational Safety and Health.* Krakow: Scientific Paper Authors and Publishers Society Universitas.
Wgner, B., ed. 2004a. *The Labour Code: A Commentary.* Gdansk: ODDK.
Wyka, T. 2003. *The Protection of a Worker's Health and Life as an Element of Employment.* Warsaw: Difin Center for Consultation and Information.

Part II

Psychophysical Capabilities of Humans in the Working Environment

2 The Physiology of Work

Joanna Bugajska

CONTENTS

2.1 INTRODUCTION

The physiology of work is a part of human physiology focusing on explaining the mechanisms that form the basis for the responses of the human body while performing work. Armed with knowledge about these mechanisms, we can predict the behaviours of an organism during the performance of various types of work under varying external conditions, which will allow us to adapt workloads and shape environmental conditions in accordance with human capacities.

The physiology of work includes the ability to perform physical exercise, and thus physical work. This depends upon the characteristics of the human body, in addition to the functioning and efficiency of many body systems and organs. The most significant systems are (1) the cardiovascular and respiratory systems, which determine the general functional capacity of humans, the ability to adapt to physical exercise, and the efficiency of thermoregulation mechanisms, and (2) the musculoskeletal system, which is characterised by both motor properties (speed, coordination, precision of movements, and flexibility) and the strength and resistance of the muscles.

Technological developments in recent decades have brought about a reduction in physical exercise requirements for most workstations. However, many of these workstations are still associated with some activities that require substantial physical exercise, which, apart from requiring high levels of energy, result in a substantial burdening of the human body, particularly the cardiovascular and musculoskeletal systems. Such workstations are found in many sectors of the economy and are associated mainly with lifting and carrying heavy loads and pushing, pulling, and handling heavy tools.

2.2 GENERAL STRUCTURE AND FUNCTIONS OF THE HUMAN LOCOMOTOR SYSTEM

Motor functions are performed by the locomotor system, which consists of (1) skeletal muscles and the nervous system, the latter controlling the functioning of the former and together comprising the active part of the locomotor system, and (2) bones, joints, and ligaments, the passive parts of the locomotor system. All of these structures are interconnected and operate together during movement. Muscles consist of cells known as muscle cells (myocytes). Both bundles of these muscles and the entire muscle are surrounded by flexible membranes (fasciae) and are attached to bones by tendons. Tendons transfer the forces generated in the muscles during contraction to the bone levers. The bones are interconnected by joints and ligaments. The structure of the joints and the way they connect the bones allow for rotation of the bones in relation to one another, and thus, for more complex movements.

2.2.1 Types of Muscle Contractions

The most significant feature of muscle tissue is its ability to contract and thus to generate force. The energy necessary for contraction is released by the decomposition of adenosine triphosphate (ATP) to adenosine diphosphate (ADP), inorganic phosphate (P_i), and one hydrogen ion (H^+), in accordance with the following equation:

$$ATP \rightarrow ADP + P_i + H^+ + \text{energy} \tag{2.1}$$

This reaction takes place in the presence of myosin ATPase, which is activated by calcium ions (Ca^{2+}) that are released into the cytoplasm from the endoplasmic reticulum as a result of stimulation of the muscle cell by a nerve impulse (Nazar 1991).

The contraction of the muscles is based on the activation of contractile elements found in the muscle cells. This process takes place in the myofibrils, which consist mainly of two contractile proteins, actin and myosin. During contraction, temporary connections are formed between the myofilaments of actin and myosin in the muscle cell using crossbridges, which are myosin projections. A hypothetical model of the

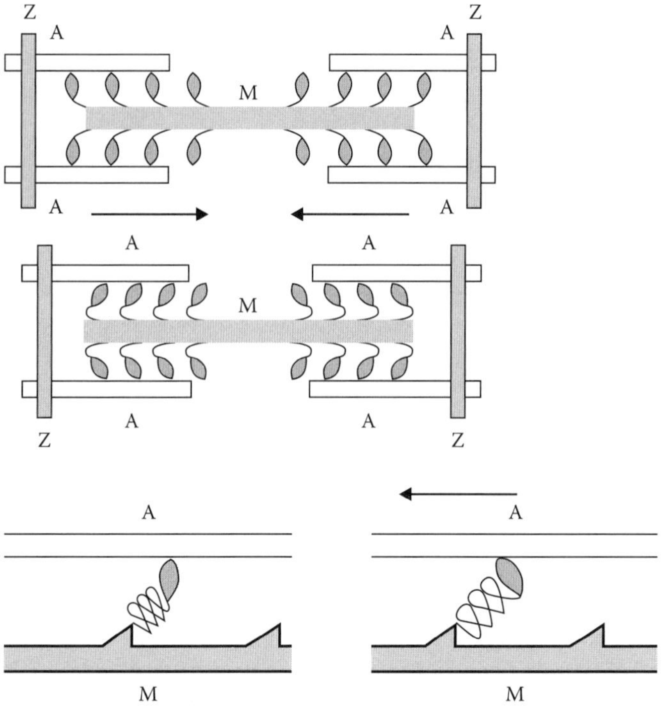

FIGURE 2.1 A hypothetical model of the contraction of myofibrils; the lower part of the drawing presents the force-generating mechanism. A, actin myofilaments; M, myosin myofilaments with the connecting crossbridges; and Z, Z-lines. (Reprinted with permission from Nazar, K. 1991. The physiology of physical efforts. In *Sports Medicine (Selected Problems)*, eds. A. Dziak and K. Nazar, 11–95. Warsaw: Polish Society of Sports Medicine.)

functioning of these crossbridges and their force-generating mechanism is presented in Figure 2.1.

Muscle contractions result from stimulation of the muscle cells by nervous impulses that are generated in the neural cells of the spinal cord (motoneurons) or the nuclei of the cranial nerves (motor nuclei). One neuron innervates a group of cells with its axon branches, which then form synapses with the muscle cells. A neuron, when connected to the muscle fibres (cells) innervated by it, forms a motor unit. The number of muscle fibres included in a single motor unit varies from about a dozen in the extraocular muscles to about 1000 in the muscles responsible for maintaining the vertical posture of the body.

There are two basic types of muscle contractions. If the length of the muscle does not change while tension develops, the contraction is known as an *isometric* or *static* contraction. If the length of the muscle changes during contraction and the tension remains unchanged, the contraction is referred to as an *isotonic* contraction. In practice, both the length and the tension of muscles typically change during contraction. Such contractions are called *auxotonic* contractions.

2.2.2 Types of Muscle Fibres

Muscle cells vary according to their functions, morphologies, and biochemical structures. These properties determine the maximum force and speed of the shortening of muscle fibres.

The functional differences between muscle fibres are based on their contraction speeds and endurance, or resistance to fatigue. Functional properties can be categorised into slow-twitch muscle fibres (ST), also known as type I fibres, and fast-twitch muscle fibres (FT), known as type II fibres. Type II fibres are further divided into two subgroups: fast-twitch red or oxidative fibres (FTa or IIa) and fast-twitch white or glycolytic fibres (FTb or IIb). In ST fibres, the time required to reach the maximum tension level during a single isometric contraction is about 80–110 milliseconds, about two times longer than the time necessary to reach maximum tension in FT fibres. On the contrary, ST fibres have a greater resistance to fatigue than FT fibres. FTa and FTb fibres have similar speeds of contraction, but FTa fibres have greater endurance than FTb fibres.

Morphological differences between the fibres are associated with their diameters (the diameter of FT fibres is greater than that of the ST fibres), their ultrastructures (FT fibres are richer in endoplasmic reticulum than ST fibres, whereas ST fibres have a greater number of mitochondria in comparison with FT fibres), and their vascular supplies (the number of capillaries surrounding the fibres is greater in ST-type fibres).

All types of muscle fibres have a similar content of energy substrates, such as ATP, phosphocreatine, and glycogen, although ST-type fibres have a higher lipid content. Differences between the various types of fibres occur in the content and activities of enzymes taking part in processes that supply energy for contraction. The activity of the enzyme that allows formation of crossbridges between actin and myosin (Ca^{2+}-dependent myosin ATPase) is greater in FT-type fibres. The activity of this enzyme is highly correlated with the speed of contraction of the muscle fibres; therefore, the difference in the activation of Ca^{2+}-dependent myosin ATPase is believed to be one of the most significant factors that contribute to the differences in the contractility of individual fibre types.

2.3 FUNCTIONAL CHANGES IN THE ORGANISM DURING PHYSICAL EXERCISE

Changes that take place in individual systems during physical exercise are subordinate to those of working muscles, which have far greater requirements to satisfy compared to the resting condition. The most significant changes take place in the cardiovascular and the respiratory systems; however, during physical exercise, changes also take place in the endocrine system, in blood volume and composition, and so on.

2.3.1 Cardiovascular System

Among the many functions performed by the cardiovascular system during physical exercise, the most significant are as follows:

- Transport of oxygen from the lungs to the working muscles and transport of carbon dioxide in the opposite direction
- Transport of energy substrates from extramuscular sources to the muscles and transport of metabolites formed in the muscles to the organs, where they are either processed further (the liver) or discharged (the kidneys)
- Discharge of excess heat generated during exercise

The scope of changes in the functions listed above depends on the intensity of exercise and changes in both cardiac activity and peripheral blood flow.

Immediately after commencement of physical exercise, the heart rate increases, which in the early stages is related to a decreased activity of the parasympathetic system, and in later stages to increased activity of the sympathetic system. During submaximal exercise that does not exceed 80%–90% $V_{O_2\,max}$, a state of balance is reached after 2–6 minutes and lasts for about 10 minutes. During more intense exercise, this state of balance is not reached. The maximum heart rate (HR_{max}) attained during maximum exercise can be calculated by the following formula:

$$HR_{max} = 220 - \text{age} \tag{2.2}$$

A measurement of the heart rate during dynamic physical exercise can be used to determine the relative workload of the employee and to assess the employee's tolerance to exercise.

Cardiac output (the amount of blood pumped by the heart in 1 minute), which is about 5–6 litres during rest, starts to increase immediately after commencement of dynamic physical exercise and reaches a state of balance after 2–6 minutes. Maximum cardiac output in untrained persons is 20–25 litres; in trained persons, it may exceed 35 litres. Such changes in the cardiac output are possible because of an increase in the heart rate and the stroke volume.

Very significant changes take place in the flow of blood through various vascular areas during exercise. The rate of blood flow through working muscles increases by 15%–20% of the cardiac output in the resting condition, and can reach up to 80% of the cardiac output. This is possible due to dilation of the resistance vessels in the muscles, which increases cardiac output, and due to a change in the general blood-flow distribution. On the contrary, the rate of blood flow through the visceral areas (the liver, the pancreas, the spleen, the intestines, and the kidneys) and through nonworking muscles decreases. Blood flow through the skin, which is controlled by the thermoregulatory system, increases, allowing for effective transfer of the heat produced during work. Blood flow through the coronary vessels increases up to four to five times during exercise. During intensive exercise, particularly when it is not preceded by a warm-up, hypoxia of the subendocardial layers of the cardiac muscle (ischaemic lesions) can be observed even among healthy persons.

The cardiovascular system responds differently during static exercise. During a static contraction, blood flow through the working muscles decreases due to the pressure exerted by the contracting muscles upon the cardiac vessels. In such situations, it is more difficult to supply both oxygen and energy substrates to the working tissues, and the metabolic products are accumulated, irritating the sensory nerve endings. The irritation of sensory nerves, in turn, results in a strong reflex action of the cardiovascular system, which is based on an increase in the heart rate and a contraction of the peripheral vessels. This reaction is proportional to the relative loading, expressed as a percentage of the maximum voluntary contraction (%MVC), and is visibly more intense even during static exercise of small muscle groups.

2.3.2 Respiratory System

Changes in the activities of the respiratory system result from the increased oxygen requirement of the working muscle cells and the increased discharge of carbon dioxide from the organism. The most significant changes in the functioning of the respiratory system during physical exercise are increased lung ventilation and improvement of the ratio of lung ventilation to blood flow through the lungs. The increase in lung ventilation takes place immediately after commencement of physical exercise and is proportional to its intensity until it reaches about 70% of the maximum intensity, that is, the intensity allowing for maximum oxygen uptake ($V_{O_2\,max}$). After exceeding this level of exercise intensity, the correlation between lung ventilation and oxygen uptake is no longer rectilinear, which means that the rate of oxygen uptake in the lungs decreases. Lung ventilation increases during exercise due to the increase in the respiratory rate and the tidal volume. The respiratory rate in adults may increase even up to 50–60 breaths per minute, with the tidal volume increasing to 3 litres, which constitutes 60% of the vital capacity of the lungs (Kozłowskai and Nazar 1999).

During physical exercise, the blood pressure in the pulmonary artery and the pulmonary blood volume increase. Blood distribution throughout the lungs evens out, and there is a better ratio of lung ventilation to blood flow throughout the lungs.

Another change that takes place during physical exercise and allows for an increased uptake of oxygen is the increase of the pulmonary diffusion capacity, which determines the ratio of oxygen diffusion between the alveoli and pulmonary blood flow.

2.4 PHYSICAL CAPACITY OF HUMANS AND PHYSIOLOGICAL CRITERIA FOR THE ALLOWANCE OF WORKLOADS

Physical capacity is a significant concept in the physiology of work because it determines a person's ability to perform physical work. The measure of physical capacity is the maximal oxygen uptake ($V_{O_2\,max}$), namely, the capacity of an individual's body to transport and utilise oxygen during submaximal exercise. Maximal oxygen uptake depends on the efficiency of the cardiovascular and respiratory systems, which enables intense and long-lasting physical exercise without experiencing changes that would disturb the homeostasis of the human body.

Physical capacity is not a constant property. The main factor influencing physical capacity in healthy humans is age. Physical capacity decreases systematically with age, mainly due to changes in the cardiovascular and respiratory systems. Many research studies have determined that the peak physical capacity level is attained at 20–25 years; from then on, it decreases in both women and men (Figure 2.2).

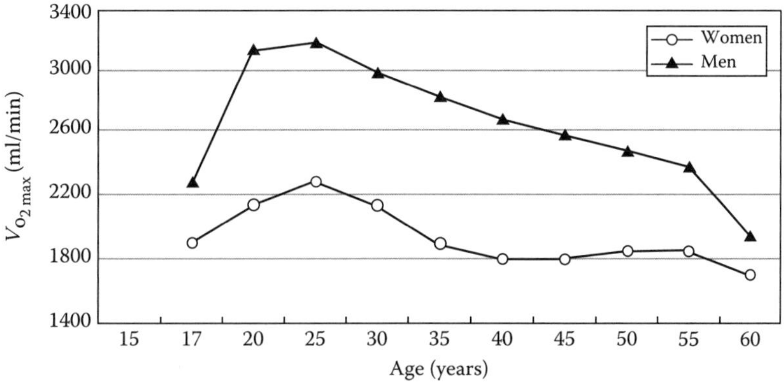

FIGURE 2.2 Maximum oxygen uptake ($V_{O_2 max}$) calculated on the basis of heart rate during submaximum exercise. (Reprinted with permission from Kozłowski, S., and K. Nazar. 1999. *Introduction to Clinical Physiology*. Warsaw: PZWL.)

In 65-year-olds, the value of $V_{O_2 max}$ is only 60%–70% of the maximum level attained in our lifetime (Shvartz and Reinhold 1990; Astrand and Rodahl 1986).

The decline of physical capacity with age is estimated to range from 0.25 to 0.80 ml/kg/min/year in men and from 0.25 to 0.40 ml/kg/min/year in women (De Zwart et al. 1995). Other research studies have determined that physical capacity decreases with age by 0.40 and 0.44 ml/kg/min/year in both men and women, respectively (Bugajska et al. 2005).

Women have a lower physical capacity than men of the same age. Women's maximum oxygen uptake is lower than men by about 20%–30% (or 17% per kilogram of body mass; Kozlowski and Nazar 1999). This is because there is generally more fat tissue in the body mass index of women than men, and fat tissue does not contribute to increased oxygen uptake during physical exercise as much as muscle tissue.

Other factors associated with work and life external environments also influence physical capacity. Systematic physical training increases the level of physical capacity. Compared to persons who are physically active, those with a sedentary lifestyle experience a more visible lowering of physical capacity level with age (Steinhaus et al. 1990).

Research has not yet determined whether the same effect can be seen in cases of people who perform many years of hard physical work. Higher $V_{O_2 max}$ values, which studies have noted among persons who perform strenuous physical work and compared with those of persons engaged in mental work, may serve as a basis for the conclusion that the selection of the type of work is correlated with the innate predispositions of the examined population. Figures 2.3 and 2.4 present the values of $V_{O_2 max}$ for women and men belonging to various age groups and performing various types of work (Bugajska et al. 2005). Among men, physical capacity ($V_{O_2 max}$) in the case of younger persons performing physical work is greater than that among those engaged in mental work; among older men, physical capacity was similar in both blue-collar and white-collar workers. On the contrary, among women, a similar physical capacity is observed in all the three groups of younger women, whereas among older women, it is lower in the case of white-collar workers.

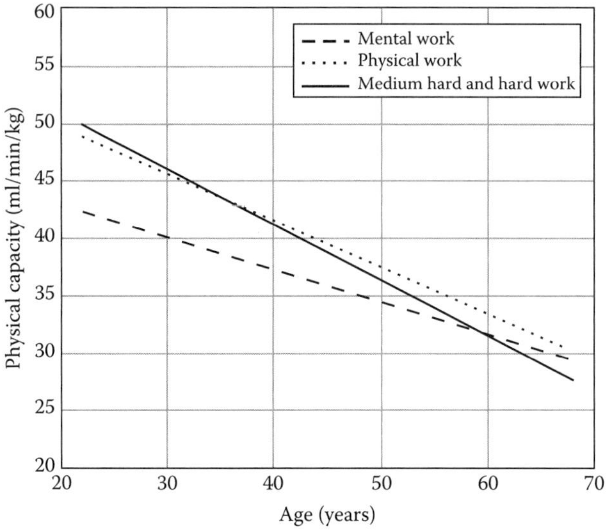

FIGURE 2.3 Maximum oxygen uptake ($V_{O_2\,max}$) in a group of men performing various types of work; a cross-sectional study ($N = 635$). (Adapted with permission from Bugajska, J., et al. 2008. *General Physical Capacity and Fitness of Occupationally Active Population in Poland.* Warsaw: CIOP-PIB.)

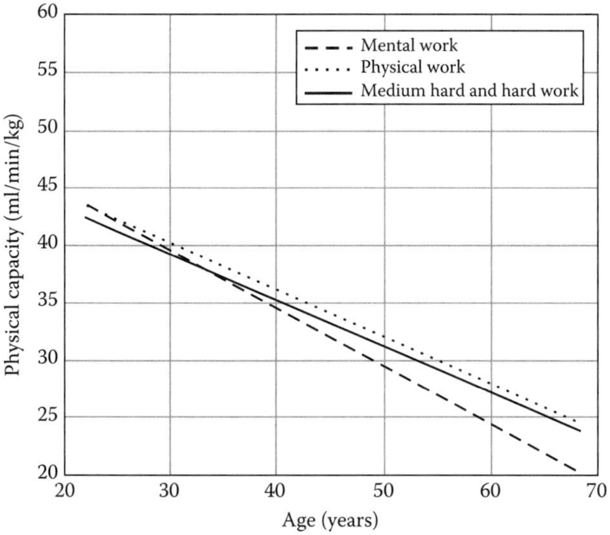

FIGURE 2.4 Maximum oxygen uptake ($V_{O_2\,max}$) in a group of women performing various types of work; a cross-sectional study ($N = 517$). (Adapted with permission from Bugajska, J., et al. 2008. *General Physical Capacity and Fitness of Occupationally Active Population in Poland.* Warsaw: CIOP-PIB.)

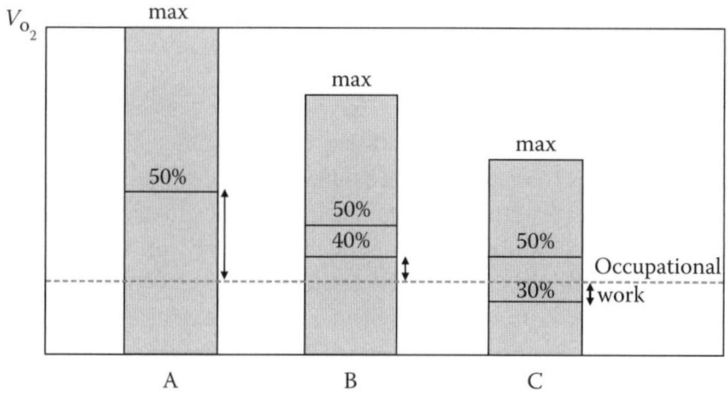

FIGURE 2.5 Physical capacity and workload tolerance. Physical capacity ($V_{O_2\,max}$): A, high; B, medium; and C, low. The dotted line stands for oxygen requirements during work. The horizontal lines on the bars indicate 50% $V_{O_2\,max}$ and are the acceptable workload level during performance of occupational work (A, 50%; B, 40%; and C, 30%). (Adapted with permission from Kozłowski, S., and K. Nazar. 1999. *Introduction to Physiology*. Warsaw: PZWL.)

Knowledge of the physical capacity of humans allows employers to determine workers' ability to perform physical work, and thus is very significant to the improvement of work conditions. Oxygen requirements increase proportionally with the intensity of exercise. By measuring the current oxygen requirement during physical exercise and comparing it with the maximum oxygen uptake ($V_{O_2\,max}$), we can numerically determine the physical workload (relative workload). Knowledge of the relative workload allows us to predict the time span we can perform a specific type of work with no fatigue, increase in the heart rate, or increase in disturbance of homeostasis. Physical exercise, even seemingly light, when performed for a long time by persons with a low level of physical capacity, may pose an excessive burden and thus may lead to negative health consequences. Figure 2.5 shows how a varying workload may require physical exercise of the same intensity among employees characterised by low, medium and high physical capacities.

Decisions concerning permission to perform work that is associated with a long-term physical workload should thus be based on an assessment of the physical capacity of the worker.

On the basis of empirical data, we can assume that for people with a low physical capacity, the physical workload should not exceed 30% $V_{O_2\,max}$; for those with medium capacity, this should be 40% $V_{O_2\,max}$; and for those with high efficiency, 50% $V_{O_2\,max}$ (Nazar 1999).

2.5 PHYSIOLOGICAL CLASSIFICATION OF PHYSICAL EXERCISE

In the physiology of work, physical exercise is understood as the work of skeletal muscles along with the functional changes that it causes in the human body. The physiological classification of physical exercise is significant, because it allows us to determine the real workload and predict the responses of individual body systems to exercise. There are many criteria for the classification of physical exercise, which can be listed as follows:

- *Type of muscle contraction*: Exercise can be either dynamic or static. During dynamic exercise, isotonic contractions of the muscles prevail (the muscle changes its length, while its tension remains the same); during static exercise, isometric contractions of the muscles prevail (the length of the muscle remains the same, while the tension increases).
- *Sizes of the muscle groups engaged in exercise*: Local exercise (work that involves one or both upper limbs) engages less than 30% of the total muscle mass; general exercise (work that involves one or both lower limbs) requires the involvement of more than 30% of the total muscle mass.
- *Time span of performance of the work*: Work could involve short-term exercise (up to 15 minutes), medium-term exercise (up to 30 minutes) or long-term exercise (more than 30 minutes; Nazar 1999).
- *Functional changes in the organism*: These include mainly changes in oxygen uptake, heart rate, and body temperature.

The actual workload during physical work is more complex than that presented in the above classifications. Apart from the type of physical exercise and its duration, there are also individual characteristics which should be used to determine workload during performance of physical work, including the physical capacity of the employee. In the physiology of work, the classification of exercise is based on the relative workload, expressed in % $V_{O_2\,max}$, which denotes the ratio of the actual oxygen consumption during work to the maximum oxygen uptake ($V_{O_2\,max}$); this ratio determines the physical capacity level. Such a classification is presented in Table 2.1.

Applying the classifications of physical exercise based on the energy expended during work (Table 2.2) allows employers to determine descriptions of workstations. Analysis of energy changes that take place during exercise involves an assessment of the arduousness of dynamic physical work, that is, whether the work is associated with substantial movement activity, which, apart from the

TABLE 2.1
Classification of the Intensity of Physical Work on the Basis of Relative Workload (% $V_{O_2\,max}$)

Intensity of Work	% $V_{O_2\,max}$
Light	<10
Moderately hard	10–30
Hard	30–50
Very hard	>50

Source: According to Kozłowski, S., and K. Nazar. 1999. *Introduction to Clinical Physiology*. Warsaw: PZWL.

TABLE 2.2
Classification of the Intensity of Physical Work on the Basis of the Effective Energy Expenditure during a Work Shift

Intensity of Work	Men				Women			
	kcal/8 h	kcal/min	kJ/8 h	kJ/min	kcal/8 h	kcal/min	kJ/8 h	kJ/min
Very light	Up to 300	Up to 1.2	1250	Up to 5	Up to 200	Up to 0.8	Up to 850	Up to 3.5
Light	300–800	1.2–2.2	1250–3350	5–10	200–700	0.8–1.8	850–2900	3.5–7.5
Moderately hard	800–1500	2.2–4.5	3350–6300	10–20	700–1000	1.8–3.0	2900–4200	7.5–12.5
Hard	1500–2000	4.5–7.0	6300–8400	20–30	1000–1200	4.0–4.8	4200–5000	12.5–20
Very hard	>2000	>7.0	>8400	>30	>1200	>4.8	>5000	>20

Source: Makowiec-Dąbrowska, T. 1999. Occupational physiology. In *Occupational Hygiene*, ed. J. A. Indulski. Lodz: IMP. With permission.

energy load, stimulates many systems, including the cardiovascular, respiratory, and thermoregulatory systems.

It is also important to determine energy consumption not only during the entire work shift, but also during the performance of various work tasks. This allows the worker to avoid those activities that are most arduous. The lowest energy expenditure is associated with work performed in a sitting position, such as office work. The energy requirement during performance of this work is usually less than 8 kJ/min—between 1.26 kJ/min for the sitting position and 7 kJ/min for typing activity.

Moderately difficult physical work includes work associated with performance of those tasks for which the energy expenditure is between 10 and 20 kJ/min. These are, for instance, activities associated with the operation of most machines and devices, as well as assembly works, which do not require substantial force. Hard work includes the packaging of objects and use of hand tools. Very hard work involves lifting of loads, loading of goods, or use of heavy tools such as shovels, pickaxes, pneumatic drills, and so on. Energy expenditure during performance of such tasks exceeds 30 kJ/min.

The intensity of static exercise is best illustrated by the relative measure, that is, the ratio of the currently developed force to the MVC of the muscle group engaged in the exercise. Physical exercise requiring a force not exceeding 15% MVC is classified as light, exercise requiring a force of 15%–30% MVC is classified as moderately hard, that requiring 30%–50% MVC is classified as hard, and work requiring a force above 50% MVC is classified as very hard (Kozlowski and Nazar 1999).

2.6 METHODS OF ASSESSMENT OF LOAD DUE TO DYNAMIC PHYSICAL EXERCISE

The workload associated with physical work, characterised by a high dynamic component, is often assessed by determining the energy expenditure during work. The energy produced by the human body during physical work is only partially converted into mechanical work (up to 25%). The remaining part of the energy produced is converted to heat. The energy expenditure is defined as the amount of energy *produced* by the organism during the performance of a work task.

The amount of energy *consumed* by the organism during the performance of work is made up of the energy expenditure associated with metabolism at rest and the energy used for the performance of the given task, that is, the energy expenditure associated with effective work.

Basic metabolism is the amount of energy consumed by an organism at both physical and emotional rest, in a state of thermal comfort, in a lying position, in a fasting state, in the morning, and after at least eight hours of sleep. This term is often used to refer to the amount of energy necessary to maintain life processes. The level of basic metabolism depends on sex, age, height, weight, and area of the body surface. In adults, basic metabolism can range from 5900 to 8000 kJ within 24 hours. Basic metabolism is higher in men, whose muscle mass is about 30% greater than women. Assessment of basic metabolism requires strict compliance with measurement conditions; therefore, metabolism at rest is usually assessed because it is easier to determine.

2.6.1 Methods of Determination of Energy Expenditure

Human energy expenditure during exercise can be assessed by direct calorimetry, which uses special calorimetric cameras and measures the amount of heat generated by the organism. This method cannot be used in real work conditions.

Indirect calorimetry is another widely used method. It measures energy expenditure based on the ratio of oxygen uptake per unit of time to the amount of energy released during metabolic processes. Indirect calorimetry is particularly useful during aerobic exercise. The amount of energy obtained from metabolic processes using 1 litre of oxygen varies depending upon the type of substance undergoing combustion; for example, during burning of glucose and fat, the amounts of energy obtained are 21.1 and 19.6 kJ, respectively. Thus, the energy equivalent of 1 litre of oxygen (necessary to convert the oxygen uptake during work to the energy expenditure) is between 19.6 and 21.1 kJ. To select the appropriate equivalent, we must determine the respiratory quotient, that is, the ratio of the amount of carbon dioxide discharged to the amount of oxygen uptake during the same period of time. Table 2.3 illustrates the values of the respiratory quotient and the corresponding equivalents of 1 litre of oxygen. Assessment of the energy expenditure using the indirect calorimetry method is thus based on measurement of the volume of oxygen uptake and carbon dioxide discharge, calculation of the respiratory quotient, and multiplication of the volume of oxygen uptake by the appropriate energy equivalent. This method is recommended by the standard EN-ISO 8996, published in 2004: *Ergonomics—Determination of Metabolic Heat Production.*

In recent years, electronic techniques have been used to measure the oxygen content in exhaled air. Laboratory and clinical apparatuses (such as Oxylog or Metabolic Measurement Cart) use oxygen sensors or electronic sensors to analyse the exhaled gases.

In industrial practices, energy expenditure is often assessed using a method that measures the exhaled (or inhaled) volume of air, or lung ventilation. The approximate

TABLE 2.3
Energy Equivalents of 1 Litre of Oxygen, Depending upon the Respiratory Quotient

Respiratory Quotient	Energy Equivalent of 1 Litre of Oxygen	
	kcal	kJ
0.70	4.69	19.62
0.75	4.74	19.83
0.80	4.80	20.10
0.85	4.86	20.36
0.90	4.92	20.61
0.95	4.98	20.86
1.00	5.05	21.12

value of energy expenditure can then be calculated based on the correlation between the ventilation and the oxygen uptake using the following formula (Datta and Ramanathan 1969):

$$E = 0.21 \cdot V_{E(STPD)} \tag{2.3}$$

where E is energy expenditure in kilojoules per minute and $V_{E(STPD)}$ is lung ventilation in litres per minute under standard temperature and pressure conditions (volume of dry gas at a temperature of 0°C and atmospheric pressure of 101.3 kPa).

This method is used in the energy-expenditure meter (MWE), which was constructed by the Central Institute for Labour Protection—National Research Institute, or CIOP-PIB, and facilitates a continuous measurement of pulmonary ventilation (Konarska et al. 1994; Koradecka and Bugajska 1999).

Measurement of energy expenditure using the MWE meter is based on the linear correlation described in Section 2.3.2 and the high correlation between pulmonary ventilation, oxygen uptake and energy expenditure. The MWE energy-expenditure meter is a portable, light, and small device that allows measurement of the energy expenditure during performance of physical work under real conditions, that is, in a situation wherein the direct determination of the oxygen uptake and the amount of carbon dioxide discharged is difficult (Bugajska 2007). This can also be determined when it is not possible to measure the energy expenditure using one of the methods listed above by reading the energy expenditure value from a table for typical activities performed in everyday life and at work. Estimating the energy expenditure using ready tables has the highest probability of error and should be used only in special cases. The differences resulting from the specific properties of the industrial sectors and the technological developments that have taken place in recent years must be taken into account.

A special way to estimate energy expenditure associated with the performance of work tasks, based on the timing and tabular methods, is Lehman's method, which takes into account the posture and type of muscle groups engaged in performance of the work. This method is divided into two stages. In the first stage, the posture during performance of the work is assessed, as well as the energy expenditure associated with maintenance of this posture. In the second stage, the main muscle groups responsible for their performance are assessed based on an analysis of the work activities, as well as the energy expenditure resulting from performance of these activities. The work-related energy expenditure is determined by summing up the results from both stages of analysis.

According to the SI system of units, energy expenditure is expressed in work units, that is, in joules (J), in joules per time unit (J/s), or in watts (W). Energy expenditure of work is often expressed also in watts per body surface area (W/m^2). However, because calorimetric units—calories (cal) or kilocalories (kcal)—have been used for many years to express energy expenditure, these units are still used to represent the level of work intensity in the industrial sector, legal provisions and textbooks. Energy expenditure, expressed in specific units, can be converted to other units as follows:

$$\text{kcal} = (\text{kJ}) \times 0.2389$$

$$\text{kJ} = (\text{kcal}) \times 4.1855$$

$$\text{W} = (\text{kcal/min}) \times 69.78$$

$$\text{kcal/min} = (\text{W}) \times 0.0143$$

$$\text{W} = \text{J/s}$$

2.7 PHYSICAL WORK AND THE EMPLOYEES' AGE

The ability to perform work changes with age, mainly due to a reduction in physical capacity and fitness and some psychophysical abilities, such as perceptiveness, reaction time, and efficiency of the sense organs. Changes in these functions have been observed from the 45th year of age. They may, of course, slow down or accelerate due to diseases, lifestyle, or the type of work performed (Bugajska 2007).

Changes associated with aging have been observed for all physiological functions and in all systems. These are as follows:

- The body composition changes; the fat-free body mass is reduced.
- The muscle mass decreases (sarcopenia); qualitative and quantitative changes occur in the skeletal muscles, based on changes in the proportion of type I and type II muscle fibres; the maximum muscle power is decreased. The strength and endurance of the muscles are also reduced, as well as the number of motor units, the mechanical properties, and the ability of the muscles to regenerate.
- The structure of the bone tissue changes; the activities of osteoblasts and—to a lesser extent—osteoclasts are reduced, particularly among women in the postmenopausal period, resulting in reduced strength of the bone tissue.
- The contraction force of the cardiac muscle in the cardiovascular system decreases, whereas the diastolic time interval increases, reactivity to catecholamines decreases, and the heart rate decreases; the vessels become less flexible and their response to adrenergic stimulation decreases.

- The volume of the respiratory tract increases, accompanied by a reduced surface for gas exchange, reduced flexibility of the lungs, and increased volume of residual air; the power and endurance of respiratory muscles decreases.
- As a result of changes in the cardiovascular and the respiratory systems, the maximum oxygen uptake ($V_{O_2\,max}$) is reduced; the extent and pace of the reduction of maximum oxygen uptake are modified significantly by physical activity (see Section 2.4).

These changes lead to a reduction in the ability to engage in physical exercise. Exercise tolerance among the elderly is visibly lower than among young people, especially during exercises that engage large muscle groups and when the intensity of exercise is high. Performance of this type of exercise, such as running, requires a substantial amount of power. On the contrary, the ability to perform exercise engaging small muscle groups is not reduced significantly among the elderly. The elderly have a lower tolerance only for physical exercise with high intensity and that involves large muscle groups (Kamińska et al. 2003; Tokarski and Kamińska 2004).

The decrease in strength associated with aging is due to the weakening of the motoneurons (Metter et al. 1997) and reduction in the size of muscle cells due to lack of physical activity (Grimby and Saltin 1983; Dutta and Hadley 1995; Porter et al. 1995). Changes in the hormonal balance (in particular, reduction in the concentrations of testosterone and growth hormone) may lead to a reduction in the muscle mass (Lamberts et al. 1997). Decreased muscle strength may also be caused by disease. Although the aging process results in diminished strength, the elderly's endurance and resistance to fatigue are similar to younger persons (Lindström et al. 1997).

As in the case of physical capacity, physical activity plays a very significant role in the changes in maximum muscle strength. Persons who engage in physical activity maintain higher levels of maximum strength and endurance than those who are not active.

Elderly persons have a lower tolerance for high temperatures, which is obvious particularly during the performance of physical work in a hot environment, during continuous work or during long-term exposure to excessive heat. This low tolerance is due to a decline in thermoregulatory mechanisms. Under the same conditions, the elderly do not start to sweat for a longer time than young people, which results in both a higher body temperature and higher average weighted skin temperature. Reduced sweating is partially due to morphological changes in the skin and the 'aging' of the sweat glands. Due to reduced sweating, the peripheral vessels of the elderly dilate to a greater extent and the heart rate increases to a greater extent, which indicates a higher physiological cost of work. Dehydration caused by heat stress is also less tolerable for the elderly.

To ensure safe work conditions and to minimise health risks among the elderly who perform physical work, it is necessary to

- Determine the intensity of physical work (measure energy expenditure).
- Determine the biomechanical factors of load on the musculoskeletal system.
- Determine the general physical capacity of the employee.
- Assess the health contraindications for performing physical exercise.

2.8 PHYSICAL WORK IN HOT AND COLD ENVIRONMENTS

The human body can maintain an internal body temperature of about 37°C, regardless of the temperature of the surrounding air. This is due to reactions controlled by the thermoregulatory centre, which maintains the balance between the amount of heat discharged or accumulated by the human body and the amount of heat produced. Maintenance of this balance is expressed by the heat balance formula.

Heat is produced in the human body as a result of the metabolism of each body cell; this heat is discharged into the external environment by (1) conduction, due to the gradient of temperatures between the cell and the surrounding area, and (2) convection, through the flow of extracellular fluid such as blood. The most significant factors that disturb the heat balance of the body are physical work and the external thermal environment. Therefore, the ability to perform physical work in cold and hot environments is another significant issue in the physiology of work.

2.8.1 Work in a Hot Environment

When working in a hot environment, there are two sources of body heat load. One of them is the heat produced by the body during metabolism. About 20%–25% of the energy produced during metabolism associated with muscle work is converted into the mechanical energy necessary to perform the work, while the remaining part is converted to heat, which must be discharged. The amount of heat produced depends on the intensity of the physical exercise. The second source of heat load while working in a hot environment is the external hot environment, which influences the human body.

In both situations, responses are triggered that prevent overheating of the body (heat stress). These responses are controlled by the thermoregulatory centres, which are located in both the anterior and in the posterior part of the hypothalamus. The most significant thermoregulatory mechanisms that allow the heat to discharge into the environment are dilation of the skin vessels and vaporisation of sweat.

Dilation of the skin vessels during exercise results from suppression of the tension of the sympathetic innervation of these vessels, which results in higher dermal blood flow. This causes an increase of the skin temperature and in the rate of heat loss between the skin and the external environment by radiation and conduction. Heat discharge to the external environment, associated with this mechanism, increases along with the difference between the temperatures of the skin and the external environment. The effectiveness of this process is reduced during physical work in a hot environment due to a decrease in the gradient between the temperatures of the skin and the surrounding air.

Sweat glands are stimulated by cholinergic sympathetic innervations, both when the internal body temperature increases during physical exercise and when the temperature of the environment is higher than that of the skin. The secretion and vaporisation of sweat, which plays a principal role in a hot environment, is a very efficient method to release heat. When the sweat glands are activated, the human body may lose a substantial amount of fluids along with electrolytes. Sweat excretion reaching up to even 7–8 litres per work shift has been observed among employees performing physical work in a hot environment.

Despite the activation of thermoregulation mechanisms allowing for heat discharge during physical exercise, the internal body temperature increases in proportion to the intensity of exercise. During long-term exercise, the body temperature stabilises after about 30 minutes. However, this occurs only temporarily and the duration, lasting for periods ranging from several minutes to about half an hour, depends on both the intensity of the physical exercise and the temperature of the external environment. During intense, long-term exercise, the body temperature may even exceed 40°C. The increase in body temperature during physical exercise is greater in dehydrated persons.

Adverse reactions to heat loads, which are a response to the vasomotor reaction and dehydration of the body as a result of sweating, include a decrease in cardiovascular blood volume, reduction of arterial blood pressure, and a general increase in the burden of the cardiovascular system.

During physical work, particularly in a hot environment, it is therefore necessary to supplement the body with fluids and electrolytes. The best source of fluid is noncarbonated mineralised water. Fluids should be delivered directly to the work area in unlimited quantities. This is particularly important in the case of older employees, because they show lower amounts of voluntary fluid intake, while the load on their cardiovascular systems is greater during physical work in a hot environment (Marszałek et al. 2005).

A decreased ability to perform physical exercise in a hot environment is the result not only of functional changes in the cardiovascular system, but also of the direct influence of temperature on the muscle metabolism.

2.8.2 Work in a Cold Environment

The body's reactions to a cold environment depend on its thermoregulatory mechanisms, which permit the maintenance of heat by both decreasing its loss through contraction of the skin vessels and increasing its production through muscle shivering, increased muscle tension, and nonshivering thermogenesis. Contraction of the skin vessels results in a reduction of the dermal blood flow, as well as a decrease in the gradient between the temperatures of the skin and the air around it, which reduces heat loss through radiation and conduction. Mechanisms that increase heat production in a cold environment are of little significance for maintenance of the body heat balance; on the contrary, they negatively influence the ability to perform physical exercise. An increase in both at-rest muscle tension and muscle shivering, resulting from the necessity to produce heat, causes muscle stiffness, reduction of muscle endurance, and less precise movements. These reactions result in lowered work ability.

During intense physical work performed in a cold environment, thermal load may also occur, triggering the thermoregulatory mechanisms of heat loss, that is, the dilation of blood vessels and activation of sweat glands. In such a situation, the employee experiences sensations of heat and often removes his clothing, resulting in the possibility of excessive cooling of the body, particularly after exercise. Therefore, persons working in cold environments need to be provided with a heated, closed room, in which they can rest during breaks.

The thermoregulatory contraction of blood vessels in the skin increases peripheral resistance, which may result in increased arterial blood pressure and load on the cardiovascular system. In persons suffering from arterial hypertension or ischaemic heart disease, exposure to low temperatures may result in aggravation of the symptoms of these diseases.

2.8.3 Principles of Organisation of Physical Work in Hot and Cold Environments

When performing work under conditions that could lead to disturbance of the heat balance, special rules should be followed to ensure employee safety. The most significant of these include the following:

- Proper qualification of employees for work, with reference to both their health and age
- Acclimation of the employee to the conditions of work
- Securing employees against both heat loss in a cold environment and overheating in a hot environment by using appropriate personal protection equipment
- Supplementing fluids and electrolytes lost as a result of sweating
- Limiting both the duration of work and exposure to conditions that disturb the heat balance

Acclimation to a hot or a cold environment refers to the complex changes taking place in the human body as a result of repetitive exposure to a hot or cold environment, leading to an increased tolerance of the body to these conditions.

In a person who is acclimated to work in a hot environment, the physiological mechanisms resulting in increased sweat secretion are triggered faster, facilitating faster discharge of heat to the environment and thus reducing the heat load under such conditions; that is, there is a lesser increase of both the heart rate and the internal body temperature. As acclimation to heat increases, sweating starts at a lower skin temperature and after a smaller increase in the internal body temperature; the sweat secreted also contains less sodium chloride.

Acclimation to a cold environment may be of either general or local nature. *General acclimation* to cold conditions means there is an increase in the basic metabolic rate and an increase in heat production as a response to the low temperature of the environment. Among persons who have been acclimated to cold conditions, a greater nonshivering thermogenesis and a thicker layer of the subcutaneous tissue are observed. However, such acclimation changes take place only under constant exposure to cold conditions. *Local acclimation* to cold conditions means there is an increased dermal blood flow and increase of the skin temperature in the areas that are directly exposed to low temperatures such as the hands, feet, and face.

Acclimation mechanisms are not retained for life. They should be stimulated after each period of inactivity at work, such as vacation or sick leave.

REFERENCES

Astrand, P.-O., and K. Rodahl. 1986. *Textbook of Physiology.* New York: McGraw-Hill.

Bugajska, J. 2007. *Older Workers—Physical Capacities and Conditions.* Warsaw: CIOP-PIB.

Bugajska, J., T. Makowiec-Dąbrowska, A. Jegier, and A. Marszałek. 2005. Physical work capacity ($V_{O_2 max}$) of active employees (men and women). In *Assessment and Promotion of Work Ability, Health and Well-Being of Ageing Workers*, eds. J. Costa and J. Ilmarinen, 1280(C):156–160. Amsterdam: Elsevier, International Congress Series.

Bugajska, J., T. Makowiec-Dąbrowska, M. Konarska, D. Roman-Liu, T. Tokarski, J. Kaminska, Marszałek, et al. 2008. *General Physical Capacity and Fitness of Occupationally Active Population in Poland.* Warsaw: CIOP-PIB.

Datta, S. R., and N. L. Ramanathan. 1969. Energy expenditure in work predicted from heart rate and pulmonary ventilation. *J Appl Physiol* 26(3):297–302.

De Zwart, B. C. H., M. H. W. Frings-Dresen, and F. J. H. Van Dijk. 1995. Physical workload and the ageing workers: A review of the literature. *Int Arch Occup Environ Health* 68:1–12.

Dutta, C., and E. C. Hadley. 1995. The significance of sarcopenia in old age. *J Gerontol* 50A:1–4.

Grimby, G., and B. Saltin. 1983. Mini-review: The aging muscle. *Clin Physiol* 3:209–218.

Kamińska, J., T. Tokarski, and J. Słowikowski. 2003. Differences in accuracy of steering in older men and women. *Acta Bioeng Biomech* 5 (Suppl. 1):224–227.

Konarska, M., B. Kurkus-Rozowska, A. Krokosz, and M. Furmanik. 1994. Application of pulmonary ventilation measurements to assess energy expenditure during manual and Massie work. In *Proceedings of the Triennial Congress of the International Ergonomics Association,* 3:316–317.

Koradecka, D., and J. Bugajska. 1999. Physiological instrumentation. In *The Occupational Ergonomics Handbook,* eds. W. Karwowski and W. S. Marras, 525–547. Boca Raton, FL: CRC Press.

Kozłowski, S., and K. Nazar. 1999. *Introduction to Clinical Physiology.* Warsaw: PZWL.

Lamberts, S. W. J., A. W. Van den Beld, and A. Van der Lely. 1997. Endocrinology aging. *Science* 278:419–424.

Lindström, B., J. Lexell, B. Gerdle, and D. Downham. 1997. Skeletal muscle fatigue and endurance in young and old men and woman. *J Gerontol Biol Sci* 52A(1):B59–B66.

Makowiec-Dąbrowska, T. 1999. Occupational physiology. In *Occupational Hygiene,* ed. J. A. Indulski. Lodz: IMP.

Marszałek, A., M. Konarska, and J. Bugajska. 2005. Assessment of work ability in a hot environment of workers of different ages. In *Assessment and Promotion of Work Ability, Health and Well-Being of Ageing Workers,* eds. J. Costa and J. Ilmarinen, 1280C: 208–213. Amsterdam: Elsevier, International Congress Series.

Metter, E. J., R. Condit, J. Tobin, and F. J. Lozar. 1997. Age-associated loss of power and strength in the upper extremities in women and men. *Gerontol Biol Sci* 52A:B267–B276.

Nazar, K. 1991. The physiology of physical efforts. In *Sports Medicine (Selected Problems),* eds. A. Dziak and K. Nazar, 11–95. Warsaw: Polish Society of Sports Medicine.

Nazar, K. 1999. The physiology of work. In *Occupational Safety and Ergonomics.* ed. D. Koradecka, 87–118. Warsaw: CIOP.

Porter, M. M., A. A. Vandervoort, and J. Lexell. 1995. Aging of human muscle; structure, function and adaptability. *Scand J Med Sci Sports* 5:129–142.

Shvartz, E., and R. C. Reinbold. 1990. Aerobic fitness norm for males and females aged 6 to 75 years, a review. *Aviat Space Environ Med* 61:3–11.

Steinhaus, L. A., R. E. Dustman, R. O. Ruhling, R. Y. Emmerson, S. C. Johnson, D. E. Shearer, R. W. Latin, J. W. Shogeoka, W. H. Bonekat. 1990. Aerobic capacity of older adults: A training study. *J Sports Med Phys Fitness* 2:163–172.

Tokarski, T., and J. Kamińska. 2004. The relationship between age and physical strength in working males. *Acta Bioeng Biomech* 6 (Suppl. 1):399–403.

3 Selected Issues of Occupational Biomechanics

Danuta Roman-Liu

CONTENTS

3.1 BEGINNING AND DEVELOPMENT OF BIOMECHANICS

Biomechanics is a science dealing with equilibria and human motions; it studies the mechanical properties of tissues, organs, and systems, as well as the mechanical motion of living organisms and its causes and results. There are external (e.g. gravity) and internal (e.g. muscle) forces that cause body motion. As a result of these forces, the position of a body (one's own body or foreign bodies) changes or body deformation occurs. The name 'biomechanics' is derived from the Greek word *mechana*, which means tool. The suffix 'bio-' indicates that this science concerns living organisms.

Biomechanics combines knowledge from the fields of mathematics, physics, and anatomy. The foundations of this science were laid long ago by mathematicians and physicians. Even Socrates, who lived 2400 years ago, thought that we would not be able to understand the world around us without understanding ourselves. A physician's son who possessed an exceptional gift for observation, Aristotle (384–322 BC) expanded upon this idea and was fascinated by the anatomy and structure of living creatures. He perceived the movements of animals as similar to those of mechanical systems. Aristotle laid the foundations of mechanics, referring to levers and forces acting on a given arm. Two hundred years later, Archimedes (287–212 BC) said, 'Give me where to stand, and I will move the Earth'; his works allowed for the development of statistics and the use of both forces and moments of forces for the mechanics of living creatures (http://www.spinalfitness.com/history.htm).

Leonardo da Vinci (1452–1519) played an unquestionable role in the development of biomechanics. He described the anatomy of humans and was likely the first person to study the stability of the spine. Da Vinci found that muscle forces acted along the link between a muscle attachment and the point of the application of force. Giovanni Alfonso Borelli (1608–1679) is considered to be the father of biomechanics; in association with Marcello Malpighi (1628–1694), he presented an analysis of spine load for the first time ever in a work entitled *De Motu Animalium*. Borelli determined the forces that were necessary to maintain static equilibrium in the different joints of the human body.

Biomechanics currently comprises a great number of different fields. Topics of research in biomechanics are widespread, from the mechanics of vegetables and fruits to complex control systems in highly developed organisms, including humans. Biomechanics is applied to describe internal phenomena, such as the forces, structure, or electrical activities of muscles, as well as external phenomena, described using the analysis of motion and force characteristics. Because biomechanics is an interdisciplinary science studying the structure of the motions of living organisms, especially humans, and mainly uses the methods of mechanics, the laws of classical mechanics are applied to biomechanics—above all, the laws of motion formulated under the name of Newton's principles of dynamics. The first principle states that a physical body will remain at rest or continue to move at a constant velocity unless an outside net force acts upon it. The second principle regards a change in the motion velocity or the momentum and states that the rate of a change in momentum is proportional to the force acting on the body and takes place in the direction of that force. The third principle states that for every action, there is an equal and opposite reaction.

One of the basic objectives of biomechanics is to describe the position of a body and the changes to this position, namely, motion. Analysis of the motion may be based on anatomy and physiology, as well as on external observation of the motion. Etienne Jules Marey (1830–1904) and Edward James Muybridge (1830–1904) were forerunners in the analysis of motion. As early as 1882, they used a prototype of a roll of film to capture photographs in burst mode (http://www.utoledo.edu/kinesiology/classess). Using special cameras, they photographed the various phases of motion of a galloping horse on wet collodion plates. These pictures were scientific proof that a galloping horse leaps from the ground with one leg while the other is lifted slightly. In addition, Marey developed a post, called Stadion Physiologique, to investigate forces of pressure acting on the ground. This was a prototype of a dynamometric platform.

3.2 OCCUPATIONAL BIOMECHANICS

The International Society of Biomechanics suggests dividing the subject of biomechanics into three subtopics: (1) engineering biomechanics, involving models and human-machine systems; (2) medical biomechanics, based on anatomy and physiology, and (3) general biomechanics, involving methodology, functional structures, control of biological systems, data collection, and biomechanics of sports and basic movements. Occupational biomechanics and its aim to investigate both the causes and the results of the external and internal forces acting on a worker is also receiving increasing attention.

More research in the field of occupational biomechanics is necessary to design ergonomic work places and work processes. Thus, occupational biomechanics is related to ergonomics, just as it is related to anthropometry, physiology, and psychology. Occupational biomechanics discusses the causes and results of load, resulting from physical work, to the human musculoskeletal system, and this biomechanics is widely used to design work places and processes. This design process should include analysis of body load due to occupational activities and its comparison with a worker's abilities, which are determined based on the biomechanical properties of the musculoskeletal system. Typical goals of occupational biomechanics are as follows:

- To investigate the structures and functions of the musculoskeletal system, which is treated as an apparatus of work; to measure the mechanical parameters (forces and moments of the forces acting on the human body; dislocations, velocities, and accelerations of body segments while working).
- To measure muscle function by electromyography (EMG).
- To perform mathematical modelling and computer simulation of work processes.
- To assess the load on the human body while it performs typical activities such as carrying weight and operating machines or computers.

3.3 POSITIONS AND MOVEMENTS OF BODY SEGMENTS

The motor apparatus of living organisms is based on mobility in adjacent body segments that are linked by joints. Movement of separate segments occurs as a result of the actions of internal forces. The total number of all movements, called the number of degrees of freedom, may be performed at specific joints—assuming that each of these movements may be performed independently—depending on the biomechanism. In humans, the number of degrees of freedom may be 240–250. As a result, the positions of different segments in relation to each other in the occupational environment have to be described using approximately 250 variables.

The body of a biomechanism may be comprised of various numbers of segments, which depend on the complexity of the model. The number of degrees of freedom also depends on the accuracy.

The positions and movements of separate human body parts are usually described with reference to three body planes: sagittal, frontal, and transverse (Figure 3.1).

The angles of abduction and adduction occur in the frontal plane. The angles of flexion and extension occur in the sagittal plane, and the angles of rotation, namely, the rotating movement towards the inside (pronation) and the outside (supination), occur in the transverse plane. As a result, the position of a body can be defined using angles of these three planes. The number of angles corresponds to the number of degrees of freedom; in other words, in the case of the most detailed human model, there would be 250 angles. The total value of all the angles equals zero in the natural position, a straight standing position, with the upper limbs hanging along the body (standing 'in attention').

The human upper limbs may be modelled using 30 degrees of freedom; however, they may also be modelled using the minimum number—seven degrees of freedom

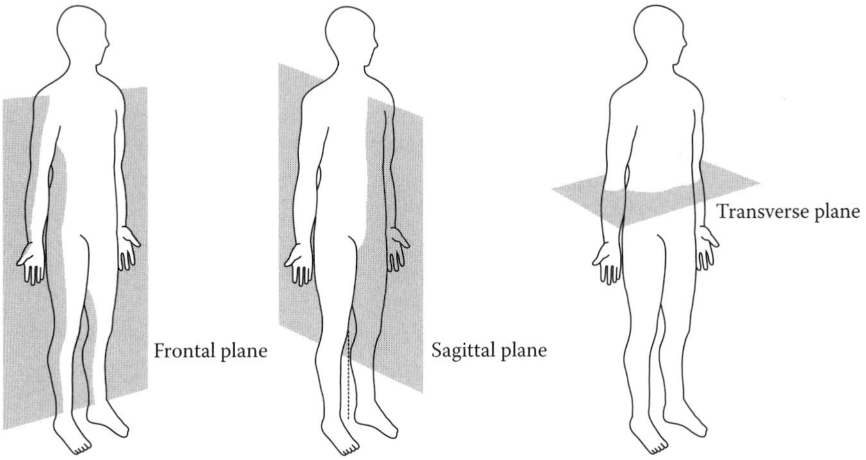

FIGURE 3.1 Frontal, sagittal, and transverse planes of motion.

(Morecki et al. 1971). In such a minimalist case, the hand is replaced with a single segment, excluding the possibilities of manipulation, which makes it possible to represent the basic force characteristics of the upper limb. A simplified model of the upper limb, expressed using a kinematic chain with seven degrees of freedom, uses seven angles, which describe how the limb deviates from a natural position in terms of flexion, extension, abduction, adduction, rotation along the arm axis, pronation, and supination. Figure 3.2 is a graphical presentation of the angles that define a position of the upper limb when the model with seven degrees of freedom is used (Roman-Liu 2003a).

Apart from the body position and motor activities, the forces connected with the execution of work are an important part of biomechanics. The forces involved in the process of human motor activity may be divided into external and internal forces, depending on their relationship to the musculoskeletal system. According to this division, external forces include forces applied on the object of work; internal forces include muscle forces. External forces can be measured experimentally using different measuring devices, such as a dynamometer. Measuring internal forces, including muscle forces, may be attempted, but it requires disruption of the body covering, that is, the use of invasive measurement techniques. Muscle force can be evaluated based on either measurements carried out by a transducer implanted between the muscle and tendon or the measurement of the EMG signal using needle electrodes injected into the belly of a muscle. Muscle force can also be assessed using surface EMG, namely, electrodes stuck on the surface of the skin over the tested muscle.

3.4 ANTHROPOMETRY IN OCCUPATIONAL BIOMECHANICS

Anthropometry is used in physical anthropology and involves comparative measurements of human body parts such as bone length, skull and head volume, body proportions, body weight, eye spacing, and so on.

The worker population has varied dimensions and body weights. Designers of machines, devices, or working postures must determine the dimensions of their

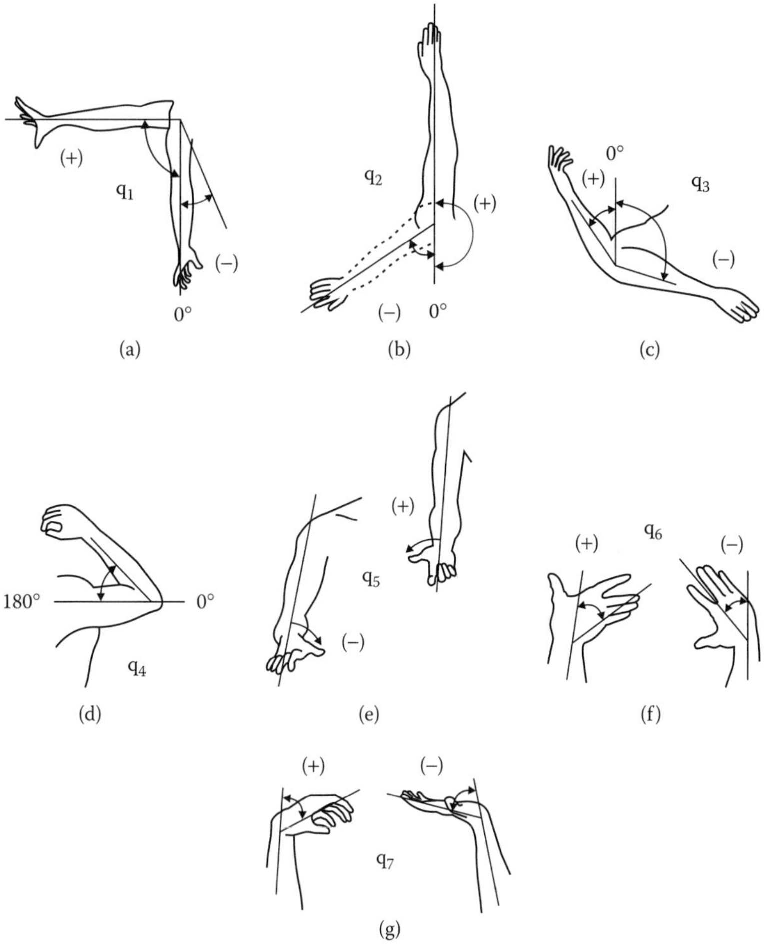

FIGURE 3.2 Graphical presentation of the angles, making it possible to define the position of the upper limb using seven angles: (a) the abduction or adduction angle in the transverse plane; (b) the angle of arm flexion or extension in the sagittal plane; (c) the angle of rotation around the arm axis; (d) the elbow flexion angle; (e) the angle of rotation around the forearm axis; (f) the angle of abduction or adduction in the wrist; and (g) the angle of extension or flexion in the wrist.

users. Hence, the dimensions of the general population must be discussed in terms of anthropometric parameters. There are somatic anthropometric parameters, such as the height, width, length, depth, and perimeters of different body parts, and functional parameters, which determine the distance and angular range of movements of different body segments.

The anthropometric parameters that describe the dimensions or measurements of different body segments are listed in standard EN 547-3 (2000). This standard also presents anthropometric data, namely, the values of anthropometric parameters for the European population. Twenty-three anthropometric features are distinguished

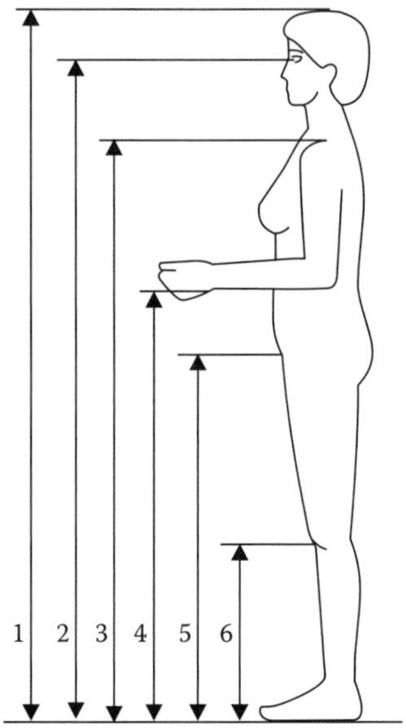

1	Body height
2	Eye height
3	Shoulder height
4	Ulnar height
5	Perineal height
6	Knee height

	Women		
	5 c	50 c	95 c
1	152.4	161.5	170.7
2	142.1	150.7	159.6
3	124.1	132.3	140.6
4	91.3	98.6	105.0
5	76.0	82.9	89.7
6	38.5	42.3	46.3

FIGURE 3.3 Anthropometric measurements of the human body in a standing body position. (Reprinted with permission from Gedliczka, A. 2001. *Atlas of the Human Body.* Warsaw: CIOP.)

therein. *Atlas miar człowieka* (Gedliczka 2001) is a compendium of current, selected information, which is useful in design and ergonomic evaluation. Information contained in the atlas includes Polish anthropometric data (somatic and functional), biomechanical data, the parameters of occupational space, and safety measurements. Figure 3.3 presents model anthropometric parameters and data for a standing position, which are also contained in the atlas.

Measurements of anthropometric parameters are subject to a normal distribution, according to the Gaussian function that describes the number of cases in terms of a value corresponding to a measured anthropometric parameter. The measurement characteristics of a population are described using the centile concept, that is, a dot on a scale, below which a determined percentage of cases can be found.

For the purpose of designing and evaluating working postures, anthropometric data are usually given in the system of values represented as the 5th and 95th centiles. These values are called extreme values and using them makes it possible to take into account differences among measurements involving 90% of the population. According to the variable distribution function, the 5th centile is a value of the measured parameter obtained in 5% of the population, whereas the 95th centile is a value obtained in 95% of the population. This is a rule of limited measurements used to design working postures.

The dimensions of the human body and dimensions of the work stand determine the spatial structure of a working posture, which means that in order to obtain optimum working posture, it is necessary to adjust the work place to the group of users. The anthropometric dimensions of a worker in relation to a working posture are of basic significance for taking into account a worker's load. A working posture may be adjusted to the individual dimensions of a user or to the dimensions of a population of target users.

The spatial structure of a working posture has the construction, shape, size, and configuration of the work stand. The worker's musculoskeletal load is mainly affected by the working posture, which is in direct contact with the work stand. A connection between the human body and the object is made through points with which the worker comes into tactual or visual contact; the work stand is a type of spatial structure, a configuration and location of control and informational elements, as well as a location for work tools. The location of the contact points should correspond to the dimensional characteristics of the worker population. The structure, along with the contact points, determines the work space and, therefore determines the position of a worker's body and its musculoskeletal load resulting from the work being done.

Criteria regarding occupational zones defined in the horizontal plane and the height of the occupational zones are used to determine the spatial occupational zones.

The horizontal plane has a normal range, a maximum range, and a forced range. The *normal range* is the range determined by subsequent positions of the middle of the hand when the forearm is being rotated in relation to the elbow joint (Figure 3.4). The *maximum range* is a less comfortable range of the upper limbs, determined by subsequent positions of the middle of the arm when the whole straightened limb is moved in relation to the shoulder. The *forced range* is reached when body movements exceed the maximum range. Works performed in the maximum and forced ranges are especially unfavourable.

The height of the occupational space, depending on the type of work, is important to configure working postures. The height of the occupational space is usually determined in relation to the ulnar height. Grandjean (1990) suggested recommended heights of working planes for different types of jobs done in a standing position. For jobs that do not require exceptional accuracy, the palms should reach 75 mm below the position of the elbows when arms are lowered freely. During manipulative jobs, the elbows should not be raised to heights greater than 100 mm above their position when the arms are lowered freely. For jobs requiring accuracy and

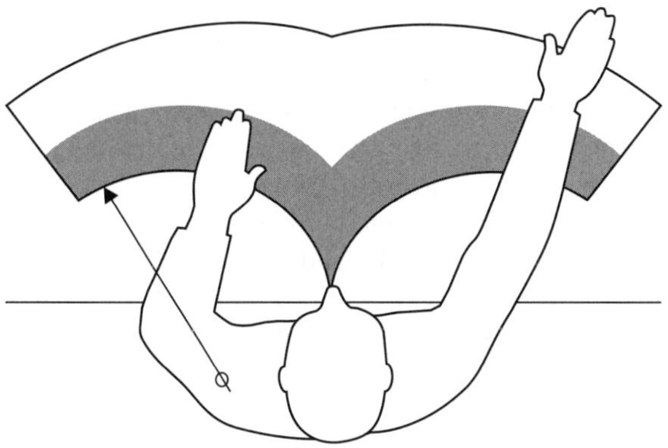

FIGURE 3.4 Occupational zones for the upper limbs in the horizontal plane.

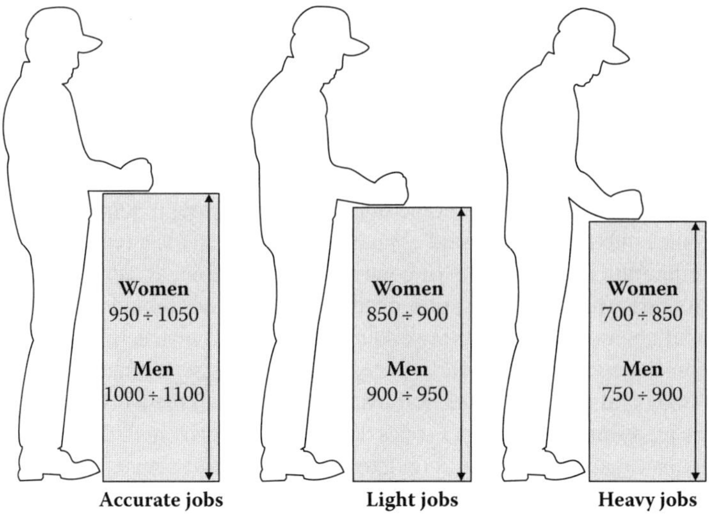

FIGURE 3.5 Occupational zones for the upper limbs in the vertical plane.

therefore exceptional visual control, occupational planes that are higher in relation to the normal position are applied (Figure 3.5).

The rule applied when designing working postures is that external dimensions—such as the location and distance of control elements—are applied according to the dimensions of the 5th centile, whereas the internal dimensions—such as the dimensions of transition openings or accesses—are applied according to the dimensions of the 95th centile (EN 547-1 2000; EN 547-2, 2000).

3.5 BIOMECHANICAL FACTORS FOR MUSCULOSKELETAL LOAD

An appropriately designed workplace should minimise the risk of musculoskeletal system disorders. The development of such disorders is strictly connected to the worker's musculoskeletal load, associated with the occupational activities that he or she performs. A worker's musculoskeletal load is influenced by biomechanical factors affecting humans, such as body segments' positions, external force (the type, direction, and value of a given force), and time factors. A time factor may be regarded as the duration for which a given body position is maintained (often when a force is applied), the frequency of repetitions of given activities, or the maximum duration of maintaining a static load or performing a repetitive task. The above three basic biomechanical factors determine the musculoskeletal load and the fatigue caused by performing given occupational activities and are always discussed together.

The body position taken during work is determined by the relationships between construction of a working posture, the necessity to perform given activities, and the anthropometric dimensions of the worker. For this reason, the design of a working posture should take into consideration the limitations associated with body dimensions and the work process. An appropriate body segment position will be provided when the spatial construction of a work stand is adjusted to the operator.

When musculoskeletal load is at a minimum, the optimum body position is the position wherein the values of the angles describing a body position equal zero; that is, the natural body position. This is a standing position with the spine straight and upper limbs hanging along the body. A body position that is uncomfortable, limited, or extremely deviated from the natural body position may cause overloading of muscles, ligaments and tendons. The greater the deviation of the body from the natural position, the greater is the load of the musculoskeletal system. Therefore, the working posture and activities of an operator should not force a worker to maintain uncomfortable positions. Such positions result in high values of compressive and shearing forces and force movements in the spine and joints of the upper and lower limbs, a direct reason for the development of traumas and disorders of the musculoskeletal system.

When the working posture is inappropriate and maintained for a longer period of time, it may cause a significant load on the muscular system. This phenomenon intensifies when a force must be applied to the components of a work stand, such as the means of work, work tools, and control elements.

The body position affects not only the load on the joints that experience the force being applied, but also the characteristics of the force. The value of the maximum force being applied is greater for some angles in the joints than for others, which means that the body position differentiates the values of the maximum force being applied, namely the maximum voluntary contraction (MVC). However, the maximum values of the force are different, depending not only on the body position, but also on the type of applied force. In the case of the upper limbs, it is possible to distinguish activities that involve the muscles of the whole upper limb, such as lifting, pulling, pushing, or rotation (pronation and supination of the forearm), as well as activities that involve only local muscles, such as the grip of the hand when the force is mainly exerted by the muscles of the hand and forearm.

Because the position of the upper limbs has a significant effect on the value of the exerted maximum force, it is necessary to consider both the type of exerted force and the position of the upper limbs in order to obtain standard values of force. Thus, a significant number of values of forces must be determined. This can be achieved by measuring forces of different types and positions for a determined population or by developing mathematical relationships that make it possible to calculate the maximum value of a specific force for any positions of the upper limbs (Roman-Liu 2003b, 2003c; Roman-Liu and Tokarski 2005).

Moreover, individual factors such as age, sex, or muscle mass are significant for defining the force characteristics. Research has shown that external force in women is about 75% of the force in men, which mainly results from differences in the muscle mass between these two populations (Balogun et al. 1991; Su et al. 1994; Hallbeck and McMullin 1993).

Apart from the external force exerted and inappropriate body positions during work, the rhythm of the work is also a cause of muscular system disorders. Static load and repetitive load are the most dangerous for the musculoskeletal system.

Static muscle tension is isometric muscle contractions in which the muscle length does not change. Constant muscle length occurs in situations in which the positions of the involved body segments do not change. From an ergonomic point of view, static muscle tension is characterised by the duration of the tension for a given level of the force. The maximum duration load is maintained is inversely proportional to the level of force being applied (Fugelvand et al. 1993). In both static constant and repetitive work, the allowable time for work performance declines exponentially as the external load increases. The load is related to the tension of the muscle group that is the most loaded in a given occupational activity. Consequently, the allowable duration of a task is usually presented as the function of load, expressed as a percentage of the maximum effort of a muscle or a group of muscles (in other words, the MVC).

Repetitive muscle tension is caused by a constant repetitive loading of the same tissues. Repetitive work, according to Kilbom (1994), is work that requires the performance of similar cycles of occupational activities several times. Similar cycles are defined in terms of time sequences, evolving muscle forces, and the spatial characteristics of movements. Repetitive work is defined by parameters of repetitive work (Roman-Liu 2003a), that is, the time characteristics of a cycle: the time period for which a given cycle lasts, the time period for which separate cycle phases last, and the number of cycle phases, as well as the parameters that characterise a force in association with the position of, for example, the upper limbs.

There are different types of repetitive work; the most basic example is a cycle consisting of a load phase and a relaxation phase (Mathiassen and Winkel 1991) Repetitive works with a larger number of working cycle phases with varying time periods and forces may occur during each of these phases. Moreover, works with an additional basic cycle may occur. A basic cycle is a cycle consisting of an element of a series of repetitive steps during a working cycle.

Silverstein et al. (1986) divided repetitive work into work characterised by cycles of high repeatability and cycles of low repeatability. She regarded cycles of high repeatability as cycles with a duration less than 30 seconds or cycles that are performed for more than 50% of the duration. A cycle of low repeatability is a cycle that lasts longer than 30 seconds and in which more than 50% of the cycle time period is devoted to performing the same sequence of activities. According to research, repetitive work, that is, work when the time period of a repeated cycle is less than 30 seconds, may be a reason for the development of musculoskeletal system disorders (Silverstein et al. 1986).

Taking into account the parallel effects of load on the musculoskeletal system, biomechanical factors such as the body position, and the exerted force and time, a working posture should be designed so that activities requiring force exertion are performed in an optimal way, taking into account positions of separate body segments, the direction, type, and value of an exerted force, its frequency of occurrence, and the duration of the action. Moreover, different muscle groups should be activated alternately during work so as not to cause static overload and fatigue of the musculoskeletal system.

3.6 ASSESSMENT OF LOAD OF THE MUSCULOSKELETAL SYSTEM BASED ON BOTH THE ACTIVITY OF THE MUSCLES AND MODELS OF A HUMAN BODY

The load on the worker's musculoskeletal system may be assessed as external load, related to the static and dynamic physical effort or internal load, or related to the reaction of the body to the development of external load.

External load can be assessed using different methods, such as the Ovako Working Posture Analysis System (OWAS; Karhu et al. 1977), Occupational Repetitive Actions (OCRA; Colombini 1998) or the Repetitive Task Indicator method (Roman-Liu 2007). These methods assess load based on parameters that describe the positions of separate body segments, the force exerted by the worker, and the time sequences of load.

Internal load and worker's fatigue can be assessed using methods such as analysis of blood pressure, analysis of energy expenditure, or analysis of the electric signal that characterises muscle contractions, namely, surface EMG.

In the last several years, surface EMG has become one of the most dynamically developing methods to assess work load and muscle fatigue. EMG is based on the recording of electrical activity from selected muscles involved in the performance of activities. Surface electrodes register signals in a noninvasive and relatively easy way, and measurements can be performed at a work place.

The EMG signal is stochastic and is within the frequency band of about 5–1000 Hz, however, the value of the upper limit is 450 Hz (above 450 Hz, harmonic components are negligible). Apart from the frequency, another parameter that characterises the EMG signal is its amplitude, which ranges from several microvolts at rest to dozens of millivolts during maximum muscle tension.

The EMG signal is the source of a great deal of information regarding processes that occur in the muscle, including muscle load and fatigue. Figure 3.6 presents a typical waveform of the raw EMG signal registered during a 1-minute timespan. A significant increase in the signal amplitude is visible, resulting from an increase in the muscle strength.

Only qualitative assessment, which indicates whether the muscle works with a greater of lower force, is possible based on the recording of the raw EMG signal. To obtain quantitative information, the signal must be processed mathematically. Based on this processing, parameters characterising the EMG signal can be distinguished, indicating the processes occurring in the muscle, such as muscle fatigue.

Muscle fatigue is the result of processes that change the ability of the muscle to retain a determined level of force and/or static body position, and is defined as decreased force-generation possibility resulting from an increased feeling of effort. Changes occurring as a result of muscle fatigue are visible in the electromyogram recording. Muscle fatigue causes an increase in the EMG signal amplitude and a

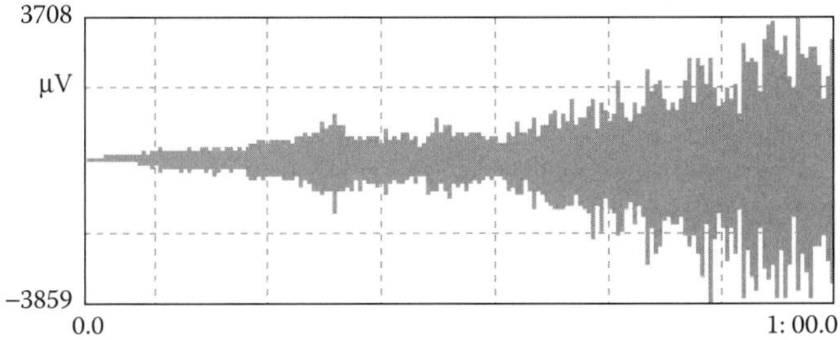

FIGURE 3.6 Raw electromyography signal registered during exertion of increasing levels of force.

shift in the power spectrum towards low frequencies; this is visible as a change in the values of the EMG signal parameters.

The analysis of the EMG signal has some limitations, especially in relation to muscle fatigue. These limitations mainly result from the need to conduct recordings during isometric muscle contraction in order to perform a spectral analysis of the EMG signal. It is especially difficult to provide such measuring conditions when analysing fatigue at various levels of the exerted force or at various positions of the upper limbs. This means that it is far more complex to examine the consequences of repetitive load than those of static load.

Rapid development in the field of technology favours the development of computer methods based on human body models of different complexities to assess the musculoskeletal system. Computer models are developed based on material models that use mathematical modelling and computer simulation. Different models of living organisms, which take into account three basic types of solutions, have been suggested: human mechanics treated as a whole, the mechanics of separate human body segments, and movement control.

The first stage of model generation is theoretical research and starts with the formulation of a mathematical description of a tested biomechanical system. Based on a formulated physical model, it is then possible to solve a system of equations, inequalities and other relationships forming a mathematical model. The last stage of this process is the presentation of results. Multibody and finite element are the most common types of models; however, mixed models are also used.

Multibody-type models examine the human motor system as a multisegment system, that is, as a biomechanism consisting of solid segments linked with biokinematic pairs (joints) and driven by the forces exerted by muscles or groups of muscles. Multibody models may be static or dynamic. If the model is for the dynamics of human body movements, then for a selected physical multisegment model, the equations of the motion are derived using the principles of mechanics and have the form of ordinary differential equations, depending on the number of variables that have been used.

An appropriate model that facilitates a study of the interactions of different factors of external load can be selected depending on the task. Two basic types of dynamic tasks can be solved using a derived mathematical model:

1. *Simple dynamics*: It is possible to track the course in time (trajectory) of transitions, velocities, and accelerations of any number of selected points in the human body, if the course in time of the forces exerted by muscles (or force moments in relation to the rotation axis in the joints) and the external forces applied to the model segments are known.
2. *Reverse dynamics*: It is possible to determine the course in time of the moments of muscle forces that caused the movement and the reaction forces in the joints by solving an equation of motion, if the trajectories of selected system points (a given motion) are known based on experimental studies.

Physical models with distributed parameters are generated by precise representation of the geometry of the human body using a great number of elements (hence the finite element method). Certain determined geometric, material, and other properties can be attributed to these elements. The mathematical description of such models is extremely complex and comprises systems of equations (from several hundred to tens of thousands of equations, depending on the size of the modelled body part), the generation and solution of which is only possible using special computer methods. Models with distributed parameters make it possible to determine both the internal forces and the distribution of stress and deformation in the human musculoskeletal system.

One example of a model developed using the finite element method (FEM) is the spatial model of the musculoskeletal system of the human trunk, which takes into account the complexity of a shape that is difficult to represent (Kamińska et al. 2004, 2006). The lumbar region of the spine is modelled extremely precisely and the differences in the structures of all lumbar vertebrae are included. As shown in Figure 3.7, this model comprises five lumbar vertebrae, six intervertebral discs from Th12/L1 to L5/S1 (anuli fibrosi with nuclei pulposi), parts of the sacral bone with the upper edge of the pelvis, parts of the trunk above the intervertebral disc Th12/L1 (modelled as a rigid body), ligaments (such as anterior, posterior, yellow and interspinous), and muscles (such as intertransverse, interspinal of the back, external and internal abdominal oblique, and straight abdominal).

This model assumes that the muscular system should perform a task with the minimum work input; this is required to formulate a mathematical model as an optimisation task. The optimisation criterion of muscle work is based on a criterion of the

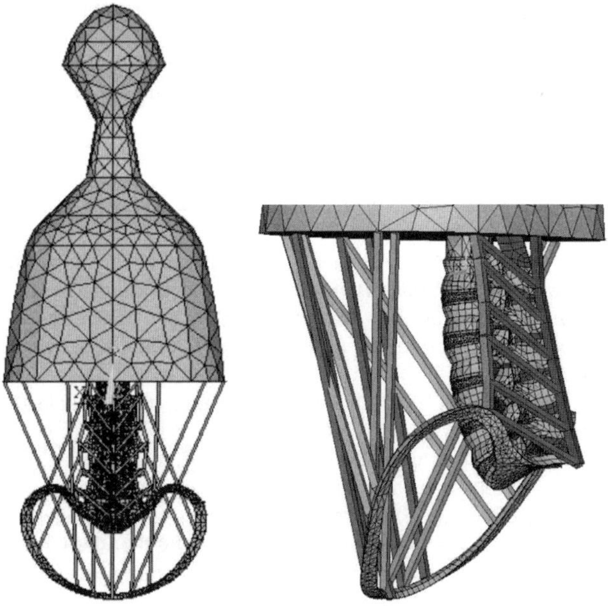

(a) (b)

FIGURE 3.7 Finite element method model of the trunk: (a) view of the model back; (b) modelled muscles.

minimum energy necessary to maintain a given position under a given external load (Dietrich et al. 1993; Zagrajek 1990; Kędzior and Zagrajek 1997) generated by the muscles and a criterion of the energy associated with the pressure in the abdominal cavity.

This model can be used to assess the load of the lumbar spine under different conditions of external load associated with the occupational activities performed. The results of the model calculations may include, among others, transitions of the model elements, stresses in intervertebral discs and/or lumbar vertebrae, and the forces acting on the modelled muscles.

REFERENCES

Balogun, J. A., C. T. Akomolafe, and L. O. Amusa. 1991. Grip strength: Effects of testing posture and elbow position. *Arch Phys Med Rehabil* 72:280–283.

Colombini, D. 1998. An observational method for classifying exposure to repetitive movements of the upper limbs. *Ergonomics* 41:1261–1289.

Dietrich, M., K. Kędzior, K. Miller, and T. Zagrajek. 1993. Statistics and stability of human spine under working conditions. In *The Ergonomics of Manual Work*, eds. W. S. Marras, W. Karwowski, and J. L. Smith, 147–150. London: Taylor & Francis.

Fugelvand, A. J., K. M. Zackowski, K. A. Huey, and R. M. Enoka. 1993. Impairment of neuromuscular propagation during human fatiguing contractions at submaximal forces. *J Physiol* 460:549–572.

Gedliczka, A. 2001. *Atlas of the Human Body*. Warsaw: CIOP.

Grandjean, E. 1990. *Ergonomics in Computerized Offices*. London: Taylor & Francis.

Hallbeck, M. S., and D. L. McMullin. 1993. Maximal power grasp and three-jaw chuck pinch force as a function of wrist position, age, and glove type. *Int J Indust Ergon* 11:195–206.

Kamińska, J., D. Roman-Liu, T. Tokarski, P. Borkowski, and T. Zagrajek. 2006. Computer simulation of spine load during backpack carrying. *Research Papers of the Department of Applied Mechanics* 26:185–190.

Kamińska, J., D. Roman-Liu, T. Zagrajek, P. Borkowski. 2004. The model of human lumbar spine. *Acta Bioeng Biomech* 6 (Suppl. 1):77–81.

Karhu, O., P. Kansi, and I. Kuorinka. 1977. Correcting working postures in industry: A practical method for analysis. *Appl Ergon* 8:199–201.

Kędzior, K., and T. Zagrajek. 1997. A biomechanical model of the human musculoskeletal system. In *Modelling and Simulation of Human and Walking Robot Simulation*, eds. A. Morecki and K. Waldron, 124–153. Wieden: Springer Verlag.

Kilbom, A. 1994. Repetitive work of the upper extremity. I. Guidelines for the practitioner. *Int J Indust Ergon* 14:51–57.

Mathiassen, S. E., and J. Winkel. 1991. Quantifying variation in physical load using exposure vs time data. *Ergonomics* 12:1455–1468.

Morecki, A., J. Ekiel and K. Fidelus. 1971. *Bionics of Movement*. Warsaw: PWN.

PN-EN-547-1. 2000. Safety of machinery. Human body measurements. Principles for determining the dimensions required for openings for the whole body access into machinery.

PN-EN-547-2. 2000. Safety of machinery. Human body measurements. Principles for determining the dimensions required for access openings.

PN-EN-547-3. 2000. Safety of machinery. Human body measurements. Anthropometric data.

Roman-Liu, D. 2003a. *Biomechanical Analysis of Repetitive Work*. Warsaw: CIOP-PIB.

Roman-Liu, D. 2003b. Maximum handgrip force in relation to upper limb location—A meta analysis. *Am Ind Hyg Assoc J* 64:609–617.

Roman-Liu, D. 2003c. Maximum force of tip pinch, lateral pinch and palmer pinch in relation to maximum handgrip force—A meta analysis. *Biol Sport* 20(4):303–319.

Roman-Liu, D. 2007. Repetitive task indicator as a tool for assessment of upper limb musculoskeletal load induced by repetitive tasks. *Ergonomics* 50(11):1740–1760.

Roman-Liu, D., and T. Tokarski. 2005. Upper limb strength in relation to upper limb posture. *Int J Indust Ergon* 35:19–31.

Silverstein, B., L. J. Fine, and T. J. Armstrong. 1986. Hand wrist cumulative trauma disorders in industry. *Br J Indust Med* 43:779–784.

Su, C., J. Lin, T. Chien, K. Cheng, and Y. Sung. 1994. Grip strength in different positions of elbow and shoulder. *Arch Phys Med Rehabil* 75:812–815. http://www.spinalfitness.com/history.htm (accessed November 11, 2009).

Zagrajek, T. 1990. *Biomechanical Modelling of the Spine with the Finite Elements Method.* Warsaw: Technical University of Warsaw.

4 Psychosocial Risk in the Workplace and Its Reduction

Maria Widerszal-Bazyl

CONTENTS

4.1 INCREASE IN THE IMPORTANCE OF PSYCHOSOCIAL RISK IN THE WORKPLACE

The past few decades have brought significant changes to the characteristics and conditions of performed work. The significance of heavy manual labour, performed in extremely tough external environments using imperfect—at least compared to the contemporary—tools and machines has decreased, while the importance of clerical work and service sectors has increased. These types of work are often performed in huge, sometimes international organisations, and usually require a level of cooperation among various categories of coworkers, supervisors, and customers. The market's globalisation and greater competition among related companies have a significant impact on the characteristics of work. This competition forces a constant growth in work intensity, as well as an organisation's continued transformation to accommodate its newest needs. It may cause employees to feel overwhelmed with work and unable to follow ongoing changes with regard to the job's insecurity. Moreover, companies striving to achieve more flexibility have been searching for new forms of employment: fixed-time employment, part-time contracts, outsourcing, and temporary employment. Such new forms of employment usually have lower security and stability.

The above situation is one in which the social and organisational layer of work is the main source of stress, rather than the physicochemical work conditions. The fourth European survey of work conditions, conducted by the European Foundation for the Improvement of Living and Working Conditions (Parent-Thirion et al. 2007), reported stress to be the second most common cause of ailments, dwarfed only by musculoskeletal disorders. Among the employees of the 27 countries of the European Union (EU), 22% have suffered stress (25% reported musculoskeletal disorders). Stress-related problems occur more often in countries that have recently joined the EU. Moreover, a comparison of Eurofund surveys from 1995, 2000–2001, and 2005 shows they are on the rise in new member states (NMS), including Poland. The relevant data are shown in Table 4.1.

As shown in the table, in 2000–2001, the level of stress was similar in both old member states (OMS) and NMS of the EU (28% each). However, in 2000–2005, this level decreased to 20% in OMS. At the same time, the levels of fatigue, irritation, and headaches also decreased. A reverse trend was found to exist in NMS. The prevalence of stress grew slightly in these countries in 2001–2005 (from 28% to 30%). Some stress-related problems also increased; for example, sleeping problems and headaches increased from 8% and 15% to 12% and 24%, respectively. Due to

TABLE 4.1
Stress and Stress-Related Symptoms: Trends over Time

Question	EU-15			CC-12	NMS-10
	1995[a]	2000[a]	2005[b]	2001[c]	2005[b]
Does your job influence your health? (% yes)	57	60	31	69	56
Stress (% yes)	28	28	20	28	30
Fatigue (% yes)	20	23	18	41	41
Headaches (% yes)	13	15	13	15	24
Irritation (% yes)	11	11	10	11	12
Sleeping problems (% yes)	7	8	8	8	12
Anxiety (% yes)	7	7	8	7	7

Sources: [a] Paoli, P., and D. Merllie. 2001. *Third European Survey on Working Conditions 2000.* European Foundation for the Improvement of Living and Working Conditions. Luxembourg: Office for Official Publications of the European Communities.
[b] Fourth European Working Conditions Survey. 2005. Statistical annex. http://www.eurofound.eu.int/docs/ewco/4EWCS/4WCS_Annex%20statistical%20annex.pdf (accessed August 8, 2008).
[c] Working conditions in the acceding and candidate countries. 2001. Polish language version. http://www.eurofound.eu.int/publications/htmlfiles/ef0306.htm (accessed August 8, 2008).
Note: CC = candidate countries.

the reversal of previous trends, the difference in stress prevalence between OMS and NMS in 2005 also grew compared to 2000–2001. In 2005, stress prevalence in NMS was markedly higher than that in the EU 15—work-related stress was reported by 20% of OMS and 30% of NMS. Other stress-related conditions (except for anxiety) were also found at higher levels in NMS. According to the same survey, around 35% of the employees in Poland suffered stress at the workplace.

4.2 WHY PSYCHOSOCIAL JOB CHARACTERISTICS ARE A SOURCE OF RISK—THE MECHANISM OF STRESS

The mechanism of stress is the psychosocial property that explains the influence of work on the organisation's health. It justifies the usage of the term 'psychosocial risks'.

4.2.1 Concept of Stress

The concept of stress was introduced in the 1930s by physiologist Hans Selye, who defined it as 'the non-specific response of the body to any demand placed upon it' (Selye 1977, 14). This concept of stress works in both physiology and psychology. Its definition has undergone variations, not only as a reaction of the organism as the author would suggest, but also as the incentive to the reaction, as well as the interaction between entities and environments (see reviews by Reykowski 1966; Heszen-Niejodek 1996; Widerszal-Bazyl 2003). On the basis of the various approaches and studies, the meaning of the term here will be as follows: Stress in the workplace is a psychophysiological response to a situation in which the environment's requirements exceed the employees' capabilities or border such limits.

Because stress is a psychophysiological response, it can be described in terms of physiology and psychology.

4.2.2 Stress at the Physiological Level

At the physiological level, the three following axes are the most characteristic of stress (Everly and Rosenfeld 1992):

1. The first axis—direct agitation of internal organs by the autonomic nervous system. Brain impulses (from the hypothalamus) agitate the autonomic nervous system, innervating the internal organs. The autonomic nervous system consists of the sympathetic and parasympathetic parts, which are antagonistic to each other. The sympathetic circuit is especially important (although not exclusively) in relation to stress mechanics. Its agitation causes, among other effects, the following symptoms:
   - Pupil dilation
   - Sweat-gland agitation
   - Heart-rate acceleration
   - Musculoskeletal blood vessel extension
   - Stenosis of the blood vessels of the skin
   - Extension of gills (acceleration of respiration rate)
   - Decomposition of glycogen and release of cellohexaose in the liver
2. The second axis—neurohormonal (also called the hypothalamic-adrenal axis). Impulses are delivered to the hypothalamus from the tonsillar body (a part of the limbic system), and then pass through the autonomic nervous system to the core of the adrenal capsule, which in turn releases adrenaline and norepinephrine, sometimes called the 'stress hormones', into the blood circulation. Their effects are often similar to those brought on by stimulation of the first axis, however, those are more delayed and tenacious. Adrenaline and norepinephrine create, among other symptoms, the following effects:
   - Increase in arterial blood pressure
   - Increase in cardiac output
   - Increase of free fatty acids, cholesterol, and triglycerides in the blood plasma
   - Increase in muscle tension
3. The third axis—hormonal (also called the hypothalamic–pituitary–adrenal axis). This axis consists of three circuits that are more or less separate. We will focus on the hypothalamus, the most substantially diagnosed and

described. The hypothalamus, through the hormonal canals, agitates the most important vessel for internal incretion, the pituitary gland. The gland then releases adrenocorticotropic hormone (ACTH) into the blood circulation, stimulating the adrenal capsule's crust to release glucocorticoids (the most recognisable is cortisol). Glucocorticoids cause the following symptoms:

- Increase in glucose production
- Increase in levels of free fatty acids in the blood
- Inhibition of immunological functions

4.2.3 Stress at the Psychological Level

At the psychological level, the emotional reactions most often caused by stress are anger, fear, and frustration. Cognitive reactions are strongly linked to emotional reactions and can be described as our feelings about the situation ultimately overwhelming and threatening us, thereby lowering our self-esteem or even causing us to find hostility all around us. The efficiency of rendition, observed on the behavioural level, varies in accordance with general emotional agitation in agreement with the classical rule of Yerkes–Dodson, which claims that the efficiency of performance is subject to agitation. With a slight increase in emotional agitation, we can observe an increase in performance efficiency (in terms of insight, memory, focus, motor skills, and intellect), whereas a large increase in emotional agitation is followed by a decrease in efficiency.

Important psychological factors come into play in the stress reaction, and must be considered in the process of restoring balance between entities and environments. Ways of restoring balance, or in other words, ways of dealing with stress in the workplace, can differ significantly. The solutions are as diverse as the threatening situations and human characters. However, various authors offer certain general categories of ways to deal with stress (Heszen-Niejodek 1996; Wrześniewski 1996). We shall cite herein the classification offered by Lazarus and Folkman (1984), who detailed four ways of dealing with stress.

1. *Seeking information*: Information is sought so that one may be more effective in reaching a goal or in dealing with failure or an existing threat more easily (e.g., one would gather information to indicate that the failure is not as big or the goal is not as attractive as first presumed).
2. *Taking direct actions*: This category includes almost an infinite number of possible actions. The actions are aimed at either changing the surroundings to better suit the entity's abilities and values or at changing the entity itself to lessen the divergence between the entity and the surroundings.
3. *Holding back actions*: Lazarus regards this category to be an important countermeasure. In the modern world, it is sometimes best not to take action to most effectively dispose off the threatening stress.
4. *Intrapsychic methods*: These are based on changes in the perception of the environment or ourselves in order to mitigate the divergence between the surroundings and the entity. Their significance lies mostly in their sedative

activity and ability to reduce feelings of emotional upset. We include defensive mechanisms such as distancing the mind (erasing from awareness the problems causing the stress), projection (assigning our emotions and traits to other people), perfunctory reactions (disregarding real motives while acknowledging opposite and fake motives), repression (disregarding thoughts or motives that may cause feelings of guilt), and many others (Grzegołowska-Klarkowska 1986).

4.2.4 Stress and Diseases

Either long-lasting or especially intense stress may bring on damage to an organ or even a whole system, thereby leading to disease. Almost every disease can be a derivative of stress to some extent. On the contrary, we cannot say that there is a definite connection between even a strong stress and any disease. Such a situation is a nondirect state and is also connected to many secondary factors, such as genetic inclination, type of the cause of stress, acquired habits of reaction under various situations, and many others (Everly and Rosenfeld 1992). All these elements together decide whether the disease occurs and, if yes, how is it going to manifest. Connections between psychosocial stress and the ailments described next are often observed.

- *Cardiovascular diseases (especially coronary heart disease, heart failure and hypertension)*: Chronically high levels of catecholamines and glucocorticoids lead to arteriosclerosis (a significant reduction in the capacity of blood vessels, including coronary vessels) and increased blood pressure, causing heart attacks and heart failure.
- *Distortion of the digestive system (especially peptic ulcers)*: The inordinate release of gastric juice caused by stress—in addition to some genetic inclinations—may cause gastric erosion.
- *Musculoskeletal disorders*: Stress causes an increase in the tension in striated muscles. If the duration of the contraction is too long, less blood flows through the muscle, while the number of metabolic by-products simultaneously increases. Consequently, pain occurs. Some headaches have a similar mechanism.
- *Decrease in immunological resistance*: The immunological system has long been acknowledged to be fairly independent of the central nervous system, while reacting mainly to intruding antigens (foreign bodies). Current research, however, clearly shows that the brain—and through it, received psychosocial signals—has an impact on the immunological system. The cause lies either in the autonomic nervous system (which innervates the organs of the autonomic nervous system such as the marrow, thymus, and lymph nodes) or in the hormones secreted. The T and B lymphocytes, playing a major part in the organism's immunological capabilities, have stress-sensitive receptors, reacting to adrenaline, norepinephrine, glucocorticoids, and other hormones. Psychosocial stress is a possible cause of a decrease in various aspects of the immunological response, such as maturation and partition of some forms of T lymphocytes. The new, exciting field devoted to the

relationship among psychic topics, the nervous system, and immunology is called psychoneuroimmunology (Maier et al. 1997; Sheridan and Radmacher 1998). A decrease in immunological resistance increases the chances for various viral, microbial, and degenerative diseases. A decrease in immunological resistance under the influence of the psychosocial stresses listed below is also perceived as an important mechanism of tumour generation (Sheridan and Radmacher 1998; Zakrzewska 1989).

- States of anxiety and depression
- Professional burnout

Professional burnout was first described in 1974 by psychiatrist H. Freudenberger and later popularised by research papers published by Ch. Maslach, a social psychologist, and her formulation of the measurement tool. In the beginning, burnout was connected to jobs related to direct services or helping other people; thus, it included nurses, teachers, doctors, social workers, and so on. The high emotional demands necessitated by such professions might have led to the first strong symptoms of stress. According to Maslach, they can be included under three main categories:

1. *Emotional exhaustion*: Subject appears to be in a depressed mood, anxious, dejected, disillusioned; may have feelings of helplessness and frustration, as well as constant fatigue and somatic ailments.
2. *Depersonalisation*: This term is used to describe a lukewarm or even a hostile position towards a charge, and means that they are treating somebody more as an object than as a sensitive person. Such effects have been acknowledged to be the outcome of an objectionable manner of dealing with stress, caused by high demands in the workplace, but mainly resulting from use of defensive techniques such as distancing. Depersonalisation causes the person to back down from tough emotional problems, as well as intellectualise the issue (think about patients in medical terms only), use professional parlance or offensive language ('they behave like animals'), and escape physical contact (dodge eye contact).
3. *Sense of lack of accomplishment at work*: The subject shows pessimism towards the possibility of helping a charge, low professional self-esteem and feelings of underestimation and failure.

Burnout is most often observed among young people with no more than 2–3 years of professional experience, especially among those who begin their work with high, usually idealistic, expectations of their role in the company and the significance of their work (Schaufeli and Enzmann 1998). Later versions of the burnout concept (Maslach and Leiter 1997) pertain to the effect not only to jobs devoted to helping or service but also to many other professions. The three symptom groups described above can be classified under three general categories: exhaustion, cynicism, and ineffectiveness. A scale of measurement of such general phenomena was developed (Schaufeli and Enzmann 1998) but has not been adopted in Poland so far. However, vast empirical resources have been gathered thus far regarding the burnout effect, and the majority of jobs described involve the service fields.

4.2.5 Consequences of Stress at the Organisational Level

Stress on employees affects the functioning of the whole company, which can lead to the following:

- *Increased absences*: According to data gathered in the United States, American industry has been losing 550 million labour days annually due to absences, with 54% of absences said to be caused by stress. Similar data were gathered in the United Kingdom; the industry there has been losing 360 million labour days annually, with half that number caused by stress in the workplace (Cooper et al. 1996).
- *Lowered productivity*: Productivity declines when stress increases. Even if the employees are present at work, the quality of their services and products is lower.
- *More accidents*: Inordinate stress levels increase the probability of a mistake and, consequently, an accident. Some experts estimate that stress is responsible for 60%–80% of accidents in the workplace (Cooper et al. 1996).
- *Higher job turnover*: Results of research over the burnout effect among nurses show that (Schaufeli and Enzmann 1998) depersonalisation was associated with job turnover 2 years later. Similarly, positive associations were found between emotional exhaustion and job turnover among teachers within 1 year, and among general practitioners within 5 years (Schaufeli and Enzmann 1998). However, the percentage of variation in fluctuation explained by the range of burnout, although significant, was not high and ranged between 1% and 5%.
- *Increase in costs related to higher morbidity rate*: Treatment costs for diseases that are a consequence of stress are best visible in United States, where an employer is directly charged with treatment of his or her employees. In European countries, such costs are more hidden, which does not mean that they are nonexistent. According to the British Health Foundation, the evaluated loss of companies employing 10,000 people due to cardiovascular disorders (where we recognise stress as an important risk factor) can be split up as follows:
  - 2.1 million pounds lost due to lowered productivity of male employees and 340,000 pounds due to lowered productivity of female workers.
  - Loss of 35 men and 7 women (due to coronary heart disease among others).
  - Truancy of 59,000 labour days for men and 14,000 for women.

The full evaluation of costs must include a whole range of other diseases we mentioned previously, significantly increasing the above numbers.

According to the British evaluation, if we scale the results to the whole country, the costs of stress in the workplace in the United Kingdom reach 10% of the gross national income (Cooper et al. 1996).

4.3 PSYCHOSOCIAL RISK FACTORS IN THE WORKPLACE

In literature, one may find various classifications of stress-causing factors present in the workplace or, in other words, classification of psychosocial risk types. Cooper and Marshall (1987) name five categories of factors: those related to work itself, to the role of the employee in the organisation, career development, relations among the employees, the organisational structure, and the work climate. Below is a different classification, which mirrors the main directions of research in the field.

4.3.1 Very High Quantitative and Qualitative Demands

The most common problems in modern work environments are too much work and pressure that the work be done quickly. According to the European Agency for Safety and Health at Work, the intensity of work is one of the major types of psychosocial risk (European Agency for Safety and Health at Work 2007). In surveys on European work conditions, conducted between 1991 and 2005, there is a continuous growth in the number of respondents who report that they work at a very high speed at least three-fourths of their time (Figure 4.1). In 2005, 46% of respondents selected this answer (in the EU-25), whereas in 1990, 35% selected this answer (for the EU-12). The same survey also shows that there is a rise in the number of workers who report working under tight deadlines around three-fourths of the time or more.

Intensive work can become a source of satisfaction and personal development when there is a significant amount of autonomy. Nonetheless, there are many epidemiological research articles showing high quantitative demands at work are

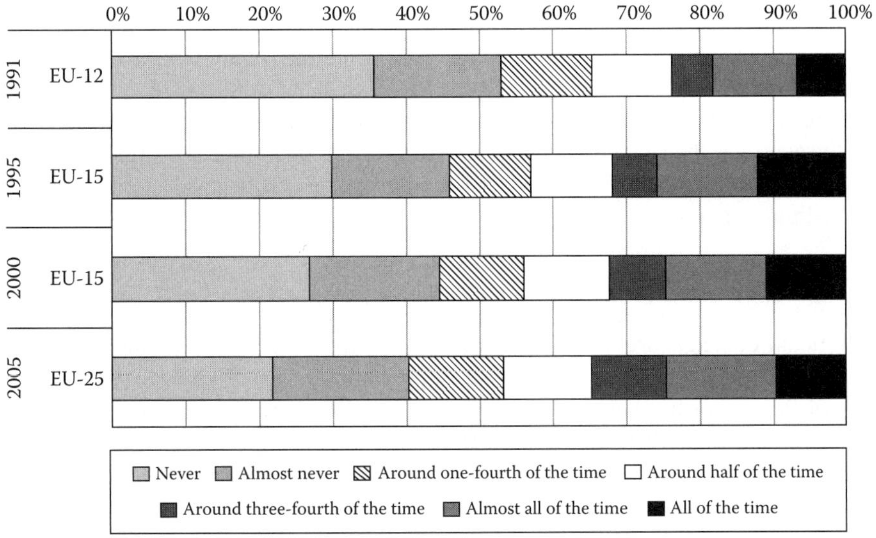

FIGURE 4.1 Percent of workers working at a very high speed.

connected with the health loss in employees. Previous and current research leads to the same conclusion. Margolis et al. (1974) surveyed a representative number of employees in the United States aged 16 and over, for a total of 1496 people. The authors found that work overload was related to the following indicators of stress: excessive use of alcohol, work absenteeism, and lower self-esteem.

A more recent study reported by Dutch researchers (de Jonge et al. 2000) involved a sample of 1700 workers; it showed that high demands were significantly related to lower job satisfaction, higher emotional burnout, a slightly higher level of depression, and more frequent psychosomatic symptoms.

Psychosocial risks can also be caused by excessive qualitative demands at the workplace, meaning work that is too hard or complex. Caplan et al. (1975) surveyed representatives from 23 countries and certified that the higher the subjective complexity of work (in terms of employee preference), the greater are the symptoms of depression among the surveyed people. Excessive qualitative demands may also involve an excessive amount of work: someone who has to perform a difficult task will have to work longer and more intensively.

4.3.2 Very Low Quantitative and Qualitative Demands

Psychosocial risks may be also caused by situations totally opposite to those mentioned previously, that is, a very low level of stimulation or satisfaction from the performed work. Such situations involve performing jobs that may seem dull, repeatable, and are usually very much below the employee's potential abilities. Risks are also related to jobs that require physical isolation, for example, in Arctic stations, in submarines, or during wardenship. Even professions commonly described as highly demanding and connected, with a high level of stimulation, may lack stimulation over a longer period of time, for example, among guards or pilots. In psychology, an effect of 'solitary altitude' is described among pilots (Terelak 1995). It occurs above certain heights and is usually recognised as losing touch with reality and a feeling of abruption from the Earth. It is accompanied by feelings of anxiety, apathy, and an urge to break out of the situation.

The current part-time professions, considered dull, not only refuse to vanish but, on the contrary, their range has been increasing due to ongoing automation and computerisation, embracing even broader social areas related to production, services, and administration.

4.3.3 Emotional Demands

Emotional demands are a significant source of stress, especially in jobs that entail frequent contact with other people, that is, predominantly in the service sector. The role of this sector has been steadily growing in all the developed economies of the world. In 2005, this sector made up 66% of the labour force in the EU (Parent-Thirion et al. 2007). Consequently, the importance of psychosocial risk, such as high emotional demands, has also been growing. Contact with other people, such as clients, students, patients, and so on, is frequently a source of negative emotions such as sadness, helplessness (e.g., when faced with suffering or an incurable disease), anger (e.g., toward a tedious client), irritation, and impatience (e.g., toward a lazy student). Generally accepted standards do not permit display of emotions; people who help are expected to be composed and friendly. Thus, they are penalised twice: not only are they more prone to experiencing negative emotions compared to those doing other jobs, but they are also expected to suppress these feelings and display emotions that are incompatible with those they are actually experiencing. A high level of emotional demand is related to a low level of mental well-being (Zapf 2002); it is conducive to professional burnout and behaviours incompatible with the standards of the organisation or profession (Bechtoldt et al. 2007).

4.3.4 Low Job Control

Job control is the possibility of influencing one's work (its methods, pace, breaks, and so on) and the conditions in which it is performed. The importance of job control has been underlined in psychology for a long time. Having control, not using it, is the most important (Glass et al. 1971).

Control of an employee in the work environment is being reassessed in two aspects—job autonomy and participation in decision making. Although these terms are similar, they refer to slightly different effects. We can use 'autonomy' if this person is in charge of the assigned tasks and their results. In other words, no one is expected to interfere with the employee's job; he or she is self-reliant. We can use 'participation' if an employee shares the responsibility of the workplace decision-making process more broadly, not only in relation to his or her duties. Limitations in both forms of control should be considered psychosocial risk factors.

As early as the 1960s, Kornhauser (1965) pointed to the negative mental effect of limited control as related to the necessity to work at a pace imposed by a machine. In his study, only 13% of workers working at a pace imposed by a machine were mentally healthy, whereas 29% of workers doing repetitive work at their own pace were mentally healthy.

The latest British research confirms the negative health effects of lack of control. The British Whitehall II project (Marmot et al. 2002) has been going on for more than 10 years, with over 10,000 civil servants as subjects. The results indicate that new cases of ischaemic heart disease (registered over a period of 5 years) are significantly more common in workers with low job control compared to workers with high job control. It is worth pointing out that the results were the same regardless of whether the worker or the supervisor was assessing the level of control (Bosma et al. 1998).

Polish studies conducted in a group of more than 300 managers at the Central Institute for Labour Protection—National Research Institute (CIOP-PIB) also proved that perceived job control was a predictor of certain symptoms of mental health, such as an increased level of anxiety and a decreased level of curiosity on Spielberger's State-Trait Personality Inventory (STPI) scale, a higher score when measured with the Beck Depression Inventory, a lower assessment of the quality of life on the Cantril ladder, and a higher frequency of somatic complaints (Widerszal-Bazyl 2003).

Jackson (1983) conducted sedulous research in the workplace. Two groups of hospital staff participated. In the experimental group, the surveyed people were given a chance to participate in the decision-making process in their department. Supervisors were trained and instructed to organise meetings with employees at least twice a month. The control group was not given such a chance. Tests conducted half a year after the experiment clearly showed that participation in decision making decreased role conflict and ambiguity (see below), which in turn led to lowered tension and higher satisfaction.

The results of research on the psychological outcome of participation in the decision-making process are not consistent. This is due to the fact that participation itself has various forms: formal and informal, voluntary and coercive, and direct and indirect. The possibility of each of these forms leading to specific results should not be ruled out. On the contrary, research seldom diversifies individual types of participation.

Spector (1986) has conducted a meta-analysis of 88 studies containing participation results. Just three (totalling $N = 213$ of respondents) measured subjective physical ailments. Their average correlation with participation equalled $r = -34$. Therefore, an average correlation of participation with bad mood equalled $r = -18$; the equation was based on research from four groups (where $N = 300$).

The relationship between participation and satisfaction is stronger. Spector (1986) reviewed 17 studies of the relationship between participation and satisfaction and observed an average correlation equal to $r = 44$.

The relationship between participation and behavioural measures of stress is not strong; however, it is noticeable. Based on his analysis and research review, Spector derived the following average correlations for participation at various levels:

- $r = 0.23$—study level (6 studies, 1343 people)
- $r = -0.20$—intention of leaving the job (4 studies, 1451 people)
- $r = -0.38$—actual job switch (3 studies, 358 people)

4.3.5 Job Insecurity

Job insecurity as related to the dynamic changes in the job market is an important stress-causing factor in many work environments. In a survey conducted in the 27 EU member states by the European Agency for Safety and Health at Work, 11.3% of 'old Europe' respondents and 25.6% of 'new Europe' respondents agreed with the following statement: 'I can lose my job within the next 6 months'. Insecurity was especially visible in Poland, with 26.6% of respondents agreeing, outpaced only by the Czech Republic and Slovenia.

Empirical research has shown a lot of evidence for job insecurity as an important risk factor for employees' health. A meta-analysis by Sverke et al. (2002) covered 37 studies on this topic conducted between 1980 and 1999; 14,888 subjects were involved. Analysis showed the correlation between job insecurity and poorer mental health to be $r = -0.24$. This relationship was also confirmed by longitudinal analyses, which unambiguously showed that job insecurity should be considered a cause of poor mental health, whereas the inverse relationship, the effect of poor mental health on the perception of job insecurity, although theoretically possible, was not statistically significant (Hellgren and Sverke 2003). The relationship between job insecurity and physical health has been analysed less frequently. Another meta-analysis of Sverke et al. (2002) covered 19 such studies with a total of 9704 subjects. It showed that the mean correlation was $r = -0.16$, which means that the greater the insecurity, the poorer is the physical health. Usually, researchers include the subjects' opinions regarding their health, which form a basis for the statement that high insecurity is related to a lower assessment of one's own health, higher frequency of somatic disorders (such as headache and back pain), and an onset of chronic diseases. It is important to point out that the longitudinal studies did not confirm the effects of job insecurity on somatic disorders experienced a year later (Hellgren and Sverke 2003). However, the authors are right in pointing out that the cause-and-effect relationship between job insecurity and physical health should be considered open for the time being because it is impossible to say whether one year is too short a period to reveal somatic disorders. A few researchers have studied the physiological indicators of health and claim that job insecurity may be related to higher systolic and diastolic blood pressures, ischaemic heart disease, and elevated body mass index (BMI*; Ferrie et al. 1998, 2001).

4.3.6 Role Conflict and Ambiguity

A person has plenty of important social roles in the workplace: a charge, supervisor, colleague, mentor, and so on. Some aspects of these roles may be important sources of psychological stress. A lot of focus has been devoted to two parameters of jobs: role clarity and role conflict. The basics of knowledge in this field were laid more than 30 years ago by Kahn et al. (1964) and there has been a lot of interest in this topic since then.

The clarity of a role can be described as a 'level of clarity in communicating goals and responsibilities to employees and a level of comprehending processes needed to achieve goals by employees' (Sawyer 1992, 130).

Usually, lack of clarity is a problem for high-level staff in an organisation (Schuler 1975). Fisher and Gitelson (1983), who conducted a meta-analysis of 42 studies on role clarity and its causes and results, observed that lack of clarity is mostly a problem to people who perform complex tasks and are well educated. In a study by Kahn et al. (1964), the lack of clarity in a role is a reason for either lower job satisfaction or leaving the job. This dependence was not too strong, although statistically noticeable.

Margolis et al. (1974) conducted a study in the United States in a representative sample of subjects ($N = 1496$); they found that unclear roles were related to depression, lower self-esteem, dissatisfaction with both life and job, and the intent to quit work. However, though these relationships are statistically significant, they are not very strong.

* BMI is an individual's body weight divided by the square of his or her height (in metres; BMI = weight/height^2).

Exhaustive research has been done regarding the importance of role clarity and its relationship to the burnout effect. A meta-analysis of 38 studies concerning this problem (Pfenning and Husch 1994; cited after Schaufeli and Enzmann 1998) clearly showed a strong correlation between the two. Lack of role clarity has a lot in common with the professional burnout effect (14% of collective variation), depersonalisation (8% of collective variation), and lack of achievement at work (10% of collective variation), thereby possessing all major components of the professional burnout effect.

Another important parameter of the professional role as a source of psychosocial risk is called role conflict. This term can be described as a 'coincidental appearance of at least two role transmissions', or demands made by the environment to the employee performing a certain performing one task certainly excludes the possibility of performing the other because the assignments are clearly adverse (Katz and Kahn 1979, 286).

Employees bordering the organisation who are in contact with other companies as well, are especially vulnerable to role conflict. Internal workstations are less vulnerable to such risk, although employees working for more than one department may feel more threatened than others.

New roles, usually developed to break stereotypes of behaviour and ways of dealing with problems in a given environment, are more at risk for role conflict. The scale of conflict depends on two factors: (1) who is transmitting the message that delivers the conflict and (2) the concerned person's position in the organisation and influence on other employees. The bigger this influence, the greater the conflicts may become. Role conflicts may lead to mental and physical health losses.

4.3.7 Mobbing

According to the Polish labour code, mobbing is an 'act or behaviour aimed at or against an employee, especially resting upon a long-lasting and systematic harassment, and threatening an employee, causing his professional self-evaluation to decrease, as well as aimed at humiliating and gibing an employee, keeping him or her away from other employees, or eliminating him or her from the team of employees' (Labour Code 2009, article 94).

Although mobbing sometimes means an act of physical aggression, it is most often seen as psychological violence. Some authors distinguish mobbing as (1) personal, such as yelling, offending, gibing, threatening, ignoring, or nasty jokes, and (2) that related directly to the work aspect: unrealistic task deadlines, too much work to do, tasks below abilities, constant control, and hiding important information. Some of the above-mentioned behaviours may be very discreet and indiscernible by other employees.

Polish labour code requires employers to counteract mobbing in their workplace (Labour Code 2009, article 94). An employee who suffers health loss due to mobbing can come forward with financial demands of his or her employer, as well as request compensation if he or she chooses to terminate the contract due to mobbing (Labour Code 2009).

Mobbing has consequence for both an employee and the organisation. A victim of abuse usually suffers high stress, which may lead to physical and mental health losses if it is long lasting. Post-traumatic stress disorder (PTSD) is also possible. PTSD is usually diagnosed if a victim of abuse lasting for at least a month constantly suffers upsetting flashbacks and feelings related to the violence, is trying to run away from thoughts, feelings, and places related to the abuse and seems to be constantly agitated. Such syndromes are often accompanied by depression, anxiety, fear, addiction, or even suicide attempts.

4.3.8 Interpersonal Relations and Social Support

We all know that good relationships with other people at work are important to our well-being. Not only should those in the work environment try to avoid conflicts, but should also actively support employees. Such support may take various forms. Usually, we can distinguish the following (House 1981) types of support:

- *Emotional support*: Showing sympathy, interest, kindness, and so on.
- *Instrumental support*: Concrete forms of support, for example, solving a difficult problem.
- *Informational support*: Delivering information crucial to solving problems.
- *Evaluational support*: Giving opinions on a person's appearance, behaviour, and speech.

Social support can have many effects on the employees' health. First, the kindness of the environment may become a shield that keeps out stress-causing factors. As a result, there are no stressful reactions, with no health consequences to follow. In a harmonious workplace, employees are unlikely to be unfairly punished.

Second, if the surrounding environment has many stress-causing factors, social support may become a buffer that shields employees from any negative impact on their health. The reason for this effect is probably 'blunting of the blade', which allows the person to deal with problems more effectively by lowering both stress and the negative impact on health.

Third, there are direct relations between an employee's well-being and social support. A person surrounded by kindness feels better and more confident even if working under the influence of other stress-causing factors.

Many studies have indicated a relationship between the amount of support received and an employee's health. An example can be found in the following Swedish undertaking:

Hoog and Eriksson (1993) conducted research on 150 female computer operators, trying to identify factors that play an important part in the occurrence of skin diseases. One such factor appeared to be their superior's support when given a lot of work. With stronger support, a lesser chance for skin disease was observed. However, given a small amount of work, no relation between the superior's support and the probability of skin disease was observed.

4.3.9 Stages of Occupational Career as a Source of Risk

Successive stages in a person's professional career can be a source of characteristic forms of stress. A considerable amount of focus has been devoted to the beginner's stage. Potential stress-causing factors at this stage are often related to shock caused by a difference between naïve expectations and the reality of work, confusion, lack of confidence, and the need to determine one's work and role in the company, strike up relations with superiors and workmates, and understand the rules and prize systems in the workplace (Burke 1988). Professional burnout is most likely to show up at this stage of the career and objectionable work conditions only fuel its strength. Cherniss (1980) names five sources of stress likely to appear in jobs related to helping other people: lack of confidence in professional skills, lack of appreciation from customers, bureaucratic procedures that aggravate work, lack of stimulation followed by a feeling of monotony, and lack of teamwork and support from colleagues.

The current focus is on older employees beyond the peak of their careers, between 50 and 65 years old. The most important stress-causing factors in this period are (Burke 1988) the need to deal with new technology, lack of promotion possibilities, likelihood of decrease in efficiency, need for additional training, discrimination, financial problems, and health problems for the employee and his or her spouse.

4.4 MODELS OF PSYCHOSOCIAL STRESS AT WORK

Discussions and studies on psychosocial risk at work have resulted in several interesting models (Siegrist 1998) that reveal the most important mechanisms of this phenomenon. The two that have had the greatest influence on contemporary thinking on psychosocial risk and have inspired the greatest number of studies are the demand–control–support (DCS) model and the effort–reward imbalance (ERI) model. A third model, demand–resources, is a synthesis of those two.

4.4.1 Demand–Control–Support Model

American researcher Robert Karasek (1979) formulated the first version of the DCS model. He assumed that two dimensions of the work environment, demand and control, have basic significance for the workers' perception of stress. Karasek understood that demand primarily involves quantitative demands* (a lot of work, tight deadlines) and demands that result from role conflict (see Section 4.3.6). Control was explained as a component of two subdimensions: skill discretion (the level of skill and creativity required on the job) and decision authority (the possibility for workers to make decisions about their work).†

* In the original version of the DC model, and in most studies devoted to it, demands were understood as predominantly quantitative. Later, however, other components of the concept of demand were incorporated into questionnaires that measured demands. These were emotional demands, cognitive demands, job insecurity, and so on (Karasek et al. 1998).

† Incorporating the complexity of the task into the concept of control has been an often-criticised element of the model and as such ignored by numerous researchers. The author of the model explained that it is possible to talk about real control only when a more complex task is carried out.

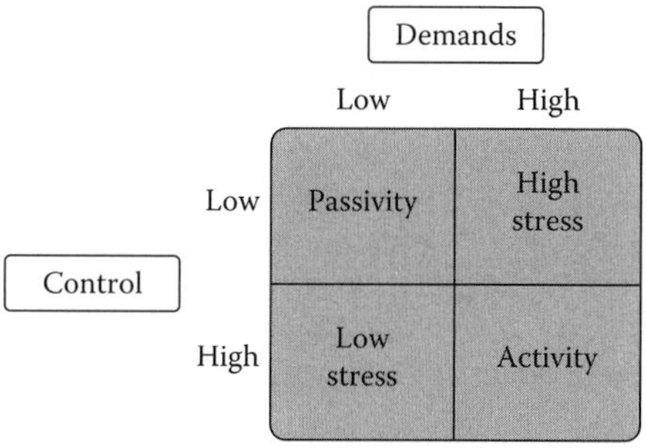

FIGURE 4.2 Karasek's demand–control model.

In his model, Karasek identified the following four main situations (Figure 4.2), which differed in two critical dimensions:

1. High demand–low control is the most stressful situation. Workers are faced with high demand, but are not given adequate latitude or means to meet those demands. This creates psychophysiological stress, which, if extensive, leads to anxiety, depression, and risk of a somatic illness.
2. High demand–high control creates conditions for personal development. Workers are faced with difficult tasks, but they can model their behaviour in such a way that they can meet their goals.
3. Low demand–low control neither stimulates activity, because demands are low, nor allows action or control. This causes passivity both in occupational life and in the way free time is spent. Seligman (1975) described similar phenomena as 'learned helplessness'. The individual has no opportunity to develop.
4. Low demand–high control is the most relaxing situation, causing the least stress. High control makes it possible to react optimally to every unchallenging demand. The risk of poor mental well-being or a psychosomatic illness is low.

Many studies have confirmed the demand-control model, proving that when demands are high and control is low, mental, cardiovascular, musculoskeletal, or immunological disorders are significantly more frequent.

An example of this type of research is an epidemiological study conducted in Stockholm (Theorell et al. 1998) involving men ages 45 to 64 who had their first myocardial infarction between 1992 and 1994. The results were as follows: considering the ratio of demand to control and forming a subgroup from the upper quartile of this index (namely, people who perceive their demands as high when compared with the possibilities of control), the risk of infarction in this subgroup is almost one and a half times higher than that in the other subgroups (relative risk [RR] = 1.4). After eliminating many possible confounding variables, such as smoking cigarettes, hypertension, cholesterol level, social class, and so on, the rate of infarction risk was slightly lower but still significant (RR = 1.3).

Johnson and Hall (1988) and Karasek and Theorell (1990) further developed the DCS model. They pointed to a third important dimension of the psychosocial work environment, which determined the level of stress—social support. According to this model, stress is highest when high demands coexist with low levels of control and social support. A few studies have confirmed this extended model (cf. reviews by De Lange et al. 2003; Widerszal-Bazyl 2003).

4.4.2 Effort–Reward Imbalance Model

German medical sociologist Johannes Siegrist (1998) created the ERI model. The model assumes that occupational work is a special process of social exchange: workers put some effort into a task; in return, they expect a reward, which can have various forms such as pay, prestige, appreciation, support from colleagues and supervisors, job security, and promotion prospects. If the reward is not proportional to the effort, there is psychophysiological stress, the accumulation of which can lead to many diseases. A significant aspect of Siegrist's conception lies in the introduction of a personal variable into the model, which he calls 'overcommitment'. Overcommitment has the following symptoms: continuous worry about work, inability to tear oneself away from work, and a tendency to work too much. If this characteristic is strong, the psychophysiological stress that results from the imbalance between effort and reward has an even greater influence on the worker's health (Figure 4.3). Siegrist's model has been tested in many studies and can be used to successfully predict a subjective state of health (Siegrist et al. 2004) and new cases of ischaemic heart disease (Bosma et al. 1998).

4.4.3 Job Demands–Resources Model

The job demands–resources (JDR) model was proposed by Demerouti et al. (2000) and Bakker et al. (2003); it is a synthesis of the DCS and ERI models. The JDR model distinguishes two categories of work conditions: demands and resources. According to Demerouti, 'job demands refer to those physical, social, or organisational aspects

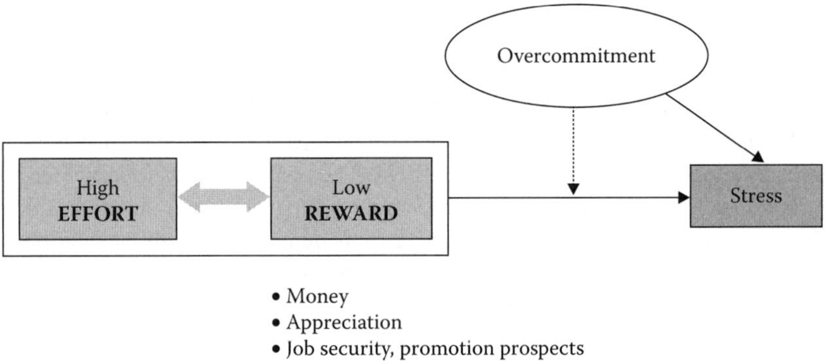

FIGURE 4.3 The effort–reward imbalance model.

of the job that require sustained physical and/or psychological (i.e. cognitive or emotional) effort on the part of the employee and are therefore associated with certain physiological and/or psychological costs. … Job resources refer to those physical, psychological, social, or organisational aspects of the job that either/or: (1) reduce job demands and the associated physiological and psychological costs; (2) are functional in achieving work goals; (3) stimulate personal growth, learning, and development' (Demerouti et al. 2000, 395).

Thus, the JDR model is a more general model, though it has much in common with the aforementioned models. The demands in this model are not quantitative demands as in the DCS model; rather, they comprise emotional demands, equipment problems, changes at work, and so on. Resources are not just control and support (as in the DCS model) or just rewards (as in the ERI model), but all of these elements together. The essence of the JDR model lies in the assumption that every organisation has its own characteristics, which can be classified into two main categories: demands and resources. According to the authors of this model, it is best to begin a study with a qualitative analysis, which will make it possible to extract the most important demands and resources in a given environment. Only then should one conduct a quantitative analysis of the influence of those dimensions on the result variables (such as health and absenteeism) and use those particular questionnaires that consider resources and demands especially significant in an environment. The problem with the JDR model is that it does not state precisely what characteristics of the work environment produce the most stress. People responsible for psychosocial work conditions may therefore not know what they should pay attention to first. However, this flaw of the model is also its advantage; the model is thus more flexible and can be adapted to specific organisations that are being analysed.

4.5 PSYCHOSOCIAL RISK MANAGEMENT

4.5.1 Basic Steps in Psychosocial Risk Management

Following the European Commission's guidance (Levi and Levi 2000), the psychosocial risk-management process should involve the following steps:

1. Define the risk factors and their severity, range of prevalence, causes, and relationship to overall health. The aim of this step is to determine—through surveys, interviews, and other techniques—which psychosocial properties of work are the most stressful in a group. It also evaluates the hardening of potential health consequences of existing psychosocial risk types.
2. Analyse the connection between risk factors and their results in terms of employees' health and organisational outcome. The goal of this step is to gain an understanding about which of the risk factors identified in step 1 are exceptionally strongly connected to entities' and organisation's health.
3. Social partners (employers and employees' representatives, experts, and other important organisations active within the company) work together to project an intervention set of rules aimed at limiting existing psychosocial risk and implementing the project.

4. Evaluate the direct and long-term consequences of enforcing sets of rules. The evaluation embraces the level of risk factors, level of physical health and well-being, quality and quantity of manufactured goods, and performed services. Economic analyses of costs and earnings related to such enforcement are also expected.

Many factors decide if applying a set of rules aimed at lowering psychosocial risks will be effective. Two factors are especially important:

1. Social support of organisational change
2. Active participation of employees in the process

In European states, labour unions are an important source of support for the improvement of work conditions. In the United States, where the number of unions is very low, a lot of pressure is exerted on superiors to support changes. Even in places where unions are strong, however, it is also crucial to receive support from superiors. With no such support, even the most rational changes have hardly any chance of success. A very important and valuable source of support is a mutual agreement between the union or unions and the company's management team. The recently signed European Framework Agreement (2004) opens a window of opportunity and provides a good formal basis for stimulating such cooperation. It obligates the main social partners (governments as well as representatives of employers and employees) to cooperate in the limitation of psychosocial risks in the workplace.

Agreements and contracts should be also formed among the lower stages of the management structure and union officials in consultation with employer's representatives. Mid- and low-level managers, who might initially support changes, can become uneasy with possible threats in later stages as an increase in the employee's decision-making power lessens the impact of managers. However, managers must understand and support changes so that the changes can be effectively applied. In other words, it is imperative to confirm the sets of rules and ideas of change at many levels of the organisation to ensure they prevail and bring results, or, as Karasek and Theorell (1990) aptly observed, that they have to come to the 'critical organisational bulk'.

Active participation from regular employees is another critical condition for the successful application of new rules and changes. The discipline of participation ergonomy particularly relies upon this participation (Noro and Imada 1991). The foundation of participation ergonomy lies in the theory that efficient work conditions (including psychosocial aspects) is not possible without employee support. The advantages of including employees in the process are as follows:

- Psychosocial risk factors occur in many fields, and their origins may vary; an effective diagnosis of such a state given by a broad and miscellaneous body usually shows factors that might be missed by a single observer or even a sole management team.
- Participation in the decision-making process, leading to solving problems, increases involvement in applying a solution previously accepted as valid.

4.5.2 Forms of Prevention

Three forms of psychosocial risk prevention are usually distinguished (e.g. Murphy 1988):

1. Limitation of the sources of psychosocial risk
2. Limitation of physical and physiological reaction to risks
3. Treatment of health losses caused by long-lasting stress resulting from psychosocial risks

Each of these forms can be aimed at either an entity or an organisation. The main goal is to strengthen resistance to stress, while the organisation benefits from the possibility of applying changes aimed at reducing or eliminating existing sources of stress or developing mechanisms that counteract the creation of stress. The primary condition to any action lies in monitoring stress at the workplace.

4.5.3 Monitoring Psychosocial Risk at Work

Monitoring psychosocial work conditions means simply finding the risk factors in the psychosocial area of work. It requires observing both work conditions and the staff's health level. The basic tools useful in such monitoring processes are checklists and questionnaires.

Checklists are simple in form; an example can be found in the 19-question list prepared by the European Foundation for the Improvement of Living and Working Conditions (Kompier and Levi 1994) to monitor the content of work tasks. The respondent answers the questions by marking *yes* or *no*. Respondents can be picked from among either employees or management teams, and have a chance to report problems in their fields through such a list.

A more advanced form of stress monitoring is applied, along with an inquiry form surveying employees' feelings and opinions regarding their work. In the Psychological Workroom of CIOP, an inquiry form for stress monitoring was developed and named *Psychosocial Conditions of Work* (Widerszal-Bazyl and Cieślak 1999).

4.5.4 Stress Prevention Directed at an Individual

Many techniques for stress prevention and increased resistance have been invented. Some of them have been known since ancient times, for example, physical workouts or meditation. Only recently have attempts been made to explain the psychophysiological mechanisms of an entity's changes due to the use of such techniques. This field is outlined next.

4.5.4.1 Healthy Lifestyle

Physical exercise is the basis of a healthy lifestyle. It not only improves the general condition of organisms (such as the cardiovascular and respiratory systems) but also improves psychological state by reducing fear, depression, and aggression and increasing self-confidence and self-esteem (Everly and Rosenfeld 1992). Stretching exercises, such as jogging, swimming, or bicycling, are believed to be the best. Effective exercise is carried out three to four times a week for a period of 30–40 minutes. Other important elements of a healthy lifestyle include proper nutrition, weight control, not smoking, and getting adequate amounts of sleep and relaxation (Schaufeli and Enzmann 1998).

4.5.4.2 Cognitive-Behavioural Techniques

Cognitive-behavioural techniques are based on the assumption that emotions—the quintessence of stress—are a derivative of cognitive judgments about reality. Thus, a change in the perception of reality may cause a decrease in negative feelings and behaviours connected with these emotions. The most accredited technique in this field is a rational-emotive therapy offered in the 1960s by Albert Ellis (1998). Its goal is to persuade a person to abandon irrational beliefs that empower stress ('I expect all people to be fond of me') and form more realistic and less stress-causing judgments. Other methods in this group are the *cognitive evaluation technique*, which teaches the subject to look at the stress-causing factor from another point of view, making it less dangerous or even revealing its positive aspects or the *cognitive endeavour technique, cognitive rehearsal*, in which the subject adapts emotionally and cognitively in an artificial surrounding to situations that normally cause anxiety, for example, refusal to perform a superior's assignment. Thanks to training, situations such as these are less stressful in real life.

4.5.4.3 Relaxation Techniques

This group embraces a wide range of techniques, each exhaustively described in literature and with plenty of practical approaches from researchers gathering information.

One of the most popular techniques is called *neuromuscular relaxation*, which is based on the gradual and methodical loosening of muscles and, therefore, decreasing mental tension. Another popular technique is progressive relaxation, developed by Jacobson. Yet another approach is called *respiratory regulation*, which is based on learning a deep, phrenic breathing. This technique requires focus and concentration on the process of breathing alone. Respiratory regulation is believed to be a reasonable way to decrease stress and increase relaxation. *Schultz's autogenic training* is based on the assumption that a human being can suggest to itself certain physiological states, for example, increasing the body heat level or heart rate. For example, if a human being suggests its arm to become increasingly warmer, an increase in the amount of blood flowing through the limb and, in consequence, a rise in temperature, are possible. Such training is based on repeating certain exercises and autosuggestions, which may also be related to other factors, such as feelings of weight, heat, rhythmic pulse, and respiration. In general, a person should only attempt to learn such techniques under the guidance of an expert and only then start to exercise on their own. *Biological feedback* is a whole set of methods based on the fact that, with help from sensitive measure gear, a customer may receive punctual feedback about their organs. In relation to information type, we can mention electromyographic feedback (providing information about tension of certain muscles), thermal feedback (providing information about temperature), and electroencephalographic feedback (providing information about brain waves). With such information, a human being can supposedly influence the functioning of his or her organs, even those commonly believed to be independent of our will.

4.5.5 Stress Prevention Directed at an Organisation

As shown by critical reviews of stress-reduction programmes in western European countries (Ivancevich et al. 1990), most stress-prevention programmes focus largely on influencing persons. Organisational changes are seldom performed. Therefore, there is a need to direct more focus onto organisational impact as a method of countering stress. Some options for organisational impact are posited next.

4.5.5.1 Optimisation of Task Content

Work-task designing is a traditional subject of interest connected with psychology and work organisation. Such enterprises were not at first directly connected to issues of stress at work but rather with productivity and lifestyle. Nevertheless, they can be seen as antistress actions and are described as such in recent literature. The 'job design' movement (Buchanan 1989) offered four techniques to optimise tasks:

1. *Job rotation*: Moving employees to new stations on a regular basis.
2. *Job enlargement*: Merging similar jobs into larger modules.
3. *Job enrichment*: Connecting simple tasks and control-and-decision elements; assigning workers to tasks with higher levels of complication.
4. *Creation of autonomous work groups*: Creating independent employee groups responsible for larger job fragments.

An example of such an approach can be found in the reorganisation conducted at the Almex Company in Sweden (Karasek and Theorel 1990), which produces ticket-validation machines. Before the change, employees worked with the help of a conveyor belt, assembling hundreds of parts to produce a product with thousands of variations. Despite the diversity of the products, every worker was able to perform only a certain amount of the work. A change in the production methods was developed at one of the biggest Swedish technical universities at the Institute for Production. Masters of Engineering there consolidated the fragmented actions into larger working modules.

The central mark of the change was the creation of autonomous groups, which consisted of 5–15 people who took full responsibility for creating whole machines. Workers now had to have higher skill levels; they needed to go through training sessions, after which they could bear a lot more responsibility than before. Teamwork brought out a tendency to help each other (instrumental support), thereby increasing the positive reactions of employees towards one another (emotional support). As a result of these changes, productivity increased and customer claims decreased. Employees from autonomous groups suffered stress symptoms less often than workers not in these groups (7% and 18%, respectively); they were comparatively seldom tired (2% and 22%, respectively), despite having more demanding jobs. Interviews showed an increase of self-confidence and self-esteem.

4.5.5.2 Increasing Employee Participation

An important cause of stress, low job control (previously discussed in Section 4.3.4), can be eliminated in two main ways: first, by increasing job autonomy, which is an important element of job enrichment; and second, by increasing employee participation in the company's management process. Participation in important decisions involving the whole organisation can be of great value for many people. Moreover, including employees in the process of work-condition optimisation can be an effective way to deal with a range of other stress-causing factors present in the work environment.

4.5.5.3 Defining Work Time Framework

An excessive number of work hours is an important cause of stress; hence, an organisation should provide a clear work time framework, eliminating overworking. This problem is often discussed with reference to the professional burnout effect, because it is commonly believed that intensive and long-lasting contact with a charge (such as patient and pupil) is a primary risk factor leading to burnout. Therefore, various 'time-related' countermeasures (Schaufeli and Enzmann 1998) should be proposed, such as global work time reduction, reduction of direct contacts with charges by granting some days free from work to rebuild mental health, granting long absences from work, assigning other interesting tasks, encouraging part-time work, and so on.

4.5.5.4 Management Development

A manager can become either a serious source of stress for his or her employees or a buffer shielding them from stress-causing factors. Managers should know the primary sources and consequences of stress in order to effectively fulfil a health-encouraging role. Managers should be familiar with stress management techniques (e.g. delegating power) and have good interpersonal skills. Managers should also have practical skills in their field. These issues are the subject of programmes for management perfection conducted in many organisations.

4.5.5.5 Development of Occupational Careers and Training of Employees

A lack of possibilities for professional development can be an important stress-causing factor at various stages of professional life. To counteract such states, an organisation should constantly optimise appropriate career-management tools, such as its recruitment system and job selection process, according to employees' preferences, and develop a promotion system and training programme, which allow workers to follow the changing demands in their surroundings. Professional career development also requires adequate planning from the employees themselves. An attractive organisation provides support in this field by organising training sessions related to career planning, which help employees to learn about their weak and strong suits, needed for successful career planning, and facilitate the aggregation of the organisation's possibilities. Some companies hire special counsellors who individually help the workers in their career planning.

4.5.5.6 Formation of Effective Organisational Behaviours

Effective organisational behaviour training is another method which can limit employee stress. It focuses on individual entities, in contrast to the focus on techniques described in Section 4.5.4; it is about creating organisational behaviours rather than forming behaviours in general. The training is thereby meant to influence an organisation, although according to Schaufeli and Enzmanna (1998), training should not be included among the ways of impacting both entities and organisations, but rather seen as interactions at the interface of the entity and the organisation. The following undertakings are included in this group:

- *Time-management training* (e.g. Fontana 1999): May help in dealing with the pressures of deadlines and time management, a common stress-causing factor, as discussed previously.
- *Interpersonal-skills training*: Especially important in service and helping professions whose employees are vulnerable to the burnout effect. Maslach and Leiter (1997) described the types of interpersonal skills that persons performing such jobs should have, such as an ability to strike up and hold eye contact, maintaining conversations on sensitive topics, distinguishing types of people (according to sex, age, descent and so on), and dealing with aggressive people. A popular form of interpersonal training is assertiveness training (Lindenfield 1995), which helps the person realise their goals without aggression.
- *Creation of a realistic image of the profession*: This interesting form of training was developed mainly as a method of preventing the burnout effect. It is not uncommon for people suffering from burnout to have begun with very high, idealistic expectations of their environment. A more realistic approach toward a profession before starting the job will be an effective countermeasure to the burnout effect. Training among nurses (Schaufeli and Enzmann 1998) has proven such an approach to be a valid one.

REFERENCES

Act of 26 June 1974—The Labour Code (Uniform text). DzU 1998, no. 21, item 94, as amended.

Bakker, A. B., E. Demerouti, and W. B. Schaufeli. 2003. Dual processes at work in a call centre: An application of the job demands-resources model. *Eur J Work Org Psychol* 12(4):393–417.

Bechtoldt, M. N., et al. 2007. Main and moderating effects of self-control, organizational justice and emotional labour on counterproductive behaviour at work. *Eur J Work Org Psychol* 16(4):479–500.

Bosma, H., et al. 1998. Two alternative job stress models and the risk of coronary heart disease. *Am J Publ Health* 88:68–74.

Buchanan, D. A. 1989. High performance: New boundaries of acceptability in worker control. In *Job Control and Worker Health*, eds. S. L. Sauter, J. J. Hurrell Jr, and C. L. Cooper, 255–273. Chichester: Wiley.

Burke, R. J. 1988. Sources of managerial and professional stress in large organizations. In *Causes, Coping and Consequences of Stress at Work*, eds. C. L. Cooper, and R. Payne. Chichester: Wiley.

Caplan, R. D., et al. 1975. *Job Demands and Worker Health.* 75–160. Publication NIOSH, Washington, DC: US Department of Health, Education and Welfare.

Cherniss, C. 1980. *Professional Burnout in the Human Service Organizations.* New York: Praeger.

Cooper, C. L., and J. Marshall J. 1987. Sources of stress in managerial and white collar work. In *Stress at Work*, eds. C. L. Cooper and R. Payne, 123–164. Warsaw: PWN.

Cooper, C. L., P. Liukkonen, and S. Cartwright. 1996. *Stress Prevention in the Workplace.* Dublin: European Foundation for Improvement of Living and Working Conditions.

De Jonge, et al. 2000. Linear and nonlinear relations between psychosocial job characteristics, subjective outcomes, and sickness absence: Baseline results from SMASH. *J Occup Health Psychol* 5(2):256–268.

De Lange, A. H., et al. 2003. The very best of the millennium: Longitudinal research and the demand-control-(support) model. *J Occup Health Psychol* 8(4):282–305.

Demerouti, E., et al. 2000. A model of burnout and life satisfaction. *J Adv Nurs* 32(2):454–464.

Ellis, A. 1998. *Short-Term Therapy.* Gdansk: Gdansk Psychology.

European Agency for Safety and Health at Work. 2007. *Expert Forecast on Emerging Psychosocial Risks Related to Occupational Safety and Health.* Luxembourg: Office for Official Publications of the European Community.

European Foundation for the Improvement of Living and Working Conditions. 2006. *Fifteen years of working conditions in the UE: Charting the trends.* http://www.eurofound.europe.eu/publications/htmlfiles/ef0685.htm (accessed August 8, 2008).

European Framework Agreement on Work-Related Stress. 2004. http://www.etuc.org/IMG/pdf_Framework_agreement_on_work-related_stress_EN.pdf

Everly, G. S., and R. Rosenfeld. 1992. *Stress: Causes, Therapy and Autotherapy.* Warsaw: PWN.

Ferrie, J. E., et al. 1998 The health effects of major organizational change and job insecurity. *Soc Sci Med* 46:243–254.

Ferrie, J. E., et al. 2001. Job insecurity in white-collar workers: Toward an explanation of associations with health. *J Occup Health Psychol* 6:26–42.

Fisher, C. D., and R. Gitelson. 1983. A meta-analysis of correlates of role conflict and ambiguity. *J Appl Psychol* 68:320–333.

Fontana, D. 1999. *Time Management.* Warsaw: PWN.

Fourth European Working Conditions Survey. 2005. Statistical annex. http://www.eurofound.eu.int/docs/ewco/4EWCS/4WCS_Annex%20statistical%20annex.pdf (accessed August 8, 2008).

Glass, D. C., B. Reim, and J. R. Singer. 1971. Behavioural consequences of adaptation to controllable and uncontrollable noise. *J Exp Social Psychol* 7:244–257.

Grzegołowska-Klarkowska, H. J. 1986. *Personality Defence Mechanisms.* Warsaw: PWN.

Hellgren, J., and M. Sverke. 2003. Does job insecurity lead to impaired well-being or vice versa? Estimation of cross-lagged effects using latent variable modeling. *J Org Behav* 24:215–236.

Heszen-Niejodek, I. 1996. Stress and coping: Main controversies. In *Man in Stress Situations*, eds. I. Heszen-Niejodek and Z. Ratajczak. Katowice: University of Silesia.

Hoog, J., and N. Eriksson. (1993) The office illness project in northern Sweden. The significance of psychosocial factors for prevalance of skin symptoms among VDT workers. A case referent study. In *Work with Display Units 92*, eds. H. Luczak, A. E. Cakir and G. Cakir. Abstractbook F-10. Berlin: Technical University of Berlin.

House, J. S. 1981. *Work Stress and Social Support.* Reading, MA: Addison-Wesley.

Ivancevich, J. M., et al. 1990. Worksite stress management interventions. *Am Psychol* 2:252–261.

Jackson, S. E. 1983. Participation in decision making as a strategy for reducing job related strain. *J Appl Psychol* 68:3–19.

Johnson, J. V., and E. M. Hall. 1988. Job strain, work place social support, and cardiovascular disease: Across-sectional study of a random sample of the Swedish working population. *Am J Pub Health* 78(10):1336–1343.

Kahn, R. L., et al. 1964. *Organisational Stress*. New York: Wiley.

Karasek, R. 1979. Job demands, job decision latitude and mental strain: Implication for job redesign. *Adm Sci Q* 24:285–308.

Karasek, R., and T. Theorell. 1990. *Healthy Work*. New York: BasicBooks.

Karasek, R., et al. 1998. The Job Content Questionnaire (JCQ): An instrument for international comparative assessments of psychosocial job characteristics. *J Occup Health Psychol* 4:322–355.

Katz, D., and R. Kahn. 1979. *Social Psychology of Organizations*. Warsaw: PWN.

Kompier, M., and L. Levi. 1994. *Stress at Work: Causes, Effects and Prevention. A Guide for Small and Medium Sized Enterprises*. Dublin: European Foundation for Improvement of Living and Working Conditions.

Kornhauser, A. 1965. *Mental Health of Industrial Worker*. New York: Wiley.

Labor Code 2009 with commentary. Warsaw: Presspublica.

Lazarus, R. S., and S. Folkman. 1984. *Stress, Appraisal and Coping*. New York: Springer Publishing Company.

Levi, L., and I. Levi. 2000. *Guidance on Work-Related Stress. Spice of Life or Kiss of Death?* Luxemburg: Office for Official Publications of the European Communities.

Lindenfield, G. 1995. *Assertiveness*. Lodz: Ravi.

Maier, S. F., R. Watkins, and M. Fleshner. 1997. Psychoneuroimmunology. *Psychology News* 1:5–35.

Margolis, B. L., W. H. Kroes, and R. A. Quinn. 1974. Job stress: An unlisted occupational hazard. *J Occup Med* 16:654–661.

Marmot, M., T. Theorell, and J. Siegriest. 2002. Work and coronary heart disease. In *Stress and the Heart. Psychosocial Pathways to Coronary Heart Disease*, eds. S. A. Stansfeld and M. Marmot. London: BMJ Books.

Maslach, Ch., and M. P. Leiter. 1997. *The Truth about Burnout*. San Francisco: Jossey-Bass Inc.

Murphy, L. R. 1988. Workplace interventions for stress reduction and prevention. In *Causes, Coping and Consequences of Stress at Work*, eds. C. L. Cooper, and R. Payne, 301–339. Chichester: Wiley.

Noro, K., and A. Imada. 1991. *Participatory Ergonomics*. London: Taylor & Francis.

Paoli, P., and D. Merllie. 2001. *Third European Survey on Working Conditions 2000*. European Foundation for the Improvement of Living and Working Conditions. Luxembourg: Office for Official Publications of the European Communities.

Parent-Thirion, A., et al. 2007. *Fourth European Working Conditions Survey*. European Foundation for the Improvement of Living and Working Conditions. Luxembourg: Office for Official Publications of the European Communities.

Pfenning, B. and Hüsch, M. 1994. Determinants and correlates of burnout syndrome: A meta-analytic approach. Master's thesis, Free University of Berlin, Institute of Pyschology.

Reykowski, J. 1966. *Personality Functioning under Psychological Stress Conditions*. Warsaw: PWN.

Sawyer, J. E. 1992. Goal and process clarity: Specification of multiple constructs of role ambiguity and structural equation model of their antecedents and consequences. *J Appl Psychol* 77:130–142.

Schaufeli, W., and D. Enzmann. 1998. *The Burnout Companion to Study and Practice: A Critical Analysis*. London: Taylor & Francis.

Schuler, R. S. 1975. Role perceptions, satisfactions, and performance: A partial reconciliation. *J Appl Psychol* 1960:683–687.

Seligman, M. E. P. 1975. *Helplessness.* San Francisco: W. H. Freeman.

Selye, H. 1977. *Stress without Distress.* Warsaw: PIW.

Sheridan, Ch. L., and S. A. Radmacher. 1998. *Health Psychology.* Warsaw: Institute of Health Psychology, Polish Psychological Association.

Siegrist, J. 1998. Adverse health effects of effort-reward imbalance at work: Theory, empirical support, and implications for prevention. In *Theories of Organisational Stress*, ed. C. L. Cooper, 190–204. New York: Oxford University Press.

Siegrist, J., et al. 2004. The measurement of effort-reward imbalance at work: European comparisons. *Soc Sci Med* 58:1483–1499.

Spector, P. 1986. Perceived control by employees: A meta-analysis of studies concerning autonomy and participation at work. *Human Relat* 39:1005–1016.

Sverke, M., J. Hellgren, and K. Näswall. 2002. No security: A meta-analysis and review of job insecurity and its consequences. *J Occup Health Psychol* 7:242–264.

Terelak, J. F. 1995. *Psychological Stress.* Bydgoszcz: Branta.

Theorell, T., et al. 1998. Decision latitude, job strain, and myocardial infanction: A study of working men in Stockholm. *Am J Pub Health* 88:382–388.

Widerszal-Bazyl, M. 2003. *Work Stress and Health: An Attempt to Verify the Robert Karasek's Model and the Demand-Control-Support Model.* Warsaw: CIOP-PIB.

Widerszal-Bazyl, M., and R. Cieślak. 1999. *Psychosocial Conditions of Work.* Warsaw: CIOP.

Working conditions in the acceding and candidate countries. 2001. Polish language version. http://www.eurofound.eu.int/publications/htmlfiles/ef0306.htm (accessed August 8, 2008).

Wrześniewski, K. 1996. Styles and strategies of coping with stress. In *Human behavior in stressful conditions*, eds. I. Heszen-Niejodek, and Z. Ratajczak, 44–61. Katowice: University of Silesia.

Zakrzewska, T. 1989. Psychosocial risk factors for cancer morbidity: Review of research. *Psychology Review* 4:1019–1039.

Zapf, D. 2002. Emotions work and psychological well-being: A review of the literature and some conceptual consideration. *Human Resourc Manage Rev* 12:237–268.

5 The Physiology of Stress

Maria Konarska

CONTENTS

5.1 INTRODUCTION

Stress is the arousal of an organism as a result of a stimulus from the *material* or *social environment* that is assessed as threatening to its physical or psychological balance to an extent that exceeds its ability to cope with the problem. Typical properties of stress include negative emotional arousal resulting from threat recognition and an excessive physiological response, including neurohormonal changes such as adrenaline and corticosteroid turnover.

Due to the broad scope of the effect of stress mechanisms at various levels—from the molecular to the social levels—it is one of the most common natural phenomena, shaping mutual relations between the environment and all living organisms, including humans, animals, and—as discovered in recent years—plants and unicellular organisms (Jenkins 1979; Ellis 1996).

Living organisms and the environment constitute a unit of components that exerts mutual influence within the framework of a continuous feedback process. Energy, in various forms, communicates environmental information: mechanical, acoustic, light, and chemical energy act as stimuli, modifying the activities of the organism. Receptor nerve cells receive these stimuli through simple nerve endings in the skin, such as pain receptors, and send them to cells equipped with complex mechanisms that process information at the level of sense organs (vision, hearing, smell, and balance). Receptors receive about 10^9 bits of information per second from the environment. Approximately 10^7 bits of information per second is transmitted back to the environment, mainly by speech, gestures, and mimicry.

Only part of the environmental information (about 10^1 bit/s) is transmitted to the consciousness. These pieces of information mostly aim to ensure the safety of the organism. For instance, pain receptors ensure a reflexive response to all stimuli if their intensity results in skin or internal organ damage.

This information is transmitted very quickly to the brain and is sensed consciously for as long as the damaging factor remains in effect. On the contrary, information received from temperature receptors is not stored in the consciousness for a long time if the temperature of the environment is comfortable—that is, if it does not indicate a situation that would pose a danger to health and life and therefore does not require triggering of thermoregulation mechanisms.

Communication with the environment is possible due to the basic properties of receptors, such as:

- *Excitability*: Allows the nerve and muscle cells to receive energy from the environment and transform of this energy into specific functions of organs and systems, for example, in the muscular layer of the internal organs or in the locomotor system.
- *Reception specificity*: Enables the nerve cells, with their diversified structure, to selectively receive information contained in various forms of environmental energy: mechanical, acoustic, and chemical. Pain receptors are an exception, as they respond to all types of stimuli if its intensity results in damage to the skin and/or internal body organs.
- *Sensory adaptation*: The transmission of information to the subconsciousness regarding stimuli that have been assessed as nonthreatening to the organism. Information from pain receptors is an exception here—it remains in the consciousness for as long as the damaging factor is exerting influence.

The brain is the main receptor for information; it manages all the systems and internal organs of the body. The nervous, endocrine, circulatory, and respiratory systems carry out the orders of the brain to ensure that all organs function while simultaneously maintaining balance of the internal processes of the organism. All of these systems cooperate with the locomotor or muscoskeletal system to carry out the orders of the brain in terms of exerting active influence upon the environment, for example, during the work processes. The environment–organism interaction follows an established pattern (Table 5.1).

TABLE 5.1
Interactions between the Environment and a Human Body

Environmental event
Perception of the event by the receptors of the sense organs
Assessment by the central nervous system
Interpretation by the brain
Modulation by personality and genes
Transmission of information by the neurohormonal system to the effector organ (heart, lungs, kidneys, muscles, and so on)
Transformation of the signal sent to the effector organ into a specific organ function
A complex physiological and behavioural response of the organism (human body), coordinated by the brain

5.2 THEORIES OF STRESS DEVELOPMENT

In the mid-nineteenth century, physician and physiologist Claude Bernard (1813–1878), in his work entitled *Lessons on the Phenomena of Life Common to Animals and Vegetables*, presented a theory summarising the basic rules of cooperation between the organism's internal environment and the external environment, as well as the unity of life processes for all living organisms. His theses, which were mostly based on intuition and careful observation, have now become particularly relevant to the most recent studies of molecular biology, which have shown similarities between cellular level stress reactions in all living animal and plant organisms.

In the 1920s, Bernard's concepts were further developed by physiologist Walter B. Cannon (1871–1945), who was the first to use the word *homeostasis* to describe the dynamical stability of the internal environment, which is maintained despite changing conditions in the external environment. In his book *Wisdom of the Body* (1963), Cannon described the results of his observations and simple analyses, indicating engagement of the entire physiological chain of reactions in response to stimulation of the nervous system under the influence of intense emotions. The reaction that resulted, which he succinctly summarised using the term 'fight or flight', is characterised by heart-rate acceleration, an increase in both the pressure and concentration of high-energy compounds (such as glucose) in the blood, dilation of the alveoli, and an increase in the number of circulating red blood cells. These processes enhance vigilance, improve reflexes, and allow quick adaptation of the muscular system to substantial effort requirements (Figure 5.1). The same changes were recorded by

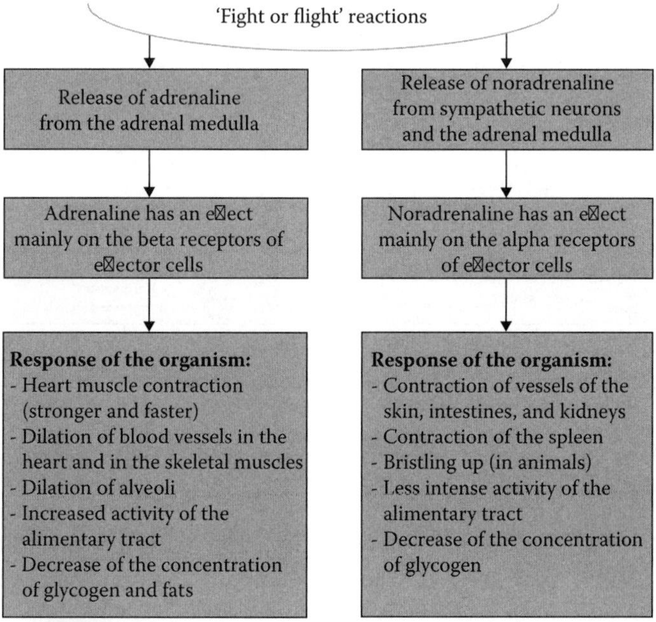

FIGURE 5.1 Diagram showing stimulation of the adrenal medulla and its physiological consequences, or the 'fight or flight' reaction.

Cannon after adrenaline injections, proving that this hormone was responsible for the preparation of the organism to respond to a stimulus.

The role of adrenaline in the mobilisation of an organism under the influence of strong emotions had already been described by Napoleon Cybulski (1895), a Polish physician and scientist, in the late nineteenth century. Discoveries that laid the foundation for knowledge of the basic mechanisms of adaptation to environmental conditions were popularised in the late twentieth century due to the efforts of a great scientist and physician, Hans Selye (1907–1982). In his research, conducted using advanced biochemical techniques, he confirmed that the reaction described by Cannon, associated with increased secretion of adrenaline, resulted in psychophysical mobilisation, which activated the organism to make an effort, similar to the agitation of sportsmen lining up at the start. Selye was the first to apply the term 'stress' to life processes, defining it as a *nonspecific response of the organism to all stimuli*; in his theory, which he called General Adaptation Syndrome (Selye 1946), he summarised his research on the effects of strong environmental stimuli. He discovered that the activating effect of adrenaline was supported by steroid hormones called corticosteroids, which mobilised the reserves of the high-energy compounds (glucose and free fatty acids) for use by the muscles and glucose for use by the brain and red blood cells. Corticosteroids intensify the brain processes of memorising and learning and influence our behaviours, including aggression level, which is significant in the processes of coping with stress. Selye also described the destructive role of excessive concentrations of corticosteroids in metabolic, immunological, and psychological processes, which lead to a limited ability to adapt to long-term stress (Figure 5.2).

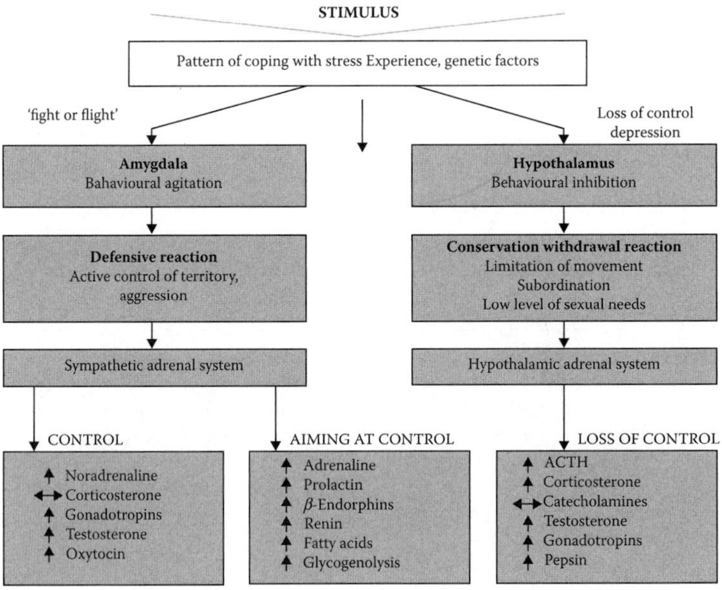

FIGURE 5.2 Two potential responses of the organism to stress: active coping with stress, that is, 'fight or flight'; and the passive reaction due to loss of the ability to cope with stress, leading to depression.

The psychosomatic consequences of long-term stress and the destructive role of corticosterone in the process has been summarised by Selye in his books, which widely and effectively popularised the issue of stress (Selye 1975a,b; 1976). Society now widely recognises adrenaline and corticosteroids as the hormones of stress that determine its intensity and consequences.

Neurophysiological research developed dynamically in the 1970s, confirming the existence of a path in the part of the brain known as the hypothalamus, which transforms the sensory perceptions from receptors into emotions and then into stress reactions. Interesting studies by endocrinologist W. Mason proved that the stress reaction is neutralised—and the adrenaline and corticosteroid concentrations in the blood do not change—in situations wherein emotions associated with a sense of threat due to pain or hunger have been blocked pharmacologically (Mason 1975). This means that typical changes caused by stress do not develop if the environmental factor is not *perceived* as a threat to the body, even if it poses a real danger (Figure 5.3). Discovery of the role of conscious perception as a condition for the emergence of somatic stress reactions has led to the establishment of stress psychology and psychophysiology as a branch of science combining the theories of physiologists and psychologists (Figure 5.4).

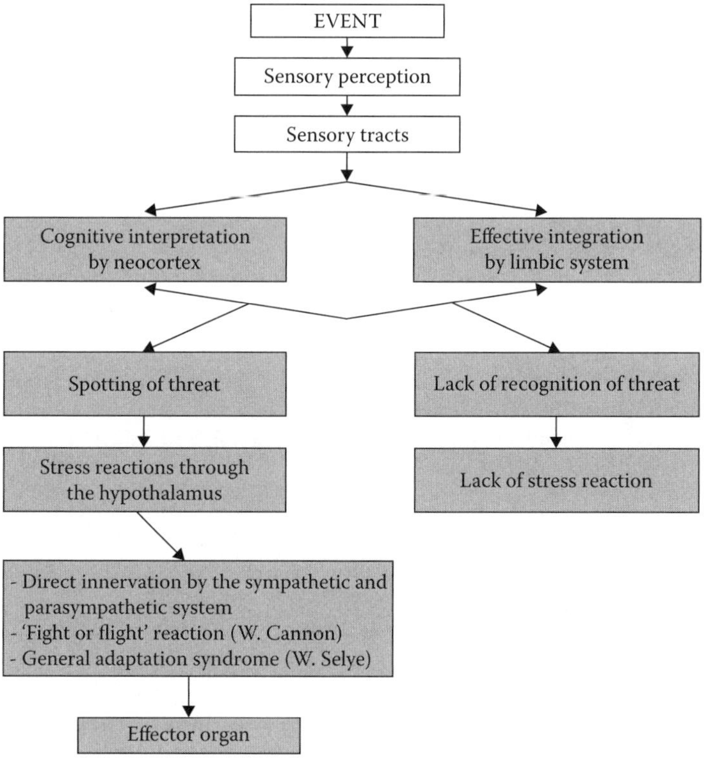

FIGURE 5.3 Reaction of the organism to stress: the role of conscious perception as a condition for the emergence of stress reactions.

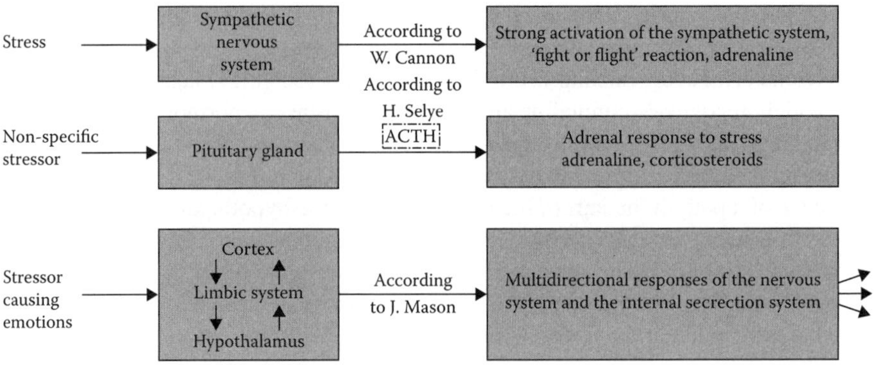

FIGURE 5.4 Development of the stress concept.

All these studies have formed the basis for a coherent psychophysiological theory of stress, which is being developed further by psychologists and biologists at deeper levels—cellular and molecular.

5.3 PHYSIOLOGICAL BASIS OF ADAPTATION TO STRESS

A high level of adrenaline in the blood has become a symbol for an interesting life that is full of extraordinary events. However, these positive emotions call for reflection on the following issue: for how long is an organism able to withstand high emotional tension? Particularly disturbing are the effects of stress upon the circulatory system. Diseases related to this system are among the main causes of death in developed societies. Analyses of adaptation mechanisms have allowed for a more comprehensive assessment of the positive and negative aspects of 'training' in ordinary and extraordinary stress situations.

Adaptation is the ability to adjust to both changing conditions and environmental stress and is one of the basic processes that ensures the survival of all living organisms. Individual and social adaptation mechanisms are interesting from the perspective of (1) ailments and health effects of this condition in everyday life and at work and (2) social and economic consequences.

Three objectives need to be achieved in the processes of adaptation to environmental stimuli (Burchfield 1979):

- Maintenance of homeostasis (physiological and psychological)
- Protection of metabolic resources and psychophysiological efficiency
- Effective protection against stress or steps to counteract it

These objectives are achieved by mechanisms that are subject to central control by the brain. The mechanisms function in the periphery (body tissues and organs) during the continuous compromise between stimulation and suppression of psychophysiological processes.

Central adaptation mechanisms are neuropsychic changes taking place as a result of stress in the central nervous system (CNS; the brain) and the associated changes of behaviour, encompassing the processes of learning to cope with stress and maintaining an active attitude throughout the process.

The stages of the learning process are (1) threat recognition by sensory receptors and organs; (2) estimation of the seriousness of the threat and prediction of the effects of its influence; and (3) determination of the possibilities of coping with it. This allows us to 'get accustomed' to stress (McCarty et al. 1992).

The active attitude of coping with stress is the processes of selection of the tactics of struggle and avoidance behaviours or prediction-based responses, including psychophysiological reactions preceding stress (Lazarus and Folkman 1984; Burchfield 1979). The processes of learning and active coping with stress are the same for all organisms, including those with a low level of CNS development. This means that the processes take place in the extracortical structures of the brain and are part of the basic neural mechanisms, which adapt organisms to interactions with the environment (Groves and Thompson 1970; Bolles 1972). The similarity of these reactions in all living organisms has allowed us to study the molecular effects of stress in simple organisms, which have a small number of nerve cells (Ellis 1996; 2007), and compare the results to more complex organisms.

The significant mechanisms of central adaptation include changes in the nervous system that lead to a decreased sensitivity to pain. This is associated with changes in the metabolism of opiate compounds (endorphins and encephalins) found in the brain, which play the role of natural anaesthetic agents.

Peripheral adaptation mechanisms protect the organs (such as the heart and the kidneys) and systems (the circulatory, respiratory, and endocrinal systems) against the excessive activity of hormones released intensively during stress (adrenaline, noradrenaline, vasopressin, corticosteroids, and renin). This is brought about by the regulation of the transmission of nervous stimulation when receptors present on the cell membrane, where hormones enter the cells of the effector organs from the blood, are nonspecifically blocked. These receptors may be blocked by hormones that are already active, and therefore their excess resulting from release into the blood during the arousal caused by stress is inactivated. This reduces the effect of the excessive amount of 'stress hormones' in the blood, limiting stimulation of the effector cells of the heart, kidneys, muscles, brain, and walls of blood vessels.

Reduction of arousal caused by stress may also take place through inhibition of the activity of the sympathetic nervous system, including prevention of excessive adrenaline secretion through the release of active inhibitors such as prostaglandins or acetylcholine from the sympathetic nervous system during the negative feedback processes.

Apart from those listed above, all systems taking part in a stress reaction also have individual mechanisms of cooperation during such a comprehensive mobilisation. For instance, adaptation of the circulatory system is ensured by baroreceptors, which regulate arterial blood pressure. This process is based upon a specific, reflexive mechanism that regulates the stroke volume, heart rate, and blood pressure in response to the arterial blood pressure and to changes in the circulating blood volume in the venous system.

The effect of all these mechanisms is to reduce the reactivity of all the systems that take part in responding to stress, that is, becoming 'habituated to stress'. Such

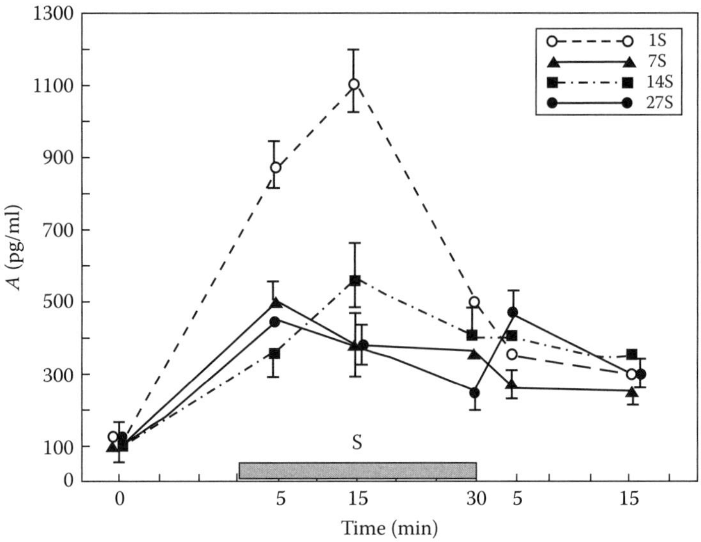

FIGURE 5.5 Adrenaline concentration in the blood during exposure to substantial stress: a 30-minute stress event (1S) and a stress event repeated for: 7 days (7S), 14 days (14S) and 27 days (27S). (Reprinted with permission from Konarska, M. 1995. *Response of the Sympathetic-Adrenal Medullary System to the Test Stress of the Chronically Stressed Rat. Habil. Diss. (abstract in English).* Warsaw: CIOP.)

a reaction has been seen in research studies in model experiments, in which release of adrenaline into the bloodstream has been analysed under the influence of a stress factor.

These studies have shown that the first contact with stress may result in an increase in the secretion of adrenaline into the bloodstream by as much as 40 times, indicating the 'operational' stimulation of the sympathetic nervous system (the 'fight or flight' reaction) and resulting in the physiological mobilisation necessary to cope with stress (Konarska et al. 1990b; Konarska 1995; Figure 5.5). However, arousal is extinguished relatively quickly, even before the stressor ceases to have an effect; this provides an opportunity to save the high-energy compounds consumed in excess under stress (glycogen, glucose, and active substances in the organism such as enzymes and hormones), and provides time for regeneration of these resources.

The ability to become accustomed to stress has been shown in a decrease in the secretion of adrenaline into the bloodstream under long-term administration of a stress stimulus (Figure 5.5). The adrenaline concentration after each subsequent exposure to a stress factor was reduced by 40%–60%, which most likely proves that the sense of threat is reduced and the negative emotions diminish (Konarska et al. 1990a, 1990b; Konarska 1995). Such long-term contact with 'familiarised' stress factors may exert a positive training effect, increasing the ability to respond to other, unknown stimuli.

The 'training' effect in relation to an active stimulus is illustrated in Figure 5.6, which shows that introduction of a novel stress factor among the adapted ('trained')

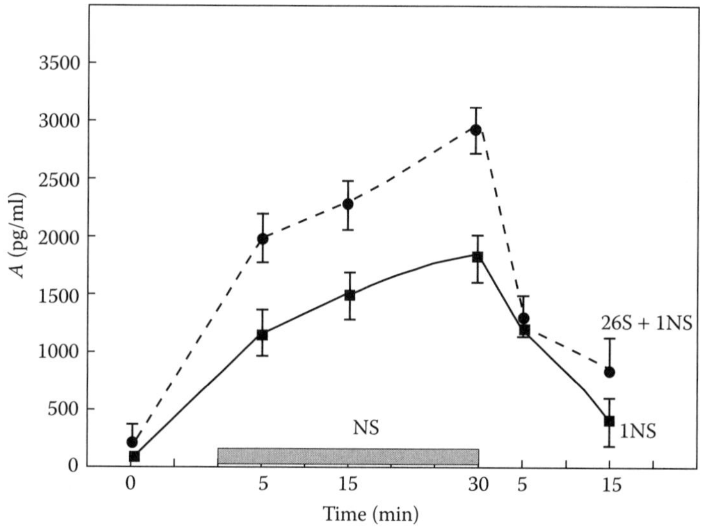

FIGURE 5.6 Adrenaline concentration in the blood during a 30-minute exposure to novel stress (1NS) and a response to the same event after previous long-term exposure to another stress situation (26S + 1NS). (Reprinted with permission from Konarska, M. 1995. *Response of the Sympathetic-Adrenal Medullary System to the Test Stress of the Chronically Stressed Rat. Habil. Diss. (abstract in English).* Warsaw: CIOP.)

individuals causes a reaction that is almost double that of the first contact with stress (Konarska et al. 1989; 1990c; 1995). The training results in an increase in the levels of active substances, including accumulation of adrenaline reserves in the adrenal glands triggered by the emergence of a new threat. In the above studies, the concentration of adrenaline and the arterial blood pressure was typical for the resting condition at all times prior to everyday exposure to stress, which fully proves an ability to maintain the stable functioning of the circulatory system.

The results of these studies illustrate some aspects of adaptation processes that occur because of the presence of many regulatory physiological mechanisms.

5.4 HEALTH EFFECTS OF STRESS

Stress causes mobilisation of the entire organism. When the mechanisms and effects of stress were being discovered in the 1980s, stress was estimated to lead to more than 14,000 physicochemical changes in an organism (Wilson and Schneide, after Asterita 1985). Attempts at summarising these changes are no longer made, because, starting from the molecular level, virtually no aspect of the organism is free from stress-induced changes. Therefore, stress is a risk factor for many diseases of very diverse origins (Johnson 1990; Dallman et al. 2004, 2007; Marmot and Shipley 1996; Walker et al. 1999; Widerszal-Bazyl 2005; Kreis and Bodeker 2007; Herbert et al. 2006; Jędryka-Góral et al. 2002; Bland et al. 2000; Nilsson 2007).

Epidemiological studies have found a distinct correlation between the stress-inducing factors of life and work and the incidence of the following conditions:

- Metabolic disorders (vascular atherosclerosis, obesity, and diabetes).
- Invasive diseases caused by both bacteria and viruses (tuberculosis and neoplastic diseases)
- Regulatory diseases (hypertension and coronary heart disease)
- Psychological disorders (depression, anxiety, insomnia, and chronic fatigue syndrome)

Under substantial stress, adrenaline and other catecholamines exert a significant influence on the reactivity of the circulatory system by increasing the heart rate and the arterial blood pressure; hyperactivity and a substantial level of variability in circulatory reactions under chronic stress, particularly in the case of genetic predisposition to hypertension, may become a direct cause for the development of 'resistant' hypertension, which is caused by hypertrophy of the blood vessel muscle tissue. The causes of hypertension include changes in the activity of hormones such as angiotensin and renin, which participate in the kidneys' control of sodium and the volume of bodily fluids.

During long-term stress, corticosteroids influence the concentration of insulin and balance of carbohydrates and lipids, leading to changes that cause metabolic diseases such as obesity, diabetes, and atherosclerosis.

Corticosteroids and catecholamines (mainly adrenaline) activate many metabolic paths through glycogenolysis, causing an increase in the production of coenzyme A molecules (Co-A; one-carbon fragments), which serve as building blocks for the synthesis of endogenous cholesterol.

In the sphere of psychogenic reactions, the role of corticosteroids in the development of chronic fatigue syndrome, conservation withdrawal reactions, depression, insomnia, and neurodegenerative diseases, such as Alzheimer's or Parkinson's disease has been confirmed (Walker 2007; Rydstedt et al. 2007a,b; Dallman et al. 2004, Dallman et al. 2007).

Studies have also confirmed a correlation between the increased secretion of corticosteroids and invasive diseases. Excessive secretion of these hormones leads to reduced immunity due to disturbances in the production of immunological factors like cytokines and antibodies, as well as a decrease in the activity of immune cells (T and B lymphocytes), phagocytes, and other cytotoxic cells. This results in a general decrease in immune system activity (Walker et al. 1999; Jędryka-Góral et al. 2002).

Since the 1980s, studies on the physiological and biomedical stress mechanisms have focused on processes caused by stress at the cellular level, initiated by the discovery of 'thermal shock proteins' (HSP family), identified by Alfred Tissiers (1917–2003) after accidental exposure of cells to a high temperature. These proteins appeared in cells in response to stress or diverse toxic factors, such as infections, toxins, hypoxia, and dehydration. The production of thermal shock proteins is a cellular-level reaction to stress, common in all living organisms—plants and animals. It confirmed the thesis of unity of the entire living world, proposed in the nineteenth century by Cannon; at the same time, it confirmed the thesis of nonspecific reaction to stress, a component of Selye's theory that had been criticised for many years.

The discovery of thermal shock proteins has enabled valuable research using simple animal and plant organisms as models, as well as tissue and cell cultures, which has drawn universal conclusions regarding the mechanisms of the emergence (and treatment) of many genetic and metabolic diseases in humans.

The common fruit fly *Drosophila melanogaster* has contributed to genetics and medicine (such research was awarded a Nobel prize in 1933, for mechanisms of inheritance of qualities, and in 1995, for control of genetic congenital defects) and today serves as a 'research model' for studies on the influence of stress on the aging processes. These are manifested by the accelerated shortening of telomeres and apoptosis, that is, programmed cell death—a process, which, when disturbed, leads to the development of neoplasms.

In neurophysiology, studies on protein aggregation in Alzheimer's and Parkinson's diseases have been conducted using the nematode *Caenorhabditis elegans* (such a study was awarded a Nobel Prize in 2006 for the study of control of information flow in the genes). This small, transparent organism has 959 somatic cells and 16,757 genes; about 51% of its genes are identical to those of humans, including those responsible for neurodegenerative processes. The result of this research is of great practical significance, as it will allow the determination of pharmacological methods to stop the basic disease mechanisms that are triggered by stress and other factors.

Many years of research have shown that prolonged stress may also be a significant risk factor for diseases, across civilisations, just as are genetic factors, age, and individual traits.

For instance, 25 years of socioepidemiological research conducted within the framework of the Whitehall or Workhealth II programmes (projects in the European Union) proved that the long-term effects of stress cause disorders of the circulatory and immune systems (Marmot and Shipley 1996; Kreis and Bodeker 2007). The societies of Central and Eastern Europe have been undergoing a process of fast socioeconomic and cultural transformation and have recently served as models for such research. The research shows that the main risk factors include socioeconomic changes resulting in higher levels of stress at work due to increased competitiveness, uncertain employment, and decreased control (Widerszal-Bazyl 2003; Cox et al. 2000). According to forecasts, there will be an epidemic increase in the incidence of diseases of the circulatory system, obesity, and depression; these will be the main causes of early deaths (Siegrist 2001; 2007) until 2020. In Russia, the increase in the death rate since 1989 has already totalled about 400,000 per year (Williams 2007), whereas in Hungary, the mortality of men aged 40–69 has increased from 12.2% to 16.2% (Kopp 2007).

All aspects of research on stress, including biomedical, psychological, and psychosocial, have been combined into a single field of research that deals with the biology of happiness and positive well-being. During a conference on stress in Budapest in 2007, Andrew Streptoe presented the results of a meta-analysis indicating that the death rate in a population of several thousand people dominated by positive emotions and pro-health behaviours was lower than the average by almost 19% (Streptoe 2007). These correlations are associated with psychobiological reactions activated by the CNS, including neuroendocrinological, immunological, inflammatory, and circulatory responses. The results of research in this field underline the need to engage in activities aimed at the promotion and shaping of pro-health behaviours.

5.5 SUMMARY

Stress is an interaction between the environment and an organism that is interpreted by the organism as:

- Threatening its physical or psychical stability
- Causing negative emotional arousal
- Leading to an excessive physiological response, including neurohormonal changes

Emotions that accompany a sense of threat are associated with a neurophysiological reaction originating in the hypothalamus of the brain and causing a multitude of psychophysiological changes. Stress has negative consequences for somatic and emotional health and leads to negative social phenomena.

Adaptation to stress, which is aimed at maintaining homeostasis in the internal environment of the organism, is one of the basic mechanisms of survival under changing environmental conditions. Psychophysiological mechanisms adapt by reducing reactivity to subsequent stress events and establishing a condition that is a compromise between agitation and suppression of the systems responding to stress. Adaptation is based on learning processes, including recognition and assessment of the quality of stimuli, assessment of the modes of coping with stress, prediction of the consequences, and determination of the possibilities of overcoming stress. Positive effects of adaptation are the stabilisation of the physiological and emotional processes and the strengthening of the immune system.

The organism's ability to adapt diminishes as a result of long-term stress with which it is unable to cope. This may lead to negative health consequences expressed as somatic and psychological disorders. Prevention of stress effects should be based on activities aimed at increasing adaptation abilities, at both the physiological and the psychological levels.

REFERENCES

Asterita, M. F. 1985. *The Physiology of Stress*. New York: Human Sciences Press.

Bland, M. L. et al. 2000. Gene dosage effects of steroidogenic factor 1 (SF-1) in adrenal development and the stress. *Endocrinol Res* 26(4):515–516.

Bolles, R. C. 1972. Reinforcement, expectancy and learning. *Psychol Rev* 79:394–409.

Burchfield, P. R. 1979. The stress response a new perspective. *Psychosom Med* 41:661–672.

Cannon, W. B. 1963. *The Wisdom of the Body*. New York: Norton (Reprinted from 1939 edition).

Cox, T., A. Griffiths, and E. Rial-Gonzalez. 2000. *Research on Work-Related Stress*. Bilbao: European Agency for Safety and Health at Work.

Cybulski, N. 1895. Further investigation of adrenal gland function. *Main Physiology Journal* 4:82–91.

Dallman, M. et al. 2007. Glucocorticoids as sculptors of adaptation. In *Book of Abstracts—2nd World Conference of Stress*, 436. Budapest.
Dallman, M. F. et al. 2004. Chronic stress-induced effects of corticosterone on brain: Direct and indirect. *Ann N Y Acad Sci* 1018:141–150.
Ellis, J. R. 1996. Discovery of molecular chaperones. *Cell Stress Chaperones* 1:155–160.
Ellis, J. R. 2007. Protein misassembly: Chloroplast, chaperones and crowding. In *Book of Abstracts—2nd World Conference of Stress*, 7. Budapest.
Groves, P. P., and R. Thompson. 1970. Habituation: A dual-process theory. *Psychol Rev* 77(5):419–450.
Herbert, J. et al. 2006. Do corticosteroids damage the brain? *J Neuroendocrinol* 18(6): 393–411.
Jędryka-Góral, A. et al. 2002. Stress—Where are we now? Does immunity play an intrinsic role? *Autoimmunity* 35(7):421–426.
Jenkins, C. D. 1979. Psychosocial modifiers of response to stress. *J Human Stress* 5:3–15.
Johnson, A. K. 1990. Stress and arousal. In *Principles of Psychophysiology*, eds. J. T. Cacioppo and L. G. Tassinary, 216–252. Cambridge, UK: Cambridge University Press.
Konarska, M. 1995. *Response of the Sympathetic-Adrenal Medullary System to the Test Stress of the Chronically Stressed Rat. Habil. Diss. (abstract in English)*. Warsaw: CIOP.
Konarska, M., R. E. Stewart, and R. McCarty. 1990c. Predictability of chronic intermittent stress: Effects on sympathetic-adrenal medullary responses of laboratory rats. *Behav Neural Biol* 53:231–243.
Konarska, M., R. Steward, and R. McCarty. 1989. Sensitization of sympathetic-adrenal medullar response to a novel stressor in chronically stressed laboratory rats. *Physiol Behav* 46:129–135.
Konarska, M., R. Steward, and R. McCarty. 1990a. Habituation of plasma catecholamine responses to chronic intermittent restraint stress. *Psychobiology* 18:30–34.
Konarska, M., R. Stewart, and R. McCarty. 1990b. Habituation and sensitization of plasma catecholamine responses to chronic intermittent stress: Effects of stressor intensity. *Physiol Behav* 47:647–652.
Kopp, M. 2007. Socioeconomic and psychosocial determinants of chronic stress in a changing society. In *Book of Abstracts—2nd World Conference of Stress*, 435. Budapest.
Kreis, J., and W. Bodeker. 2007. *Indicators for work-related health monitoring in Europe*. http://enwhp.org/ (accessed August 8, 2008).
Lazarus, R. S., and S. Folkman. 1984. *Stress, Appraisal and Coping*. New York: Springer.
Marmot, M. G., and M. J. Shipley. 1996. Do socioeconomic differences in mortality persist after retirement? 25 year follow-up of civil servant from the Whitehall study. *Br Med J* 313:1177–1180.
Mason, J. W. 1975. A historical view of the stress. *Human Stress* 1:22–35.
McCarty, R., M. Konarska, and R. E. Stewart. 1992. Adaptation to stress: A learned response? In *Stress Neuroendocrine and Molecular Approaches*, Vol. 2. eds. R. Kvetnansky, R. McCarty, and J. Axelrod, 521–533. Philadelphia: Gordon and Breach Science Publishers.
Nilsson, P. M. 2007. Early ageing as a consequence of adverse psychosocial stress exposure. In *Book of Abstracts—2nd World Conference of Stress*, 441. Budapest.
Rydstedt, L. W. et al. 2007a. The long-term impact of perceived stress on sleep quality. In *Book of Abstracts—2nd World Conference of Stress*, 412. Budapest.
Rydstedt, L. W., J. Deveroux, and M. Sverke. 2007b. Comparing and combining the demand-control-support model and effort-reward imbalance model to predict long-term mental strain. *Eur J Work Organ Psychol* 16(3):261–278.
Selye, H. 1946. The general adaptation syndrome and the diseases of adaptation. *J Clin Endocrinol* 6:117–230.
Selye, H. 1975a. Confusion and controversy in the stress field. *J Human Stress* 1:6–12.

Selye, H. 1975b. *Stress without Distress*. New York: McGraw-Hill.

Selye, H. 1976. *Stress of Life*. Rev. ed. 1936, New York: McGraw-Hill.

Siegrist, J. 2001. Long-term stress in daily life in a socioepidemiologic perspective. *Adv Psychosom Med* 22:91–103.

Siegrist, J. 2007. Work stress and health in the cultural context of globalisation. In *Book of Abstracts—2nd World Conference of Stress*, 436. Budapest.

Streptoe, A. 2007. The biology of happiness and positive well-being. In *Book of abstracts—2nd World Conference of Stress*, 9. Budapest.

Walker, A. 2007. Stress, quality of life and multiple chronic disease: Patterns emerging from a large national sample, Australia. In *Book of Abstracts—2nd World Conference of Stress*, 411. Budapest.

Walker, J. G. et al. 1999. Stress system response and rheumatoid arthritis: A multilevel approach. *Rheumatology* 38:1050–1057.

Widerszal-Bazyl, M. 2003. *Work Stress and Health: An Attempt to Verify the Robert Karasek's Model and the Demand-Control-Support Model*. Warsaw: CIOP-PIB.

Williams, R. 2007. Psychosocial stress management programs in transforming societies. In *Book of Abstracts—2nd World Conference of Stress*, 436. Budapest.

Part III

Basic Hazards in the Work Environment

6 Harmful Chemical Agents in the Work Environment

Małgorzata Pośniak and Jolanta Skowroń

CONTENTS

6.1 INTRODUCTION

Due to the high level of industrialisation in the contemporary world, humans are increasingly exposed to dangerous chemicals in their work and living environments. Chemicals cause many diseases, including insufficient respiratory ability, inflammatory skin conditions, psychoneurological disorders, and neoplastic diseases. Studies conducted by the European Foundation of Living and Working Conditions in Dublin on the work conditions in the countries of the European Union (EU) in 2005–2006 showed that 18% of workers were exposed to or came in contact with chemical agents, including 20.5% through the respiratory tract (Parent-Thirion et al. 2007). In 2007, more than 74,000 deaths due to chemical exposure were recorded in the EU-27 member states (Takala et al. 2009; Musu 2007). One in three occupational diseases recognised every year in the 15 EU countries is caused by exposure to chemical agents (Eurostat 2004).

According to the European Agency for Safety and Health at Work (2006) in Bilbao, about 16 million chemicals are available on the market and 10,000 are manufactured in amounts exceeding 10 tons per year.

In the last few decades, sciences that investigate the effects of chemical substances on human health and the environment have seen rapid development. Still, sufficient data to classify these chemicals according to their toxic and physicochemical properties are available for only 14% of commonly used compounds. Research units active in industrial toxicology therefore attempt to identify the hazards to human health posed by chemical substances and preparations present in the work and living environments.

An important element for the safe use of chemicals in the work environment is the occupational exposure limit (OEL)—the main criterion for assessment of occupational exposure. The American Conference of Governmental Industrial Hygienists (ACGIH), a nongovernmental organisation of industrial hygienists and safety specialists, created one of the first systems for establishing and revising OELs. ACGIH was formed in 1938 and released its first list of OELs in 1941 (Paustenbach 2000). In Europe (Commission Decision 95/320/EC), the Scientific Committee for Occupational Exposure Limits to Chemical Agents (SCOEL) has been developing indicative occupational exposure limit values (IOELVs) for chemical agents in the work environment since 1995. Directives 98/24/EC, 91/322/EWG, 2000/39/EC, and 2006/15/EC of the Environmental Working Group European Commission contain the current lists of IOELVs for 104 chemical substances and binding occupational exposure limit values (BOELVs) for 10 substances (98/24/EC, 99/38/EC, directive 2003/18/EC). France, Germany, Sweden, and Finland have national systems for establishing admissible concentrations of chemicals in the workplace air.

Poland has had such a system since 1983—the Interdepartmental Commission for Maximum Admissible Concentrations and Intensities of Agents Harmful to Health in the Working Environment is its core element. The commission develops new

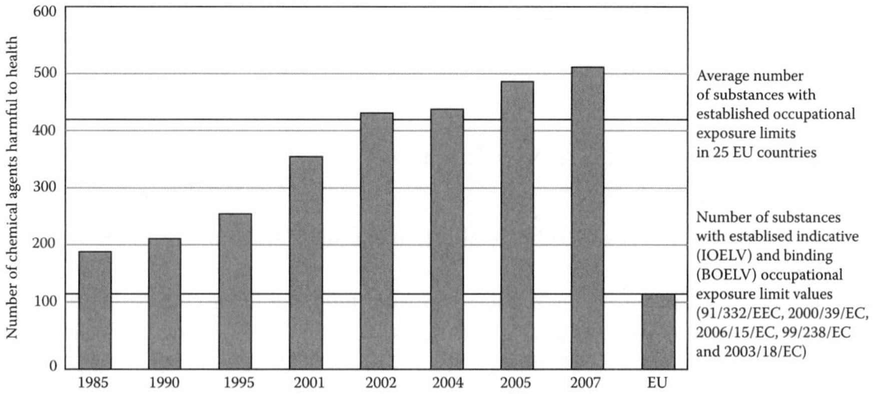

FIGURE 6.1 Establishment and verification of the maximum admissible concentrations in workplace air in Poland (1985–2007) and in the European Union (2007) for chemical agents harmful to health.

values and systematically revises the existing values of maximum admissible concentrations (MACs) of chemical and aerosol agents in accordance with the provisions of the EU directives (Council Directive 89/391/EEC, Council Directive 99/38/EC, Council Directive 98/24/EC) and national regulations (labour code, together with the executive provisions).

In the more than 20 years Poland has been using the national system, more than 300 new items have been added to the list of MACs (Figure 6.1). The list currently in force in Poland includes 514 chemical substances and aerosols, along with their OELs, which are legally binding in all national economy sectors.

An integral part of the documentation of MACs for chemical substances involves the use of analytical methods that measure concentrations of substances at the workplace. The Polish Committee for Standardisation has developed these methods over more than 40 years as standards for air-purity protection at the workplace (Koradecka 1999).

6.2 GENERAL CHARACTERISTICS OF CHEMICAL AGENTS HARMFUL TO HEALTH

6.2.1 Basic Definitions

Chemical substances are chemical elements and their compounds either in their natural state or obtained by a manufacturing process, including by using any additives necessary to preserve their stability and any impurities derived from the process used; they do not include solvents that may be separated without affecting the stability of the substance or changing its composition. *Preparations* are mixtures or solutions composed of two or more substances (Registration, Evaluation, Authorisation, and Restriction of Chemicals [REACH] regulation).

Dangerous substances and *dangerous preparations* are substances and preparations that pose a hazard to human health or the environment. These are classified into at least one of the following categories: highly toxic, toxic, harmful, corrosive, irritant, sensitising, carcinogenic, mutagenic, toxic for reproduction, flammable, extremely flammable, explosive, oxidative, and dangerous to the environment.

6.2.2 Routes of Absorption of Chemicals into the Body

Chemical substances occur as gases, vapours, liquids, or solid bodies. They are absorbed into the body mostly through the respiratory tract and skin or from the gastrointestinal tract.

Substances such as gases or vapours form a homogeneous mixture with air. Two-phase system solid-gases or liquid-gases are called *aerosols*. Aerosols can be divided into *dusts, fumes*, and *mists*. Dusts are suspensions of solid particles in the air, arising from mechanical disintegration. Fumes are also suspensions of solid particles in the air, arising from processes such as combustion, calcination, or sublimation. Mists are suspensions of liquid droplets in the air formed during liquid spraying and fume condensation. Aerosols of oil fumes arise during machining and solvent fumes during spray painting. These substances are absorbed in the respiratory tract mainly as gases, vapours, and aerosols.

The *respiratory tract* is the anatomical part that governs the process of respiration, that is, brings in oxygen and excretes the products of metabolic changes, such as carbon dioxide. The major function of the respiratory system is to exchange gas between the external environment and the organism's circulatory system. In humans and other mammals, this exchange facilitates the oxygenation of blood by removing carbon dioxide and other gaseous metabolic wastes from the circulation. The respiratory tract is divided into upper and lower respiratory tracts; the skin also plays a minimal role in gas exchange. The upper respiratory tract includes the nasal passages, pharynx, and larynx. The lower respiratory tract includes the trachea, right and left bronchi, and lungs. The trachea divides into the left and right air tubes, called bronchi; the bronchi in turn connect to the lungs. Within the lungs, the bronchi branch into smaller bronchi and even smaller tubes called bronchioles. Bronchioles end in tiny air sacs called alveoli, where the exchange of oxygen and carbon dioxide actually takes place. Each lung contains about 300–400 million alveoli. The lungs also contain elastic tissues that allow them to inflate and deflate without losing shape and are encased by a thin lining called the pleura. This network of alveoli, bronchioles, and bronchi is known as the bronchial tree (Aleksandrowicz 1994).

The respiratory tract is covered with an epithelium, the type of which varies down the tract. The respiratory tract also contains smooth muscle, elastin and cartilage, and glands in parts of the tract that produce mucus.

There is no cartilage or smooth muscle in the bronchi. Each of the segmental bronchus supplies a bronchopulmonary segment. A bronchopulmonary segment is a division of the lung separated from the rest of the lung by a connective tissue septum. The left lung is divided into two lobes and the right into three lobes.

The bronchial tree continues branching until it becomes terminal bronchioles, which lead to the alveolar sacs. Alveolar sacs are made up of clusters of alveoli, similar to individual grapes in a bunch. The individual alveoli are tightly wrapped in blood vessels, and the gas exchange actually occurs here. Deoxygenated blood from the heart is pumped through the pulmonary artery to the lungs, where oxygen diffuses into the blood and is exchanged for carbon dioxide in the haemoglobin of erythrocytes. The oxygen-rich blood returns to the heart via the pulmonary veins to be pumped back into the systemic circulation.

The alveoli consist of an epithelial layer and extracellular matrix surrounded by capillaries. There are three major alveolar cell types in the alveolar wall.

Each alveolus is wrapped in a fine mesh of capillaries that covers about 70% of its area. The blood brings carbon dioxide from the rest of the body to be released in the alveoli; then, the blood in the alveolar blood vessels takes up the oxygen from the alveoli to be transported to the cells in the body. An adult alveolus has an average diameter of 0.2–0.3 mm (Pabst and Putz 2007). Together, the lungs contain approximately 1500 miles (2400 km) of airways and 300–500 million alveoli, with a total surface area of about 70 m^2 in adults. An adult at rest breathes in about 5 litres per minute and during work about 20 litres or more.

Gases and vapours are absorbed into the body directly through the respiratory tract, depending on the physical activity. Fumes in the liquid phase can be absorbed directly by the alveolar epithelial layer. Aerosols (dusts and fumes) are not totally absorbed—some dusts can be eliminated with mucus, coughed up with sputum, or swallowed.

The contribution of the respiratory-tract segments (nasopharyngeal, tracheobronchial, and pulmonary) to aerosol absorption depends on the particles' aerodynamic diameter. Particles less than 7 μm (the respirable fraction) are the most dangerous, because they are most easily deposited in the lungs and are also easily distributed throughout the body (Figure 6.2).

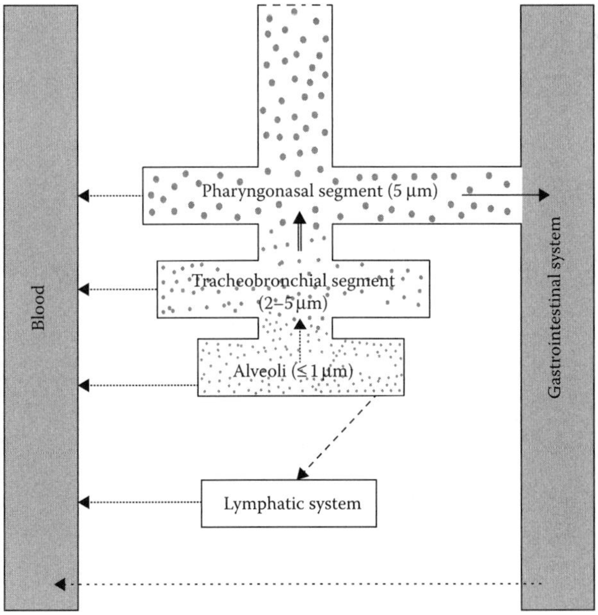

FIGURE 6.2 Absorption and distribution of chemicals in the lungs.

The *skin* is the outer covering of the body; it protects the body against the environment. Some chemicals destroy the skin, but others can be absorbed through it. There are two ways chemicals can be absorbed through the skin:

- Diffusion through the skin layers (the transepidermal routes)
- Diffusion through skin appendages—hair follicles, oil glands, or sweat glands (the transfollicular routes)

Organic chemicals with a high partition coefficient in oil or water emulsions and a small degree of ionisation can be absorbed by passive diffusion (the transepidermal route). Aliphatic and aromatic hydrocarbons, nitric compounds and amines, phenols, phosphor-organic insecticides, carbon disulphide, and tetraethyl lead can be absorbed this way.

The role of absorption of organic solvents or pesticides through skin is often underestimated.

The transfollicular route has a less important role in absorption, because skin appendages occupy only a small surface of the skin—from 0.1% to 0.01%. Electrolytes, metals, and their organic compounds are absorbed through this route.

Absorption of chemicals through the skin from the workplace air is not very important for most substances, apart from phenol, nitrobenzene, and aniline. The absorption rate through the skin for these substances can increase to about 30% of the volume present in the work environment. Different factors can change the absorption efficiency; among them are the state of the skin, anatomic differences, age, temperature, and humidity (Czerczak and Kupczewska 2002).

Absorption of chemicals through the gastrointestinal tract can occur in different sectors, from the mouth through the oesophagus and stomach to the intestines (Figure 6.3). Chemicals are absorbed from the gastrointestinal tract when they are in

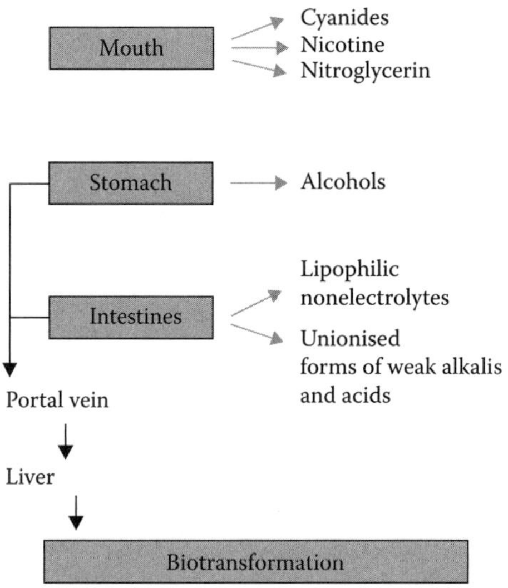

FIGURE 6.3 Absorption of chemicals in the gastrointestinal tract.

a nonionic form. Weak acids are absorbed mainly in the stomach, while weak bases are absorbed in the intestines. Dangerous chemicals undergo resorption very quickly in the intestines because the intestinal mucosa is physiologically adapted to absorption. Metals can also be absorbed by the gastrointestinal tract, but the mechanism and efficiency of this process are not well known. Knowledge of the chemical form of the metal is important, for example, methyl mercury is absorbed by the gastrointestinal tract with an efficiency of 90%–100%, whereas inorganic mercury compounds are absorbed with only 10% efficiency and metallic mercury is not absorbed in this way. The absorption efficiency also depends on the person's diet. Low levels of calcium, ferric ions, and proteins in the diet cause an increase in the absorption of toxic metals such as lead and cadmium.

6.2.3 Combined Effects of Chemicals

In industrial conditions, situations in which the same population is exposed to different substances or preparations are the rule rather than the exception. When humans are exposed to two or more chemicals, the combined toxic effects may be one of the following types:

- *Independent*: The mechanisms of the substances are different and therefore induce different effects. The combined effects are the same as if the person was exposed to each substance individually.
- *Additive*: The mechanisms of the substances are the same and the volume of the effects is quantitatively equal to the sum of the effects caused by each substance individually.
- *Synergistic*: The toxic effects of one substance are intensified by the other.
- *Strengthened (potentialisation)*: The chemical substance itself does not cause a harmful effect, but it intensifies the harmful effects of another substance.
- *Antagonistic*: The toxic effect of a given substance is reduced or cancelled by another substance (Skowroń et al. 2000).

6.3 EFFECTS OF HARMFUL CHEMICALS IN THE WORK ENVIRONMENT ON HUMANS

6.3.1 Types of Poisoning

Poisoning occurs when exposure to chemical substances or preparations exceeds the limits of human capacity, that is, when the body cannot remove these substances either by digestion or excretion. Poisoning can be divided into the following groups according to the exposure and duration:

- *Acute*: Harmful effects are induced by a substance absorbed into the body in a high dose or on one occasion; occurs in a relatively short period of time (within 24 hours).

- *Chronic*: Harmful effects occur after prolonged exposure to low doses or concentrations of substances present in the industrial or living environment, such as contamination of air, water, or food.

Occupational poisonings are usually chronic. The body's response to chemicals depends on the chemicals' physicochemical properties and absorption route, the health, sex, and age of the exposed person, and the condition of both the endocrine and immune systems, in addition to external factors such as temperature, exposure period, and humidity.

6.3.2 Effects of Harmful Chemicals on Humans

The effects of chemicals on the human body are divided into local (corrosive and irritant), systemic, and remote (genotoxic, carcinogenic, embryotoxic, fetotoxic, and teratogenic) effects.

6.3.2.1 Corrosive and Irritant Effects

Local effects of chemical substances occur at the site of contact with the substance, such as on the skin or in the respiratory tract.

In the occupational environment, acute poisonings are often considered accidents at work. Corrosive substances or preparations may induce burns of the skin and mucosal membranes. Direct eye contact with irritant chemicals may cause health effects of different intensities, ranging from temporary stinging and lacrimation to permanent corneal damage. The rate and intensity of the damage depend on the quantity of the substance to which the eyes were exposed and the time elapsed before the administration of first aid. Acids, alkalis, and solvents are examples of irritant substances.

When a chemical come in contact with the skin, damage to the protective skin layer may occur and, consequently, the skin will become dry, rough, and sore. This condition is called *irritant* or *toxic eczema*. The substances that cause this type of skin lesion are called *primary irritants*. Alkalis, acids, organic solvents, soaps, and washing agents are the most important primary irritants. Chemicals irritate the respiratory tract and cause inflammatory lesions in different parts of the tract. The gases and vapours of some substances, such as hydrogen fluoride, hydrogen chloride, ammonium, formaldehyde, and acetic acid, usually cause lesions in the nose, pharyngonasal cavity and larynx. Substances such as chlorine, arsenic trichloride, and phosphorus trichloride cause lesions in the upper respiratory tract and bronchi. Irritant chemicals cause cough and sneezing and can provoke spasms of the epiglottis and bronchi in high concentrations. Phosgene or nitrogen oxides directly cause lesions of the lung, which may be the cause of pulmonary oedema and pleural inflammation. Such inflammation may occur immediately after exposure or within a few hours after exposure, that is, after the latency period. In this case, the clinical symptoms of poisoning include cough, cyanosis (oxygen-deficiency symptom), and coughing up large amounts of mucus (Seńczuk 2007).

Substances that irritate the respiratory tract also damage its defence mechanisms, decreasing immunity and increasing susceptibility to asthma and pulmonary oedema.

The effects of irritating substances on the respiratory tract depend upon a number of factors, including the concentration of the substance, duration of exposure, and individual susceptibility. Another important factor is the manner of breathing—through the mouth or through the nose. People who are exposed to irritants over a long period often have an impaired sense of smell and cannot detect the presence of these substances in the workplace air.

6.3.2.2 Sensitisation

Allergic contact eczema appears in workers who are in contact with substances that provoke sensitising effects. Allergic reactions of the skin are often similar to symptoms of inflammation: itching, burning, erythema, and papular, vesicular or exfoliated epidermis, mainly on the hands, forearms, and face. Allergic contact eczema has recently been the most frequently occurring occupational skin disease. Chemical allergies cause about 70% of skin inflammation. The course of allergic contact eczema depends on the type of contact with the sensitiser. Allergies resulting from permanent, long-term contact with such substances are resistant to treatment. Only completely preventing contact with the allergen allows the person to recover. Aromatic amines, turpentine, epoxy resins, triethyltetramine, rubber, chromium VI, nickel, cobalt, tetracycline, formalin, aniline dye, and essential oils are the most frequent substances that produce allergies.

Asthma is caused by sensitising changes in the respiratory tract arising from exposure to allergens in the occupational environment. The characteristic signs of asthma are a cough, especially at night, and respiratory difficulties such as gasping and shortness of breath. Occupational asthma is caused by both macro- and microparticles in the work environment, often in concentrations that do not exceed the exposure limits or exceed them minimally (Górski 1997). Macroparticles include flour and its impurities, animal allergens, resin (especially colophony), antibiotics (especially beta-lactam antibiotics), latex, grains of oil plants, and detergents. Microparticles include isocyanines, platinum salts, nickel, chromium, cobalt, aluminium, dyes, persulphates, henna, disinfection agents, and acid anhydrides.

An allergy specialist can conduct many types of examinations to diagnose occupational asthma, such as the skin-patch test and testing for antibodies. To prevent allergies, contact with sensitising substances should be avoided by: (1) eliminating such substances from the environment, (2) using personal protective equipment, or (3) changing the work location (Górski 1997).

6.3.2.3 Systemic Actions

The human body is made up of many tissues and organs forming different types of systems. Chemicals can have various effects on these systems that lead to morphological and functional changes in some organs or groups of organs.

Exposure to neurotoxic substances can disturb the process of external inhibition or the stimulation of nerve conduction in the central and peripheral nervous systems. It can lead to functional disturbances in the central (encephalopathy) or peripheral (peripheral neuropathy) nervous systems. Depression or stimulation of the nervous system may also occur, which indicates a correlation of the toxic action of the substance with the adaptation mechanism. This can lead to normalisation of the activity of this system. The rate of these changes and the time needed to elicit them depend upon the concentration or dose of the neurotoxic substance. Small functional changes in the central and peripheral nervous systems are often a serious result of exposure to lower concentration of chemicals. Chloroalkanes, carbon disulphide, and mercury vapours and lead are substances that affect the nervous system. Alkanes (hexane), aliphatic ketones, and carbon disulphide show a strong action on the peripheral nervous system. Exposure to phosphoro-organic compounds, such as parathion, can obstruct and destroy the functioning of the central and peripheral nervous systems.

Nitro compounds (nitrotoluene), chloroalkanes, and chloro- and bromobenzene derivatives affect the liver. The symptoms of liver injury caused by such substances, such as a yellow colour of the eyes, are sometimes misdiagnosed as liver inflammation.

Substances such as carbon tetrachloride, ethylene glycol, and carbon disulphide affect the kidney's ability to eliminate toxic substances from the body. Other substances, such as cadmium, lead, turpentine, methanol, toluene, and xylene affect the kidneys more slowly.

Attention recently has been focused on the harmful effects of chemicals on the reproductive systems of both women and men. Toxicology has focused on the mechanisms of the toxic effects of substances on the entire reproductive cycle, including the production of reproductive cells in both sexes, fertilisation of the ovum, its implantation, embryo or organogenesis, foetal development, delivery, and growth until sexual maturity. Fertility impairment in men may be connected with exposure to chemicals such as ethylene dibromide, benzene, anaesthetic gases, chloroprene, lead, organic solvents, and carbon disulphide. Miscarriages may follow exposure to anaesthetic gases, ethylmercury oxide, glutaraldehyde, chloroprene, lead, organic solvents, carbon disulphide, and vinyl chloride.

6.3.2.4 Carcinogenic Effects

Exposure to some chemicals can cause an uncontrolled growth of cells, leading to cancerous lesions. Most carcinogenic substances are nonthreshold compounds, that is, those for which safe exposure levels cannot be determined.

Cancerous lesions may occur many years after the first exposure to the chemical substance. This period is called the *latency period* and can range from 4 to 40 years. Cancers that develop due to occupational exposure can be localised in different places in the body, and are not necessarily limited to those areas that were in direct contact with the chemical substance. Substances such as arsenic, asbestos, chromium, and nickel may cause lung cancers. Chromium, nickel, isopropyl oils, wood dust, and dust from tanned leather can cause neoplasms of the nasal cavity and nasal sinuses. Bladder cancer is associated with exposure to benzidine, 2-naphthylamine, or dust of tanned leather, whereas skin cancer is associated with exposure to arsenic, coal tar, and petroleum derivatives containing polycyclic aromatic hydrocarbons. Vinyl chloride may cause cancerous lesions in the liver and benzene in the bone marrow (Szadkowska-Stańczyk and Szeszenia-Dąbrowska 2001).

Exposure limits for certain carcinogenic substances are not established in many countries because safe exposure levels cannot be determined. Instead of proposing an exposure limit, agencies or organisations determine the risk caused by a definite exposure level. Government agencies and national or international organisations use the concept of *acceptable risk* for establishing or proposing admissible exposure levels for carcinogenic substances. The level of acceptable risk depends on commonly accepted social and economic criteria; in developed countries, three interest groups, made up of representatives of employees, employers, and state administrators whose task is to perform law-enforcement surveillance, make decisions regarding this topic (Czerczak 2004b).

Enterprises having carcinogenic or mutagenic substances should strive to eliminate them from their technological processes or maintain their concentrations at levels below the maximum admissible values, preferably at a very low level. Exposure to nonthreshold carcinogenic substances should also be minimised.

6.4 ADMISSIBLE CONCENTRATIONS OF CHEMICAL AGENTS HARMFUL TO HEALTH IN THE WORK ENVIRONMENT (POLAND, EU)

An admissible concentration of a substance in the air exists for all harmful chemical agents. Workers may be exposed to chemicals without any adverse health effects at levels below these concentrations. The Polish list of exposure limits for chemical substances in the work environment has the following criteria:

- *Maximum admissible concentration (MAC-TWA)*: The TWA concentration, for a conventional 8-hour workday and a work week as defined in the Labour Code, to which workers may be exposed during their whole working life, without any adverse effects on their health (also when retired) or that of the next generations.
- *Maximum admissible short-term concentration (MAC-STEL)*: The average concentration to which workers may be exposed without any adverse health effects if the exposure does not last longer than 15 minutes, does not occur more than twice during a workday, and occurs at intervals not shorter than 1 hour.
- *Maximum admissible ceiling concentration (MAC-C)*: A concentration that should not be exceeded even for an instant because of the threat to workers' health or life.

The proposed values of MACs, prepared by scientists, are in the form of documentation (criteria document containing information on the identification of harmful agents, their physicochemical properties, level of occupational exposure, and the consequences of human and animal exposure. They also include toxicokinetic and toxicodynamic data, the dependence of the toxic effect on the exposure level, a justification of the proposed value of exposure limit, methods for its determination in the workplace, and recommendations for prevention of exposure. Experts make use of original bibliographic materials available from computer databases, such as Toxline, Medline, and Chemical Abstracts. They also use documentation prepared in the United States, Germany, and Sweden, materials from both the World Health Organisation (WHO) and the International Agency for Research on Cancer (IARC), and some specialised periodicals and unpublished documents. The Interdepartmental Commission for MAC and MAI (maximum admissible intensity) adopted socially acceptable levels of occupational risk for carcinogenic agents in the range of 10^{-5}–10^{-3}. This means that the representatives of social partners and state administrators accept the possibility that the incidence of one cancer per 1000 or one cancer per 100,000 individuals exposed to a carcinogenic substance of a definite concentration may increase proportionally with exposure. The Team of Experts for Chemical Agents scientifically determines the risk characteristics for substances with proven carcinogenic effects and defines the MAC values for different risk levels. A commission then adopts the proposed MAC values for a given level of acceptable risk (Czerczak 2004a).

The proposed values of admissible concentrations together with the relevant documentation are presented at the session of the Interdepartmental Commission and are submitted as a proposal to the Minister of Labour. Once approved, the MAC values for chemical and aerosol agents harmful to the health are published in *Dziennik Ustaw* (DzU; *Polish Journal of Laws*) as a legally binding decree on exposure limits for all sectors of the national economy (Skowroń 2007a).

The MAC values and the intensities for agents harmful to health serve as guidelines for the designers of new and modernised technologies and products, as criteria for the assessment of work conditions, and as the basis for prevention activity in various enterprises.

Documentation of admissible OELs is published in the Interdepartmental Commission's *Podstawy i Metody Oceny Środowiska Pracy* (*Principles and Methods of Assessing the Working Environment*) and indexed in Chemical Abstracts and in the CISDOC database. Knowledge of the data contained in the documentation on the effects of harmful agents on humans is indispensable in establishing appropriate medical prevention efforts and taking corrective actions to improve work conditions (Skowroń 2007a).

Twenty-seven EU member states (the EU-27) legally regulate exposure limits and list admissible values for specific chemical substances. In most countries, the Minister of Labour establishes these values and issues the relevant decrees. The MACs of chemical substances in the workplace air are determined based on a scientific assessment of their toxic effects, but these decisions are sometimes of a political nature. These values are mostly TWA concentrations for 8-hour workday exposure at a level that ensures there are no adverse effects on workers' health. In most EU-27 countries, these values are similar and are linked with other national regulations on occupational safety and health (OSH).

The process of harmonising the exposure limits in the EU countries began with the establishment of IOELVs by the Scientific Committee on Occupational Exposure Limits (SCOEL; Commission Decision 2006). IOELVs are health-based, nonbinding values that are determined using the latest data and available measuring techniques. They determine the threshold exposure levels, below which exposure is not expected to lead to adverse effects. IOELVs are necessary for the employer to determine and assess risk in accordance with Article 4 of directive 98/24/EC (Council

Directive 1998). For every chemical agent that has indicative admissible OELs established at the community level, the member states determine the national OEL, taking into account the community's admissible value (Commission Directive 1991; Commission Directive 2006). They can define its nature in accordance with national law and practice (European Commission, Report EUR 19253 EN 1999).

Initially, when the EU was composed of only 15 countries (the EU-15), small countries such as Cyprus, Malta, Portugal, or even Italy adapted their lists of admissible values to the first list of IOELVs issued by the SCOEL (Commission directive 2000/39/EC). Lithuania also incorporated this list into its national list while it was in a transitional period before joining the EU. In some countries, the number of substances for which exposure limits were to be established was considerably reduced for different reasons, such as a lack of sufficient data, continued discussion of the risk assessment, or low exposure to these substances in the workplace (Netherlands, Germany, and United Kingdom).

Determination of IOELVs for carcinogenic substances by the SCOEL depends on the type and mechanism of their carcinogenic effect, that is, on whether or not the substance produces genotoxic effects. Thus, carcinogenic substances have been divided into the following groups (Bolt and Huici-Montagud 2008):

- *Group A—nonthreshold genotoxic carcinogens*: Risk assessment involves a linear nonthreshold model of extrapolation of test results from animals (high doses) to humans (low doses); for example, 1,3-butadiene, vinyl chloride, and dimethyl sulphate.
- *Group B—genotoxic carcinogens*: The existing data are not sufficient to apply the linear nonthreshold (LNT) model; for example, acrylonitrile, benzene, naphthalene, and wood dusts.
- *Group C—genotoxic carcinogens*: A practical threshold can be set based on existing data; for example, formaldehyde, vinyl acetate, nitrobenzene, pyridine, crystalline silica, and lead.
- *Group D—nongenotoxic and non–DNA-reactive carcinogens*: A threshold can be set based on nonobserved adverse effect level (NOAEL); for example, carbon tetrachloride and chloroform.

SCOEL sets the OELs for compounds in groups C and D (European Commission, Report EUR 19253 EN 1999; Skowroń 2007b).

For some nonthreshold carcinogenic substances for which IOELVs cannot be set, BOELVs have been adopted. These are established based on currently available scientific data, socioeconomic conditions, and the possibility of achieving these values in industry. Contrary to IOELVs, which are implemented into EU law by the Council Directive, BOELVs are introduced by the decision of the European Commission and European Parliament. For substances for which BOELVs are set, member states then establish appropriate national values, which may be at the same level but not above the level established by the EU. Binding values have been set for the following substances: asbestos (actinolite, anthophyllite, chrysotile, grunerite, crocidolite, and tremolite), benzene, hardwood dusts, lead and its inorganic compounds, and vinyl chloride monomer (directives 98/24/EC, 99/38/EC, 2003/18/EC; Table 6.1).

TABLE 6.1
Binding Occupational Exposure Limit Values for Chemical Substances as Established by the European Union

Chemical Name	CAS	TWA (mg/m^3)	TWA (ppm)	TWA (fibres/ml)	Notation[a]	Directive
Asbestos actinolite	77536-66-4	–	–	0.1	–	2003/18/EC
Asbestos anthophyllite	77536-67-5	–	–	0.1	–	2003/18/EC
Asbestos chrysotile	12001-29-5	–	–	0.1	–	2003/18/EC
Asbestos grunerite (amosite)	12172-73-5	–	–	0.1	–	2003/18/EC
Asbestos crocidolite	12001-28-4	–	–	0.1	–	2003/18/EC
Asbestos tremolite	77536-68-6	–	–	0.1	–	2003/18/EC
Benzene	71-43-2	3.25	1	–	Skin	99/38/EC
Hardwood dust	–	5	–	–	–	99/38/EC
Lead and its inorganic compounds	7439-92-1	0.15	–	–	–	98/24/EC
Vinyl chloride monomer	75-01-4	7.77	3	–	–	99/38/EC

Note: CAS = chemical abstract service number; TWA = time-weighted average concentration measured or calculated for an 8-hour workday exposure; mg/m^3 = milligrams per cubic metre of air at 20°C and 101.3 kPa; ppm = parts per million by volume in air (ml/m^3).
[a]A skin notation identifies the possibility of significant uptake through the skin.

The procedure the EU follows to establish IOELVs for chemical substances includes the following:

- Establishment of a list of priority substances for which IOELVs should be set by the Directorate General for Employment, Social Affairs, and Equal Opportunities
- Adoption of the priority list of chemical substances by SCOEL
- Evaluation of recent published scientific data on the toxicology of the chemicals by SCOEL
- Preparation of a draft recommendation summary document (SCOEL/SUM) by SCOEL
- Six months of public consultation through the contact points
- Collation of comments about the document and the proposed IOELVs; amendments to the document
- Publication of the document together with a justification of the values by the commission
- Development of a proposal for a directive based on the SCOEL recommendation by the commission

- Consultation with the Advisory Committee on Safety and Health at Work about the commission's proposal
- Adoption of the directive by the Technical Committee when establishing IOELVs for chemical substances or by the Council and European Parliament for BOELVs (Employment and Social Affairs 2008; Skowroń 2007b)

6.5 ADMISSIBLE CONCENTRATIONS IN THE BIOLOGICAL MEDIUM OR BIOLOGICAL LIMIT VALUES

Biological monitoring measures the amount of a substance or its metabolites contained in a biological medium and assesses its biological effects. Biological limit values (BLVs) are reference values used to assess potential hazards to the workers' health. These values are established based on available scientific data. Concentrations of a substance equivalent to its BLV should not cause adverse health effects in people exposed in the workplace (8 hours per day, 5 days per week). Toxic substances or their metabolites are measured mainly in blood, urine, and exhaled air. Biological monitoring makes it possible to assess exposure to a substance absorbed from the living and work environments through various routes, such as the respiratory tract, the digestive system, or the skin; hence, large individual differences are observed in the exposure indices (Jakubowski 1997).

SCOEL determines BLVs based on the following items: (1) currently available scientific data from studies on humans (epidemiological or volunteer studies); (2) extrapolating the IOELVs established for a substance, taking into account the toxicokinetics of the substance (if the IOELV for a substance is established because of its corrosive effects, the BLV may be based on systemic adverse effects; then these values will not correspond to each other); and (3) the results of the substance's effects (e.g. acetylcholinesterase inhibition in serum); the BLV can be directly derived from suitable studies in humans (European Commission, Report EUR 19253 EN 1999).

In Poland, persons exposed to lead in the work environment must be tested for lead in the blood. The recommendations of the Interdepartmental Commission for MAC and MAI include conditions for taking samples of the biological medium and an interpretation of results for 29 chemical agents (Jakubowski 2004; Augustyńska and Pośniak 2007).

6.6 INFORMATION SOURCES ON HAZARDS RELATED TO THE USE OF CHEMICAL AGENTS IN THE WORKPLACE

6.6.1 Safety Data Sheets for Chemical Substances and Preparations

Chemical substances and preparations are potentially hazardous for human health and life, not only in the work environment, but also in the natural environment. Comprehensive information about the dangerous properties of chemical substances and preparations, the type and dimension of hazard they pose, and the rules to be followed when handling them allows for rational and efficient prevention in the workplace and for the protection of people and the natural environment outside the workplace. Essential information about chemical substances and preparations, intended for the general public, is presented in the form of safety data sheets.

Chemical substances and preparations with explosive, oxidising, flammable, highly toxic, toxic, harmful, corrosive, irritating, sensitising, carcinogenic, and mutagenic properties, as well as those that can impair reproduction and are destructive to the environment, are classified as dangerous.

Guidelines for the compilation of safety data sheets for a substance or preparation are included in Annex II to the regulation (EC) No. 1907/2006 (REACH).

A safety data sheet contains the date of issue and the following items:

- Identification of the substance or preparation and of the company
- Identification of hazards
- Composition or information on ingredients
- First-aid measures
- Firefighting measures
- Accidental release measures
- Handling and storage
- Exposure controls or personal protection
- Physical and chemical properties
- Stability and reactivity
- Toxicological information
- Ecological information
- Disposal considerations
- Transport information
- Regulatory information
- Other information

Toxicological information includes the relevant derived no-effect level (DNEL), at which the agent does not cause adverse effects in different groups of exposed populations (e.g. workers, consumers, and small children, reflecting the likely routes, duration, and effects of exposure) and a predicted no-effect concentration (PNEC), at which the agent does not cause adverse effects in the environment. These are required in safety reports for chemical substances introduced to the market in amounts exceeding 10 tons/year (Gromiec 2008).

The supplier of a substance or a preparation shall also provide the recipient with a safety data sheet compiled in accordance with the following regulations:

- If a substance or preparation meets the criteria for classification as dangerous
- If a substance is persistent, bioaccumulative, and toxic or is very persistent and very bioaccumulative (vPvB) in the environment according to criteria laid down in Annex XIII of the REACH regulation
- If a substance is included in the list established in accordance with Article 59(1) for reasons other than those referred to in point 1, that is, the substance meets the criteria for classification as carcinogenic of category 1 or 2

in accordance with directive 67/548/EEC or the substance meets the criteria for classification as mutagenic of category 1 or 2 in accordance with directive 67/548/EEC.

6.7 EU POLICY REGARDING CHEMICAL AGENTS—REACH REGULATION

6.7.1 REACH Regulation Provisions

The December 2006 REACH regulation no. 1907/2006 of the European Parliament and of the Council of 18 is the governing act of the new European chemicals policy, which ensures a high level of environment and human health protection, especially workers' health, and enhances the competitiveness of the European chemical industry. The regulation set specific duties and obligations for manufacturers, importers and downstream users of substances to be used for their own reference and in preparations and articles. The biggest shortcoming of the previous regulations was an unnatural division between new and existing substances; this meant that 99% of new agents on the market are not tested, and therefore the information on most of them is not sufficient. The REACH system will increase the interest in test results and their use in industry; this should translate to an increase in innovation and thereby in the competitiveness of the European chemical industry against other great world economies (Miranowicz-Dzierżawska 2007). The REACH regulation began in all EU countries on 1 June 2007.

6.7.2 Main Elements of REACH

The REACH system is based on the following four pillars:

1. *Registration*: Each substance manufactured or imported into a community in quantities exceeding 1 ton/year must be registered with the European Chemicals Agency. The level of detail and the date the registration must be prepared depend on the total yearly tonnage of the substance. Substances should be registered over an appropriate period of time as it is shown below, depending on the quantity marketed:
   - Above 1000 tons/year + substances of very high concern (SVHC)—over 3.5 years
   - 100–1000 tons/year—over 6 years
   - 1–100 tons/year—over 11 years
   - Chemical substances can also be preregistered, allowing transition periods for the manufacturers and importers. At preregistration, the registrant must submit the following basic information to the agency:
     - The identity of the substance as specified in Section 2 of Annex VI of the regulation
     - The identity and contact details of the producer or importer who should be contacted
     - The prospective deadline for registration or tonnage band

- These data must be submitted between 1 June 2008 and 1 December 2009. The producer must complete preregistration. Substances exceeding 1 ton/year require submission of a registration dossier. Substances exceeding 10 tons/year require a chemical assessment, and a chemical safety report must be prepared if the substances are classified as vPvB or persistent, bioaccumulative and/or toxic (PBT; Miranowicz-Dzierżawska 2008).

2. *Evaluation*: Both the industry (within registration obligations) and the competent authorities in the member states will evaluate the substance's potential risk. The agency will also evaluate the registration dossier, perform preliminary verification of the technical dossier, and send it to the competent authorities in the relevant country for a detailed evaluation.
3. *Authorisation*: Certain substances will require authorisation from the European Commission. The authorisation procedure covers SVHC because some agents have very serious and often irreversible effects on humans and the environment, regardless of their tonnage. These substances are classified as carcinogenic, mutagenic, or harmful to reproduction or PBT or vPvB.
4. *Restriction*: Some dangerous substances will be subject to restrictions on their manufacture, marketing and use. The list of substances subject to these restrictions is specified in Annex C to the regulation, and will be established and modified as necessary at the request of the European Commission or member states.

The regulation also established a new institution, the European Chemicals Agency, based in Helsinki, which will be responsible for the implementation of the REACH system at the community level.

REACH introduces substantial changes to the system for developing and communicating information about chemical substances. The most important changes are as follows:

- Shifts the obligation to prove that a substance is dangerous from the state authorities to the manufacturer or importer.
- Creates a uniform legal system for placing chemical substances on the market.
- Updates information about chemical substances according to an established schedule.
- Imposes an obligation to evaluate the chemical substance to intermediate users.
- Disseminates information about registered chemical substances.

6.8 TESTING AND MEASURING CHEMICAL SUBSTANCES IN THE WORKPLACE AIR FOR OCCUPATIONAL EXPOSURE ASSESSMENT

Supervision of the hygiene conditions of a work environment and assessment of occupational exposure to chemical substances are the most important elements of workers' health protection. When the concentrations and absorbed doses of harmful airborne substances are known, the health effects of exposure can be predicted and preventive measures applied early to reduce the occupational risk. The principal obligation of an employer should be the systematic identification and assessment of hazards connected with chemicals in the work environment.

6.8.1 Methods for Measuring Chemical Agents in the Workplace Air

The concentration of chemical substances in the workplace air should be measured by accredited laboratories in accordance with regulations on testing and certification. The measurements should use determination methods characterised by selectivity, adequate sensitivity, and precision. This can be methods specified in Polish and international standards or other equivalent methods that are consistent with the general requirements of the European standard PN EN 482 2006.

Measurement procedures for the quantitative determination of chemical substances in workplace air involve taking and analysing air samples. The chemical substances should be quantitatively separated from air, and their concentrations should be measured (ideally at the level 0.1 MAC and at least 0.25 MAC). The specifics of the air samples have a considerable influence on the results of the measurements and the assessment of exposure to harmful chemical substances in the work environment, and therefore should be taken with great precision regarding the volume of air and quantitative identification of the tested substances.

The method of taking the air sample is selected depending on the substance's properties and the analytical techniques used. The person selecting the appropriate methods should know the technological process and physicochemical properties of the substances they are hoping to quantify. The main methods of taking air samples are the isolation method and the integration method.

The *isolation method* is used to take samples of air containing gases and vapours of low-boiling liquids. It involves taking a definite amount of air in a special container, for example, gas pipettes, cylinders, glass or plastic syringes, or even plastic bags. Although this method is simple, its applications are limited due to the small amount of air taken, which is often not sufficient to determine the MAC. Moreover, adsorption of the container walls can cause losses of the substance being measured. This method should not be used for taking air samples that contain reactive or easily adsorbing materials.

In the *integration method*, a definite amount of air flows through sorbents that quantitatively absorb the tested substances. Air samples are enriched by the absorption of substances into solutions or by adsorption of solid sorbents. The kit for taking samples includes the following:

- An absorber, which may be an impinger filled with an absorbing solution, a tube filled with a solid absorber, or a filter placed in a fitting
- An air pump to ensure a constant flow rate
- A rotameter

For liquid sorbents, the determined substances either dissolve in the sorbent or enter into the chemical reaction, resulting in a colour change. The absorption efficiency depends on the sorbent type and absorption conditions such as the velocity and volume of airflow and the column height of the liquid in the impinger temperature. For chemisorption, that is, absorption as a chemical reaction between the substance and the surface liquid sorbent, the airflow velocity should be adapted to the speed of this reaction. Prolonging the time the two phases are in contact and enlarging the contact surface by using washers with sintered glass increases the absorption efficiency.

While using solid sorbers to separate the substances from the air, the substances are adsorbed on the sorbent surface. Different materials with well-developed surfaces may be used as sorbents, for example, activated carbon, silica gel, aluminium oxide, molecular sieves, a mixture of aluminium oxide and magnesium oxide, Florisil, modified diatom soil (e.g. chromosorbs), and plastics (Amberlite XAD resins, tenax, resins, chromosorbs of series 100, and porapak).

Depending on the physicochemical properties of the substance being separated from the tested air, suitable solid sorbents can be used. The sorption ability of the substance depends mainly on its polarity, particle size, and vapour pressure. Polar compounds adsorb more easily on polar sorbents such as activated carbon, and nonpolar compounds on nonpolar sorbents such as silica gel.

Adsorbing tubes are most frequently used for taking air samples. They are usually made of glass and contain the sorbent in two layers (10 and 100 mg) separated by dividers of glass fibre or polyurethane foam and closed with plastic plugs.

Substances adsorbed on solid sorbents in adsorbing tubes are separated quantitatively, using a process reverse to adsorption, desorption. Two types of desorption may be used in the analysis of air: desorption with a suitable solvent (eluent) and thermal desorption into a gaseous phase. However, desorption with a solvent is commonly used in standardised methods to determine harmful substances, using gas chromatography as a measuring technique. The procedure should ensure quantitative desorption with the highest possible efficiency. Usually the efficiency is lower than 100% and therefore it is necessary to determine the desorption coefficient, which should be taken into account when calculating the content of determined substances in the air sample.

Chemisorption is also used to separate and determine the substances on solid sorbents. In these cases, the solid sorbents are covered with special reagents that react with the tested substances during absorption and form special products that are the basis for further determination.

Taking air samples on solid sorbents has many advantages: it allows enrichment and long-term storage of samples, easy transport, and easy service. The integration method to separate air contaminants uses pumps that allow constant airflow through the sorbents. Different types of pumps can be used for this purpose, but membrane pumps powered by a battery or main supply are most frequently used. Portable battery pumps are the most convenient, as they enable airflow with different velocities—from 0.02 to several litres per minute, for at least 12 hours. Nonexplosive battery pumps are also produced for use in explosive conditions.

6.8.2 Strategy of Measurements for Inhalation Exposure Assessment

Proper air sampling using the above-mentioned methods is necessary for occupational exposure assessment of chemical agents, and should be performed in accordance with existing rules that govern the assessment of the compatibility of work conditions with MAC-TWA, MAC-STEL, and MAC-ceiling (MAC-C).

Air samples should be taken in each worker's breathing zone during his or her entire stay at a workstation, using individual dosimetry. Taking air samples with this method requires suitable equipment, either samplers with individual battery pumps or passive dosimeters without pumps. Laboratories without such equipment can use stationary measurements, in which air samples are taken at permanent measurement points, as close to the workstations as possible. Samples must be taken randomly, and the time of each sampling depends on the determination method that is being used.

In occupational exposure assessment, personal samplers are placed in the worker's breathing zone for at least 75% of the work shift. The assessment should cover all workers whose jobs involve contact with harmful substances. To assess the compatibility of work conditions with MAC values, two to five samples should be taken continuously using pumps with a sampler, and one sample should be taken using a passive dosimeter.

At least two 15-minute samples should be taken when the concentration is the highest, to determine the 'short-term concentration' and its compatibility with MAC-STEL values. For stationary measurements, the rules for taking air samples depend on the *timing* of the tasks.

If the chemicals in the work environment cause acute irritation, act quickly, or have a disagreeable smell, and have MAC-C values established, then air samples should be taken as quickly as possible. Continuous monitoring of substance concentrations may be performed at stationary posts localised in the workers' area or by individual analysers placed directly on the work clothing.

Because there are no devices for making continuous measurements, the samples should be taken in a short amount of time, or, the measurements should be taken at regular intervals and in all periods of expected high concentrations. The time span between successive measurements depends upon the homogeneity of the technological process and the changeability of concentrations. The time spans are as follows:

- 30 minutes for nonhomogenous technological processes of high concentration changeability
- 1 hour for technological processes of low concentration changeability

The concentration should not exceed the MAC-C in samples taken as mentioned above or in other samples taken for exposure assessment.

Devices such as portable analysers and tube indicators can be used for this type of measurement, rather than using methods that yield results only after the sample has been analysed.

Gas and absorption chromatography are most frequently used to determine the chemicals in air samples taken at workstations. Visible, infrared, and ultraviolet spectrophotometry and high-performance liquid chromatography are used less often.

6.8.2.1 Assessment of Occupational Inhalation Exposure

The main criteria for assessing inhalation exposure are the results of the measurement of chemical substance concentrations in the workplace air and the relevant

MACs. Exposure is assessed based on these results and also by using calculated exposure indicators, that is, TWA concentration X_g, the upper and lower limit of the confidence interval (UC and LC) for a real-time average or TWA concentration (PN-Z-04008-7 2002; Gromiec 2005; PN-EN 689 2002).

If the TWA concentration or the upper limit of the confidence interval of real-time average does not exceed the MAC-TWA value, then work conditions are considered to be compliant with the requirements. If the MAC-TWA value is in the confidence interval for the geometric mean of the measurement results, then the data are not sufficient to determine if work conditions comply with exposure limits. Additional tests of the work environment should then be conducted, and measurements should be made over 30 days on two randomly chosen work shifts, taking at least five air samples at each shift. If the results of more than half of the samples are over the MAC-TWA, the work conditions are considered harmful. If the results for half of the samples are equal to or below MAC-TWA, the work conditions are considered safe.

If several chemical substances are present simultaneously in the workplace air, then combined exposure should also be assessed by summing up the toxic effects of all chemicals.

Work conditions are considered safe if, for both samples taken intentionally in the periods of highest emission of substances and samples taken randomly, the concentration in any of the 15-minute air samples taken does not exceed the MAC-STEL for the given substance.

Work conditions are considered harmful in the following cases:

- The concentration in any 15-minute sample is higher than the MAC-STEL.
- A concentration equal to the MAC-STEL is in the work environment for more than 15 minutes or occurs more than twice.
- The interval between two 15-minute periods when the concentration is equal to the MAC-STEL is shorter than 1 hour.

Work conditions can be considered safe if the determined concentration does not exceed the MAC-C for the given substance. If a chemical agent harmful to workers' health is present in a workplace, the employer must carry out tests and periodical measurements of this agent within the following time frames:

- At least every 2 years if, in the last test, the concentration of the harmful agent was 0.1–0.5 inclusive of the MAC-TWA set out in the regulations.
- At least once a year if the concentration of the harmful agent is between 0.5 and 1.0 inclusive of the MAC-TWA set out in the regulations.

When a carcinogenic or mutagenic agent occurs in workplace air, the employer must measure this agent in accordance with the regulations:

- In all cases when the conditions of using the agent have changed
- At least once every 3 months if the concentration of carcinogenic or mutagenic agent is between 0.5 and 1.0 inclusive of the MAC-TWA value

- At least once every 6 months if the concentration of carcinogenic or mutagenic agent is between 0.1 and 0.5 inclusive of the MAC-TWA value

If the MACs of an agent harmful to workers' health are exceeded, the employer must determine the causes and introduce the necessary technical, technological, or organisational changes immediately.

Periodical measurements of harmful chemical agents are not obligatory if the results of the last two measurements do not exceed 0.1 MAC-TWA and no changes in the technological process that could affect the concentration of the harmful agent are foreseen. This also applies to measurements of carcinogenic or mutagenic agents.

6.8.3 Strategy for Dermal Exposure Assessment

Dermal exposure occurs when a substance is in contact with the epidermis, can be absorbed dermally, and shows systemic effects and/or causes local effects, that is, effects on the skin's surface. Contact of chemical substances with the skin may lead to irritation, rashes, acne, and even ulceration or subcutaneous haemorrhage, burns and damage of the skin's protective properties due to long-term exposure.

Dermal exposure causes dynamic interaction between environmental contaminants and the skin. Unlike inhalation exposure, dermal exposure is not currently considered in general occupational risk assessment, although the harmful effects and occupational skin diseases it can cause are documented in literature and in the results of epidemiological and statistical studies (Marquart et al. 2006).

In accordance with recommendations of European directives and national regulations, the assessment of exposure to chemical substances in the work environment should take into account many parameters of dermal exposure. These include the quantities of the chemical substances deposited on the skin, the surface area of exposed skin, the duration of the deposition on the skin, and the amount of the substance that is absorbed through the skin under normal occupational exposure conditions.

Methods to measure the concentrations of chemical substances and assess dermal exposure increasingly have been the subject of research. CEN TC 137 has dealt with this problem as well and has started developing standards. *Workplace exposure—Measurement of dermal exposure—Principles and methods* proposes principles and methods for measuring occupational dermal exposure (presented in Table 6.2).

Selecting a method by which to measure chemical substances on the skin or on work or protective clothing depends on physicochemical properties and quantity of the deposited substance. However, basing the method on the determination of mass will fail when the results of the measurement are used to calculate and evaluate the quantity of chemical substance absorbed into the body, mainly due to the lack of limit values for chemical substances deposited on the skin.

Surrogate skin and biomonitoring of the chemical substances or their metabolites in out-breathing air of the worker, in urea, blood, or in other biological materials can

TABLE 6.2
Measurement Methods for the Assessment of Dermal Exposure to Harmful Chemical Agents

Principles of Measurement	Sampling Medium
Interception method: absorbent or relevant dosimeters are attached to an operator's clothing or skin at various locations on the body prior to exposure. Following exposure, the dosimeters are removed and the amount of agent they retain is determined by an appropriate analytical method	Patches of filter paper, α-cellulose paper, cotton, rayon, flannel, filter paper impregnated with lanolin, aluminium foil, polypropylene or polyurethane foam may be used as a sampling medium Gloves, whole working suit
Removal method: contaminants are removed from the skin area or layer with an adhesive tape or by washing and the amount of chemical substance is then determined by an appropriate analytical method.	Strip of adhesive tape Wash water Rinsing water
In-situ method: the quantity of the agent or an added chemical tracer on the clothing or skin is measured to determine dermal exposure to an agent.	Added chemical tracer, such as fluorescent material or visible dye, present directly on the clothing or skin

Source: CEN/TS 15279. 2006. Workplace exposure. Measurement of dermal exposure: Principles and methods.

be used for dermal exposure assessment. Forecasting using statistic and deterministic methods may also be useful in the estimation of chemicals deposited on the skin (Warren et al. 2006).

6.9 ESTIMATION AND CONTROL OF OCCUPATIONAL RISK CAUSED BY EXPOSURE TO HARMFUL CHEMICAL AGENTS IN THE WORK ENVIRONMENT

Occupational risk assessment is the process of identifying agents that pose a hazard to the workers' health and examining the work conditions in which exposure to these agents occurs. It is a continuous process, and its main objectives are to acquire detailed knowledge about the harmful chemical agents that can cause damage and to encourage the employer to undertake risk control measures.

Chemical substance risk assessment is a multistage process. Each stage should be conducted very carefully and responsibly. If possible, risk assessment should be based on measurements of chemical concentrations in the workplace.

In accordance with the provisions of European directives and national regulations on workers' safety and health protection against chemical agent–related risks, employers must identify any chemical agents present in the work environment that pose a hazard to the workers and assess the occupational risk caused by these agents. If the employer starts a new activity involving the use of such agents, they must establish whether an agent that may pose a hazard to the workers' health will be present in the work environment, assess the occupational risk, and, if necessary, undertake preventive measures before the works are started.

When determining the admissibility of risk arising from chemical agents in the workplace, the employers should, in accordance with directive 98/24/EC, consider the following data and circumstances:

- The dangerous properties of the chemical agent
- The information available on the effects of the chemical agent on human health and the environment as well recommendations for its safe use, and above all, the information included in the safety data sheets
- The routes by which the agent enters the body during occupational exposure (inhalation, skin contact, or dermal ingestion)
- MACs in the work environment and BLVs, if established
- The efficiency of personal and collective protective equipment and other preventive measures
- The opinions of industry physicians and the outcomes of medical examinations
- The conditions of work with chemical agents

Combined risk should be assessed in workers exposed to several chemical agents. Occupational risk assessment should also cover periods when the risk increases, for example, during the repair and service of machines and when equipment used at work stands is being assessed. The risk should be reassessed if any modifications have been introduced in the composition of agents or in the course of technological processes or if progress in medical science regarding the health effects of those agents has been made.

Risks connected with chemical agents result from both direct contact with the agent and as an effect of energy formed due to chemical transformations that accompany a fire or explosion. Because of these two mechanisms, employers should use methods that allow assessment of the occupational risk connected with chemical agents and of the likelihood of an accident resulting from their presence.

6.9.1 Principles of Quantitative Risk Assessment of Inhalation Exposure

The Polish standard PN-N-18002 (2000) sets out general principles of risk assessment, including the risk connected with exposure to chemical agents. The standard recommends that, where possible, the occupational risk should be estimated based on quantities that characterise the risk, that is, for chemical substances, MACs (MAC-TWA, MAC-STEL, MAC-C) and the results of measurements of chemical substance concentrations in the work environment.

There is little probability of adverse health effects resulting from workers' exposure to these substances when exposure indicators are below or equal to exposure limits. Ceiling concentrations, when exceeded, are most dangerous because of the threat to the workers' health or life. These substances may cause acute mucosal inflammation of the eyes and upper respiratory tract even during short-term exposure; therefore, they cannot exceed MAC-C values.

There are three levels of occupational risk from exposure to harmful chemical substances; national legislation has established exposure limits (Table 6.3; PN-N-18002 2000; Pośniak and Skowroń 2007).

The principles adopted for occupational risk assessment do not apply to carcinogenic or mutagenic substances (directive 2004/37/EC), from which there is a risk of cancer at the 10^{-5} or 10^{-3} probability when their concentration is at the MAC-TWA level. When these substances are present in the work environment, the health risk is high for all workers if exposure indicators are equal to or higher than 0.1 MAC-TWA. When concentrations in the air are lower than 0.1 MAC-TWA, the risk is medium.

Because there are special regulations for young persons and pregnant women, risk assessment for these groups sometimes diverges from the adopted principles. Jobs in which these groups are exposed to chemical substances mentioned in the lists are prohibited for these groups. If these groups are exposed to chemicals such as carcinogenic agents, pesticides, or substances harmful to reproduction, the occupational risk is considered high regardless of the concentrations in the workplace air. Exposure of pregnant and breast-feeding women to organic solvents at concentrations exceeding one-third of their MACs is also considered high risk.

Risk assessment based on the principles presented in Table 6.3 can only be made for about 1000 harmful chemical agents for which the OELs are determined by the national regulations. This number is very small, as there are about 30,000 chemical agents commonly used in the EU. The employer should therefore establish his or her own criteria of occupational risk admissibility for nonthreshold chemical agents, taking into account the opinions of OSH experts, his or her own experience, and the workers' opinions and experiences. This task is very complicated and difficult to accomplish.

TABLE 6.3
Quantitative Risk Assessment

Risk Level	Risk-Assessment Criteria
Low risk (L)	Exposure indicators for the assessment of work conditions are compliant with maximum admissible concentrations–time-weighted average (MAC-TWA) and, additionally, the MAC–short-term exposure limit (MAC-STEL) or ceiling concentration (MAC-C) does not exceed 0.5 of this value
Medium risk (M)	Exposure indicators are equal to or higher than 0.5 MAC-TWA, MAC-STEL, or MAC-C, but do not exceed these values
High risk (H)	Exposure indicators are higher than MAC-TWA, MAC-STEL, or MAC-C, but do not exceed these values

6.9.2 Principles of Qualitative Risk Assessment—Inhalation Exposure

Neither the EU legislation nor the national legislation specifies detailed principles for the assessment of risk connected with chemical agents in the work environment. There are many methods of risk assessment available to obtain reliable results. One is proposed in the "Practical guidelines of a nonbinding nature on the protection of the health and safety of workers from the risks related to chemical agents at work", developed by the Advisory Committee on Safety and Health at Work of the European Commission DG Employment and Social Affairs (2005). There are two independent methods for the assessment of risk connected with the following:

- Exposure (through inhalation or the skin) to chemical agents
- Possibility of an accident resulting from the presence of chemical agents

Three variables are taken into account to assess the occupational chemical risk connected with inhalation exposure:

- The intrinsic hazard of the substance—the risk category is assessed according to the R phrases that describe the risk
- The tendency of the substance to pass into the environment
- The quantity of the substance used in the assessed operation

The main hazard posed by a substance is a very important characteristic in occupational risk assessment. Based on the phrases that describe the type of risk (R phrases), which are placed on the label or set out in the data sheet of the product, chemical agents are classified into five risk categories: A, B, C, D, and E. Table 6.4 gives

TABLE 6.4
Main Risk Categories of Chemical Substances

	Risk Category E
Number of R Phrase	**Name of R Phrase**
R42	May cause sensitisation by inhalation
R42/43	May cause sensitisation by both inhalation and skin contact
R45	May cause cancer
R46	May cause heritable genetic damage
R49	May cause cancer by inhalation
Mutagenic category 3, R68	Possible risk of irreversible health effects

Source: Directorate General for Employment, Social Affairs and Equal Opportunities. 2005. "Practical guidelines of nonbinding nature on protection of the health and safety of workers from hazards related to chemical agents at work". Luxembourg: Advisory Committee on Safety, Hygiene and Health Protection at Work of European Commission Employment and Social Affairs.

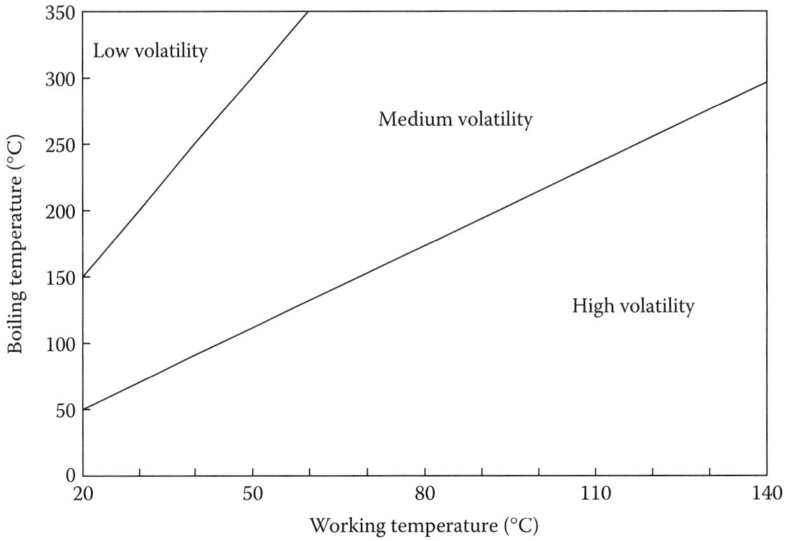

FIGURE 6.4 Classification of liquids with reference to their volatilities. (Reprinted from Directorate General for Employment, Social Affairs, and Equal Opportunities. 2005. "Practical guidelines of nonbinding nature on protection of the health and safety of workers from hazards related to chemical agents at work". Luxembourg: Advisory Committee on Safety, Hygiene, and Health Protection at Work of European Commission Employment and Social Affairs.)

classification criteria for the E category, based on the risks resulting from the toxic properties of chemical agents.

The tendency of the substance to pass into the environment should also be considered when assessing occupational risk due to chemical exposure. Chemical agents can be classified into three categories, based on their volatility and working temperature (Figure 6.4), which determine the capacity of liquids to evaporate and solids to form dusts. The agents can have a high, medium, or low tendency to pass into the environment.

The third variable in risk assessment is the quantity of the substance. Depending on the mass or volume of chemical agents used, the quantity is classified as small (grams or millilitres), medium (kilograms or litres), or large (ton or cubic metres).

A foreseeable level of occupational risk is determined based on these variables and using the specific principles given in Table 6.5 as an example for one risk category. Four risk levels are considered, and each of them is linked to an appropriate prevention strategy.

Occupational risk assessment as it is connected to exposure to chemical agents is relatively simple, while allowing for all factors that can influence the volume of harmful effects to the workers' health. Risk assessment at all workstations or workplaces where harmful chemical substances and preparations are used or manufactured is also possible due to these assessment practices.

TABLE 6.5
Determination of Risk Levels

Hazard Level C

Quantity Used	Volatility or Dust Generation			
	Low Level of Volatility or Dust Generation	Medium Volatility	Medium Level of Dust Generation	High Level of Volatility or Dust Generation
Small	1	2	1	2
Medium	2	3	3	3
Large	2	4	4	4

Source: Directorate General for Employment, Social Affairs, and Equal Opportunities. 2005. "Practical guidelines of nonbinding nature on protection of the health and safety of workers from hazards related to chemical agents at work". Luxembourg: Advisory Committee on Safety, Hygiene, and Health Protection at Work of European Commission Employment and Social Affairs.

6.9.3 Preventive Measures

When the identified risk is not acceptable, appropriate actions should immediately be taken to lower the concentrations of harmful chemical substances in workplace air to safe levels. The technical measures most often used to reduce the risk include the installation of new ventilation systems or the modernisation of existing ventilation systems, hermetisation, automation, and robotisation of technological processes, isolation of workstations, and proper storage of chemicals. The appropriate technical solution depends on the process type and available financial means.

Until the chemical risk is reduced by the use of appropriate technical and organisational measures, workers should use properly selected personal protective equipment.

It is very difficult to achieve the lowest possible concentrations of harmful substances in workplace air, and it is often possible only after a thorough technical modernisation. Therefore, there are no universal solutions in this area, only general actions that should be taken to eliminate of the source of hazard either indirectly or directly can be indicated—the workers must be isolated from the hazards.

The most effective way to eliminate hazards is to discontinue using the toxic substance or substituting it with a less toxic one. Chemicals should be selected at the design phase of the technological process, after a careful analysis of the toxic and physicochemical properties of raw materials or intermediate products. Such an analysis may often lead to the use of substances that do not pose a hazard to the workers' health while maintaining the parameters necessary for the technological process in question.

Isolating workstations limits the emission of harmful chemical substances. Workstations where the worker is exposed to harmful substances should be placed in separate, specially protected rooms. Screening the sources that generate toxic substances is a variant of isolation.

In industry, the hermetisation of processes is important for technical prevention and is used to limit exposure to chemical substances. It protects the equipment in which processes connected with the generation or uses of these harmful agents occur so that the passing of these agents into the workplace air is impossible. Hermetisation should cover production units in which the reactions take place and the transport of chemicals. For liquid substances, a pipeline transport with an automatic hermetised feeder can be used.

Automatisation and robotisation of technological processes are optimal technical solutions that completely eliminate workers' exposure to chemical substances. Limiting the worker's role to general supervision over control equipment protects him or her against the harmful effects of emitted substances.

Introducing automated production processes such as industrial robots also contributes to increase in effective use of production resources and an improvement in product quality. Industrial robots can be used to perform certain operations instead of humans, for example if the harmful effects of chemical agents on the worker cannot be limited or when the physical load of the worker is too high when performing these operations manually. Robotisation of processes is the most promising technical solution to protect against chemical hazards; therefore, industrial robots should be popularised and workers should be trained to use them.

Commonly used collective protective equipment includes general and local mechanical ventilation installations and devices equipped with suitable systems for cleaning air from vapours and gases (sorbents) and from solid and liquid particles (air filters). General principles concerning room ventilation are set out in relevant regulations on OSH requirements.

The aims of mechanical ventilation, which involves continuous or periodical indoor air exchange, are as follows:

- Improving the condition and composition of workplace air in accordance with hygienic requirements for human health protection and technological requirements necessary to obtain products with specific properties.
- Controlling indoor air environment parameters such as concentrations of chemical contaminants and dusts, temperature, humidity, and speed and direction of airflow.

Indicators of air exchange, which are necessary to obtain the correct parameters of indoor air, are closely related to the intended use of the room. The parameters should be based on requirements determined by the type of the and the technological process binding laws and recommendations set out in relevant standards.

If the emission source cannot be enclosed or encased completely, partial casings or local ventilation installations equipped with suction nozzles and hoods (stationary or mobile) can be used. These installations are connected to the central air purification installation or individual filtration and ventilation devices. Local ventilation prevents the propagation of chemical contaminants and dusts in the entire room and should be supported by general ventilation systems.

If the use of collective protective equipment does not ensure the required indoor air purity, the worker should be provided with personal protective equipment selected according to the type of chemical substances and preparations present in the workplace.

Respiratory protective equipment is most often used because the majority of chemical agents are absorbed into the body through the respiratory tract. The selection of suitable respiratory protective equipment should follow a thorough identification and assessment of occupational risk. The information obtained from this assessment will determine the relation between the concentration of chemicals in the work environment and the MACs, which is the basis for selecting the type and protective class of equipment. There are two main groups of respiratory tract hazards (Majchrzycka 2002):

1. *Contaminated air*: An occurrence of harmful substances in the form of aerosols, gases or vapours
2. *Oxygen deficiency*: Oxygen concentration below 17% in the air

Respiratory protective equipment is therefore divided into two main groups:

1. Filtering devices
2. Breathing apparatus

In filtering equipment, inhaled air flows through the filtering element, which removes the contaminants. Filters or gas filters are used depending on the forms of the contaminant—gas, vapour, or aerosol. Each type of filtering device can appear in either of the following versions: with no assisted airflow, as a powered filtering device, or as a power-assisted filtering device. Indications concerning selection of respiratory protective equipment in chemical substances hazard are also given in the producer's instructions. The following items should be used for short-term exposure and when the exposure limits are only slightly exceeded:

- Filtering half masks for aerosol contaminants or pollutants
- Half masks with combined filters for aerosol and vapour contaminants or pollutants as well as for gases of chemical substances contaminants or pollutants

The following items should be used for contact with carcinogenic substances or when the MAC value is considerably exceeded and the harmful substance is irritant or sensitising:

- A filter, gas filter, or combined filter with a half mask
- A filter, gas filter, or combined filter with a full mask

The following items should be used in emergency situations, when long-term work is necessary, or when exposure limits are considerably exceeded:

- Powered or power-assisted combined gas filters incorporating relevant face pieces, such as helmet, hood, or full-face mask

- Breathing apparatus as compressed air line open- or closed-circuit breathing apparatus with full-face mask

Apart from properly selected equipment type, the protective class of the substance should also be selected properly, taking into account the following:

- Type, properties, and state of the substance
- Concentration and intensity of effects on human skin
- Exposure time and ambient conditions
- Type of job, which determines the clothing construction and material type (e.g., coated materials of different thicknesses made from chemically resistant polymers)

The most frequently used clothing are light suits, garments, aprons, and clothing made from textiles, coated or impregnated nonwovens, or foils.

At a relatively low chemical intensity, the following types of clothing protect against liquid chemicals:

- Clothing protecting against sprayed liquid—type 4 (PN-EN 465 2003)
- Clothing protecting against liquid splashes—type 6 (PN-EN 13034 2005)

Handwear is another group of equipment protecting against chemicals. Workers in contact with chemical substances should use tight gloves made from natural rubber, synthetic rubber (polychloroprene, butylene, polyacrylonitrile), or plastics (PCV, PVA, viton). Polymer gloves with textile support have better utility properties and greater mechanical resistance.

The type and concentration of the substance, duration of contact, type of exposure (e.g., splashing with drops of the substance or possibly of pouring onto the hands, contact with one or several chemical substances or mixtures), type of work, and related risks of mechanical damages to the glove material must be taken into account when selecting gloves to protect against chemical agents.

REFERENCES

Aleksandrowicz, R. 1994. *Small Anatomic Atlas*. Warsaw: PZWL.

Augustyńska, D., and M. Pośniak, eds. 2007. *Harmful Agents in the Working Environment—Admissible Values*. 3rd ed. Warsaw: CIOP-PIB.

Bolt, H. M., and A. Huici-Montagud. 2008. Strategy of the Scientific Committee on Occupational Exposure Limits (SCOEL) in the derivation of occupational exposure limits for carcinogens and mutagens. *Arch Toxicol* 82:61–64.

CEN/TS 15279. 2006. Workplace exposure. Measurement of dermal exposure: Principles and methods.

Commission Decision 2006/573/EC of 18 August 2006 appointing members of the Scientific Committee for Occupational Exposure Limits to Chemical Agents for a new term of office. OJ L 228, 22.08.2006, 22.

Commission Directive 2000/39/EC of 8 June 2000 establishing a first list of indicative occupational exposure limit values in implementation of Council Directive 98/24/EC on the protection of the health and safety of workers from the risks related to chemical agents at work. OJ L 142, 16.6.2000, 432.

Commission Directive 2006/15/EC of 7 February 2006 establishing a second list of indicative occupational exposure limit values in implementation of Council Directive 98/24/EC and amending Directives 91/322/EEC and 2000/39/EC. OJ L 38, 9.2.2006, 36.

Commission Directive of 29 May 1991 on establishing indicative limit values by implementing Council Directive 80/1107/EEC on the protection of workers from the risks related to exposure to chemical, physical and biological agents at work (91/322/EEC). OJ L 177, 5.7.1991, 22.

Council Directive 1999/38/EC of 29 April 1999 amending for the second time Directive 90/394/EEC on the protection of workers from the risks related to exposure to carcinogens at work and extending it to mutagens. OJl L 138, 01.06.1999, 0066–0069.

Council Directive 98/24/EC of 7 April 1998 on the protection of the health and safety of workers from the risks related to chemical agents at work (fourteenth individual Directive within the meaning of Article 16(1) of Directive 89/391/EEC). OJ L 131, 5.5.1998, 11–23.

Council Directive of 12 June 1989 on the introduction of measures to encourage improvements in the safety and health of workers at work (89/391/EEC). OJ L 183, 29.6.1989, 1–8.

Czerczak, S. 2004a. The principles of establishing MAC values of harmful chemical compounds in the working environment. *Principles and Methods of Assessing the Working Environment* 4(42): 5–18.

Czerczak, S. 2004b. Classifications of chemical carcinogenic factors—A review. *Occupational Safety* 1:9–14.

Czerczak, S., and M. Kupczewska. 2002. Assignment of skin notation for maximum allowable concentration (MAC) list in Poland. *Appl Occup Environ Med* 55:795–804.

Directive 2003/18/EC of the European Parliament and of the Council of 27 March 2003 amending Council Directive 83/477/EEC on the protection of workers from the risks related to exposure to asbestos at work. OJ L 97, 15.4.2003, 48.

Directive 2004/37/EC of the European Parliament and of the Council of 29 April 2004 on the protection of workers from the risks related to exposure to carcinogens or mutagens at work (Sixth individual Directive within the meaning of Article 16(1) of Council Directive 89/391/EEC. OJ L 158, 30.4.2004, 50–76.

Directorate General for Employment, Social Affairs, and Equal Opportunities. 2005. "Practical Guidelines of Nonbinding Nature on Protection of the Health and Safety of Workers from Hazards Related to Chemical Agents at Work". Luxembourg: Advisory Committee on Safety, Hygiene, and Health Protection at Work of European Commission Employment and Social Affairs.

Employment and Social Affairs. http://ec.europa.eu/employment_social/health_safety/occupational_en.htm (accessed October 13, 2008).

European Agency for Safety and Health at Work. 2006. TC WE 2006. Task TC WE 3.3: To prepare a policy overview of occupational exposure limits for publication on the agency's dangerous substances web area. Unpublished work.

European Commission. 1999. Methodology for the derivation of occupational exposure limits: Key documentation. Scientific Committee Group on Occupational Exposure. Report EUR 19253 EN.

Eurostat. 2004. *European Business—Facts and Figures 2004*. Luxembourg: European Commission.

Górski, P. 1997. *Recommendation for Occupational Asthma Diagnosis and Prevention*. Lodz: IMP.

Gromiec, J. 2005. *Guidelines for the Measurement and Assessment of Environmental Agents Harmful to Health*. Warsaw: CIOP-PIB.

Gromiec, J. 2008. Problems concerning the integration of 'derived-no-effect-levels' (DNELS) into occupational safety and health regulations. *Occupational Medicine* 59(1):65–73.

Jakubowski, M. 2004. Biological monitoring of occupational chemical exposure. *Occupational Medicine* 55(1):13–18.

Jakubowski, M., ed. 1997. *Biological Monitoring of Exposure to Chemical Agents in the Working Environment*. Lodz: IMP.

Koradecka, D., ed. 1999. *Occupational Safety and Ergonomics*. Vols. 1 and 2. Warsaw: CIOP.

Majchrzycka, K. 2002. Respiratory protective devices. *Occupational Safety* 3:8–9.

Marquart, H., et al. 2006. Default values for assessment of potential dermal exposure of the hands to industrial chemicals in the scope of regulatory risk assessments. *Ann Hyg* 50:469–489.

Miranowicz-Dzierżawska, K. 2007. REACH—A new EU regulation that increases chemical safety. *Principles and Methods of Assessing the Working Environment* 3(53):5–17.

Miranowicz-Dzierżawska, K. 2008. REACH regulation—Preparing a chemical safety report. *Occupational Safety* 9(444):10–55.

Musu, T. 2007. Evolution of the European Chemical Industry. http://www.eesc.europa.eu/sections/ccmi/docs/documents/03_05_2007_Dresden/MUSU.pdf.

Pabst, R., and R. Putz. 2007. *Human Anatomic Atlas: Trunk, Internal Organs, Legs*. 3rd ed. Wroclaw: Urban & Partner.

Parent-Thirion, A., E. F. Macias, and G. Vermeylen. 2007. *Fourth European Working Conditions Survey*. Dublin: European Foundation for the Improvement of Living and Working Conditions.

Paustenbach, D. J. 2000. The history and biological basis of occupational exposure limits for chemical agents. In *Patty's Industrial Hygiene*, 5th ed., ed. R. L. Harris, 1903–2000. New York: John Wiley & Sons.

PN-EN 13034. 2005. (U) Protective clothing against liquid chemicals—Performance requirements for chemical protective clothing offering limited protective performance against liquid chemicals (Type 6).

PN-EN 465. 2003. Protective clothing. Protection against liquid chemicals. Performance requirements for chemical protective clothing with spray-tight connections between different parts of the clothing (type 4 equipment).

PN-EN 482. 2006. Workplace atmospheres. General requirements for the performance of procedures for the measurement of chemical agents.

PN-EN 689. 2002. Workplace atmospheres. Guidance for the assessment of exposure by inhalation of chemical agents and measurement strategy.

PN-N-18002. 2000. Occupational safety and health management systems. General guidelines to evaluating occupational risk.

PN-Z-04008-7. 2002. Air purity protection. Sampling methods. Principles of air sampling in workplace and interpretation of results.

Pośniak, M., and J. Skowroń. 2007. Harmful chemical substances. In *Occupational Risk: Methodology of Assessment*. 3rd ed. ed. W. M. Zawieska, 37–91. Warsaw: CIOP-PIB.

Regulation (EC) No 1907/2006 of the European Parliament and of the Council of 18 December 2006 concerning the Registration, Evaluation, Authorisation and Restriction of Chemicals (REACH), establishing a European Chemicals Agency, amending Directive 1999/45/EC and repealing Council Regulation (EEC) No 793/93 and Commission Regulation (EC) No 1488/94 as well as Council Directive 76/769/EEC and Commission Directives 91/155/EEC, 93/67/EEC, 93/105/EC and 2000/21/EC. OJ L 396, 30.12.2006, 1.

Seńczuk, W. 2007. *Toxicology*. Warsaw: PZWL.

Skowroń, J. 2007a. The activity of the Interdepartmental Commission for Maximum Admissible Concentrations and Intensities for Agents Harmful to Health in the Working Environment in 2006. *Principles and Methods of Assessing the Working Environment* 1(51):155–62.

Skowroń, J. 2007b. Carcinogenic and mutagenic agents in Polish and UE legal regulations. *Principles and Methods of Assessing the Working Environment* 4(54):5–43.

Skowroń, J., et al. 2000. Combined effects of some toxic inorganic solvents on the organism. *Occupational Safety* 10:1–5.

Szadkowska-Stańczyk, I., and N. Szeszenia-Dąbrowska. 2001. Lung cancer due to occupational exposures: A review of epidemiological evidences. *Occupational Medicine* 51(1):27–34.

Takala, J., M. Urrutia, P. Hämäläinen, and K. L. Saarela. 2009. The global and European work environment: Numbers, trends, and strategies. *SJWEH Suppl* (7):15–23.

Warren, N. D., et al. 2006. Task-based dermal exposure models for regulatory risk assessment. *Ann Hyg* 50:491–503.

7 Dusts

Elżbieta Jankowska

CONTENTS

7.1 INTRODUCTION

Dusts are among the most significant harmful agents present in the work environment. Human exposure to dusts can cause mechanical injury of the mucosa or skin, allergic reactions and diseases, pneumoconiosis, and cancer.

Production plants and service outlets emit the greatest amount of harmful dusts to the work environment; their concentrations and harmful effects on human health depend on the work processes. Although dust concentrations in office space air are much lower than in industrial facilities, concentrations and particle size distributions are still significant in the context of 'sick building syndrome' (SBS; Jankowska et al. 2004a; Jankowska et al. 2004b).

The most recent research focuses on the harmful effects of dusts containing very small particles, that is, particles with nanometric dimensions. These particles are emitted in processes that have been used for a long time, such as welding, soldering, burning of diesel motor oils, and smoking. Today, emphasis is put mainly on the harmful effects of newly generated particles with dimensions below 100 nm that are emitted during their production or application in nanotechnological processes. These particles may induce an inflammatory reaction in the alveoli and lead to blood clotting. Nanoparticles enter the human body mainly through the respiratory system, but they can also be found in the liver (Kreyling et al. 2002; Oberdörster et al. 2002).

Reducing the risk of occupational diseases resulting from dust exposure is one of the most important ways to ensure occupational safety and health. Employees' protection against the harmful effects of dusts in the work environment requires the following:

- Determination of the type, concentration, and other basic parameters of dusts
- Assessment of employees' exposure to harmful dusts
- Assessment of the occupational risks of the employees exposed to harmful dusts
- Application of collective protection against dusts, eliminating or limiting the dust exposure of employees, and, if this is not possible, the application of appropriate personal protection equipment

7.2 DEFINITIONS

Particle: A small, separate part of a solid or liquid substance (PN-ISO 4225 1999).

Nanoparticle: Nano-object with all three external dimensions in the nanoscale (ISO/TS 27687 2008).

Particle aerodynamic diameter: The diameter of a sphere of a density 1 g/cm^3 with the same terminal velocity due to gravitational force in calm air as a particle under the prevailing conditions of temperature, pressure, and relative humidity (ISO 7708 1995; EN 481 1993).

Dust: A general term that refers to solid particles of varying size and origin, which remain suspended in a gas for a period of time (PN-ISO 4225 1999).

Dust emission: Liberation of dust into the atmosphere. The point or area from which such liberation takes place is known as the 'emission source' (PN-ISO 4225 1999).

Dust concentration: The mass or number of particles of a solid substance in a unit volume of gas, grams per cubic metre, or the number of particles per cubic metre.

Breathing zone: The space from which humans directly inhale air. This is a radius of about 3 dm (decimetres) around the face; its centre lies in the middle of the line joining both the ears. The base of the hemisphere is the plane running through this line and the top of the head and the larynx (PN-ISO 4225 1999).

Exposure index: A number describing employees' exposure to a harmful substance, calculated based on the substance's concentration in the air and compared to the appropriate admissible value (PN-ISO 4225/Ak 1999).

Maximum admissible concentrations (MAC-TWA): The time-weighted average concentration for a conventional 8-hour workday and a work week, as defined in the Labour Code, to which workers may be exposed during their whole working life without any adverse effects on their health (even when retired) or on the next generations (Regulation of the Minister of Labour and Social Policy of 29 November 2002).

Filter: A device on which particles are deposited in a porous medium.

Collective protection against dust: The simultaneous protection of a group of people against harmful dusts at work, applied as technical solutions in workplaces, machines, and devices (Regulation of the Minister of Labour and Social Policy of 26 September 1997).

Equipment for personal protection against dust: Equipment used by the employee for protection against harmful dusts in the work environment, including all accessories and additional elements designated for this purpose (Regulation of the Minister of Labour and Social Policy of 26 September 1997).

General mechanical ventilation: Ventilation of the entire room or a complex of rooms.

Local mechanical ventilation: Ventilation of a specific space in a room, a workstation, or a production device.

Filtration and ventilation device: The device used to remove pollutants from the air at a workstation.

7.3 DUSTS IN THE WORK ENVIRONMENT

Dusts emitted at workstations can have various properties depending upon the emission source. The degree of human exposure and a selection of collective and personal protection equipment can be estimated based on the following parameters:

- Dust concentration (concentration of various size fractions of dust particles)
- Particle size (coarse-, medium-, and fine-dispersed dusts)
- Particle shape (fibrous and nonfibrous dusts)
- Chemical composition and crystal structure (e.g. crystalline silica or amorphous silica)
- Explosive properties of the dust (e.g. wood dusts or flour dusts)

Dust emission sources in workplaces include the following:

- Technological processes (in typical production spaces)
- Personnel handling machines and equipment (e.g. in 'clean rooms')
- Air supplied from the outside by forced air ventilation unit or by infiltration (in production spaces, office spaces, or in 'clean rooms')
- Nanotechnological processes (in rooms in which nanoparticles are produced or used for production)

7.3.1 Production Spaces

During the production process, emissions usually occur from point sources. The size distributions of the emitted particles and the concentration of harmful dusts in the workstations depend upon the type of technological process (Szymczykiewicz 1973).

The particles generated in production processes like the grinding of brittle materials (milling, cutting, or crushing of solids) are dispersed and suspended in the air and usually form a polydispersive aerosol (containing particles of various sizes). Particles originating from the solidification of vapours of metals or other compounds, which assume the form of solids at room temperature, constitute the dispersed phase of regular particles and form a monodispersive aerosol (containing particles of similar sizes).

During the production process, particles of various sizes usually are present in the air (polydispersive aerosols). The particles' suspension period in the air (suspended particles) and their ability to become suspended near their settlement location (settled dust) depends upon factors such as the size, shape, and specific gravity of the particles and the velocity, humidity, and temperature of the air.

The most significant dust-generating technological processes, such as milling, crushing, sieving, transportation, mixing of loose materials, sharpening, cutting, polishing, and smoothing, emit substantial quantities of dust particles that may be inhaled by employees and may cause occupational diseases. Figures 7.1 and 7.2 present the mass and number concentration of some dust particles emitted at various workstations in the furniture industry and at a construction site, respectively.

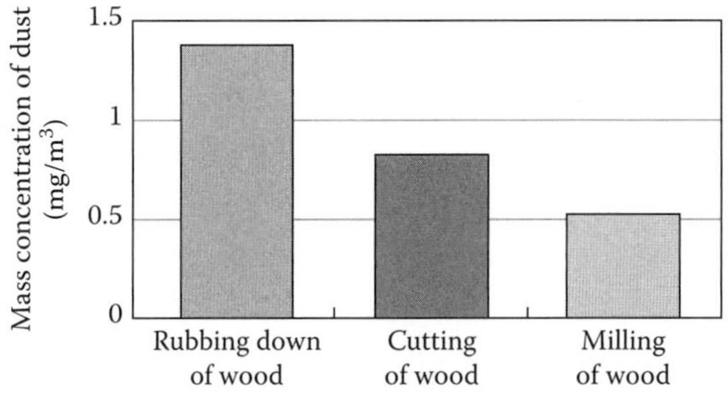

FIGURE 7.1 Total dust concentration at workplaces in the furniture industry; measurements conducted using the filtration-gravimetric method.

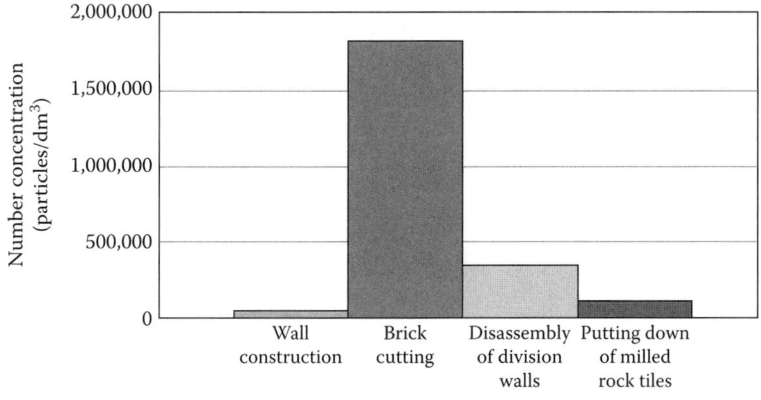

FIGURE 7.2 Number concentration of dust particles within the range of 0.3–3 μm at construction site workstations; measurements conducted using the particle counting method.

Local ventilation systems are usually used to capture dust from point source emissions (e.g. production spaces and some service outlets), along with auxiliary general mechanical ventilation systems, in work spaces in which point sources of dust are present. However, some dust emitted from point sources, particularly fine and very fine particles, disperse throughout the room, which may lead to dust exposure (usually at concentrations higher than that in the outdoor air) among not only the employees occupying the dust-emitting workstations, but also those who are at other workstations or pass through the room. Attention should be given to rooms in which point-emission dust sources are located and local ventilation systems are absent, because they generate significant concentrations of particles in the room as both as suspended and settled dust, depending upon the production process.

7.3.2 Office Spaces

The dust concentration in office space air is lower than that in typical industrial production spaces and usually lower than the concentration in the outdoor air (Hämeri et al. 2003; Jankowska et al. 2004b; Jankowska and Jankowski 2005; Figure 7.3). The concentration and size distribution of particles entering the room are determined by the room ventilation method (natural or mechanical). The quality of the air supplied to the rooms is controlled by using appropriate air filters, cleaning the ventilation systems, and so on.

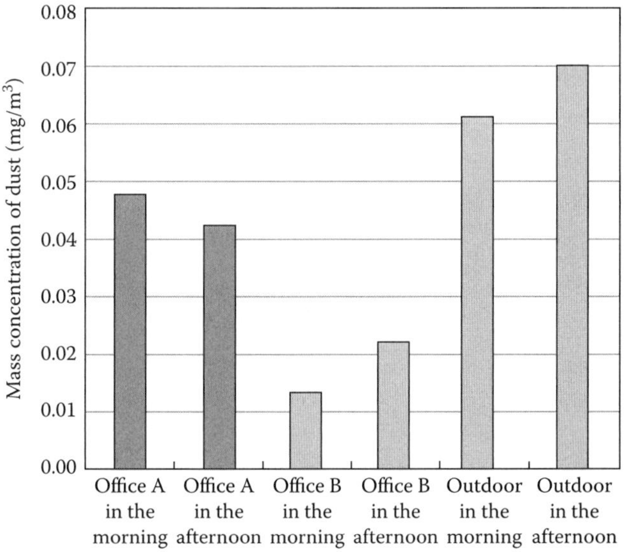

FIGURE 7.3 Mass concentration of dust in the air outside the building and in office spaces A and B; measurements conducted using the photometric method. (Adapted with permission from Jankowska, E., et al. 2004b. *An Ordered Targeted Project.* [Unpublished work.])

The presence of a large number of people in a room (Jankowska 2006; Figure 7.4), the finishing material and equipment room, the type of work performed (e.g. administrative work), and smoking (Jankowska et al. 2007; Figure 7.5) may also be internal emission sources, mainly of fine and very fine particles. Although the particle concentrations in office spaces are much lower than those in industrial production spaces, the influence of particles from external and internal emission sources and the

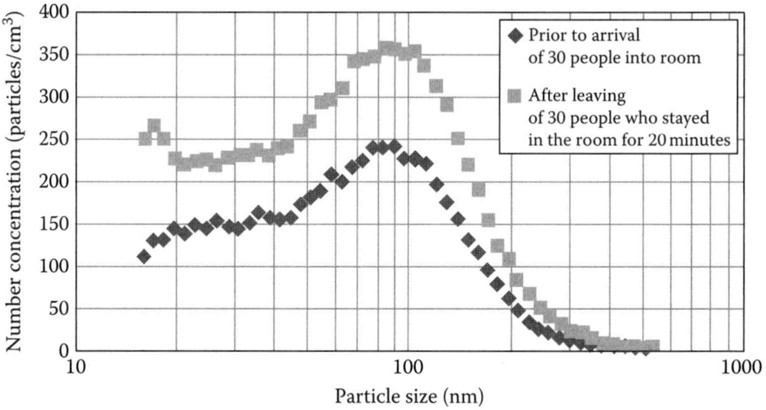

FIGURE 7.4 Influence of people upon the number concentrations and size distributions of particles in an office room; measurements conducted using the particle counting method. (Reprinted from Jankowska, E. 2006. Influence of human activities on fine particles suspended in indoor air. In *Proceedings of the Seventh International Aerosol Conference*. St. Paul, MN, 792–3.)

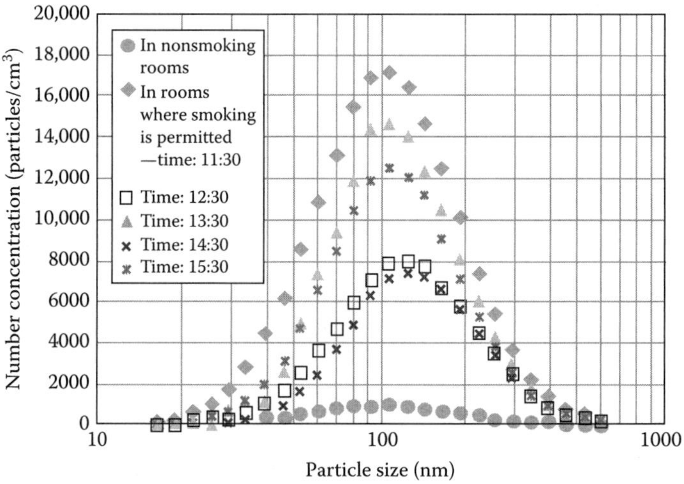

FIGURE 7.5 Number concentrations of particles in the air for smoking and nonsmoking areas; measurements conducted using the particle counting method. (Adapted with permission from Jankowska, E., M. Pośniak, and E. Walicka. 2007. *Occupational Safety* (10):15–17.)

air parameters (velocity, humidity, and temperature) upon the concentration and size distribution of the particles is significant. Dusts are believed to be one of the most harmful agents present in the office spaces, causing SBS, which affects a number of people in postindustrial society.

7.3.3 Nanotechnological Processes

The production and use of nanoparticles in work processes has resulted in intense worldwide research to assess the harmful effects of nanoparticles on humans. The analysis of exposure to nanoparticles is significant to the protection of human health because these particles may be suspended in the air for a long time even after the completion of the process, posing a threat not only to persons occupying the workstations directly associated with the process, but also to those near these workstations, as well as the general population, if, for example, improper air filters are used in the exhaust mechanical ventilation systems. Nanoparticles may settle in the alveoli of the human respiratory system and cause diseases, and, due to their size, may also get into the blood of exposed persons, causing other diseases that have not yet been recognised.

Knowledge on the emission of nanoparticles during various stages of their production or application is very limited; therefore, research in this field is a significant priority. Large plants producing or using nanoparticles in technological processes protect against their harmful effects mainly by air-tight sealing of processes or effective mechanical ventilation, whereas employees of most small- and medium-sized companies are usually unprotected against the harmful effects of nanoparticles.

In nanotechnological processes, such as those used in typical production processes, large amounts of particles are emitted to the air during, for example, mixing of loose raw materials, which are made up of very fine particles of up to several dozen nanometres in size; there is a substantial increase in their concentration in the air upon the completion of the process even after mixing nanoparticles for 2 minutes (Figure 7.6). This is because these particles diffuse into the air and

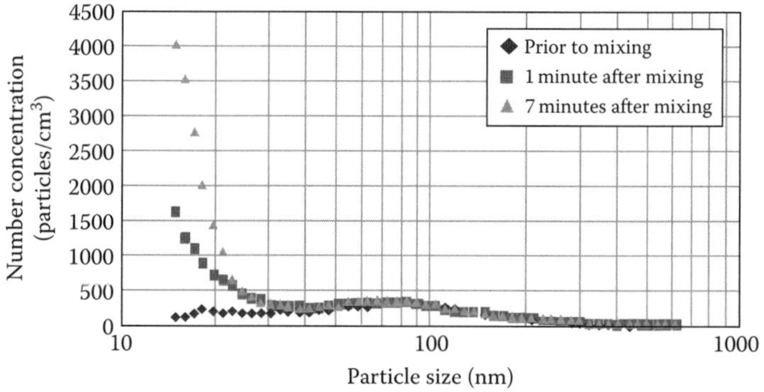

FIGURE 7.6 Size distributions of particles present in the air before and after mixing raw materials containing nanoparticles; measurements conducted using the particle counting method.

may accumulate in the breathing zones of the employees as late as 7 minutes after completion of the process.

7.4 METHODS OF DETERMINATION OF DUST PARAMETERS

The quantities of dust entering the human respiratory system from the air depend upon the dust's properties, the velocity and direction of the air's movement in the surrounding area, breathing frequency, and whether the particles are inhaled through the mouth or the nose. These parameters are measured using various research methods depending on the size of dust particles emitted at the workstations. At typical production spaces, dust samples are collected using the filtration-gravimetric method or the microscopic method. In office spaces and in nanotechnological processes, dust samples are examined mainly using optical methods, such as photometers and particle counters.

The basic parameters of the air are determined during the analysis of the concentration and size distribution of dust particles. These include temperature, humidity, and velocity, which may significantly influence the movement and number of particles suspended in the air. Analyses should also determine the basic harmful chemical substances in the dust—particularly in dust fractions that enter the area of gaseous exchange.

7.4.1 Production Spaces

In Poland, the rules for collecting air samples in the work environment and the interpretation of the results obtained from their analysis are specified in the standard PN-Z-04008-7 (2002/Az12004). Measurement strategies and guidelines for the assessment of exposure are provided in the standard PN-EN 689 (2002).

To estimate exposure to dust, dust particles suspended in the air in the individual's breathing area (the total dust) and the dust fraction that settles in the human alveoli (the respirable dust or respirable fibres) is measured using the filtration-gravimetric method (PN-91/Z-04030 2005; PN-91/Z-04030 2006) or the microscopic method (PN-89/Z-04202 2002). *Total dust* is the set of all particles surrounded by the air within a given air volume. *Respirable dust* is the set of particles passing through the preliminary selector, which is characterised by permeability based on the dust particle size. It is expressed by the lognormal probability function with the average aerodynamic diameter value of 3.5 ± 0.3 μm and a standard deviation of 1.5 ± 0.1. *Respirable fibres* have a length of more than 5 μm, a maximum diameter of less than 3 μm, and a length-to-diameter ratio of > 3 (Regulation of the Minister of Labour and Social Policy of 29 November, 2002). Air samples may be collected at a specific fixed point of the work environment using stationary equipment or using individual devices worn by the employee and equipped with a dust sampler placed within the employee's breathing zone as close to the respiratory system as possible. Individual dosimetry is preferred, as it best reflects the employee's exposure to harmful dust throughout the entire working day.

The European standard (EN 481 1993; PN-EN 481 1998) and the international standard (ISO 7708 1995; PN-ISO 7708 2001) recommend that exposure to dust at

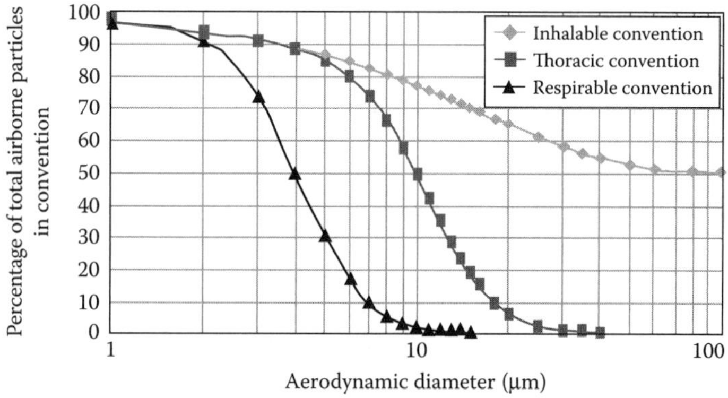

FIGURE 7.7 The inhalable, thoracic and respirable conventions as percentages of total airborne particles. (Adapted with permission from PN-EN 481. 1998. Workplace atmospheres—Size fraction definitions for measurement of airborne particles; and PN-ISO 7708. 2001. Air quality—Particle size fraction definitions for health-related sampling.)

production spaces is measured using methods that ensure selection at the entry due to aerodynamic effects. The following dust fractions of specific aerodynamic diameters of suspended particles in the air are collected from workstations (Figure 7.7):

- *The inhalable fraction*: The mass fraction of total airborne particles inhaled through the nose and mouth
- *The thoracic fraction*: The mass fraction of inhaled particles penetrating beyond the larynx
- *The respirable (alveolar) fraction*: The mass fraction of the inhaled particles penetrating to the unciliated airways

7.4.2 Office Spaces and Nanotechnological Processes

Filtration-gravimetric methods cannot be used to measure the concentration of dusts in the air in office space and nanotechnological processes because the results are not accurate and the dusts measured cannot be related to the strictly defined fractions of suspended particles in the air. To determine the concentration and size distribution of these particles, optical methods, including photometers (for measuring mass concentrations of dust) and optical and condensation counters (for measuring the number concentrations and size distributions of dust particles), are used. Various photometers and particle counters that allow for measurements within a very broad range of particle sizes from nanometric diameters (condensation counters) to several dozen micrometers (photometers, optical counters) are currently available.

Although the concentration and size distribution of particles determined using photometers and various particle counters cannot be directly compared, these devices gather multidimensional data on the parameters of dust suspended in the air within a specified range of particle sizes and indicate the type of particle fractions present in the air. These measurements are particularly significant to determine the presence of fine dust particles and nanoparticles in the air.

7.5 ASSESSMENT OF OCCUPATIONAL RISK ASSOCIATED WITH DUST EXPOSURE

The assessment of occupational risk associated with dust exposure is a complex process consisting of the following stages:

- Estimation of dust exposure through:
  - Identification of the dust type present at the workstation
  - Estimation of the concentration and other significant parameters of the dust (e.g. its chemical composition)
  - Calculation of the dust exposure index in relation to the daily working time
- Comparison of the obtained exposure index values with the admissible concentration values
- Assessment of the occupational risk associated with dust exposure
- Specification of risk admissibility

7.5.1 Production Spaces

For dust at typical production workstations, the MAC-TWA values are usually defined and specified in the legal provisions (Regulation of the Minister of Labour and Social Policy of 29 November 2002). Exposure to dust can be estimated by comparing the exposure indices—the dust concentrations related to the daily working time—with MAC-TWA values. This will serve as a basis for occupational risk assessment. Various methods and scales can be used to estimate the occupational risk. Table 7.1 presents an estimation of occupational risk associated with dust exposure according to a three-scale model, as recommended by the standard PN-N-18002 (2000).

High risk is not permissible. If a high occupational risk is associated with work that is already being performed, action must be taken immediately to reduce the risk level (e.g. by using protective equipment). Work should not begin or resume until the occupational risk is decreased to an admissible level.

TABLE 7.1
Estimation of Occupational Risk Associated with Dust Exposure Using a Three-Scale Model

$W >$ MAC-TWA	High risk
MAC-TWA $\geq W \geq$ 0.5 MAC-TWA	Medium risk
$W <$ 0.5 MAC-TWA	Low risk

Note: W = exposure index value; MAC-TWA = maximum admissible concentration.

Source: PN-N-18002. 2000. Systems of management of occupational health and safety. General guidelines for assessment of occupational risk.

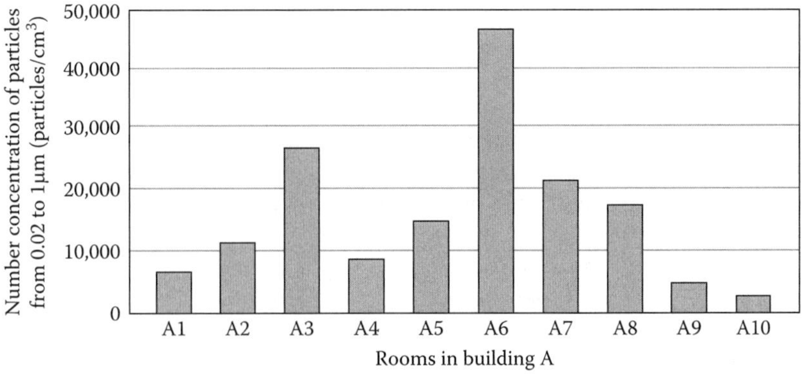

FIGURE 7.8 Number concentrations of particles between 0.02 and 1 µm in size in the air of 10 office rooms in building A; measurements conducted using the counting method. (Adapted with permission from Jankowska, E., et al. 2004b. *An Ordered Targeted Project.* [Unpublished work.])

Medium risk is permissible. However, the occupational risk level should still be reduced.

Low risk is preferable; the occupational risk should be maintained at this level.

7.5.2 Office Spaces and Nanotechnological Processes

Dust exposure in office space is usually estimated by analysing and comparing the mass concentration values or number concentration values, specified for individual rooms using photometer or counters to measure size fractions of particles similar to those determined by filtration-gravimetric methods, that is, sets of dust particles of a size under 10 µm or under 4 µm, as well as fine particles of the size below 1 µm in the surrounding air (Figure 7.8).

Research conducted at present assesses the real threat resulting from human exposure to nanoparticles.

7.6 COLLECTIVE PROTECTION AGAINST DUST

Collective protection against dust includes general mechanical ventilation systems, local mechanical ventilation systems, and devices equipped with air filters.

The purposes of ventilation systems, based upon continuous or periodical air exchange in rooms, are as follows:

- To improve the condition and composition of the air at the workstations in accordance with specified requirements for human health protection and the technology necessary to manufacture products with specific properties.
- To regulate air parameters in rooms, including pollution concentrations, temperature, humidity, velocity, and direction of air movement.

For industrial buildings where substantial amounts of dusts are emitted at individual workstations, the most beneficial solution in local exhaust ventilation (LEV) is to ensure the air-tight sealing of processes, that is, air-tight encasing of the pollution emission sources. When fully encasing the technological processes is not possible, partial encasing or local ventilation systems are used, which are connected to the dust extraction systems or to filtration–ventilation equipment. The types of LEV systems according to the degree of encasing of the pollution emission source are illustrated in Figure 7.9 (Mierzwiński et al. 2004).

Each LEV system should be accompanied by a general mechanical ventilation system. The type of LEV system depends on both the location of the emission source and the type of dust emitted. The direction and velocity of dust dispersion are also significant.

In office space, general room ventilation systems are usually used; the system layout significantly influences the air quality, including the concentrations of particles. Appropriate air flow should be ensured particularly in large open-space offices. Although the air streams are supplied to rooms as required by legal provisions, improper distribution of ventilation openings, for example, a large number of openings located close to windows and walls, may lead to draughts at the workstations

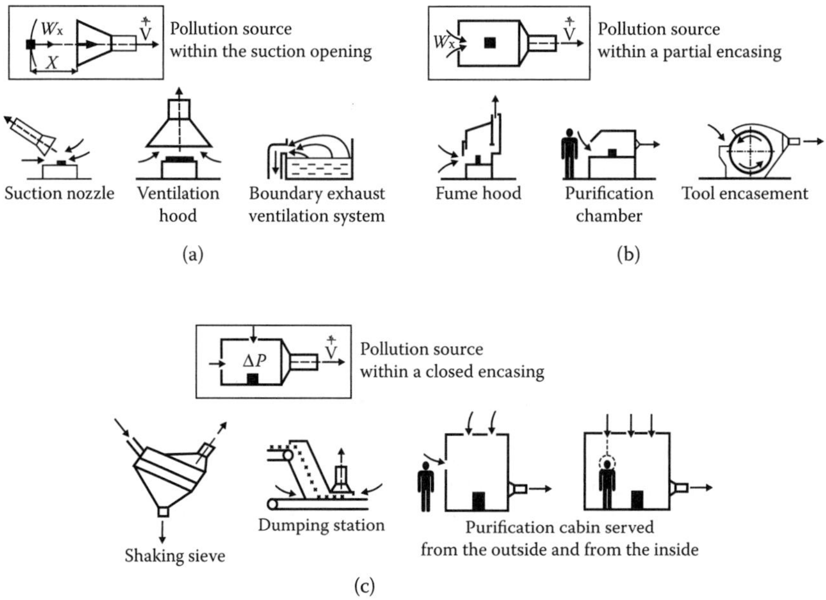

FIGURE 7.9 Types of local exhaust ventilation systems according to the degree of encasing of pollution emission source: (a) within the suction opening, (b) within a partial encasing, and (c) within a closed encasing. (Adapted with permission from Mierzwiński, S., et al. 2004. Research task no. II-5.04 carried out within the framework of the multiannual programme entitled: Adjustment of work conditions in Poland to the standards of the European Union. [Unpublished work.])

located nearby, while proper air flow is not ensured in the central part of the room, resulting in higher dust concentrations.

In both general ventilation systems and local ventilation equipment, purification systems (single or multistage) equipped with the appropriate air filters will ensure the required quality of the air is extracted from or supplied to the rooms.

The basic indicators of air filter usability are filtration efficiency and pressure drop; these parameters depend upon dust properties, air properties, and the structural parameters of the filter.

Filtration efficiency is the capacity of the filter to remove from the air pollution particles of a specified size range. The *pressure drop* (pressure difference) largely influences the selection of equipment responsible for the air movement during its flow through the filtration media.

The required level of purity for the air leaving the system (for specific pollutant particle sizes or the filtration–ventilation device) is significant for the selection of the number of filtration stages in the purification system and their classes, which are determined based on the appropriate standardised testing methods. Air filters are classified based on their filtration efficiency, which is determined using the appropriate test aerosols.

If collective protection against dust does not ensure the required level of air purity in the workplace, then appropriate personal protection equipment should be selected.

REFERENCES

EN 481. 1993. Workplace atmospheres—Size fraction definitions for measurement of airborne particles.

Hämeri, K., et al. 2003. Fine aerosols indoors and outdoors in downtown Helsinki. *J Aerosol Sci* II:1359–1360.

ISO 7708. 1995. Air quality—Particle size fraction definitions for health-related sampling.

Jankowska, E. 2006. Influence of human activities on fine particles suspended in indoor air. In *Proceedings of the Seventh International Aerosol Conference*. St. Paul, MN, 792–793.

Jankowska, E., and T. Jankowski. 2005. Influence of ventilation on dust and carbon dioxide concentration in office air. In *European Aerosol Conference*. ed. W. Maenhaut, 702.

Jankowska, E., et al. 2004a. Indoor environmental quality in office buildings. *Ergonomics* 26(4):305–319.

Jankowska, E., et al. 2004b. *An Ordered Targeted Project.* (Unpublished work).

Jankowska, E., M. Pośniak, and E. Walicka. 2007. Contamination of the air with cigarette smoke in office rooms. *Occupational Safety* 10(433):15–17.

Kreyling, W. G., et al. 2002. Translocation of ultrafine insoluble iridium particles from lung epithelium to extrapulmonary organs is size dependent but very low. *J Toxic Environ Health* 65:1513–1530.

Mierzwiński, S., et al. 2004. Research task no. II-5.04 carried out within the framework of the multiannual programme entitled: Adjustment of work conditions in Poland to the standards of the European Union. (Unpublished work).

Oberdörster, G., et al. 2002. Ultrafine particles in the urban air: To the respiratory tract-and beyond? *Environ Health Perspect* 110(8):440–441.

Ordinance of the Minister of Labour and Social Policy of November 29th, 2002 on the highest permissible concentrations and levels of agents harmful to health in the work environment. 2002. DzU no. 217, item 1833; amended by: DzU 2005, no. 212, C 1769; DzU 2007 no. 161, item 1142.

Ordinance of the Minister of Labour and Social Policy of September 26th, 1997, on the general provisions of occupational health and safety (Uniform text). 1997. DzU 2003, no 169, item 1650; amended by: DzU 2007, no. 49, item 330; DzU 2008, no. 108, item 690.
PN-89/Z-04202. 2002. Air purity protection. Tests for asbestos. Determination of the number concentration of respirable asbestos fibres in work places by optical microscopy.
PN-91/Z-04030. 2005. Air purity protection. Tests for dust. Determination of total dust in work places by filtration-gravimetric method.
PN-91/Z-04030. 2006. Air purity protection. Tests for dust. Determination of respirable dust in work places by filtration-gravimetric method.
PN-EN 481. 1998. Workplace atmospheres—Size fraction definitions for measurement of airborne particles.
PN-EN 689. 2002. Workplace atmospheres—Guidance for the assessment of exposure by inhalation to chemical agents for comparison with limit values and measurement strategy.
PN-ISO 4225. 1999. Air quality—General aspects. Vocabulary.
PN-ISO 4225/Ak. 1999. Air quality—General aspects. Terminology (national annex).
PN-ISO 7708. 2001. Air quality—Particle size fraction definitions for health-related sampling.
PN-N-18002. 2000. Systems of management of occupational health and safety. General guidelines for assessment of occupational risk.
PN-Z-04008-7:2002/Az1. 2004. Air purity protection—Sampling methods—Principles of air sampling in work place and interpretation of results.
Szymczykiewicz, K. 1973. Industrial dust. Warsaw: CRZZ.

8 Vibroacoustic Hazards

Zbigniew Engel, Danuta Koradecka, Danuta Augustyńska, Piotr Kowalski, Leszek Morzyński, and Jan Żera

CONTENTS

8.1 INTRODUCTION

Vibroacoustics is a field of science that deals with vibration and acoustic processes observed in nature, engineering, machines and equipment, means of transport, and communication, that is, in the entire work and life environment. It is useful in reducing vibroacoustic disturbances to the minimum level attainable with the current knowledge and technology.

Vibroacoustic hazards related to exposure to noise and vibration in a work environment lead to its degradation. According to a survey conducted in 2005 by the European Foundation for the Improvement of Living and Working Conditions, 41.6% of employees in Poland and 30% of employees in the 25 European Union member states complained about noise in their work environment that was perceptible for at least one-fourth of their work time (Parent-Thirion et al. 2007).

Humans have been interested in issues related to noise since the dawn of time. Analysis of Sumerian clay tablets and other documents shows that—in a sense—we had declared war on noise even before the discovery of steam, electricity and combustion engines. In Poland, the first measurements of urban noise were conducted as early as 1933. The year 1948 is considered the beginning of the history of noise control in Poland. Early studies on the subject mostly concerned acoustic insulation in buildings.

Before discussing the issues of vibroacoustics, some definitions are given below.

Oscillation motion or—in short—*oscillation* (*vibration*) is a process that periodically changes any physical quantity specific to this process. This chapter focuses on mechanical system vibrations. A mechanical system is a set of interconnected solid bodies; an acoustic system is an area filled with gas, or sometimes liquid, and limited by surfaces (such as room walls). Mechanical and the acoustic systems often form a whole unit, referred to as a *vibroacoustic system.*

Vibration has three correlated vibratory quantities: displacement, velocity, and acceleration.

Another term applied to vibroacoustic hazards is *noise*. Noise is all adverse, unpleasant, strenuous, or harmful oscillations of the elastic medium, acting through the air on the hearing organ and other parts of the human body.

Vibroacoustic parameters can be determined experimentally. The physical parameters of a vibroacoustic signal can be measured, and their influence on the psychophysiological properties of humans can be determined by research. A vibroacoustic signal is a rich source of information on the technical condition of equipment and machines found in workplaces, apartments, and so on.

Vibroacoustic signals are usually complex periodical oscillations consisting of many basic harmonic oscillations. The characteristics of a vibroacoustic signal are shown in Figure 8.1.

The average value of a given signal $a(t)$ is defined by the expression:

$$A_{av} = \frac{1}{T}\int_0^T |a|\,dt \quad (8.1)$$

where T is the oscillation period in seconds.

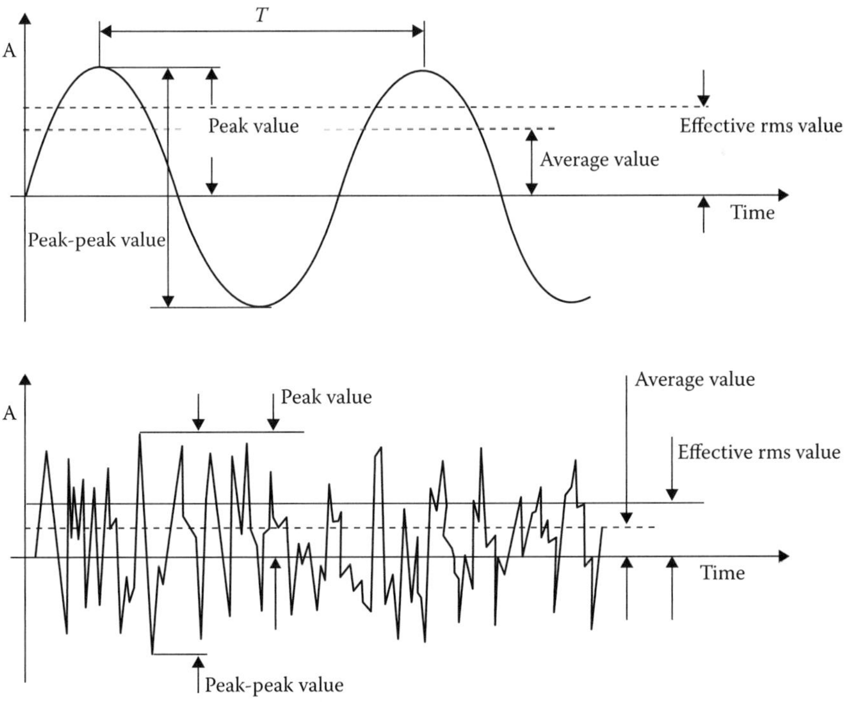

FIGURE 8.1 Parameters of a vibroacoustic signal.

The root mean square A_{rms}, also known as the quadratic mean, is expressed by the formula:

$$A_{rms} = \sqrt{\frac{1}{T}\int_0^T a^2(t)\mathrm{d}t} \qquad (8.2)$$

A_{peak} is the maximum value reached by the measured signal during observation. The difference between the maximum and the minimum value of the vibroacoustic signal amplitude is termed 'peak-to-peak value' ($A_{peak\text{-}peak}$). For a symmetric signal, such as harmonic oscillation, it is twice the peak value (Figure 8.1). The correlation between the discussed values can be presented as follows:

$$A_{rms} = F_f A_{av} = \frac{1}{F_{peak}} A_{peak} \qquad (8.3)$$

where F_f is the form factor and F_{peak} is the crest factor.

For sinusoidal signals, these factors are

$$F_f = \frac{\pi}{2\sqrt{2}} = 1.11 \quad F_{peak} = \sqrt{2} = 1.41$$

In the mechanical and acoustic oscillation theory, a logarithmic scale is often applied due to the wide range of oscillation amplitudes. The level of the vibroacoustic quantity (measured in decibels [dB]) applied in this scale is expressed by the following formula:

$$N = 20\log\frac{A}{A_0} \qquad (8.4)$$

where N is the level of the vibroacoustic quantity, A is the quantity measured, and A_0 is the reference value for a given vibroacoustic quantity (Table 8.1).

Any environment, including the work environment, has many vibroacoustic hazard sources, which can be classified based on various aspects. Figure 8.2 presents a classification of vibroacoustic hazard sources based on physical causes of disturbances. These sources originate mainly in the machines and equipment used in industry and transport.

8.2 VIBRATION

8.2.1 Basic Terms and Quantities

From the perspective of occupational safety and health protection, mechanical oscillation (vibration) is a set of phenomena observed at workstations, constituting the transmission of energy from the vibration source to the human body through various body parts that come into contact with the oscillating source while performing work tasks.

TABLE 8.1
Reference Values for Vibroacoustic Levels

Quantity	Level	Reference Value
Displacement	$L_d = 20\log\left(\frac{x}{x_0}\right)$ dB	$x_0 = 10^{-12}$ m = 1 pm
Velocity	$L_v = 20\log\left(\frac{v}{v_0}\right)$ dB	$n_0 = 10^{-9}$ m/s = 1 nm/s
Acceleration	$L_a = 20\log\left(\frac{a}{a_0}\right)$ dB	$a_0 = 10^{-6}$ m/s² = 1 μm/s²
Force	$L_F = 20\log\left(\frac{F}{F_0}\right)$ dB	$F_0 = 10^{-6}$ N = 1 μN
Sound pressure (air)	$L_p = 20\log\left(\frac{p}{p_0}\right)$ dB	$p_0 = 2\cdot 10^{-5}$ Pa = 20 μPa
Sound pressure (media other than air)	$L_p = 20\log\left(\frac{p}{p_0}\right)$ dB	$p_0 = 10^{-6}$ Pa = 1 μPa
Sound power	$L_N = 10\log\left(\frac{P}{P_0}\right)$ dB	$P_0 = 10^{-12}$ W = 1 pW
Sound intensity	$L_I = 10\log\left(\frac{I_a}{I_0}\right)$ dB	$I_0 = 10^{-12}$ W/m² = 1 pW/m²
Sound energy	$L_E = 10\log\left(\frac{E}{E_0}\right)$ dB	$E_0 = 10^{-12}$ J = 1 pJ

Long-term occupational exposure to vibration may lead to many disorders and cause permanent pathological changes. The type of changes and the pace of their development depend to a great extent upon the point of penetration on the human body. In this regard, two types of mechanical vibration can be distinguished:

1. Whole-body vibration influences the human body in general, penetrating through the legs, pelvis, the back, or the sides.
2. Hand-arm vibration acts on the human body through the upper limbs.

Vibration acceleration is usually measured to assess the influence of mechanical vibration upon the human body; it provides the best characterisation of the energy-related aspects of the vibration process. This quantity is associated with the adopted admissible values relating to daily (8-hour) exposure, which are defined for health protection in many domestic and international legal provisions. Therefore, in

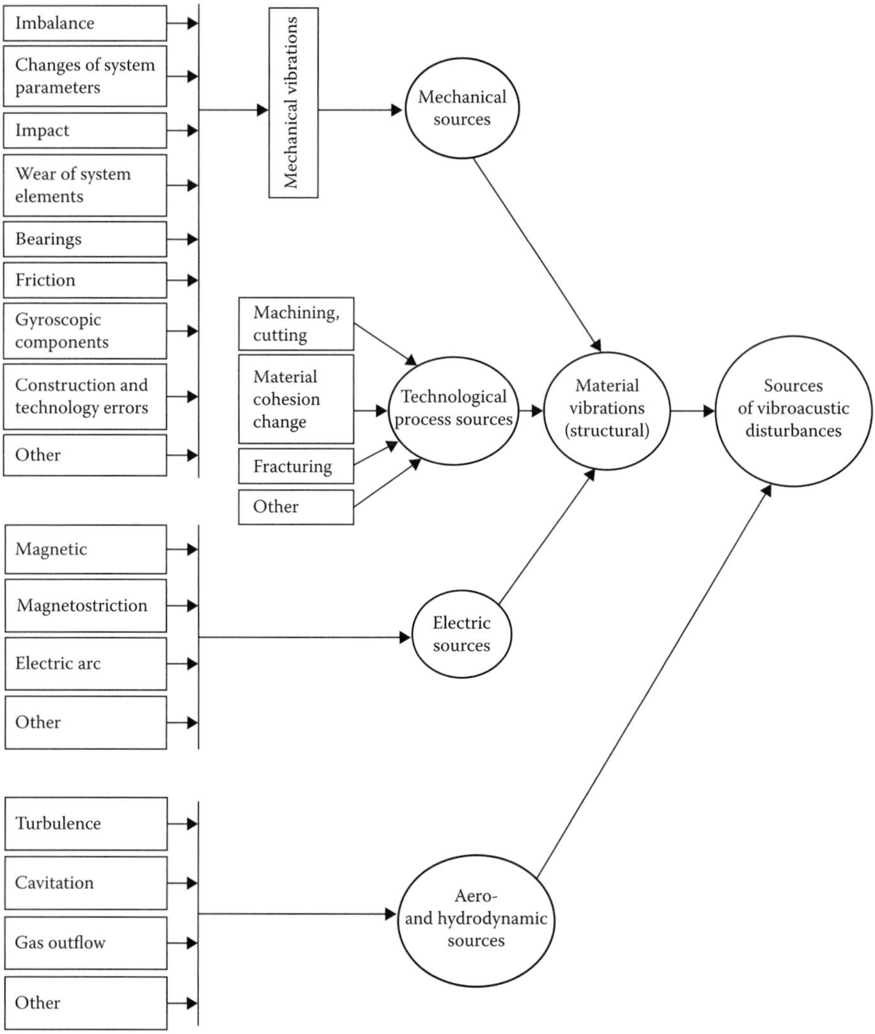

FIGURE 8.2 Sources of vibroacoustic disturbances in industry.

Section 8.2, vibration acceleration is taken into account in the analysis of mechanical vibration.

The vibration signal may contain a single component of a specific frequency (sinusoidal vibration); however, in practice, vibration is most often composed of many sinusoidal components or even a sum of their infinite number. In simple sinusoidal vibration, the value of instantaneous acceleration of vibration is expressed as follows:

$$a = a_{\text{peak}} \sin\left(2\pi \frac{t}{T}\right) = a_{\text{peak}} \sin(2\pi ft) \tag{8.5}$$

where a_{peak} is the is the maximum (peak) acceleration value in metres per second squared (m/s^2), t is the time in seconds, T is the vibration period in seconds, and $f = \frac{1}{T}$ is the frequency in hertz (Hz).

Apart from the instantaneous acceleration value and the peak value, the following terms are used to describe vibration (see Section 8.1):

- Root mean square (rms) value of acceleration a_{rms} (in m/s^2)
- Average value of acceleration a_{av} (in m/s^2)

Vibration spectrum analysis is performed to assess the vibration impact on the human body. The rms value of acceleration is determined for each frequency; the set of these values constitutes the vibration spectrum.

The response of the human body to vibration depends upon many factors, including the spectral composition of the vibration signal recorded at a given workstation. Diversified responses of the human body to vibration, depending upon its frequency, are accounted for by introducing the rms value of weighted acceleration ($a_{w,rms}$, in m/s^2).

The weighted values of acceleration are determined by applying specially adapted frequency characteristics for correction filters (Figure 8.3). The weighted value of acceleration is used to measure the employee's daily exposure to vibration.

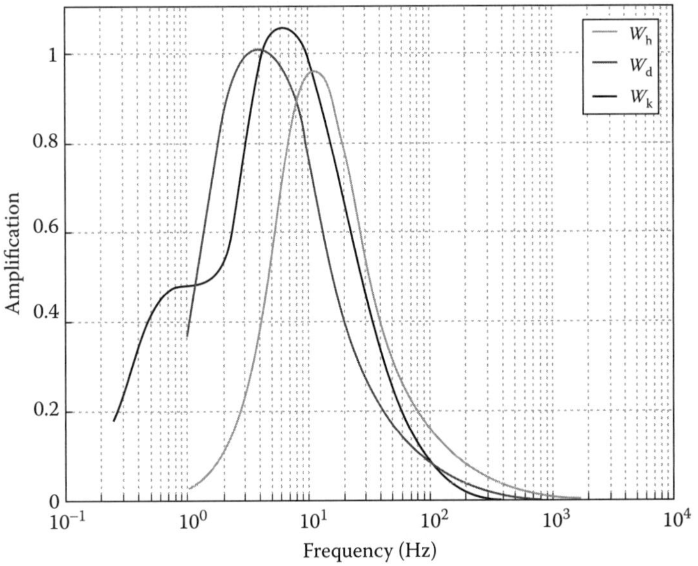

FIGURE 8.3 Frequency characteristics of correction filters used during the measurement of mechanical vibrations in the work environment: W_h for hand-arm vibration acting in directions x, y, z; W_d for whole-body vibration acting in directions x, y (horizontal); W_k for whole-body vibration acting in direction z (vertical).

8.2.2 Impact of Vibration on the Human Body

Mechanical vibration transferred from oscillatory systems to the human body may have a negative impact on individual systems and organs, or it may cause vibration of the whole body or its parts and its cellular structures. Long-term exposure to vibration may lead to many disorders and cause permanent and irreversible pathological changes; these changes depend upon the type of vibration to which an individual is exposed—whole-body or hand-arm vibration.

8.2.2.1 Hand-Arm Vibration

Exposure to vibration transferred to the organism through the upper limbs—hand-arm vibration—results mainly in pathological changes in

- The circulatory system
- The nervous system
- The osteoarticular system

Epidemiological research has shown a strict causal relationship between pathological changes in these systems and work conditions. Therefore, a set of these changes, known as *vibration syndrome*, is considered an occupational disease in many countries, including Poland.

The vascular form of vibration syndrome is the most frequent and is characterised by changes mainly in peripheral blood circulation in the form of paroxysmal vasospastic disorders of the upper extremities (white fingers). Paroxysmal vasospasms are indicated by one or more fingertips turning pale. This form of vibration syndrome is often confused with Raynaud's disease.

Vasospastic disorders are often accompanied by nervous disorders and osteoarticular changes. Changes in the nervous system caused by exposure to hand-arm vibration mainly include pallesthesia, apselaphesia, loss of sense of temperature, and associated problems such as numbness or tingling of fingertips. If exposure to vibration continues, these changes develop further, decreasing the ability to work and perform other life functions.

Osteoarticular changes in the hand are caused mainly by hand-arm vibration of frequency below 30 Hz. These include the deformity of articular spaces, calcification of articular capsules, and changes in the periosteum and the woven bone, leading to the emergence of characteristic cysts in the small bones, particularly in the wrist. Operators of manual vibration tools are the main group exposed to hand-arm vibration.

8.2.2.2 Whole-Body Vibration

The negative consequences of occupational exposure to whole-body vibration are mostly related to

- The skeletal system
- The internal organs

In the skeletal system, pathological changes emerge mainly in the lumbar segment of the spine, and less often in the cervical segment. Spinal pain syndrome resulting from pathological changes occurring in persons occupationally exposed to whole-body vibration has been recognised as an occupational disease—like vibration syndrome caused by hand-arm vibration—in some countries, such as Belgium and Germany.

Internal organ disorders caused by whole-body vibration are mainly due to the resonance excitation of individual organs (the frequency of natural vibrations of most organs is between 2 and 18 Hz). This causes adverse changes in the functioning of the alimentary system organs, mainly the stomach and the oesophagus; however, tests conducted on large groups occupationally exposed to whole-body vibration show that disorders are also found, for example, in the vestibulocochlear organ, the organs of the thorax, the female reproductive system, and the nasopharyngeal cavity.

The main groups of employees exposed to whole-body vibration are car drivers, tram drivers, engine drivers, and construction and road machine operators. In these cases, vibration is transmitted to the body from the vehicle seat through the pelvis, the back, and the sides. However, occupational exposure to whole-body vibration often affects persons who handle fixed equipment and machines operated in various work areas in a standing position. In such cases, vibrations penetrate the employee's body through the feet from the vibrating base of the workstation. The effects of these vibrations are similar to those transmitted by seats.

The biological effects of hand-arm and whole-body vibrations on the human body are usually accompanied by functional effects, which include the following:

- Prolonged motor response time
- Prolonged visual response time
- Movement coordination disturbances
- Excessive fatigue
- Irritation
- Sleep disorders

Adverse functional changes lead to a decrease in the effectiveness and quality of the work performed, sometimes making it impossible to work at all.

Because mechanical vibration and its effects are common in the work environment, it should be measured to assess occupational risk and to reduce the risk level.

8.2.3 Methods of Measurement and Criteria for the Assessment of Vibration

8.2.3.1 Methods of Determination of Machine Vibration Emission Values

Information concerning the machine vibration emission is valuable for designers, manufacturers, users, and inspection authorities, and is necessary for the comparison of vibration emissions generated by various machines and for the assessment of vibrations in order to comply with the requirements of occupational health and safety.

According to the European and domestic regulations (Directive 2006/42/EC):

- Machinery must be designed and constructed to reduce risks resulting from vibrations produced by the machinery to the lowest level, taking into account the technical progress and the availability of means for reducing vibration, particularly at the source.
- Instructions must provide the following information regarding the vibrations transmitted by the machinery to the hand-arm system or to the whole body:
  - The total vibration value to which the hand-arm system is subjected, if it exceeds 2.5 m/s^2; if this value does not exceed 2.5 m/s^2, this must be mentioned.
  - The highest rms value of the weighted acceleration to which the whole body is subjected if it exceeds 0.5 m/s^2; if this value does not exceed 0.5 m/s^2, this must be mentioned.

Methods for determining the vibration emission values for machines are specified in the appropriate standards such as PN-EN 1032, PN-EN 1033, PN-EN 20643, and the standards of the series PN-EN 28662. The methods of declaring and verifying machine vibration emission values are specified in the standard PN-EN 12096.

8.2.3.2 Methods of Measurement and Criteria for the Assessment of Vibration at Workstations

In Poland, according to the regulations of the Minister of Labour and Social Policy concerning the maximum admissible concentrations and intensities for agents harmful to health in the work environment, mechanical vibration at workstations are divided into daily exposure (8 hours) and short-term exposure (lasting 30 minutes or less). The daily exposure value $A(8)$ for hand-arm vibration is expressed as the 8-hour equivalent of the vector value of weighted acceleration, specified for three directional components a_{hwx}, a_{hwy}, a_{hwz}. The daily exposure $A(8)$ for whole-body vibration is expressed as the 8-hour equivalent of the vibration acceleration value, calculated as the highest rms value of weighted acceleration among the determined values of the three directional components, taking into account the appropriate coefficients $1.4a_{wx}$, $1.4a_{wy}$, a_{wz}.

The daily and short-term exposure values permissible for employee health protection are provided in Table 8.2.

The regulations of the Minister of Labour on occupational health and safety related to exposure to noise or mechanical vibration (directive 2002/44/EC) introduces an additional criterion for the assessment of daily vibration exposure—the *daily exposure action value* (Table 8.3). If the daily exposure action values are exceeded, the employer must undertake actions to limit the vibration exposure.

The admissible values for vibration at workstations for juveniles and pregnant women (provided in Table 8.4) have been defined in separate legal provisions (Regulations of the Council of Ministers).

Poland's general methods for the measurement and assessment of mechanical vibration are defined in the Regulations of the Minister of Health concerning tests and measurements of agents harmful to health in the work environment.

TABLE 8.2
Maximum Admissible Values of Exposure to Mechanical Vibration

Vibration Type	Admissible Values for Daily Exposure to Mechanical Vibration	Admissible Values for Short-Term Exposure to Mechanical Vibration
Hand-arm vibration	$A(8)_{adm} = 2.8\ m/s^2$	$a_{hv,30\ min,adm} = 11.2\ m/s^2$
Whole-body vibration	$A(8)_{adm} = 0.8\ m/s^2$	$a_{w,30\ min,adm} = 3.2\ m/s^2$

TABLE 8.3
Daily Vibration Exposure Action Values

Vibration Type	Daily Exposure Action Values
Hand-arm vibration	$A(8)_{imp.} = 2.5\ m/s^2$
Whole-body vibration	$A(8)_{imp.} = 0.5\ m/s^2$

TABLE 8.4
Admissible Values of Vibration Exposure for Juveniles and Pregnant Women

Vibration Type	Exposure Type	Admissible Values for Juveniles	Admissible Values for Pregnant Women
Hand-arm vibration	Daily	$A(8)_{adm} = 1.0\ m/s^2$	$A(8)_{adm} = 1.0\ m/s^2$
	Short-term	$a_{hv,30\ min,adm} = 4.0\ m/s^2$	$a_{hv,30\ min,adm} = 4.0\ m/s^2$
Whole-body vibration	Daily	$A(8)_{adm} = 0.19\ m/s^2$	Work prohibited
	Short-term	$a_{w,30\ min,adm} = 0.76\ m/s^2$	

The detailed methods are described in the appropriate Polish standards PN-EN ISO 5349-1, PN-EN ISO 5349-2, and PN-EN 14253 + A1.

Vibrations are measured under typical conditions at a given workstation while the employee performs typical activities, such as operating a tool, machine, or device. For each separate (*i*th) activity of the employee, measurements are recorded with regard to the weighted vibration acceleration values in three mutually perpendicular directions: $a_{wx,i}$, $a_{wy,i}$, $a_{wz,i}$ for whole-body vibration (Figure 8.4) and $a_{hx,i}$, $a_{hy,i}$, $a_{hz,i}$ for hand-arm vibration (Figure 8.5).

8.2.4 Current Situation of Vibration Hazard and Sources in the Work Environment

According to data from the years 1995–2007 from the Central Statistical Office, about 3%–4% (approximately 20,000–40,000) of employees in the principal sectors

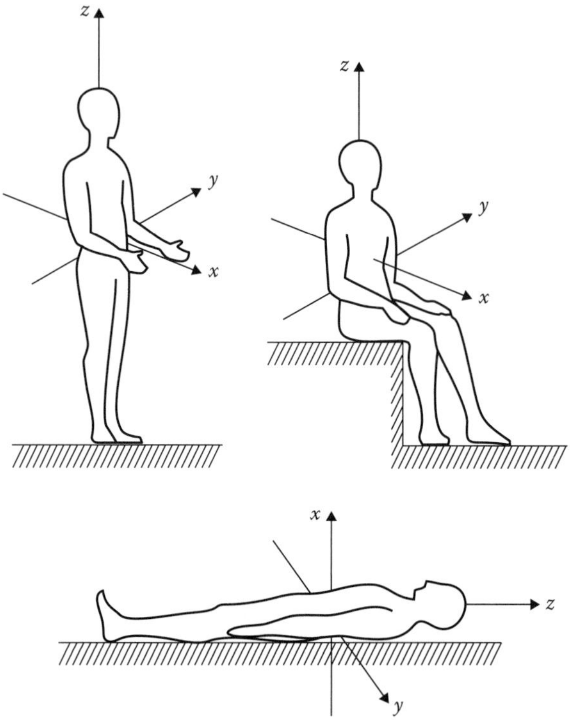

FIGURE 8.4 The anatomical (mobile) coordinate reference system used for measuring whole-body vibration. (Adapted with permission from ISO 2631-1. 1997. Mechanical vibration and shock—Evaluation of human exposure to whole body vibration. I: General requirements.)

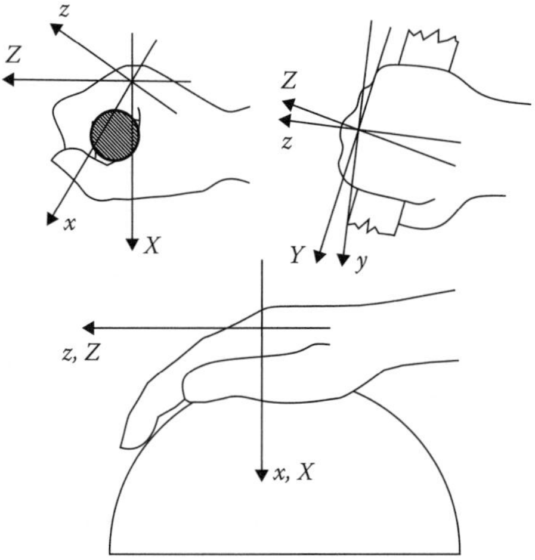

FIGURE 8.5 The mobile- and fixed-coordinate reference system used for measuring hand-arm vibration. (Reprinted with permission from PN-EN ISO 5349-1. 2004. Mechanical vibration—Measurement and evaluation of human exposure to hand-transmitted vibration. Parts 1 and 2.)

of the national economy who are exposed to strenuous and harmful agents in their work environments are exposed to vibration exceeding the permissible values. The highest health hazards are observed in industrial processing, construction, mining, and transportation. However, the data of the Central Statistical Office concerning these hazards are not complete, as they do not cover all employees (the data are gathered only from business entities employing more than nine persons) and do not include private farming.

According to a survey conducted in 2005 by the European Foundation for the Improvement of Living and Working Conditions in the 25 member states of the European Union (including Poland), 25% of employees report exposure to vibration for at least one-fourth of their working time.

According to the introduced divisions of mechanical vibration, vibration sources in the work environment can be divided into two groups:

1. Whole-body vibration sources
2. Hand-arm vibration sources

Whole-body vibration sources include the following:

- Floors, landings, and platforms in production buildings and other facilities upon which workstations are located; the primary sources of vibration in this case are mobile, portable, or fixed devices and machines, operated inside or outside, which cause vibration of the floor on which the operator is standing (Figure 8.2).
- Oscillating platforms.
- Seats and floors of transport vehicles (cars, tractors, buses, trams, trolleys, trains, ships, planes).
- Seats and floors of construction machines (for earthworks, foundation works, soil compacting).

The main sources of hand-arm vibration are as follows:

- Pneumatic, hydraulic, or electric hand-held impact tools (e.g. pneumatic drills, moulding hammers and concrete tampers, riveting hammers, hammer drills, wrenches)
- Electric- or diesel-powered hand-held rotary tools (drills, grinders, chainsaws)
- Hand-operated control levers of machines and vehicles
- Technological process sources (hand-held or hand-guided components processed by grinding, smoothing, polishing, etc.)

Some hand-held tools, such as hammers and mechanical saws (listed among typical hand-arm vibration sources), can generate very high vibration acceleration. Such vibration may be transmitted through the shoulders to the torso and head, leading to resonance excitation in the internal organs. Thus, hand-held tools are also whole-body vibration sources.

8.2.5 Methods of Preventing, Eliminating, and Limiting Vibration Exposure

Methods of prevention, elimination, and limitation of vibration exposure generally can be divided into technical and organisational or administrative methods. The *technical methods* include

- Elimination or limitation of vibration at the source
- Limitation of vibration along its propagation route
- Automation of technological processes and remote control of vibration sources
- Active control of vibration

The vibroactivity of the sources may be reduced by changing their structure, proper assembly, or fixing the machines to the floor (foundation engineering).

Using expansion joints separating the foundations of machines and devices from the surrounding area, various forms of vibroisolation materials (mats, washers, special vibration isolators), and personal protection equipment such as antivibration gloves can reduce the vibration along its propagation route.

Automation of technological processes and remote control of vibration sources allow the employees to be away from the areas exposed to mechanical vibration and thus reduce the threats to health due to vibration.

The active methods discussed next are the most effective for vibration control.

8.2.5.1 Reduction of Vibroactivity of the Sources

The vibroactivity of sources can be reduced in various ways:

- *Altering the internal structure of the vibration source*: Vibration may occur due to defective or imprecise construction of a machine or mechanism, for instance, rough finishing, resulting in an imbalance of rotating parts and an increase in the vibration level proportional to the spinning frequency (rotating speed). Improper assembly of the cooperating components, despite their proper geometry has similar effects. Vibration of the entire machine can be reduced by the reducing clearances to a minimum, improving balancing, and eliminating mutual impact of cooperating components.
- *Fixing the machine to the floor (foundation engineering)*: Fixing machines to the floor significantly influences the propagation of vibration to the environment. Dynamically complex foundations must be designed; they may change the resonance frequency due to an increase in their mass when they are rigidly fixed to the machine (i.e. the vibration source).
- *Changing the parameters of the vibration exciting force*: Cempel (1989) discussed the possibility of vibration reduction by changing the parameters of the exciting force. This method can reduce vibration resulting from short impacts by generating broad-frequency spectrum signals. Vibration can be substantially reduced by eliminating microimpacts (collisions between

masses) and replacing them, for instance, with a rotation of the cooperating surfaces.

- *Introducing an additional system*: An additional system is applied when the frequency of the source vibration is close to its resonance frequency. This additional mass can alter the vibration frequency substantially, significantly reducing the vibration quantities. For example, the dynamic eliminator, which consists of a mass component and a flexible component, can be applied. Adequately selected eliminator components compensate for the exciting forces when the force excitation frequency is equal to the resonance frequency of the additional system. The eliminator mass vibration is out of phase with the source vibration, thus effectively reducing the source vibration.

8.2.5.2 Reduction of Vibration Propagation

Vibroisolation of sources can be achieved using methods that reduce vibration propagation between the source and the receiver by applying certain structures, such as an expansion joint separating the machine foundation from the surrounding area or various vibroisolation materials like mats, washers, and vibroisolators. Vibroisolation materials, regardless of the shape and type, function based on the transformation of the vibration energy into internal material friction, and then into heat. Therefore, vibroisolation is very effective in reducing higher-frequency vibration, but it does not have much effect on low frequencies.

Adequate materials can be selected in accordance with the vibration parameters and based on the specification data for homogeneous, or even more preferably, layered, materials. The shape of the vibroisolation component, usually made of slightly flexible viscoelastic materials such as rubber, cork, or polyurethane, also substantially increases the effectiveness.

Vibroisolation can be used in the construction of personal protection equipment that is used by employees to protect their hands. Antivibration gloves effectively reduce vibrations transmitted to the hands of the operators from hand-held vibration tools within the maximum possible frequency range. Designing these gloves is difficult due to the changeability of the parameters based on the conditions of use. For instance, the tool pressure not only changes over time but also depends on the tool operation method; its distribution on the surface of contact with the source is also diversified. Therefore, the ability to select the proper vibroisolation material is significantly limited in practice.

Methods for testing antivibration gloves and the requirements they should meet are specified in the standard PN-EN ISO 10819. Antivibration gloves are tested at the National Research Institute of the Central Institute for Labour Protection according to its laboratory research standard, which fulfils the requirements of the standard mentioned above.

The results of research conducted so far in Poland and other countries show that antivibration gloves are most efficient in the frequency range of 200–1250 Hz.

Improperly designed devices for protection against vibration may not only fail to reduce the exposure of employees at workstations, but may even increase exposure; moreover, they may negatively affect the condition of the machine that generates the vibration or the condition of nearby devices. More details on methods for reducing mechanical vibration levels can be found in the extensive literature on the subject (Cempel 1989; Engel 2001; Gryfin 1990; Harris and Piersol 2002; Osiński 1997).

8.2.5.3 Organisational and Administrative Methods

Reducing vibration exposure through organisational and administrative methods includes mainly the following:

- Shortening the vibration exposure time per work shift.
- Designating special rooms for rest.
- Relocating persons particularly sensitive to vibrations to other job positions.
- Training employees to increase their awareness of the threats associated with exposure to vibration and the safe operation of machines and tools.
- Administering preventive medical tests, aimed mainly at the relocation of employees whose functional systems are critical and may worsen due to vibration. Persons who are already exposed to vibration at work should be subjected to periodic tests. The scope and frequency of preliminary, periodical, and preventive medical tests for employees exposed to various factors at work, including vibration, are defined in the regulations of the Minister of Health and Labour Policy of Poland.

Organisational and administrative methods should be applied particularly when technical methods cannot limit the exposure. Technical methods aimed at reducing vibration levels at work should take into account both vibration sources and their transmission routes. When selecting such methods, their cost in relation to the degree of vibration reduction that will be attained should also be considered.

8.2.5.4 Active Control of Vibration

Passive systems for vibration reduction are less effective in a low frequency range, in broadband frequency, and when the equipment operating conditions are highly changeable. Active systems can be used as an alternative in such situations. Active systems are equipped with additional external power sources that are controlled to supply or absorb energy at specific points of the vibrating device. These include reference sensors, control systems, and the controlled components (pneumatic, hydraulic, electric, electromagnetic, piezoelectric, etc.).

Vibration reduction systems equipped with additional controlled energy sources can be divided into

- Active systems
- Semiactive systems
- Hybrid systems

In active systems (Figure 8.6a), external energy sources are applied. They have a large capacity to generate additional forces, ensuring direct compensation for the disturbance excitations. In semiactive systems (Figure 8.6b), vibration is deadened by passive components and low-capacity external energy sources are used to change the

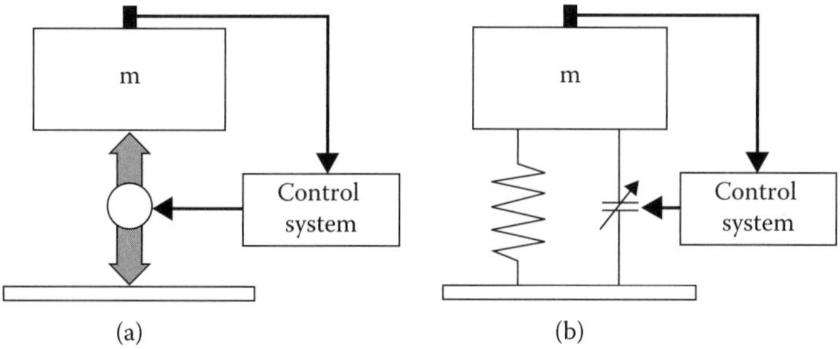

FIGURE 8.6 Diagram of vibration reduction systems: (a) active and (b) semiactive. (Adapted with permission from Engel, Z. 2001. *Protecting the Environment against Vibration and Noise*. Warsaw: PWN.)

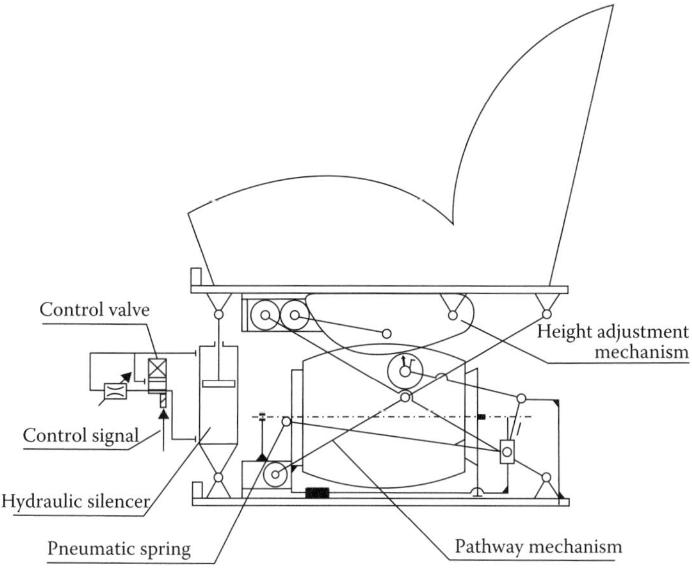

FIGURE 8.7 Example of semiactive suspension of a driver's seat. (Adapted with permission from Engel, Z. 2001. *Protecting the Environment against Vibration and Noise*. Warsaw: PWN.)

parameters of passive components such as the elastic forces and the damping forces. Hybrid systems are combinations of active and semiactive systems.

Adaptive systems, a significant class of the active systems of vibration reduction, adapt themselves to changes in the parameters of the controlled object; for example, when they are under the influence of ambient temperature, the rigidity of some of the system components is altered. In these systems, adaptive digital filters are commonly used.

Practical applications for active vibration reduction methods include

- Active suspension systems in seats for machine and vehicle operators (Figure 8.7)
- Active suspension systems in road and track vehicles

- Systems for active vibration reduction in steel load-bearing structures, such as beams or trusses
- Systems for active vibration reduction in machine and device enclosures, such as steel plates
- Systems for active vibration reduction of the structural components of airplanes
- Systems for active vibration reduction of rotors
- Systems for active vibration reduction of buildings (protection against seismic vibration)

8.3 NOISE

8.3.1 Basic Terms and Quantities

Noise is defined as all adverse, unpleasant, irritating, annoying, or harmful sound that reaches the hearing organ and influences other senses and parts of the human body. In physical terms, sound is a mechanical vibration of an elastic medium (gas, liquid, or solid). The vibration may be considered to be an oscillatory movement of the particles of the medium in relation to the equilibrium condition, resulting in a change in the medium static (atmospheric) pressure.

This change of pressure—or disturbance from equilibrium—is transmitted as a subsequent local condensation and rarefaction of the medium's particles in the space surrounding the vibrating source, generating an acoustic wave.

The difference between the instantaneous pressure value in a medium during the propagation of an acoustic wave and the static (atmospheric) pressure value is referred to as *sound pressure, p*, expressed in pascals (Pa; Figure 8.8).

Due to the wide range of changes in the sound pressure—from 2×10^{-5} to 2×10^{2} Pa (Figure 8.9)—the logarithmic scale is widely used; the *sound pressure level,* L_p, is expressed in decibels, according to the following equation:

$$L_p = 10 \log \frac{p^2}{p_0^2} \tag{8.6}$$

where p is the sound pressure rms value in pascals and p_0 is the reference sound pressure, equal to 20 μPa, a conventional value equivalent to the sound pressure at which tones (sinusoidal vibrations) of a frequency of 1000–5000 Hz can be heard.

The sound pressure level weighted by A-weighting frequency characteristics of the sound level meter is called the A-weighted sound level, while the sound pressure level weighted by C-weighing frequency characteristics is known as the C-weighted sound level. A-, B-, and C-weighted frequency characteristics (Figure 8.10), like the time weightings of the meter, are related to the characteristic of the ear known as isophonic curves, or equal loudness contours. At present, A-weighted and C-weighted characteristics are applicable regardless of the sound pressure level value.

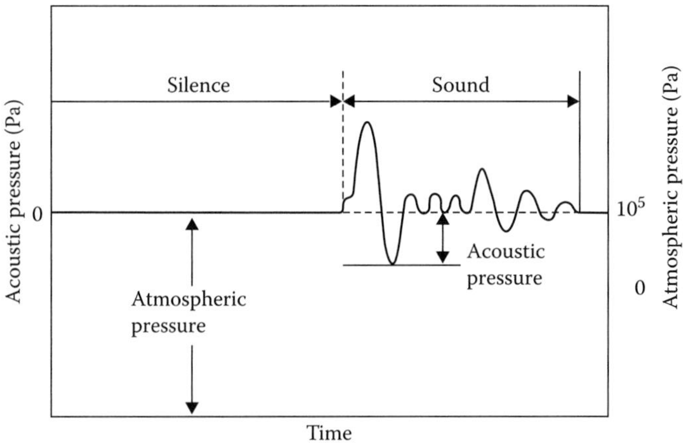

FIGURE 8.8 Sound pressure and atmospheric pressure.

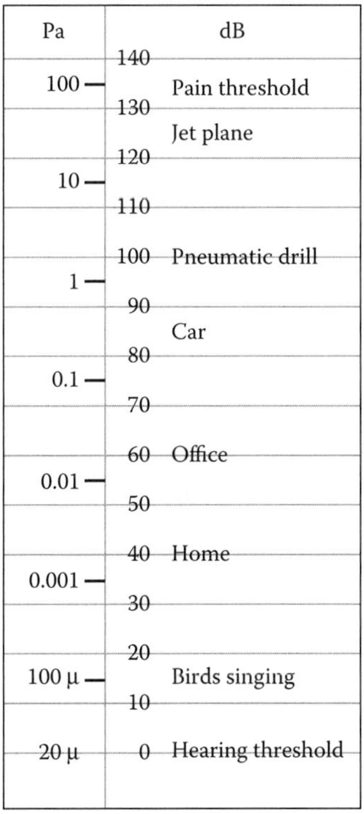

FIGURE 8.9 Sound pressure values for selected noise sources.

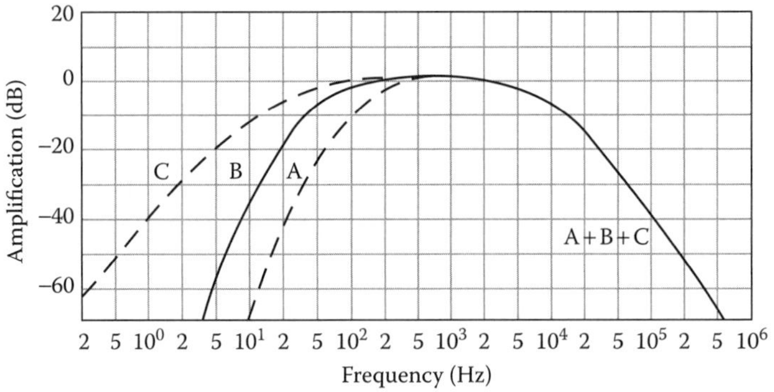

FIGURE 8.10 Frequency characteristics of the sound level meter.

Recently, the concept of G-weighted sound pressure level has also been applied (PN ISO 7196). This quantity characterises infrasound noise (see Section 8.3.6).

The maximum A-weighted sound level, that is, the maximum observed rms value of the A-weighted sound level, and the peak C-weighted sound level, that is, the level of the maximum C-weighted instantaneous sound pressure, are the basic quantities, besides the so-called A-weighted equivalent continuous sound level, used for the formulation of hearing protection criteria in the work environment.

The A-weighted equivalent continuous sound level (the quantity used to characterise noise that varies with time) is defined as the time-averaged value of the A-weighted sound level during a stated time interval T. It is expressed in decibels by the following formula:

$$L_{\mathrm{Aeq},T} = 10\log\left[\frac{1}{T}\int_0^T\left(\frac{p_{\mathrm{A}}(t)}{p_0}\right)^2 \mathrm{d}t\right] \tag{8.7}$$

where T is the observation time in seconds and p_{A} is the instantaneous value of the A-weighted sound pressure in pascals.

When determined for the noise exposure time equal to the normalised working time (i.e. an 8-hour working day or the working week), the A-weighted equivalent continuous sound level is known as the A-weighted noise exposure level normalised to an 8-hour working day (daily noise exposure level), $L_{\mathrm{Ex,8h}}$, or the A-weighted noise exposure level normalised to a nominal working week (weekly noise exposure level), $L_{\mathrm{Ex,w}}$, and it is expressed by the formulas (in dB):

$$L_{\mathrm{Ex,8h}} = L_{\mathrm{Aeq},T_e} + 10\log\frac{T_e}{T_0} \tag{8.8}$$

$$L_{\mathrm{Ex,w}} = 10\log\left[\frac{1}{5}\sum_{i=1}^{n}10^{0{,}1(L_{\mathrm{Ex,8h}})_i}\right] \tag{8.9}$$

where L_{Aeq,T_e} is the A-weighted equivalent continuous sound level determined for T_e in decibels, T_e is the effective duration of the working day, T_0 is the reference duration = 8 hours = 28,800 seconds, i is the subsequent working day in the analysed week, and n is the number of working days in the analysed week, which may be different than five.

Instead of the noise exposure level normalised to the working day or week, the daily or weekly noise exposure can be applied. This quantity is the acoustic energy dose supplied to the organism during a defined time interval and is expressed by the formula (in Pa²·s):

$$E_{A,T_e} = \int_0^{T_e} p_A^2(t)\mathrm{d}t \tag{8.10}$$

where p_A is the instantaneous value of A-weighted sound pressure in pascals and T_e is the effective duration of the working day in seconds.

The relationship between noise exposure and the noise exposure level is as follows:

$$E_{A,T_e} = 1.15 \cdot 10^{-5} \cdot 10^{0,1L_{Ex,8h}} \tag{8.11}$$

The exemplary noise exposure values and the corresponding noise exposure level values are given in Table 8.5.

The basic quantities needed to assess employee exposure to noise in the work environment are the noise exposure level normalised to an 8-hour daily or weekly working time, the corresponding daily or weekly exposure to noise, the maximum

TABLE 8.5
Relationship between Noise Exposure (*E*) and Noise Exposure Level (*L*)—Exemplary Values

E_{A,T_e} (Pa²·s)	E_{A,T_e} (Pa²·h)	$L_{Ex,8h}$ (dB)	E_{A,T_e} (Pa²·s)	E_{A,T_e} (Pa²·h)	$L_{Ex,8h}$ (dB)
1.15×10^3	0.32	80	14.5×10^3	4.03	91
1.45×10^3	0.40	81	18.2×10^3	5.07	92
1.82×10^3	0.51	82	22.9×10^3	6.39	93
2.29×10^3	0.64	83	28.9×10^3	8.04	94
2.89×10^3	0.80	84	36.4×10^3	10.12	95
3.64×10^3	1.01	85	45.8×10^3	12.74	96
4.58×10^3	1.27	86	57.6×10^3	16.04	97
5.76×10^3	1.60	87	72.6×10^3	20.19	98
7.26×10^3	2.02	88	91.3×10^3	25.42	99
9.13×10^3	2.54	89	115×10^3	32.00	100
11.5×10^3	3.20	90	–	–	–

Source: PN-N-01307. 1994. Noise. Admissible noise values in the working environment: Requirements relating to measurements.

A-weighted sound level, and the peak C-weighted sound level. These quantities also help to determine the admissible values for noise exposure in accordance with European and international regulations and standards.

Other quantities of acoustic phenomena that must be determined for effective noise control are as follows:

- Speed of sound (c), that is, the velocity of acoustic wave propagation, is the rate at which the acoustic wave travels through the medium. For instance, in air at a temperature of 20°C under normal atmospheric pressure, the acoustic wave velocity is 340 m/s.
- Particle velocity (v) is the velocity of oscillation of the medium particles while the acoustic wave passes through (usually below 1 m/s).
- Period of oscillation (T) is the smallest time interval after which the same conditions of the phenomenon (oscillation or disturbance) is repeated.
- Frequency (sound frequency, f) is the number of vibration periods in a unit of time.
- Wavelength (λ) is the distance between two subsequent points measured in the direction of propagation of a disturbance in which the vibrations are in phase (i.e. the distance covered by the wave front during a single period). The acoustic wavelength is expressed by the formula: $\lambda = c/f$, in metres, where c is the sound speed in metre per second and f is the frequency in s^{-1}. For the range of audible frequencies (f = 16–16,000 Hz), the acoustic wavelengths are λ = 21–0.021 m, respectively.

Noise is usually the sum of a large number of sinusoidal vibrations. The frequency distribution of these vibrations is called the frequency spectrum of noise.

The propagation of an acoustic wave in a medium is associated with the transmission of acoustic energy, which has the following quantities:

- *Sound power of the source*: The measure of the amount of energy (in watts [W]) radiated by the source in a unit of time
- *Sound intensity*: A vector quantity, defined as the value of the stream of energy (in watts per metre squared) flowing through the surface area in the same direction as the acoustic velocity at this point

As in the case of sound pressure, a logarithmic scale and decibels are used due to the wide range of changes in sound power and sound intensity. Acoustic power level and sound intensity level are used to describe the transmission of acoustic energy. The sound power level is the basic quantity of noise emission from the source. It is therefore used for the assessment of noise generated by machines. It is usually the A-weighted sound power level.

Apart from the terms and quantities discussed, other terms related to various methods of noise classification should also be familiar.

Noise is divided according to its influence on people into *annoying*, which has no permanent effects on the human body, and *harmful*, which has permanent effects or poses specific risks to humans. It can be steady or variable (interrupted) in time.

For example, impulse noise is variable in time, as it rises sharply to high levels and is of a short duration.

Based on the frequency range, noise is categorised as follows:

- *Infrasonic noise*, with a spectrum of components of infrasound frequencies from 1 to 20 Hz
- *Audible noise*, with a spectrum of components of audible frequencies from about 16 to 16,000 Hz
- *Ultrasonic noise*, with a spectrum of components of high audible frequencies and low ultrasound frequencies, that is, from 10 to 40 kHz

There are other noise classifications as well, based, for instance, on its causes or sources. For example, aerodynamic noise is caused by a flow of air or other gases, while mechanical noise results from friction and impacts of solids, mainly machine parts. Noise can also be classified by the environment in which it is observed. Noise in an industrial environment is known as industrial noise; noise in residential areas, public spaces, and recreational areas is known as community noise; and noise generated during transport is known as transport noise.

8.3.2 Effects of Noise on Humans

The adverse effects of noise on humans are associated mainly with the hearing organ, however, nonauditory effects can also be observed in the entire body. The effects of noise on hearing are the best known and documented, whereas nonauditory effects have not been fully examined.

8.3.2.1 Effects of Noise on the Hearing Organ

At low sound levels, hearing is limited by the *hearing threshold* (Figure 8.11). Sound pressure level values that just exceed the threshold are the barely-heard tones of specific frequencies. When the hearing organ is exposed to excessive noise, the hearing threshold is raised. This effect may be *reversible*, a temporary threshold shift, or *permanent*, after many years of dangerous exposure, and may occur in different frequency ranges. In both cases, the symptom is difficulty in hearing. A temporary threshold shift diminishes after the noise ceases; a permanent threshold shift is irreversible and does not show any recovery over time.

A temporary threshold shift starts after sufficiently long exposure to noise at more than 75–80 dB (A) and increases quickly with an increased noise level. The increase is in both the size of temporary threshold shift in decibels and the recovery time needed to restore normal hearing. Subjectively, temporary threshold shift is considered a temporary loss of hearing sensitivity. It may occur due to hearing fatigue after long exposure to noise or due to an acute acoustic trauma caused by a significant sudden or impulsive noise. A temporary threshold shift is the first sign that the surrounding noise may pose a threat to hearing.

A permanent threshold shift is an irreversible damage to the hearing organ. The risk of permanent hearing damage begins when the sound level exceeds 80–85 dB (A). Weaker stimuli do not damage the hearing organ even after long and continuous exposure.

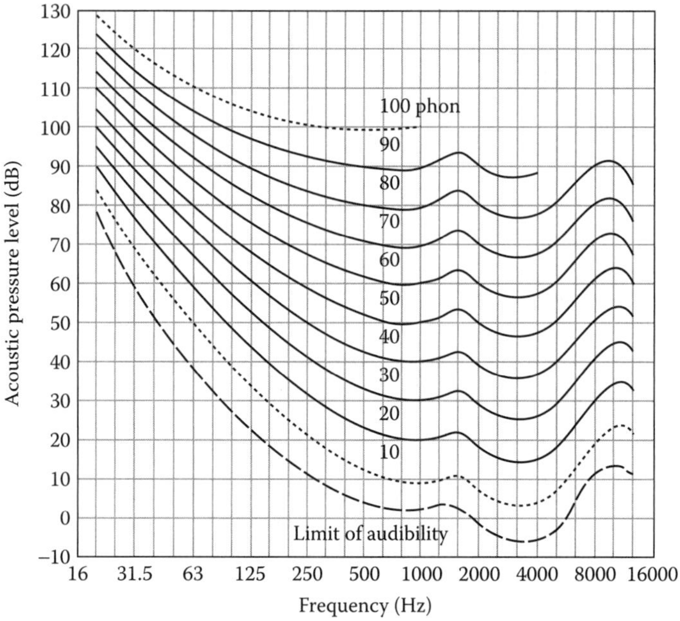

FIGURE 8.11 The area of audibility.

Hearing damage is usually an effect of long-term exposure to noise, as the effect on the hearing organ is accumulated over time. Hearing loss is assumed to be related to the amount of acoustic energy received by the hearing organ within a specified time interval. Moreover, continuous exposure to noise is more harmful than interrupted exposure, because hearing regeneration can begin even during short interruptions.

Elevation of the hearing threshold is diagnosed using audiometric tests, which can identify hearing threshold changes of only several decibels that may not be noticed by the patient. Small changes do not noticeably impair hearing ability, understood as a decline in speech perception ability. These changes are typical in the early period of noise-induced hearing damage, which advances slowly and consistently. However, such changes in the hearing threshold also can be caused by various other factors, such as biological ageing (called presbycusis) or diseases of the hearing organ.

Audiometric tests also reveal the development of permanent hearing loss (Figure 8.12). The type of hearing loss depends to a great extent upon its cause. In chronic noise exposure, hearing loss often begins in a high frequency range (4–6 kHz) and then extends to lower frequencies.

Bilateral noise-induced permanent hearing loss is expressed by an increase in the hearing threshold by at least 45 dB in the better ear. It is calculated as the arithmetic averages for audiometric frequencies of 1, 2, and 3 kHz and is a criterion for identifying and diagnosing occupational hearing loss.

Impulse noise, in which the sound pressure increases quickly to high levels, is particularly destructive to hearing. Research has shown that hearing damage induced by

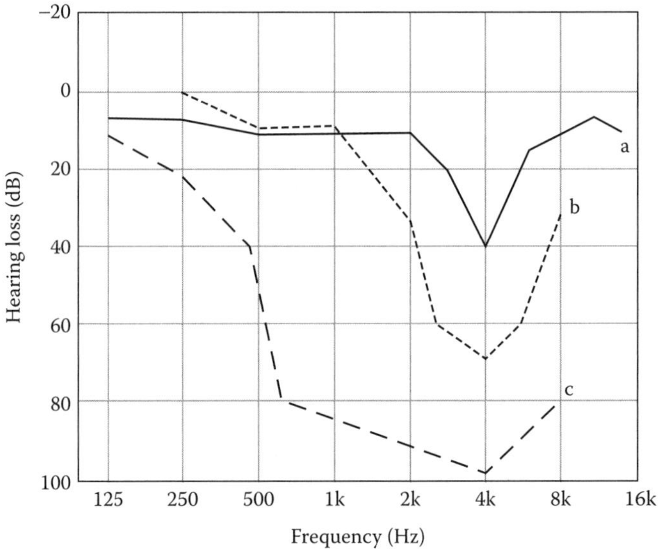

FIGURE 8.12 Examples of hearing loss: (a) mild hearing loss (——), (b) moderate hearing loss (- - - -), and (c) extensive deafness (– – – –).

impulse noise depends upon the peak sound pressure. The hearing organ's protective mechanisms, such as the stapedius reflex (acoustic reflex), have a high level of inertia (a delay of about 100 ms), and thus cannot limit the amplitude and acoustic energy of impulse noise. This may lead to a substantial overload of the hearing organ. A combination of impulse and steady noise of a high sound pressure level is also considered to be particularly damaging to the hearing organ.

Medium- and high-frequency noise is more dangerous than noise with a low-frequency maximum energy. This is due to the frequency-dependent sensitivity of the ear, which is the highest in the frequency range of 3–5 kHz (Figure 8.11). Individual susceptibility to noise varies depending upon genetic factors and past diseases.

Hearing damage may also be due to sudden damage to the hearing organ resulting in acoustic trauma or acute acoustic trauma. A sudden noise in a broad frequency range causes acoustic trauma, and a sudden increase in the noise level to above 130–140 dB causes acute acoustic trauma. Tones of specific frequencies above 1–2 kHz have been found to be dangerous and increase the risk of hearing loss. These frequencies are caused by the development of electronic sound generation devices, which often use tonal sounds.

Acoustic trauma is often caused by shots and explosions. Despite having a short exposure time, shots of a peak sound pressure level exceeding 160 dB can lead to permanent changes in hearing. Noise causes irreversible damage mainly to the hair cells of the inner ear. In extreme cases, explosions resulting in a shock wave of a peak sound pressure level exceeding 180 dB may cause perforation and defects of the tympanic membrane and damage to the auditory ossicles in the middle ear.

Acoustic trauma always causes a temporary shift in the hearing threshold, which, in the case of acute trauma, may lead to a permanent hearing damage. The hearing threshold elevation may be symmetric or it may dominate one ear, depending upon the type of exposure. Unilateral threshold elevation in one ear may occur due to a directional noise impact, which affects mainly the ear on the side of the noise source. This is usually impulse noise from a source positioned laterally to the employee.

An abnormal perception of loudness is a common effect of hearing damage. When the hearing threshold is elevated due to substantial hearing damage, sounds at a level of 50–60 dB fall below the hearing threshold or are barely audible, whereas the loudness of high-level sounds (100–110 dB), at the so-called hearing discomfort threshold, does not change; sounds at this high level produce same loudness in persons with hearing impairment and those with normal hearing. Thus, a hearing loss at a level of about 50 dB results in a double-narrowing of the range of audible sounds from about 100 to 50 dB. Narrow range of audible sounds in persons with hearing loss results in loudness recruitment, that is, a very quick loudness increase with an increase in the sound pressure level. Loudness recruitment results in hindered speech perception because pronouncing consonants and vowels is associated with a large change in the instantaneous sound level. A person suffering from hearing loss cannot hear some of the silent consonants. After increasing the signal intensity, they cannot tolerate the sound of vowels, which become too loud. Loudness recruitment is partially balanced by the use of compressing amplifiers in hearing aids (amplification decreases as the signal level from the microphone increases).

The undamaged hearing system is a sound spectrum analyser with high frequency selectivity. For instance, at 1 kHz frequency, we can distinguish between tones that differ by 0.2 Hz. Hearing damage results in the loss of frequency selectivity and a substantial increase in the mutual masking of sounds, even if their frequencies are very different. This increase in masking makes it impossible to perceive speech when many people are speaking at the same time. Due to a lack of frequency selectivity and increased masking, persons suffering from hearing loss prefer conversation with no background noise and with a single interlocutor, because they cannot understand speech under noisy conditions or when many people are talking simultaneously.

Hearing damage, particularly unilateral, may result in the same sounds being heard differently by the healthy and the damaged ear. Persons with hearing loss cannot differentiate between sounds generated quickly one after another (i.e. changes of sound in time), which limits their speech intelligibility. Finally, both unilateral and asymmetric hearing loss may result in a limited localisation of sound sources. Currently, there are no methods to improve the frequency selectivity of the damaged hearing organ, to improve the perception of sound in time, to eliminate increased masking, to restore proper sound localisation, or to limit disturbances in two-ear hearing.

Hearing loss is not the only consequence of the noise overload of a hearing system. Exposure to noise may also result in a substantial increase in the frequency and intensity of *tinnitus*. Tinnitus is the occurrence of whistling and buzzing sounds, usually of a frequency above 1 kHz, which are not physically present but are subjectively heard and are generated at various levels of the hearing system. Tinnitus is intensified in persons older than 50 years. Exposure to excessive noise also causes *hyperacusis*, that is, hypersensitivity to sound. A person with hyperacusis finds it difficult to tolerate sounds and perceives them as irritating even if they are soft by normal standards.

Permanent cochlear hearing loss resulting in a loss of the ability to understand speech (i.e. the social hearing capability) is a permanent disability for which there is no rehabilitation. This has long topped the list of occupational diseases.

8.3.2.2 Nonauditory Noise Effects

Nonauditory noise effects are not yet fully recognised. The anatomical connection between the hearing nerve tract and the cortex allows hearing stimuli to influence other brain centres and the endocrine system and, as a result, the condition and functioning of many internal organs. Sound stimuli may thus influence all the functions of an organism. This has been confirmed by reflexive actions of the respiratory system, the circulatory system, the alimentary tract, and many other organs in response to noise.

Examples of nonauditory physiological responses include motor reflexes, such as muscle contractions, which change the body posture after an unexpected signal such as an explosion or a shot, and the reactions of other systems, such as the reduction of the respiratory rate, contraction of the peripheral blood vessels, and decreased intensity of intestinal peristalsis.

Experiments have shown that physiological disorders may emerge when the sound level exceeds 75 dB. Strong acoustic stimuli exceeding 110–120 dB may impact the functioning of the sensory organs, for instance, vision, balance, or tactile sensation. Weaker acoustic stimuli of 55–75 dB may result in the loss of concentration, making work difficult and decreasing efficiency. Thus, nonauditory noise effects are a general response to noise as a stressor.

8.3.3 Measurement and Assessment of Quantities of Noise in the Work Environment

Based on the purpose of the assessment, noise measurement methods are divided into

- Methods of measurement of noise emission by machines
- Methods of measurement of noise in places occupied by people (at workstatizons)

8.3.3.1 Machine Noise Measurement Methods

These methods are used to determine the quantities of machine noise emission, analysed as separate noise sources under established experimental and operational conditions (Augustyńska et al. 2000; Engel and Pleban 2001). Pursuant to the European directives and the corresponding domestic regulations, these quantities include the emission sound pressure level at a workstation or at other specified positions and the sound power level. Selection of these quantities depends upon the noise emission values. The sound power level should be determined when the averaged A-weighted emission sound pressure level of emission at the machine workstation exceeds 85 dB.

Methods for determining the sound power level based on the measurement of sound pressure levels or sound intensity are specified in the appropriate Polish series of standards PN-EN ISO 3740 and PN-EN ISO 9614. Methods for determining emission sound pressure levels are specified in the series of standards PN-EN ISO 11200.

These methods differ mainly in their attainable accuracy in various research environments, such as special reverberation rooms and anechoic rooms or ordinary use rooms. They can be applied for acoustic tests of machines within the framework for assessment of their compliance with EU directives. They can also be used by designers and manufacturers to determine noise emission values and by users to verify the noise emission declarations.

8.3.3.2 Methods of Measurement and Assessment of Noise at the Workstation

These methods are used to determine noise exposure at workstations and in other places occupied by people. The noise measurement results are used mainly to compare the existing acoustic conditions with the requirements specified in the applicable standards and occupational health and safety regulations, as well as to assess, select, plan, and implement methods of noise reduction.

Methods for the measurement and assessment of noise at workplaces are specified in applicable regulations and standards—the regulations of the Minister of Labour and Social Policy concerning the maximum admissible concentrations and intensities in the work environment for agents harmful to the health, the regulations of the Minister of Labour on occupational health and safety as related to work associated with exposure to noise or mechanical vibrations (implementing directives 2003/10/EC and 2002/44/EC), the regulations of the Minister of Health concerning testing and measuring agents in the work environment that are harmful to the health, and the Polish standards PN ISO 9612, PN ISO 1999, and PN-N-01307.

Measurements of noise quantities can be taken during a full working day in selected typical exposure periods (the sampling method) or during the performance of specific tasks and activities. For the measurement of all types of noise (steady, unsteady, and impulse), class 1 personal sound exposure meters or class 1 or 2 integrating sound level meters should be used, fulfilling the requirements of PN-EN IEC 61252 and PN-EN IEC 61672-1. Assessment of noise exposure is based on a comparison of the measured or determined quantities of noise with the admissible values.

8.3.3.3 Admissible Values of Quantities of Noise

8.3.3.3.1 Admissible Values of Quantities of Noise for Hearing Protection

The noise exposure level normalised to a nominal 8-hour working day ($L_{\mathrm{Ex,8h}}$) should not exceed 85 dB; the corresponding daily noise exposure ($E_{\mathrm{A},Te}$) should not exceed 3.64×10^3 Pa^2·s. If the noise exposure level varies from day to day, the noise exposure level normalised to a nominal working week ($L_{\mathrm{Ex,w}}$) should not exceed 85 dB; the corresponding weekly exposure ($E_{\mathrm{A,w}}$) should not exceed 18.2×10^3 Pa^2·s.

The maximum A-weighted sound level (L_{Amax}) measured with time-weighting S (slow) should not exceed 115 dB. The peak C-weighted sound level (L_{Cpeak}) should not exceed 135 dB.

These values are applicable to all three types of noise (steady, unsteady, impulse). With regard to the averaging time value, noise can be categorised using the following methods:

- Long-term averaging during an 8-hour working day (28,800 seconds) or a working week (noise exposure level or noise exposure)
- Short-term averaging (of 1 second) of particular significance for the assessment of unsteady and short-term noise, lasting several seconds (maximum A-weighted sound level)
- Observation of the instantaneous value, that is, the peak C-weighted sound level, which is particularly significant for the assessment of impulse noise

The values provided are applicable if other detailed regulations do not define any lower values (e.g. in workstations occupied by juveniles, $L_{Ex,8h} = 80$ dB; in workstations occupied by pregnant women, $L_{Ex,8h} = 65$ dB).

8.3.3.3.2 Admissible Noise Values According to the Employee's Possibility of Fulfilling the Basic Tasks—The Criterion of Annoyance of Noise (PN-N-01307)

The equivalent A-weighted continuous sound level during the occupation of a workstation by an employee ($L_{Aeq,Te}$) should not exceed the values presented in Table 8.6. The maximum A-weighted sound level and the peak C-weighted sound level should not exceed these values.

TABLE 8.6
The Equivalent A-Weighted Sound Level A, $L_{Aeq,\,Te}$, during Occupation of the Workstation by the Employee

Workstation	Equivalent A-Weighted Sound Level $L_{Aeq,Te}$ (dB)
In direct control cabins not equipped with telephone communication means; at laboratories with noise sources; in rooms with machines and counting devices, typing machines, or teleprinters; in other rooms of similar designation	75
In dispatch, observation, and remote control cabins equipped with telephone communication means used in the process of control; in rooms for performance of precision works; in other rooms of similar designation	65
In administrative rooms, design office rooms, theoretical study rooms, data processing rooms, and other rooms of similar designation	55

Source: Data from PN-N-01307. 1994. Noise. Admissible noise values in the working environment: Requirements relating to measurements.

8.3.3.3.3 Noise Exposure Action Values—Values That Require Specific Actions (Directive 2003/10/EC and Domestic Legal Provisions)

Noise exposure action values are

- Noise exposure level: $L_{Ex,8h}$ or $L_{Ex,w} = 80$ dB
- Peak C-weighted sound level: $L_{Cpeak} = 135$ dB

The following activities should be undertaken: inform employees about noise exposure and methods of noise elimination or reduction and provide employees with personal hearing protection equipment.

8.3.4 Sources of Noise Exposure at Work

According to the Central Statistical Office data for 1995–2007, nearly 40% (about 300,000–400,000) of employees employed in hazardous conditions work under excessive noise conditions at levels exceeding 85 dB. Mining, textile manufacturing, metal manufacturing, construction, furniture production, wood production, and metal products manufacturing have the worst occurrence of excessive noise conditions. The main sources of noise at workstations are the machines and devices listed in Table 8.7.

8.3.5 Methods of Preventing, Eliminating, and Limiting Noise Exposure

State-of-the-art methods for preventing, eliminating, or limiting noise exposure are based upon the simultaneous application of technical, administrative, and organisational solutions, which are selected based on a detailed analysis of the acoustic conditions at workstations with excessive noise.

The most significant technical solutions that limit noise include the following:

- Use of low-noise technological processes.
- Mechanisation and automation of technological processes including remote and automatic control devices for noisy machines.
- Construction and application of low-noise or noiseless machines, devices, and tools.
- Proper layout of the plant and adaptation of rooms, taking into account the acoustic aspects.
- Use of noise control devices such as silencers, enclosures, screens, and sound-absorbing materials and systems (Figure 8.13).
- Use of structure-related sound insulation (vibration isolators and vibroisolated foundations of machines and devices).

The application of low-noise technological processes, as well as their mechanisation and automation, is among the most significant requirements faced by designers and managers of production processes and manufacturing plants. Manufacturers should also use low-noise machines, devices, and tools.

Only at the design and construction stage of a plant can all of the available noise control measures be used. Noise reduction after the construction of the plant and the

TABLE 8.7
The Main Sources of Noise and Exemplary Technical Solutions Limiting the Noise Level

Noise Source	A-Weighted Noise Level (dB)	Noise-Limiting Technical Solutions
Machines generating energy (combustion engines, compressors, electrical machines, transformers, and power generators)	98–130	Silencers, enclosures
Pneumatic tools and engines (drills, grinders, hammers, riveters, coarse files, and tampers)	90–120	Change of technology (precision casting, electric processing, chemical processing, mechanisation and automation)
Plastic-forming machines (presses, hammers, wire drawing machines, levelling machines, and rolling mills)	92–120	Change of technology (rolling, pressing or electric compression)
Machines for grinding, crushing, sieving, cutting, cleaning (vibrating sieves, tipping grates, cast cleaning drums, and sandblasting machines)	96–111	Automation and remote control, sound-insulating cabins
Metal machining tools (automatic lathes, wheel lathes, multicut lathes, revolving lathes, grinders, milling machines, and drills)	92–105	Change of technology (electrochemical processing)
Wood machining tools (chainsaws, shapers, milling machines, circular saws, and band saws)	92–108	Use of robots, enclosures
Textile industry machines (looms, spinning machines, yarn-twisting machines, stretcher machines, winders, lace machines, cards, knitting machines, and hosiery-making looms)	93–114	Enclosures, sound-absorbing treatments, automation, introduction of shuttleless loom technology
Turbomachines (fans, valves, reducing gears, jet pumps, burners, injectors, and oven nozzles)	98–120	Silencers, enclosures
Internal transport devices (conveyors, feeders, distributors, and cranes)	98–112	Enclosures, automation

installation of machines and devices or during the full-scale operation stage is not only much more difficult, but also much less efficient, leading to only a slight reduction of noise at a relatively high cost. Moreover, the necessary alterations or adaptations are not always favourable to the technological process applied.

Table 8.7 presents exemplary technical solutions applicable to the machines, devices, and tools that are the main sources of noise at workstations. These include technological changes, mechanisation, automation, and use of sound protection devices, such as silencers and enclosures.

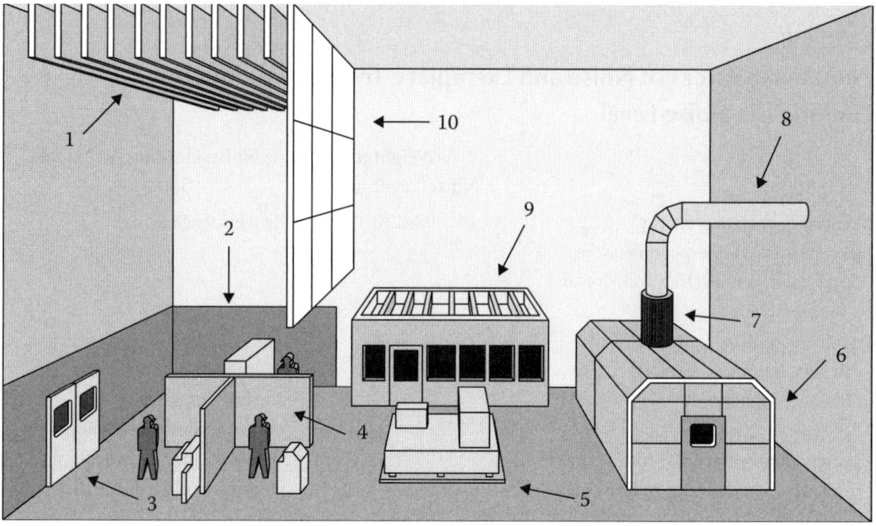

FIGURE 8.13 Noise control measures at work stations: (1) absorbers, (2) sound-absorbing materials, (3) soundproof doors, (4) screens, (5) vibration isolation, (6) enclosure, (7) silencer, (8) sound isolated pipeline, (9) sound-insulating cabin and (10) sound-absorbing system. (Adapted with permission from Kaczmarska A. 1997. Methods of low-frequency noise reduction in industrial cabins. Ph.D. disertation. Warsaw: CIOP.)

8.3.5.1 Technical Means of Noise Limitation

8.3.5.1.1 Replacement of a High-Noise Technological Process with a Low-Noise Process

The loudest production processes can often be replaced with those that are less noisy, for example, hammering may be replaced by rolling and pressing, while manual tool processing may be replaced by electrical and chemical processing or by the use of mechanical tools.

8.3.5.1.2 Mechanisation and Automation of Technological Processes

The mechanisation and automation of technological processes and sound-isolating control cabins for the service personnel are among the most modern and effective methods for eliminating exposure to noise, vibration, and other harmful agents (including dusting, high temperatures, injuries). Most cabins used in industry ensure noise reduction at a level of 20–50 dB within a frequency range above 500 Hz (Kaczmarska and Augustyńska 1999).

8.3.5.1.3 Construction and Application of Low-Noise Machines, Devices, and Tools

The alteration of technological processes and the introduction of mechanisation and automation require long periods for implementation and cannot be applied to small-scale or nonstandard production processes. Reducing noise generated by devices and tools is a very efficient method for eliminating noise sources. Sound emissions can be limited using the following methods (Engel 2001):

- Reducing excitation (i.e. minimising vibration excitation forces and limiting their spectrum), for example, by a careful balance of machine components, a change in the rigidity and structure of the system, or a change in friction resistance values.
- Changing the aerodynamic and hydrodynamic conditions, for example, by the alteration of the geometry of inlets and outlets of power utilities and their flow rates.
- Reducing the radiation efficiency coefficient, for example, by the alteration of the dimensions of the components radiating vibroacoustic energy, a change of materials, or the isolation of plates in the system.

8.3.5.1.4 Appropriate Planning, Arrangement, and Installation of Machines and Equipment

Designers of industrial plant buildings should follow these rules:

- Buildings and rooms in which silence is required (e.g. labs, offices, concept work rooms) should be separated from buildings and rooms in which noisy production processes take place.
- Machines and devices should be grouped according to the degree of noise generated, in separate rooms whenever possible.

Noise in a given room may be intensified by an improper room layout, including an excessively dense arrangement of machines. The recommended minimum distance between machines is 2–3 m.

8.3.5.1.5 Silencers

Acoustic silencers decrease noise in the ducts through which air or gas flows (ventilation systems, inlet and outlet systems of turbomachines, such as compressors, blowers, turbines, and combustion engines; Figure 8.14; Engel et al. 1993).

Modern acoustic silencers do not reduce machine capacity. They establish a substantial resistance to highly noisy, unsteady flows, while allowing for steady flows that ensure the transport of air or gas. Popular silencers of this kind include reflective silencers, that is, acoustic wave filters, as well as absorption silencers containing sound-absorbent materials.

Reflective silencers reflect and interfere with acoustic waves and work well at low and medium frequencies. They are applied at locations in which flow rates and temperatures are high, for example, in combustion engines, blowers, compressors, and sometimes in fans.

Absorption silencers prevent the transmission of acoustic energy along the ducts by absorbing most of it through sound-absorbent materials. They work well at medium and high frequencies and are widely used in ventilation ducts. In practice, often two types of silencers should be applied together, since many industrial noise

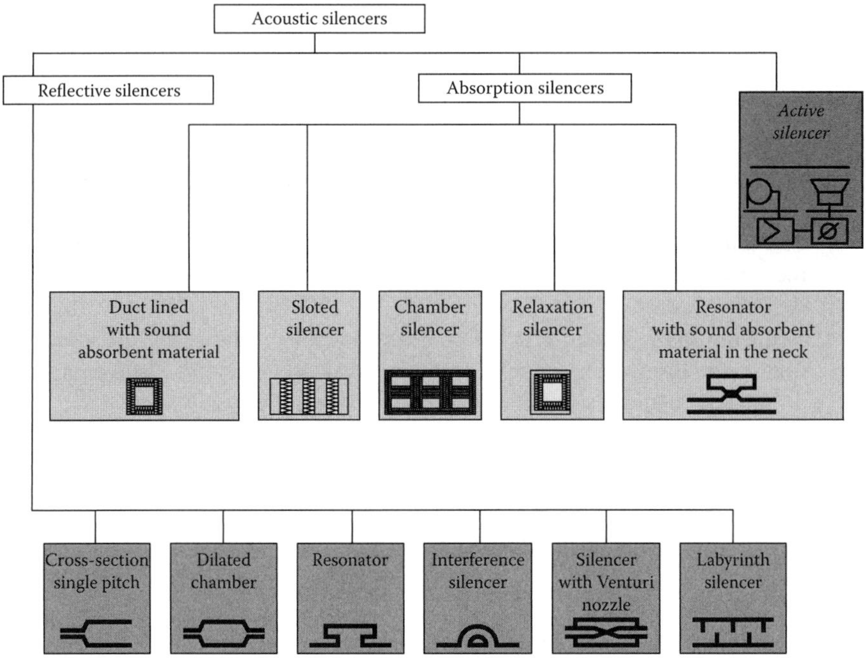

FIGURE 8.14 Classifications of acoustic silencers. (Adapted with permission from Engel, Z., J. Sadowski, B. Szudrowicz, and I. Żuchowicz-Wodnikowska, eds. 1993. *Guidelines for Computer-Aided Sound Protection of Work Stations in Industrial Halls.* Warsaw: CIOP.)

sources emit energy in a broad frequency band encompassing both the infrasound and audible range.

Active silencers are a separate group of silencers and are discussed in Section 8.3.5.3.

8.3.5.1.6 Enclosures

Noise can be eliminated at its source by enclosing the noise-generating machine in whole or in part (Figure 8.15). The enclosure of the machine should suppress sounds emitted by the source of noise as effectively as possible; at the same time, it should not hinder the normal work and handling of the machine. Machine enclosures are usually equipped with sound-absorbing isolation walls made of steel sheets lined with suppression or sound-absorbent materials. Enclosures with multilayered walls are also used.

Properly constructed enclosures can reduce the A-weighted sound level by 10–25 dB. Partial enclosures are much less effective, amounting to a reduction of only about 5 dB. Ventilation holes and other openings, which are necessary for technological purposes, also reduce the effectiveness of enclosures. Ventilation holes should be equipped with an appropriate silencer, for example, lining the duct with a sound-absorbing material.

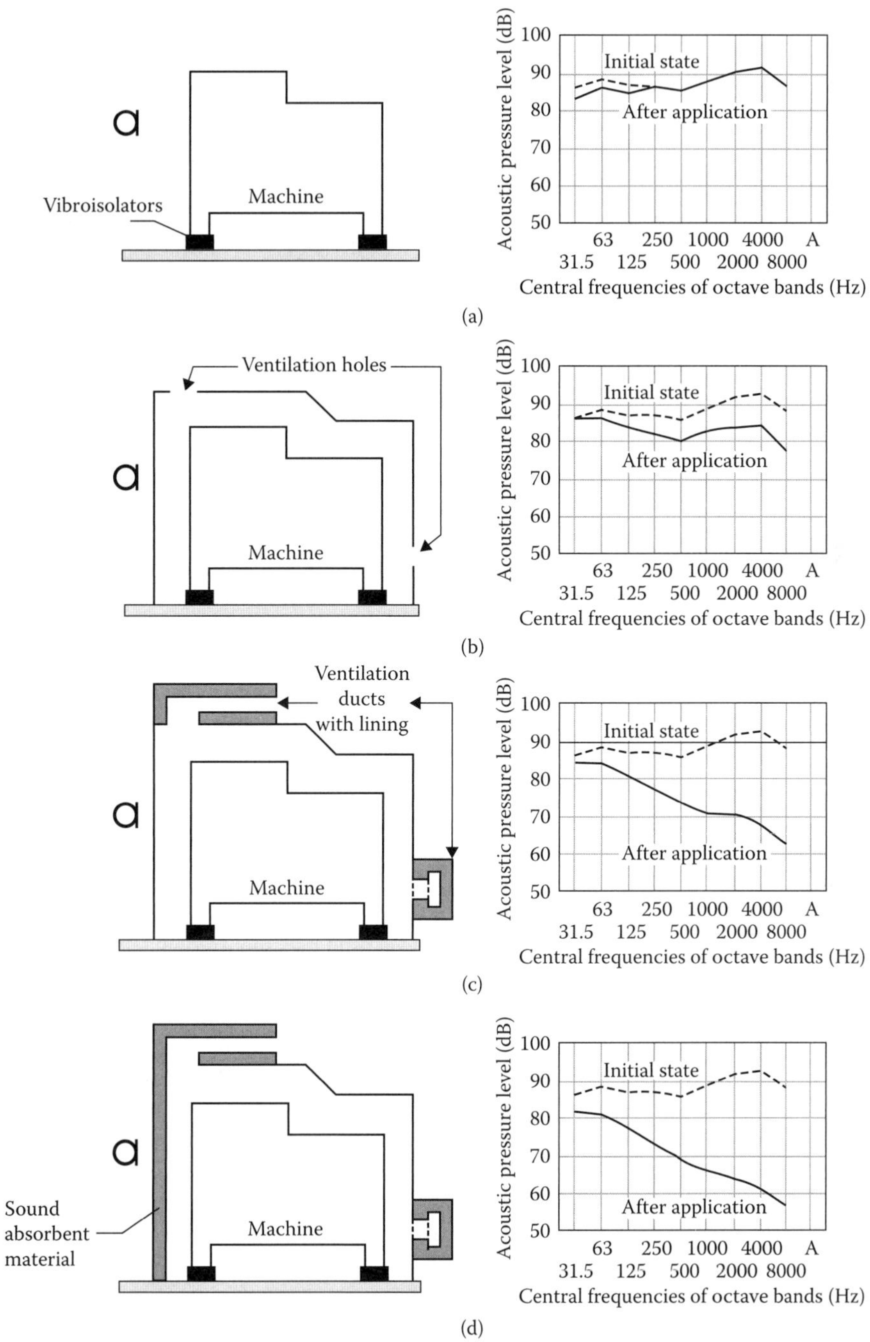

FIGURE 8.15 Typical noise reduction achieved for different machine and enclosure configurations: (a) source of sound in open space, (b) a screen in open space, (c) a screen in a room with a sound-reflective ceiling, and (d) a screen in a room with an acoustically-adapted ceiling. (Reprinted with permission from PN-EN ISO 11690-2.)

8.3.5.1.7 Screens

Screens are used to shield selected workstations in order to suppress the noise emitted by other machines as well as the noise generated by the workstation and emitted to the outside. To ensure maximum efficiency, the screen should be placed as close as possible to the source of noise or the workstation.

The basic components of a screen are the inner isolation layer (usually a metal sheet of proper thickness) and the outer sound-absorbent layers (mineral wool or glass wool panels, covered with perforated metal sheets). While using a screen in a room, the entire acoustic system should be taken into account to make sure the screen interacts with other components responsible for the reduction of the energy of reflected waves (such as sound-absorbing materials and systems).

Properly used screens reduce noise by about 5–15 dB at a distance of about 1.5 m behind the screen, along the axis perpendicular to its surface.

8.3.5.1.8 Sound-Absorbing Materials and Systems

Sound-absorbing materials and systems applied to the room walls and ceiling increase sound absorption in the room. The reflected wave sound level is thus reduced, resulting in the general reduction of the noise level in the room.

The most commonly used sound-absorbing materials are porous materials, which include textiles, minerals and glass, wool materials and mats, porous panels and wall coatings, plastic porous plates and mats, and coatings applied under pressure. The sound-absorbing materials or systems should be selected in such a way that the maximum sound absorption coefficients are within the frequency ranges to which the maximum noise spectrum components belong. In practice, only rooms with a low original absorption can properly suppress noise, by 3–7 dB. Ready-to-use sound-absorbing systems, such as ceilings, division walls, panels, and screens, are currently available on the market.

8.3.5.1.9 Administrative and Organisational Solutions

Technical solutions to eliminate or limit noise should be supported by administrative and organisational activities. These include mainly the following:

- Applying work breaks and limiting working time at the noisy workstations: This method is not very effective, as the reduction of the exposure time by half decreases the noise exposure level by only 3 dB. Giving employees breaks from work with noise exposure is more effective, as it allows the employees to rest away from high levels of noise and reduces the discomfort of using hearing protectors.
- Arranging so-called silence oases, or soundproof resting rooms, for employees operating noisy equipment.
- Planning so that tasks generating a high noise level are performed during the second and third shifts, thus limiting the number of persons exposed to the noise.
- Conducting preventive medical tests, including hearing tests for all newly hired employees and periodical control tests for all employees exposed to noise; general tests conducted every 4 years; otolaryngological and

audiometric tests conducted once a year for the first 3 years of work under noise exposure, and then every 3 years.
- Relocating persons who are particularly sensitive to vibrations and those diagnosed with hearing loss to other, less arduous job locations.

However, the basic organisational and technical solution is the automation of production processes, which allows employees to stay outside the zone of exposure to noise and other harmful agents.

8.3.5.2 Hearing Protectors

The use of hearing protectors is the simplest and quickest way to protect the hearing organ against the effects of noise exposure. However, hearing protectors should be used only if all available technical and organisational methods for limiting noise at the workplace have been applied and the permissible values of noise are still exceeded.

Hearing protectors can be divided into noise reduction earmuffs and earplugs. Earmuffs can be worn separately or fixed to protective helmets. Earplugs can be disposable or reusable. Noise-reduction earplugs can be factory-made, formed by the user, or customised by the manufacturer for the needs of individual users, taking into account specific features of the employee's external ear canal. Hearing protectors can be equipped with electronic systems, which are divided into three groups: hearing protectors with active noise reduction, level-dependent hearing protectors, and hearing protectors equipped with wired or wireless communication systems.

Properly selected hearing protectors protect the hearing organ so that the A-weighted sound level under the protector should not exceed 80 dB. Hearing protectors for specific workplaces are selected based on sound pressure levels in octave frequency bands or A-weighted and C-weighted sound levels measured for these workplaces, as well as the acoustic parameters of the hearing protectors, such as sound attenuation and H, M, L and the signal-to-noise ratio (Kozłowski 2007).

In addition to the proper selection of hearing protectors, the employer must educate employees on the proper use of these protectors in accordance with the information provided by the manufacturer. This information includes instructions for fitting the hearing protectors and recommendations regarding the duration and method of their use and storage.

8.3.5.3 Active Noise Control

Passive methods of noise reduction are expensive and less effective when applied to low-frequency noise. Active noise reduction methods, which are increasingly popular, can be used for such low-frequency noise. Active methods use additional sound sources (e.g. loudspeakers), referred to as secondary sources, to reduce the noise generated by primary noise sources. Secondary sources should be controlled to generate a secondary acoustic wave of the same amplitude and a reversed phase in relation to the primary acoustic wave. The interference of this secondary acoustic wave with the primary acoustic wave leads to noise level reduction (Figure 8.16).

Accurate control of the secondary source is very important in order to achieve sufficient noise reduction. Therefore, information on the noise signal and other parameters of the active noise reduction system (e.g. the transfer functions of individual

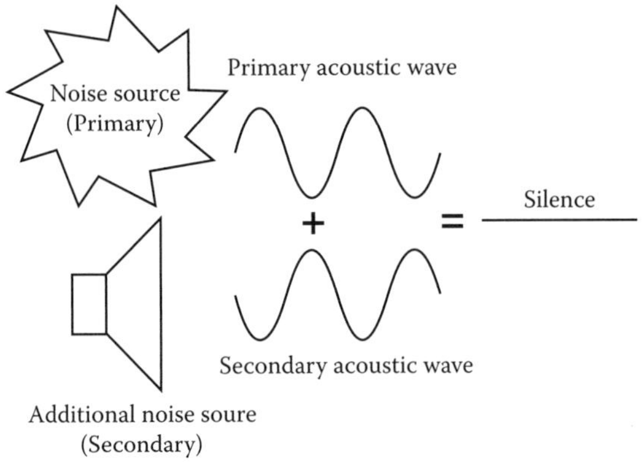

FIGURE 8.16 The principles of active noise reduction.

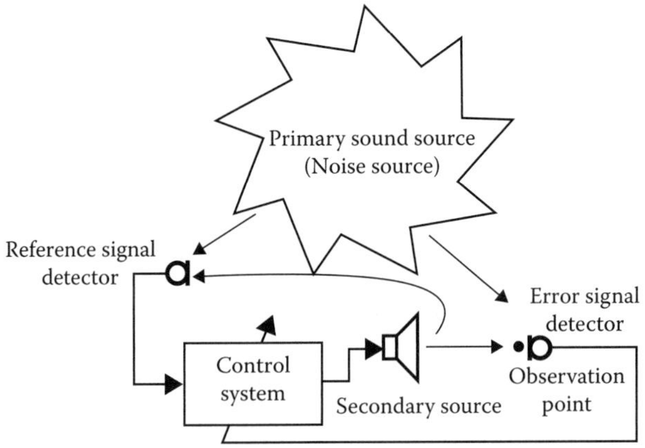

FIGURE 8.17 Diagram of an active noise reduction system.

acoustic paths of the system) should be taken into account while controlling the secondary source. A diagram of a basic active noise reduction system is shown in Figure 8.17. The purpose of this system is to reduce noise at a certain point in space, known as the observation point. Information on the noise reaches the system controlling the secondary source from the reference sensor (e.g. a microphone). The remaining information, such as the characteristics of the secondary source or the characteristics of the acoustic path between the secondary source and the observation point, should be determined using experimental methods or calculations. This is often very difficult, for instance, when these characteristics change over time; therefore, an adaptive system is usually used to control the operation of the secondary source.

The adaptive system automatically calculates all of the necessary information on the active noise reduction system and uses it in the control process; however, it requires an additional measuring component (an error sensor), which provides the control system with data on the effects of its operation, enabling the modification of the control system parameters.

Active noise reduction systems may be single-channel systems (containing one secondary source, a reference sensor, and an error sensor) or multichannel systems (containing many secondary sources, error sensors, and reference sensors). A processor-based electronic digital system, which operates as an adaptive digital filter, is an example of a typical control system. The signal that controls the secondary source is obtained by the digital filtration of the signal from the reference sensor.

Specific applications for active noise reduction systems, which are currently proposed only as customised solutions adapted to specific needs, include the following:

- Reducing noise in waveguides: For example, ventilation systems (Figure 8.18) and outlet structures of combustion systems can attain a substantial level of noise reduction, up to 30–40 dB or more. Depending upon the perpendicular dimensions of a waveguide and the applied active reduction system solution, the range of operation of the system may increase to exceed 1 kHz.
- Increasing the attenuation of hearing protectors for low-frequency noise: Applying miniature microphones and speakers controlled by analogue circuits inside the hearing protector cups increases attenuation by an additional 10–20 dB within a frequency range below 500 Hz.
- Establishing silence zones around selected areas: Silence zones can be established, for example, surrounding the employee (machine operator) or the head of an airplane passenger (active headrest). The size of the silence zone depends upon the noise frequency, the number of secondary sources, and their distribution. The noise reduction level attained ranges from several to more than 20 dB.

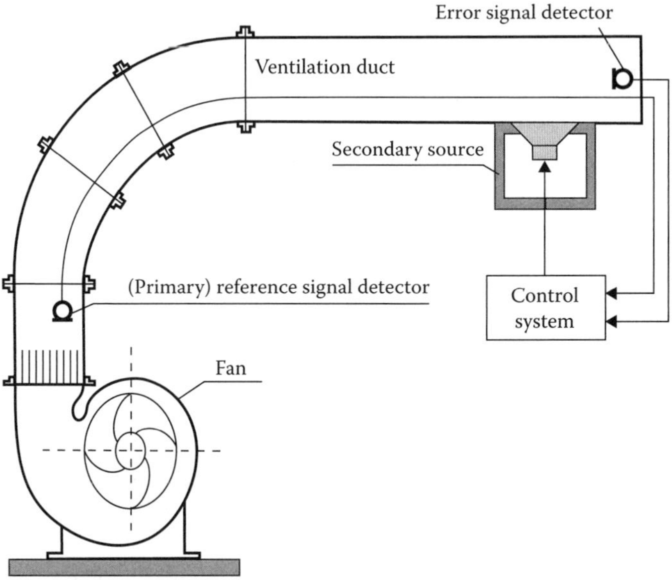

FIGURE 8.18 Active noise reduction system in a ventilation system. (Reprinted with permission from Engel, Z., G. Makarewicz, L. Morzyński, and W. M. Zawieska. 2001. *Active Noise Reduction Methods*. Warsaw: CIOP.)

- Reducing the noise generated by power transformers: The noise generated by a power transformer can be reduced along the selected propagation directions. Depending upon the number and distribution of secondary sources, the noise reduction level attained may range from several to more than 10 dB.
- Reducing the warning signal level inside priority vehicles at the drivers' ears: The use of light, active noise reduction earphones almost completely eliminates the noise reduced by the signalling system. The driver also retains the ability to receive acoustic signals from the road, which is significant for traffic safety.

More detailed descriptions of the methods of noise control can be found in the extensive literature on this subject (Crocker 2007; Engel 2001; Engel and Sikora 1998; Engel et al. 2001; Kaczmarska and Augustyńska 1999; Kaczmarska et al. 2001; Augustyńska and Zawieska 1999; Ver and Beranek 2006).

8.3.6 Infrasonic Noise

8.3.6.1 Characteristics of Infrasonic Noise

Infrasonic noise (infrasound) is the sound or noise of a spectrum within the frequency band of 1–20 Hz (PN ISO 7196, PN-N-01338). Infrasounds, despite the common belief that they are inaudible, are received by the human body by a specific auditory tract, mainly by the hearing organ. Their audibility depends on the sound pressure level. However, humans have a high level of individual variability with regard to hearing perception, particularly at the lowest frequencies. The higher the infrasound hearing thresholds, the lower is their frequency; for instance, for a frequency of 6–8 Hz, the infrasound hearing threshold is about 100 dB, and for the frequency of 12–16 Hz, it is about 90 dB. The average value of the G-weighted hearing threshold (PN ISO 7196) results in a hearing perception level of 102 dB.

Apart from the specific auditory tract, vibration receptors also perceive the infrasounds. Here, the perception thresholds are about 20–30 dB higher than the hearing thresholds. When the sound pressure level exceeds 140–150 dB, infrasounds may lead to permanent, harmful changes in the human body. A resonance of the internal structures and organs of the body may emerge, subjectively sensed from 100 dB as an unpleasant feeling of internal 'vibration'. Along with 'ear squeeze', this is one of the most typical symptoms of persons exposed to infrasounds. Occupational exposure to infrasounds mainly causes an *annoying effect*, even if the hearing threshold is only slightly exceeded. This effect is characterised by a subjective feeling of excessive fatigue, discomfort, drowsiness, and disturbed balance, as well as disorders of psychomotor performance and physiological functions. Research shows that these symptoms are temporary and regress after the infrasound source is eliminated. Many authors propose applying the hearing perception level as the criterion for the annoyance of infrasounds in the work environment.

The main sources of infrasonic noise in the work environment, particularly in industry, are low-speed turbomachines (compressors, fans, engines), power devices (mills, boilers, funnels), smelting furnaces (particularly arc furnaces), and foundry equipment (moulding machines, tipping grates).

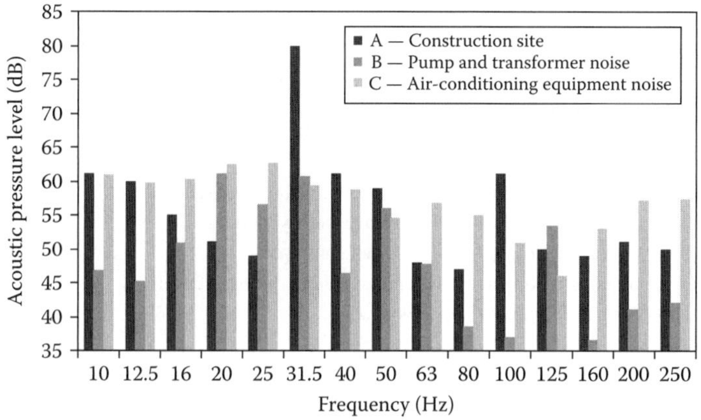

FIGURE 8.19 Noise spectrums at an office workstation. (Reprinted with permission from Zawieska, W. M., ed. 2007. *Occupational Risk. Methodological Principles of Assessment.* Warsaw: CIOP-PIB.)

Infrasonic noise sources are also found in office work environments. They may include the technical infrastructure and other equipment in the building or outside, such as air-conditioning and ventilation equipment, IT network devices, transformers, and pumps located in the building, machine rooms, and operating elevator unit engines. Moreover, such noise also reaches the work environment from the outside, generated mainly by communication traffic, heavy road traffic and track vehicles, and nearby construction sites. This noise is often the cause of employee complaints due to its annoying nature.

Exemplary noise spectrums containing infrasound components at an office workstation are presented in Figure 8.19.

8.3.6.2 Methods of Measurement and Assessment of Infrasonic Noise

Infrasonic noise at workstations is characterised by the following:

- The equivalent continuous G-weighted sound pressure level normalised to a nominal 8-hour working day or the 40-hour average weekly working time as specified by the labour code (an exception is the case of irregular impact of infrasound noise on the human body on various weekdays; $L_{\mathrm{Geq,8h}}$ or $L_{\mathrm{Geq,w}}$)
- The equivalent continuous G-weighted sound pressure level during occupation of the workstation by the employee ($L_{\mathrm{Geq},Te}$)

Values that constitute the criterion of annoyance of infrasonic noise are applied based on the current knowledge on the effects of infrasounds on the human body and are defined by the standard PN-N-01338 (Table 8.8).

The permissible values of infrasonic noise at workstations occupied by juveniles and pregnant women ($L_{\mathrm{Geq,8h}}$ or $L_{\mathrm{Geq,w}}$ = 86 dB) are defined in the regulations issued by the Council of Ministers. Methods for the measurement of quantities of infrasound noise are defined in the standard PN-N-01338.

TABLE 8.8
Equivalent Continuous G-Weighted Sound Pressure Level—Criterion of Annoyance

Quantity Assessed	Value (dB)
The equivalent continuous G-weighted sound pressure level, normalised to a nominal 8-hour working day or the 40-hour average weekly working time as specified by the labour code, $L_{Geq,8h}$ or $L_{Geq,w}$, for all employees	102
The equivalent continuous G-weighted sound pressure level during occupation of a workstation designated for conceptual work, $L_{Geq,Te}$	86

8.3.6.3 Methods of Limiting Infrasound Noise Exposure

The same requirements and principles applied in other types of noise can be applied for the prevention of infrasound noise exposure. However, protection against infrasounds is more complex due to the substantial length of infrasound waves (17–340 m), against which traditional walls, compartments, screens, and acoustic absorbers are not effective barriers. In some cases, infrasound waves are intensified by the resonance in a room, structural component, or an entire facility. The best protection against infrasounds is eliminating them at the source, that is, in the machines and devices. Other solutions include

- Applying silencers to the air or gas inlets and outlets of turbomachines.
- Providing proper foundations, with vibroisolation, for machines and devices.
- Reinforcing wall and building structures in the case of resonance.
- Applying heavy (bricked) soundproof cabins for operators of machines and devices.
- Applying active noise-control methods based on active sound absorption and compensation.

8.3.7 Ultrasonic Noise

8.3.7.1 Characteristics of the Agent

Ultrasonic noise contains components of high audible frequencies and low ultrasound frequencies from 10 to 40 kHz. Ultrasound components may penetrate the human body through the hearing organ or the entire body surface. There is not a lot of research on the effects of ultrasonic noise on the hearing organ, because under industrial conditions ultrasounds are usually accompanied by audible noise, so it is difficult to determine whether the changes in the hearing ability of the tested persons are caused by the effect of audible components, by ultrasound components, or by both types acting simultaneously. However, there is an increasingly popular belief that nonlinear phenomena occurring in the ear under the influence of ultrasounds cause subharmonic components often with the same sound pressure level as the basic ultrasound component. This phenomenon results in hearing loss within the range of the frequency of subharmonic ultrasounds.

Apart from an adverse effect on hearing, headaches and vertigo, disturbed balance, nausea, drowsiness during the day, and excessive fatigue are caused by the negative impact of ultrasounds on the vestibular organ in the internal ear.

Research on nonauditory effects has shown that occupational exposure to ultrasound noise exceeding 80 dB within the audible high-frequency range (up to 20 kHz) and above 100 dB within the ultrasound low-frequency range (above 20 kHz) causes vegetative-vascular disorders.

The main sources of ultrasonic noise in the work environment are technological ultrasound devices with low frequencies, in which ultrasounds are generated intentionally as a factor necessary to perform specific technological processes. These devices include ultrasound washers, ultrasound welders, and ultrasound wire-eroding machines and soldering irons.

Ultrasonic noise may also be emitted by high-speed machines, such as metal-processing machines, some textile industrial machines, and pneumatic devices, in which an outflow of compressed gases emits ultrasonic noise.

8.3.7.2 Methods of Measurement and Assessment of Ultrasonic Noise

Ultrasonic noise at the workstations is characterised by the following:

- Equivalent continuous sound pressure levels in third-octave bands of centre frequencies from 10 to 40 kHz, normalised to a nominal 8-hour working day or average weekly working time as specified in the labour code (an exception is cases of irregular impact of ultrasound noise on the human body on various weekdays)
- Maximum sound pressure levels in third-octave bands of centre frequencies from 10 to 40 kHz

The levels specified here should not exceed the maximum admissible values, specified in Table 8.9. Lower values should be applied to workstations occupied by juveniles and pregnant women, specified in Table 8.10.

TABLE 8.9
Maximum Admissible Values of Ultrasonic Noise for All Employees

Central Frequency of Third-Octave Bands (kHz)	Equivalent Continuous Sound Pressure Level in Third-Octave Bands of Centre Frequencies from 10 to 40 kHz, Normalised to a Nominal 8-Hour Working Day or Average Weekly Working Time as Specified in the Labour Code (dB)	Maximum Sound Pressure Level (dB)
10, 12.5, 16	80	100
20	90	110
25	105	125
31.5, 40	110	130

TABLE 8.10
Admissible Values of Ultrasound Noise at Workstations Occupied by Juveniles and Pregnant Women

Central Frequency of Third-Octave Bands (kHz)	Equivalent Continuous Sound Pressure Level in Third-Octave Bands of Centre Frequencies from 10 to 40 kHz, Normalised to a Nominal 8-Hour Working Day or Average Weekly Working Time as Specified in the Labour Code (dB)		Maximum Sound Pressure Level (dB)
	Juveniles	Pregnant Women	
10, 12.5, 16	75	77	100
20	85	87	110
25	100	102	125
31.5, 40	105	107	130

Methods for ultrasonic noise measurement are specified in the procedure for ultrasonic noise testing, published in the quarterly *Principles and Methods of Assessing the Working Environment* issued by the Central Institute for Labour Protection—National Research Institute and in the standard PN-N-01321.

8.3.7.3 Methods of Limitation of Ultrasonic Noise Exposure

The same requirements and principles applied in the case of noise can be applied for the prevention of ultrasonic noise exposure. However, medical tests should be performed more frequently—every 2 years.

Due to the short wavelength of ultrasounds in the air (1–2 cm), it is relatively easy to limit their adverse impact on humans, for example, by air-tight sealing and enclosure of noise sources and by the use of personal hearing protection equipment, which also effectively protects employees against high-frequency noise.

REFERENCES

Act of June 26th, 1974—Labour Code. DzU 1998, no. 21, item 94, as amended.

Augustyńska, D., and W. M. Zawieska, eds. 1999. *Noise and Vibration Protection in the Working Environment*. Warsaw: CIOP.

Augustyńska, D., D. Pleban, W. Mikulski, and P. Tadzik. 2000. *Assessment of Machine Noise Emission. Requirements and Methods*. Warsaw: CIOP.

Cempel, Cz. 1989. *Applied Vibroacoustics*. Warsaw: PWN.

Crocker, M. J. 2007. *Handbook of Noise and Vibration Control*. New York: J. Wiley.

Directive 2003/10/EC of the European Parliament and of the Council of 6 February 2003 on the minimum health and safety requirements regarding the exposure of workers to the risks arising from physical agents (noise). OJ L 42, Vol. 38:15 Feb 2003.

Directive 2002/44/EC of the European Parliament and of the Council of 6 February 2003 on the minimum health and safety requirements regarding the exposure of workers to the risks arising from physical agents (vibration), OJ L 177, Vol. 45: 6 July 2002.

Directive 2006/42/EC of the European Parliament and of the Council of 17 May 2006 on machinery, OJ L 157, Vol. 24:9 Jun 2006.

Engel, Z. 2001. *Protecting the Environment against Vibration and Noise*. Warsaw: PWN.
Engel, Z., and D. Pleban. 2001. *Machinery Noise—Sources, Assessment*. Warsaw: CIOP.
Engel, Z., and J. Sikora. 1998. *Acoustic Enclosures*. Krakow: AGH.
Engel, Z., G. Makarewicz, L. Morzyński, and W. M. Zawieska. 2001. *Active Noise Reduction Methods*. Warsaw: CIOP.
Engel, Z., J. Sadowski, B. Szudrowicz, and I. Żuchowicz-Wodnikowska, eds. 1993. *Guidelines for Computer-Aided Sound Protection of Work Stations in Industrial Halls*. Warsaw: CIOP.
Gryfin, M. J. 1990. *Handbook of Human Vibration*. London: Academic Press.
Harris, C. M., and A. G. Piersol, eds. 2002. *Harris' Shock and Vibration Handbook*. New York: McGraw Hill.
ISO 226. 2003 (E). Acoustics—Normal equal-loudness-level contours.
ISO 2631-1. 1997. Mechanical vibration and shock—Evaluation of human exposure to whole-body vibration. I. General requirements.
Kaczmarska A. 1997. Methods of low-frequency noise reduction in industrial cabins. Ph.D. disertation. Warsaw: CIOP.
Kaczmarska, A., and D. Augustyńska. 1999. *Low-Frequency Noise Reduction in Industrial Cabins*. Warsaw: CIOP.
Kaczmarska, A., D. Augustyńska, Z. Engel, and P. Górski. 2001. *Industrial Safety Devices Against Infrasound and Low-Frequency Noise*. Warsaw: CIOP.
Kozłowski, E. 2007. Noise. In *Selection of Personal Protective Equipment*, ed. K. Majchrzycka and A. Pościk, 203–214. Warsaw: CIOP-PIB.
Ordinance of the Minister of Economy and Labour of August 5th, 2005, on occupational health and safety with regard to works associated with exposure to noise or mechanical vibrations. DzU no. 157, item 1318.
Ordinance of the Minister of Labour and Social Policy of September 26th, 1997, on the general occupational health and safety provisions. DzU 2003 no. 169, item 1650 (uniform text), amended by: DzU 2007, no. 49, item 330; DzU 2008, no. 108, item 60.
Ordinance of the Minister of Labour and Social Policy of November 29th, 2002 on the maximum admissible values concentration and intensities for agents harmful to health in the working environment. DzU 2002 no. 217, item 1833; amended by DzU 2005, no. 212, item 1769; DzU 2007, no.161, item 1142.
Ordinance of the Minister of Environment of June 14th, 2007, on the admissible noise levels in the environment. DzU 2007, no. 120, item 826.
Ordinance of the Minister of Health and Social Welfare of May 30th, 1996, on conducting of medical examinations of employees, the scope of preventive employee healthcare and medical certificates issued for the purposes specified by the Labour Code. DzU 1996, no. 69, item 332, amended by DzU 2001, no. 128, item 1405.
Ordinance of the Minister of Health of August 1st, 2002 on the method of documenting of occupational diseases and their effects. DzU no. 132, item 1121.
Ordinance of the Minister of Health of April 20th, 2005 on tests and measurements of agents harmful to health in the working environment. DzU 2005, no. 73, item 645, as amended.
Ordinance of the Council of Ministers of September 10th, 1996, on the list of works prohibited to women. DzU 1996, item 114, item 545, amended by DzU 2002, no. 27, item 1092.
Ordinance of the Council of Ministers of August 24th, 2004 on a list of works prohibited to juveniles and the conditions of their employment for performance of some of these works. DzU no. 200, item 2047; amended by DzU 2005, no. 136, item 1145.
Ordinance of the Council of Ministers of June 30th 2009 on the occupational diseases. DzU 2009, no.105, item 869.
Osiński, Z., ed. 1997. *Vibration Reduction*. Warsaw: PWN.
Parent-Thirion, A., et al. 2007. *Fourth European Working Conditions Survey*. Luxembourg: Office for Official Publications of the European Communities.

Pawlaczyk-Łuszczyńskia, M., J. Koton, M. Śliwińska-Kowalska, D. Augustyńska, M. Kameduła. 2001. Ultrasonic noise. *Principles and Methods of Assessing the Working Environment* 2(30):88–95.
PN-N-01321. 1986. Ultrasonic noise. Admissible sound pressure levels at work places and methods of measurements.
PN-N-01307. 1994. Noise. Admissible noise values in the working environment. Requirements relating to measurements.
PN-EN 28662. 1998. Hand-held portable power tools—Measurement of vibrations on the handle. General requirements (series of standards).
PN-EN ISO 11200. 1999. Acoustics—Noise emitted by machinery and devices—Guidelines for the use of basic standards for the determination of emission sound pressure levels at work station and at other specified position.
PN-ISO 1999. 2000. Acoustics—Determination of occupational noise exposure and estimation of noise-induced hearing impairment.
PN-EN ISO 9614. 2000. Acoustics—Determination of sound power levels of noise sources using sound intensity (series of standards).
PN-EN 1033. 2000. Hand-arm vibration—Laboratory measurement of vibration at the grip surface of hand-guided machinery—General.
PN-EN 10819. 2000. Mechanical vibration and shock—Hand-arm vibration—Method for the measurement and evaluation of the vibration transmissibility of gloves at the palm of the hand.
PN ISO 7196. 2002. Acoustics—Frequency—Weighting characteristic for infrasound measurement.
PN-EN 12096. 2002. Mechanical vibration—Declaring and verification of vibration emission values.
PN-EN ISO 3740. 2003. Acoustics—Determination of sound power levels of noise sources.—Guidelines for the use of basic standards.
PN-EN ISO 5349-1. 2004. Mechanical vibration—Measurement and evaluation of human exposure to hand-transmitted vibration. Part 1 and Part 2.
PN-EN 14253+A1. 2008. Mechanical vibration—Measurement and evaluation of occupational exposure to whole-body vibration with reference to health—Practical guidance.
PN-EN ISO 9612. 2009. Acoustics: Determination of occupational noise exposure.
PN-N-01338. 2009. Infrasonic noise. Reference values of sound pressure levels at work stations. Requirements relating to measurements and assessment.
PN-EN ISO 20643. 2009. Mechanical vibration—Hand-held and hand-guided machinery—Principles for evaluation of vibration emission.
PN-EN 1032. 2009. Mechanical vibrations—Testing of mobile machines to determine the vibration emission values.
Ver, I. L., and L. L. Beranek. 2006. *Noise and Vibration Control Engineering*. 2nd ed. New York: J. Wiley.
Zawieska, W. M., ed. 2007. *Occupational Risk. Methodological Principles of Assessment*. Warsaw: CIOP-PIB.

9 Electromagnetic Hazards in the Workplace

Jolanta Karpowicz and Krzysztof Gryz

CONTENTS

9.1 INTRODUCTION

Electromagnetic fields (EMFs) have existed since the dawn of history. Geostatic magnetic fields, the electric fields of Earth's atmosphere, and low-frequency magnetic fields exist in the natural environment. The development of science and technology at the turn of the nineteenth and twentieth centuries has led to a massive use of electric energy in all areas of the economy and in households, as well as to the use of electromagnetic waves of various frequencies in wireless communications (Figure 9.1). Artificial EMFs are common and humans are exposed to complex EMFs composed of various frequencies. Workers who operate these devices may be exposed to relatively high-level EMFs; the conditions of their exposure should be controlled.

The ubiquitous presence of EMFs in the work environment requires employees, inspectors and workers to identify the sources and characteristics of the fields they generate, assess the severity of workers' exposure in the context of occupational health and safety and reduce identified hazards where necessary. The principle of avoiding unnecessary exposure also means limiting EMF exposure wherever possible. The highest priority should be given to technical activities aimed at eliminating or reducing EMF levels in the vicinity of devices. Identifying and assessing electromagnetic hazards when designing devices and designing and organizing the

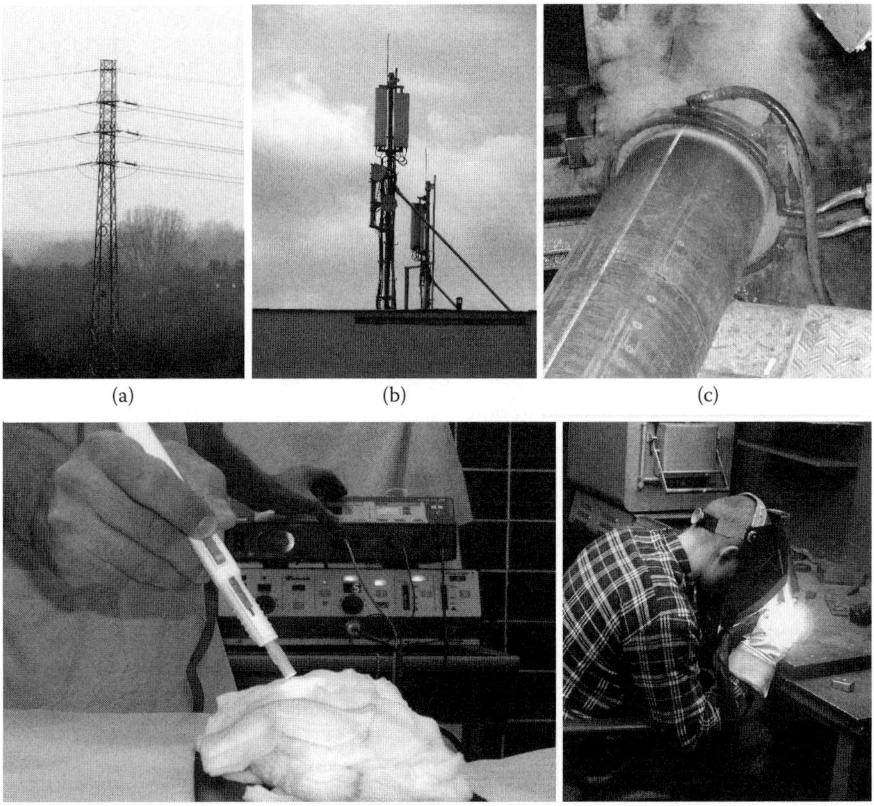

(a) (b) (c) (d) (e)

FIGURE 9.1 Examples of electromagnetic field sources: (a) a high-voltage power line; (b) antennas of mobile phone base stations; (c) an induction heater; (d) an electrical surgery device; and (e) an arc welding device.

workplace to reduce these levels are the most efficient methods to do this, in terms of both reducing the costs of protective measures and improving their effectiveness.

9.2 GENERAL DESCRIPTION

The frequency of time-varying EMFs is unlimited. EMFs of frequencies exceeding several megahertz (MHz) are also called electromagnetic radiation. The electromagnetic spectrum (Figure 9.2) includes the fields and radiation of various frequencies and biophysical properties, such as EMFs and optical radiation (nonionising), as well as X-ray, gamma, and cosmic radiation (ionising radiation). In occupational health and safety, the term *electromagnetic field* is used to describe static electric and static magnetic fields (invariable in time) and time-varying fields of frequencies less than 300 GHz (gigahertz), that is, fields produced by sources emitting waves of a length exceeding 1 mm. Such radiation cannot be directly perceived by human senses and does not cause the ionisation of the medium of its propagation.

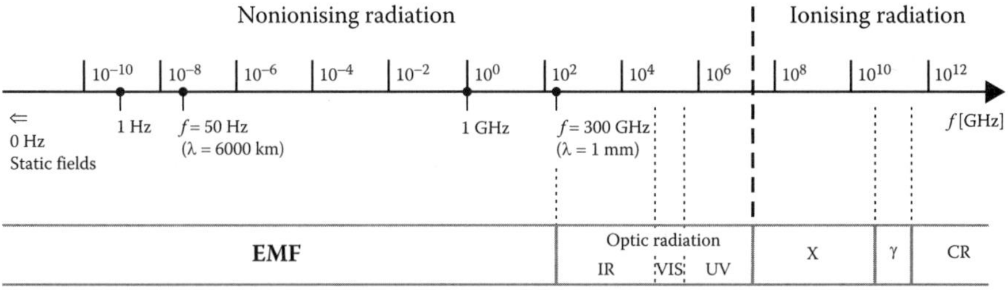

FIGURE 9.2 Frequency spectrum of electromagnetic fields and radiation. IR = infrared radiation, VIS = visible radiation, UV = ultra-violet radiation, X = roentgens's ray, γ = gamma radiation, CR = cosmic radiation.

The properties of the EMF are usually analysed with regard to its two components: the electric field and the magnetic field. The magnetic field is present around the moving electric charges (creating an electric current) or as a result of the magnetisation of some materials. The electric field exists around both the moving and motionless charges. The exposure level is usually described by the electric field strength (E), magnetic field strength (H), and frequency (f) of those fields present in the workplace (affecting the worker's body). The level of electromagnetic hazard depends on the polarisation and spatial distribution of the field in relation to the human body and the relationship between the strengths of the electric field and the magnetic field.

The energy of electric fields affects both moving and motionless charges, whereas that of magnetic fields affects only the moving charges. The static electric field is present around motionless charges, and the static magnetic field exists in the vicinity of conductors of direct current (DC) or permanent magnets.

9.3 DEFINITIONS AND RELATED TERMS

Wavelength (λ): This is the distance between two points in space, in which the wave is in the same phase of its oscillations (e.g. the distance between two adjacent peaks). The wavelength expressed in metres can be estimated by dividing 300 by the field frequency expressed in MHz; for example, for fields of frequency 100 kHz (i.e. 0.1 MHz), the wavelength (λ) is 3 km.

Electric field strength (E): This is a vector quantity representing the force affecting a unit electrical charge situated at a given point in the electric field. Its unit is volts per metre (V/m). For example, the electric field strength in a parallel plate capacitor, expressed in V/m, can be determined by dividing the difference in electrical potential on its plates, expressed in volts, by the distance between the plates, expressed in metres. An electric field strength of 25,000 V/m (i.e. 25 kV/m) exists in the centre of the plates of an air-spaced capacitor, separated by a distance of 1 cm and supplied by 250 V.

Magnetic field strength (H): This is a vector quantity representing the force affecting a unit electric charge moving at unit speed in the direction perpendicular to that vector measure. Its unit is ampere per metre (A/m). A magnetic field is also characterised by magnetic flux density (*B*), expressed in tesla (T). In the air, a magnetic field of 1 A/m is characterised by a magnetic flux density of about 1.25 μT. For example, the magnetic field strength, expressed in A/m, in the surroundings of a straight electric cable, can be determined by dividing the current, expressed in amperes, by double the π constant (about 3.14) multiplied by the distance from the cable, expressed in metres. At a distance of 15 cm from a single cable with a 100 A current, the magnetic field strength is about 106 A/m and the magnetic flux density is about 132 μT.

Near field (or near field area): A field directly adjacent to the field source, up to a distance approximately equal to one wavelength (if the size of the source is small compared with the wavelength), in which the electric and magnetic field strengths are independent of each other.

Far field (or far field area): A field that extends to distances exceeding approximately three wavelengths from the field source (if the size of the source is small compared with the wavelength), where the orientations of the electric and magnetic fields are perpendicular to the direction of radiation propagation and perpendicular to each other (i.e. they create a 'plane wave').

Specific absorption rate (SAR): The rate at which energy is absorbed in the unit mass of an exposed body, expressed in watts per kilogram (W/kg). The SAR, averaged over the whole body, is a widely accepted measure of the thermal effects of exposure to radio-frequency fields. Local SAR is used to assess the level of energy absorbed in small parts and resulting from specific exposure conditions. Local SAR should be assessed, for example, for a grounded person exposed to a radio-frequency field of a frequency in the low megahertz (MHz) range or persons exposed to near field in the immediate surroundings of the source.

Current density (J): The current in a voluminous conductor (e.g. the human body) flowing through a unit cross-sectional area, perpendicular to the direction of current flow. The unit is ampere per square metre (A/m^2).

Permissible values of internal measures of exposure: The parameters derived to protect workers against adverse consequences of thermal effects and the effects of induced currents during exposure to EMF, expressed as limit values of the SAR and the density of the induced current (*J*).

Permissible values of external measures of exposure: The parameters measurable in the real-work environment, expressed as electric field strength (*E*), magnetic field strength (*H*), magnetic flux density (*B*), and power density (*S*). If the permissible values of external measures are defined in order to fulfil the requirements regarding the permissible values of internal measures of exposure in the most unfavourable exposure conditions (e.g. exposure to a homogeneous field), then exceeding the permissible values of external measures is not always equivalent to the excess of the permissible values of internal measures (i.e. in the case of local exposure to a heterogeneous fields of a relatively high level).

Values related to parameters of EMFs are often presented using the following submultiple and multiple units:

n – Nano	$\times 10^{-9}$	(×0.000 000 001)
μ – Micro	$\times 10^{-6}$	(×0.000 001)
m – Milli	$\times 10^{-3}$	(×0.001)
	$\times 10^{0}$	(×1)
k – Kilo	$\times 10^{3}$	(×1,000)
M – Mega	$\times 10^{6}$	(×1,000,000)
G – Giga	$\times 10^{9}$	(×1,000,000,000)

9.4 MECHANISM OF INFLUENCE OF AN ELECTROMAGNETIC FIELD ON THE HUMAN BODY

An EMF may affect the body of an exposed person directly or indirectly due to the absorption of field energy by exposed objects and its influence on the human body (ICNIRP 1998; IEEE 2002, 2005; Reilly 1998; WHO 1993, 2006, 2007).

Most indirect effects are caused by contact currents, which flow through the body of a person who touches a metal object with a different electrical potential. The difference of potentials results from exposure to EMF. This phenomenon, which is linked with hazards of serious burns, needs attention in fields of frequencies less than 100 MHz, whereas in the case of fields of frequencies below 100 kHz, it can stimulate electrically sensitive tissues such as muscles or nerves and cause pain (ICNIRP 1998; IEEE 2002, 2005; Reilly 1998; WHO 2007). The intensity and the spatial distribution of contact currents depend on the frequencies of the EMF, the dimensions of the exposed object, and the area of contact between the object and the human body.

An EMF also may be hazardous to people due to an impact on the technical infrastructure, as currents induced by EMFs in devices may cause interference with automatic control devices or detonations of electrically controlled explosion equipment. The ignition of flammable or explosive materials by sparks produced by the flow of an induced current or the discharge of an electrostatic charge (WHO 2006; IEEE 2002) can also cause fires and explosions. Induced or contact currents flowing through the body may also interfere with the function of active medical implants (such as pacemakers) and mechanical implants in the body.

EMFs are not usually perceived by human senses. In some situations, however, EMFs can be sensed directly; for example, in strong magnetic or electric fields of low frequencies (several or tens of hertz), a person may have visual sensations, called magneto- or electrophosphenes. Exposure to pulsed microwave fields can cause hearing sensations, called the Frey effect (Reilly 1998; IEEE 2005).

The direct effects of EMF exposure include stimulation of electrically sensitive tissues as a result of the flow of currents induced directly in the body (this is the most significant interaction mechanism for EMFs of frequencies not exceeding hundreds of kilohertz) and heating of tissues, including serious burns, caused by the energy of field absorbed in (this mechanism is the most significant for EMFs of frequencies exceeding 1 MHz; ICNIRP 1998; Reilly 1998; WHO 1993). The effects of the current induction and rise of temperature in an exposed body are the basis for internal measures of exposure, whereas electric or magnetic field strengths, representing the level of fields in which a person is present, are called external measures of exposure (Directive 2004/40/EC; ICNIRP 1998; Karpowicz et al. 2006).

The health consequences of various interactions of EMFs with the human body are not yet established (Karpowicz and Gryz 2007a; WHO 1987, 1993, 2006, 2007). The following mechanisms of human interaction with EMF have been found (Reilly 1998):

1. Established mechanisms of human interaction with EMF:
   - Synapse activity alteration by membrane polarisation (phosphenes)
   - Peripheral nerve excitation by membrane depolarisation
   - Muscle cell excitation by membrane depolarisation (skeletal)
   - Electroporation
   - Resistive (joule) heating
   - Audio effects by thermoelastic expansion (Frey effect)
   - Magnetohydrodynamic effects
2. Proposed mechanisms of human interaction with EMF:
   - Soliton mechanism through cell membrane proteins
   - Spatial or temporal cellular integration
   - Stochastic resonance
   - Temperature mediated alteration of membrane ion transport
   - Plasmon resonance
   - Radon decay product attractors
   - Rectification by cellular membranes
   - Ion resonance
   - Ca^{2+} oscillations
   - Nuclear magnetic resonance
   - Radical pair mechanism
   - Magnetite interactions

Other undesirable effects can also appear and significantly reduce working ability. Such effects, related for example to movements in the high-level static magnetic field, can include vertigo, magnetophosphenes, nausea, a metallic taste in the mouth, and difficulties with eye–hand coordination (Karpowicz et al. 2007; WHO 2007).

Occupational exposure to EMFs extended over many years may affect health and the ability to work. So far, results of investigations have not excluded the possibility of adverse health effects from chronic exposure, especially to EMF of high levels. Possible adverse health effects from EMF exposure include the development of tumours or malfunctions of the cardiovascular, nervous, and immunological systems. Research continues in this respect.

At present, the scientific background for the assessment of health risk from such exposures is not adequate, especially in relation to exposure to static fields and intermediate frequency fields. Further investigation is a research priority. The wide use of technical or organisational measures to reduce workers' exposure to the lowest possible level is necessary. There are many examples of such measures with a high efficiency and low cost.

9.5 METHODS OF ASSESSING THE ELECTROMAGNETIC FIELD

Investigations of EMFs in work environments are conducted to identify the sources of fields, which can produce a potential hazard to workers, and to assess the severity of the hazards caused by such fields. The investigations are conducted by making measurements or calculations of the parameters of the fields affecting the workers and of the technical objects present in the work environment (Gryz and Karpowicz 2000, 2008; Karpowicz and Gryz 2007a). These measurements are mostly used to assess the level of the electric or magnetic fields present in the workplace (Figure 9.3). The measurements of induced or contact currents may be a complement to electric and magnetic field measurements (Figure 9.4).

To obtain standardised and repetitive results, the quantities of the unperturbed field (i.e. not interfered with by human presence) are used to determine the EMF affecting the human body or the environment. Therefore, measurements are taken at the location of normal work activities, but in the absence of workers.

Measurement devices are composed of exchangeable measurement probes with magnetic or electric field sensors and a battery-powered monitor containing an indicator. Electric dipoles are usually used as the electric field sensors, whereas the magnetic fields are measured by multiturn coils or Hall sensors.

The calculations that are usually performed when assessing electromagnetic hazards in the work environment are focused on the internal measures of exposure effects (J and SAR) in the worker's body (Figure 9.5; Karpowicz and Gryz 2007a). This may lead to being able to determine the distribution of external measures of exposure as well (e.g. when the unperturbed field cannot be properly measured or when there are problems with the assessment of EMF of highly heterogeneous spatial distribution in the workplace; Figure 9.6; Karpowicz and Gryz 2007a).

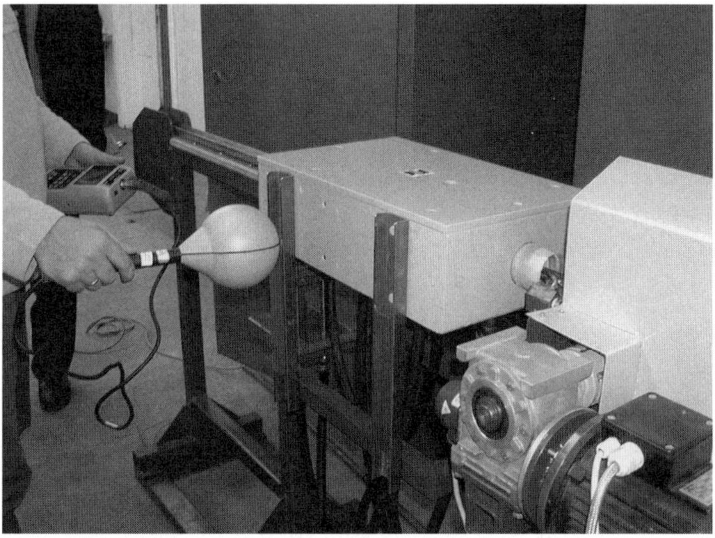

FIGURE 9.3 Assessing the worker's exposure based on the magnetic field strength in the operation of an induction heater.

FIGURE 9.4 Assessing the worker's exposure based on the induced or contact current in the workplace while touching conductive objects exposed to an electromagnetic field.

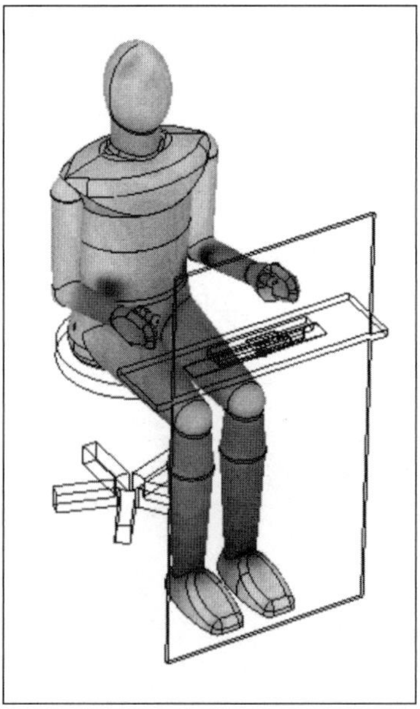

FIGURE 9.5 Assessing the worker's exposure by calculating internal measures of exposure; an example of simulation of the specific absorption rate (SAR) distribution in the body of a worker sitting in front of a dielectric heater (darker grey represents higher values).

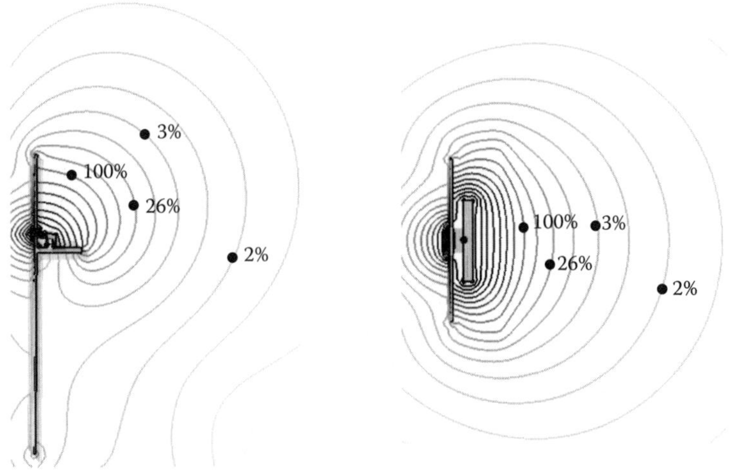

FIGURE 9.6 Assessing the level of the worker's exposure by calculating external measures of exposure; an example of spatial distribution of the electric field strength near a dielectric heater.

Computer simulations use digital models that realistically represent exposure conditions in the workplace being investigated. This would include the geometry of the investigated area, the objects present there, the electrical properties of the materials, and models of the worker's body. To accurately determine the distribution of internal measures in a human body, the model should take into account details of the anatomy. The model can be simplified, and a homogeneous model (with simple geometric forms, such as cylinder, ellipsoid, or sphere) can be used where integral parameters are studied, for example, SAR averaged over the whole body or contact and induced current.

Workers' exposure level should be assessed in the highest risk conditions. That is, the worst-case assessment in which the worker is affected by the fields of the highest level should be performed by measurements or calculations. Alternatively, the possibility of exposures higher than as investigated should be considered in analysing the obtained results.

9.6 PARAMETERS REPRESENTING THE EXPOSURE LEVEL

Two types of quantities are used to assess the level of exposure to EMFs (Gryz and Karpowicz 2000; Karpowicz 2007; Karpowicz and Gryz 2007a,b; Karpowicz et al. 2006; Koradecka et al. 2006):

1. *Internal measures of exposure* (J and SAR) cannot be measured in a work environment, but their permissible values define the maximum level of exposure to EMFs.
2. *External measures of exposure* (E, H and B) can be measured in a work environment. They determine whether the exposure level in the workplace should be controlled and if preventive measures should be undertaken.

These are supplemented by permissible values of contact and induced currents that are related to the internal measures but measurable in a work environment. The limits for permissible internal measures of exposure should never be exceeded.

Widely accepted international documents providing guidelines on the permissible level of workers' exposure were published by ICNIRP, IEEE, and the European Parliament (Directive 2004/40/EC; ICNIRP 1998; IEEE 2002, 2005). Many countries use slightly modified values of the above-mentioned limit values for their national occupational safety and health (OSH) policy (e.g. in Poland; Karpowicz et al. 2006). In the European Union OSH legislation system, the highest priority should be given to national legislation. International guidelines should be used for research or for OSH engineering where there is no national legislation.

Limits on the density of induced current refer to the head and torso (Table 9.1). Limits on the SAR ratio are defined as values averaged for the entire body (0.4 W/kg), local values in the head and torso (10 W/kg), and separate, local values in the limbs (20 W/kg), for frequencies between 100 kHz and 10 GHz (Directive 2004/40/EC; ICNIRP 1998) or for frequencies between 100 kHz and 3 GHz (IEEE 2005). The permissible values for induced and contact currents are specified in Tables 9.2 and 9.3 for assessing the exposure level of limbs.

Internal measures, correlated with the effects of exposure inside the body, were used in creating these guidelines to determine the limits of external measures, which can ensure compliance with limits of internal measures even in the most unfavourable conditions of exposure (i.e. in the worst case of exposure to constant homogeneous fields, with the highest level of interaction of the worker's body with the field, for example, the exposure of a grounded worker in an electric field of vertical polarisation). Permissible values of external measures may be exceeded in a workplace if it is demonstrated that no risk exists when permissible internal measures are exceeded. Analysing this requirement usually involves assessing the level of internal measures based on the results of numeric simulations and is of little practical significance. Therefore, the current system is to manage health and safety in the work environment containing EMFs sources by investigating and assessing the levels of external

TABLE 9.1
Permissible Values (RMS) of Current Density in Head and Torso

Frequency Range	Current Density in Head and Torso, J (mA/m²)
Up to 1 Hz	40
1–4 Hz	$40/f$
4–1000 Hz	10
1 kHz–10 MHz	$f/100$

Note: f = frequency in Hz.
Source: Data from Directive 2004/40/EC; and ICNIRP. 1998. *Health Phys* 74(4):494–522.

TABLE 9.2
Permissible Values of Induced Current

Document	Frequency	Induced Current (mA)
IEEE Std C.95.6, 2002; IEEE Std C 95.1, 2005	Up to 3 kHz	6 (both feet) 3 (each foot)
	3–100 kHz	2*f* (both feet) 1*f* (each foot)
	0.1–110 MHz	200 (both feet) 100 (each foot)
Directive 2004/40/EC; ICNIRP 1998	10–110 MHz	100 (limb)

Note: *f* = frequency in kHz.

TABLE 9.3
Permissible Values of Contact Current

Document	Frequency	Contact Current (mA)
IEEE Std C.95.6, 2002; IEEE Std C95.1, 2005–grasp	Up to 3 kHz	3
	3–100 kHz	1*f*
	0.1–100 MHz	100
IEEE Std C.95.6, 2002; IEEE Std C95.1, 2005–contact	Up to 3 kHz	1.5
	3–100 kHz	0.5*f*
	0.1–110 MHz	50
Directive 2004/40/EC; ICNIRP, 1998–contact	Up to 2.5 kHz	1.0
	2.5–100 kHz	0.4*f*
	0.1–110 MHz	40

Note: *f* = frequency in kHz.

measures of exposure in the workplace. Relatively high uncertainties of EMF-related measurements and simulations, reaching 10%–50% in the case of measurements or more in the case of calculations, should be taken into account in the EMF-related OSH-management system.

Because bioelectrical parameters of the human body depend on the field frequency, permissible values of external measures also depend on the frequency (Figure 9.7).

Examples of EMF-related regulations can be found in OSH legislation in Poland. Detailed provisions are available from various Web pages (e.g. http://www.ciop.pl/EMF) or research papers (e.g. Gryz and Karpowicz 2000; Karpowicz 2007;

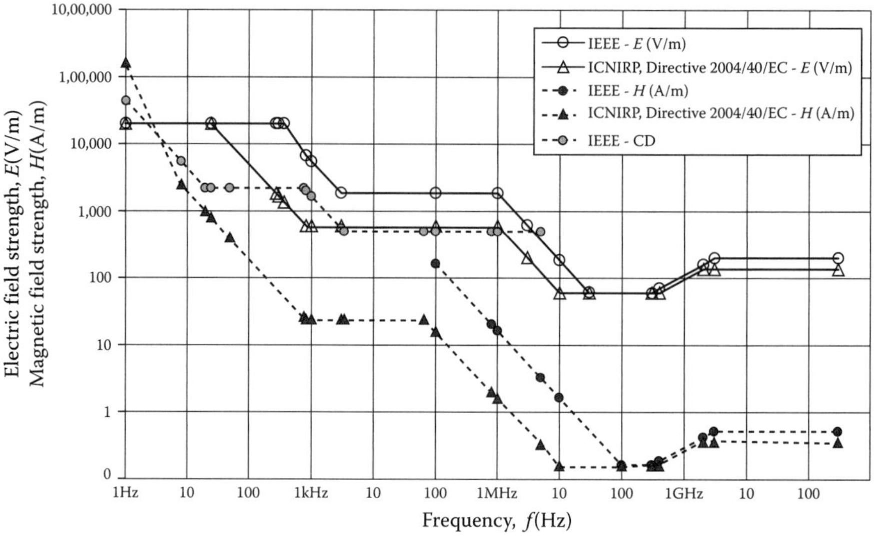

FIGURE 9.7 Permissible values of external measure, electric and magnetic field strength.

Karpowicz and Gryz 2007b; Karpowicz et al. 2006, 2008; Koradecka et al. 2006), in which the following limits are specified:

- Level of exposure permissible during an 8-hour shift, in the case of occupational exposure, expressed in electric and magnetic field strengths (external measures)
- Principles for shortening the duration of exposure to fields of higher levels, expressed as the exposure factor
- Threshold of prohibited exposure, acceptable for an exposure of only a few minutes per shift, expressed in electric and magnetic field strength
- Conditions in which it is permissible to work in fields at a prohibited level: (1) when using a protective suit, (2) when low-frequency magnetic fields affect the limbs only
- Permissible exposure to pulse microwave fields
- Definition of occupational and nonoccupational exposures, related to various groups of workers and various conditions and permissible levels of their exposure

For EMF, permissible exposure conditions were stated to ensure that the influence of a field on a worker over the period of his or her work life, up to 40 years duration, should not cause adverse changes in his or her health or the health of future generations.

Measurements should be performed in accordance with the standards harmonized with particular documents regarding the permissible conditions of exposure in the workplace. Polish standard PN-T-06580:2002, harmonized with the abovementioned regulations on occupational exposure limitation, defines the terminology and principles of measurement and assessment of workers' exposure to EMFs. The maximum permissible level of fields affecting workers are specified to ensure that the permissible values of internal measures are not exceeded in typical, realistic conditions of exposure in the workplace, and when such exposure does not exceed the threshold of prohibited exposure.

International guidelines available are as follows: measurement standard IEEE C95.3, 2002, harmonized with exposure limitations from standards IEEE C95.1, 2006 and IEEE C95.6, 2002. A detailed measurement methodology is not available for ICNIRP recommendations, and standards harmonized with the requirements of directive 2004/40/EC are being prepared by CENELEC, leaving practical problems calling for interpretation of the directive provisions (Hansson Mild et al. 2009).

9.7 WORKER'S EXPOSURE TO ELECTROMAGNETIC FIELDS

Any electrical device is a source of EMF, which can be generated intentionally or as a side effect of its operation. The most popular sources of EMFs encountered in a work environment are devices and installations used to distribute electric power, wireless communication and radars, electrothermal devices for thermal processing of metal or dielectric elements, and therapeutic or diagnostic medical equipment. An estimated several million workers are subject to exposure to EMFs throughout Europe, but such data can be significantly underestimated because of the lack of relevant registers.

Most workers are exposed to weak EMFs, which can occur near the following:

- Electrical appliances available also for the use of the general public, for example, office or computer equipment that is not located directly on the body of the worker
- Lighting equipment, except for some types of specialised radio frequency–energised lighting
- Wireless or cordless phone systems and base stations, except when work is perfomed directly on the active antennae
- Power supplying devices, such as high-voltage power lines, transformer stations, and switch yards, except for workers performing work on the living devices (e.g. at a live high-voltage power line)
- Medical devices, except magnetic resonance imaging (MRI) scanners, physiotherapy diathermies, and electrical surgery devices

Exposure to strong EMFs can occur near the following:

- Electrothermal devices, such as arc furnaces for melting scrap steel, induction furnaces, and heaters for thermal processing of steel elements (e.g. in hardening, forging)
- Dielectric heaters and presses for joining plastic elements
- Resistance welders for connecting metal elements
- Industrial magnetisers and demagnetisers

- Radio- and telecommunication devices, such as radio and television broadcasting antennas and radars
- Medical and laboratory devices, such as physiotherapy diathermies, electrical surgery devices, MRI scanners, NMR spectrometers, and magnetic therapy devices
- Electrochemical devices, such as electrolytic vats
- Magnetic separators for capturing metal elements from loose and fragmented materials

9.8 GOOD PRACTICES FOR PREVENTING ELECTROMAGNETIC HAZARDS

To avoid the aforementioned adverse effects of chronic exposure, EMF affecting employees should be minimised using the available technical and organisational measures, regardless of the exposure level. To eliminate the exposure of workers to strong EMFs in the work environment, it is compulsory to apply diverse technical or organisational protective measures. The most effective good practices can be determined by reviewing the solutions and tools used in various industries.

Occupational and nonoccupational exposures are differentiated in the assessment of EMFs (Karpowicz and Gryz 2007a). *Occupational exposure* is exposure to high-level fields prohibited for the general public and results from operating devices generating strong EMFs (Figure 9.8). *Nonoccupational exposure* is exposure to weak EMFs in the area far from the source of a strong EMF. Such exposure is not restricted, regardless of the status of the worker or public members. Particular protection is extended to young workers and pregnant women, for whom only nonoccupational exposure is allowed. The level of nonoccupational exposure is harmonized with the level of exposure acceptable for the general population, specified by regulations for public environments.

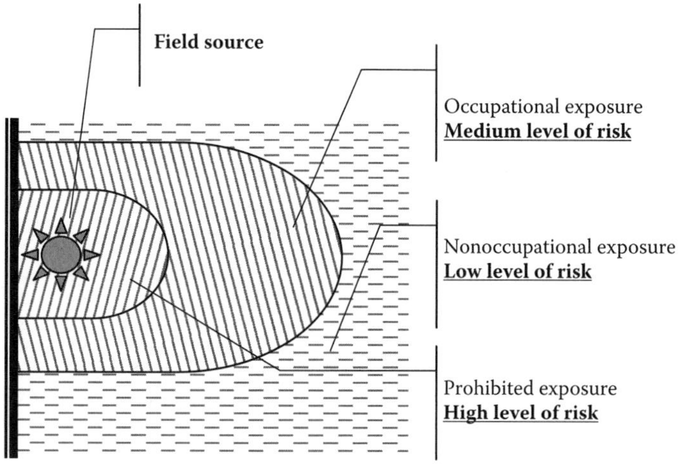

FIGURE 9.8 Principles for evaluating the occupational risk of exposure to electromagnetic fields.

Identification of groups of workers who are subject to occupational exposure and analysis of whether these groups can be minimised are key to the prevention of electromagnetic hazards in the work environment.

Occupational exposure is allowed if the sources of EMFs are identified and marked (Figure 9.9) and workers have been informed about possible hazards in their surroundings. They also must be subject to medical surveillance (especially where excessive exposure has been found), and periodic training must be conducted on the principles of working safely in EMFs. Labelling EMF sources and the areas of high-level EMFs is necessary to warn workers of the presence of a hazard, particularly people with increased sensitivity, such as pregnant women or workers with medical implants.

Persons with a higher sensitivity to EMFs, for example, those with electronic implants, may require increased protection even in fields of nonoccupational exposure level. Following international recommendations, persons with active implants should not be in static magnetic fields exceeding 0.5 mT (millitesla) or in time-varying fields of a power frequency of 50 Hz, exceeding 100 μT or 1 kV/m (ACGIH 2009, IEEE 2002, WHO 2007). They should not stay near the sources of radio-frequency radiation, for example, in the vicinity of antitheft gates; they should not use a cordless or cellular phone handset at a distance less than 15 cm from the pacemaker.

Before assessing the level of workers' exposure, the field source and the profile of electromagnetic hazards around it should be identified. Devices can be considered sources of weak EMFs if the exposure level of workers near the devices is much less than the limits stated by the regulations on the permissible exposure of workers. For example, electrical appliances for common office or home use, including office and

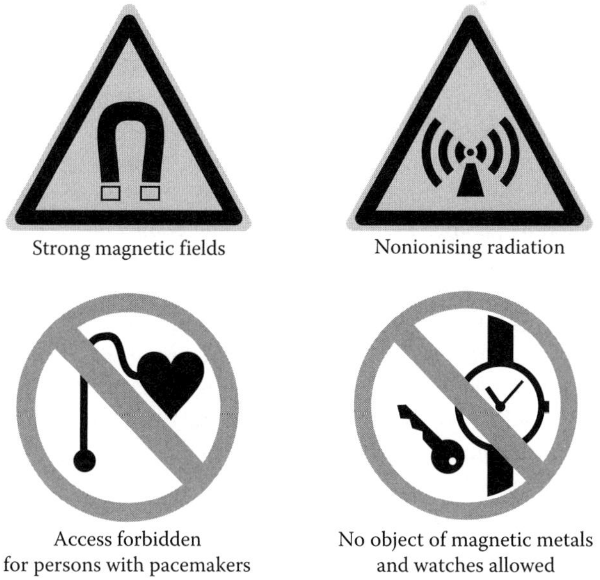

FIGURE 9.9 Warning signs of electromagnetic field (EMF) sources or EMF-related hazards.

computer devices meeting European requirements of exposure to the general public (Recommendation 1999/519/EC) and that are not present directly on the worker's body, do not require periodic inspection of the worker's exposure level. According to provisions of directive 2004/40/CE, the assessment of EMFs in the area accessible to the general public is not required in the case workers are affected by abovementioned EMF sources only. If workers stay close to such devices, they may be exposed to stronger fields. The most effective method to avoid such exposure is to organise the workplace in such a way that the sources of the fields are not in direct proximity of the workers' body.

While planning the installation or construction of a new device, electromagnetic safety requirements and information on the levels of fields that may be generated around it during its operation should be analysed. The personnel directly operating the devices should not be within the areas of strong fields, and the workers must be exposed to the least possible levels. Workers should be as far as possible from the EMF-emitting devices, or either the devices or the workplaces near the devices should be shielded.

Measurements of fields at the time of installing and test operation allow determination of the spatial distribution of the field level around the devices. These results are necessary for the proper assessment of workers' exposure, which covers both the environment of the use of EMF-emitting device and the nature activities performed by workers while operating that device. They also help in formulating recommendations and principles for safety. A detailed description of the principles of limiting the exposure of workers can be found at the Web page of CIOP-PIB (http://www.ciop.pl/EMF) and in quoted publications on the topic.

9.9 COLLECTIVE PROTECTIVE EQUIPMENT

Collective protection measures against EMFs include two kinds of shielding: localising (shielding of the EMF source) or protective (shielding of the workplace). Localising shielding is preferred, as it reduces the field level in a larger area, ensures more protection of the people within the surroundings of the source, and prevents the possibility of accidental exposure. The effectiveness of the shielding depends on various physical phenomena that take place in the shielding material (usually a metal mesh or sheet) and its geometric shape and galvanic connections. The field is partly attenuated at the surface due to surface reflection. The part that is not reflected at the surface is attenuated when it passes through the shield, due to an absorption loss. Further attenuation may happen when the field reflects towards the inside of the shield and multiple reflections (called internal reflection) occur inside it (Figure 9.10).

The efficiency of shielding is a function of the frequency and type of field and the current technical condition of the shield; the technical condition of the shield must be controlled and the efficiency of the shielding must be periodically measured.

EMF exposure can also be reduced by changing the technology that is used, for example, by reducing the output power of the device to the minimum level allowing normal operation. Similarly, the using manipulators and automatising the device operation so that the workplace is far from the area of strong EMFs protect workers from electromagnetic hazards. Changes in the device design, such as reducing the

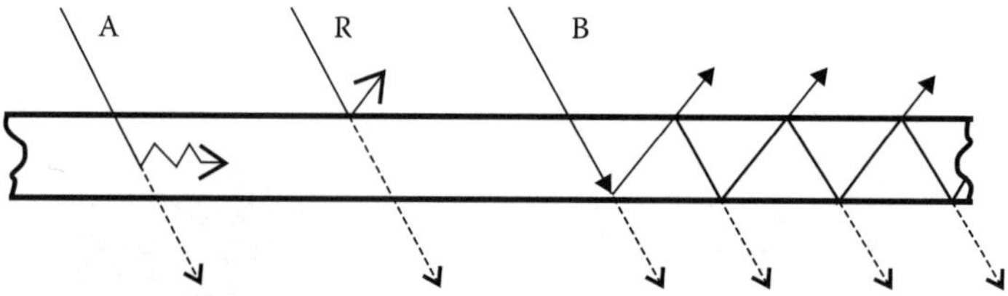

FIGURE 9.10 Interaction of an electromagnetic field with a shield. A = absorption; R = external reflection; B = internal reflection.

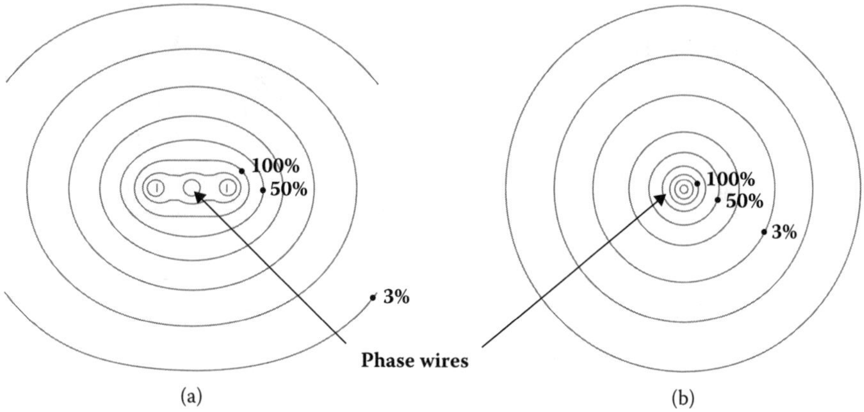

FIGURE 9.11 Magnetic flux density around three-phase wires: (a) 50 cm distance between wires; (b) 3 cm distance between wires. Calculations performed for equal current, in each of the phase wires, the phase shift is 120° and the centre of the coordinates is located in the axis of the central wire.

sizes of elements that generate EMFs or keeping them closer to each other, can be alternatives to electromagnetic shielding. For example in three-phase electric installations, when closing cables and bus bars on each other, magnetic fields generated by such a set of cables might be significantly lowered as a result of the interaction of vector quantities shift in phase (Figure 9.11).

9.10 PERSONAL PROTECTIVE EQUIPMENT

The use of protective clothing, which shields workers from EMF, is an alternative method for reducing the level of exposure. This option is the best for situations in which the workers must work in strong EMFs and in direct proximity to the field source, which cannot be switched off during the workers' presence, for example, when performing repairs or maintenance on broadcasting or radar devices. In such situations, protective clothing might protect workers from hazardous, excessive exposure to electromagnetic radiation (Figure 9.12). However, the main problem is the limited availability of clothing that is efficient enough to reduce many kinds of EMF exposures.

FIGURE 9.12 Suits protecting workers against electromagnetic field exposure, designed by CIOP-PIB researchers.

The use of dielectric seats, switches, handles, carpets and so on insulates the workplace or worker and reduces the current flowing through the worker in contact with the EMF-exposed metal structure or resulting from capacitive coupling between the field source, human body, and the ground.

9.11 SUMMARY

The assessment and reduction of electromagnetic hazards in the work environment can be perfomed due to the requirements of health and safety regulations, as activities undertaken to improve the OSH management system in a company or as required by a voluntary policy to protect workers from hazards in an occupational environment. In any case, proper identification of and distinction between the EMF sources that should be subject to preventive measures and those that pose significant hazards is key to effectively managing the resources dedicated to developing a safer and more friendly work environment.

Due to the widespread presence of EMF in the work environment, one key action is to identify the groups of workers on whom the preventive measures should primarily focus, that is, the workers who are subject to so-called occupational exposure due to operating sources of elevated EMF. These workers should be trained to safely operate field sources, and should have no health counterindications (found during medical examination). The exposure level in their workplaces should be controlled periodically.

Other workers not involved in the operation of EMF sources, young workers, and pregnant women must be subject to nonoccupational exposure, which means they should not work in the vicinity of devices which emit strong EMFs.

REFERENCES

ACGIH. 2009. American Conference of Governmental and Industrial Hygienists, TLVs and BEIs: Based on the Documentation of the Threshold Limit Values for Chemical Substances and Physical Agents and Biological Exposure Indices, Cincinnati, OH, pp. 152–163.

Council of the European Union Recommendation of 12 July 1999 on the limitation of exposure of the general public to electromagnetic fields (0 Hz to 300 GHz). 1999/519/EC. OJ EC, L 199/59 (available from the European legislation web portal: http://www.eur-lex.europa.eu).

Directive 2004/40/EC of the European Parliament and Council of April 29th, 2004, on the minimum health and safety requirements regarding the exposure of workers to risks arising from physical agents (electromagnetic fields). Eighteenth detailed directive within the meaning of Art. 16.1 of Directive 89/391/EEC. OJ, No L 184, 2004 (available from the European legislation web portal: http://www.eur-lex.europa.eu).

EN 50499. 2008. Procedure for the assessment of the exposure of workers to electromagnetic fields.

Gryz, K., and J. Karpowicz. 2000. Electromagnetic fields in working environment. *Series: Occupational Safety and Health Management*, ed. D. Koradecka. Warsaw: CIOP.

Gryz, K., and J. Karpowicz. 2008. Principles for assessing of electromagnetic hazards related to induced and contact currents phenomena. *Principles and Methods of Assessing the Working Environment* 4(58):137–171.

Hansson Mild, K., T. Alanko, G. Decat, et al. 2009. Exposure of workers to electromagnetic fields. A review of open questions on exposure assessment techniques. *Int J Occup Saf Ergon* 15(1):3–33.

ICNIRP (International Commission on Non-ionizing Radiation Protection). 1998. Guidelines for limiting exposure to time-varying electric, magnetic, and electromagnetic fields (up to 300 GHz). *Health Phys* 74(4):494–522.

IEEE Std C95.1. 2005. Standard for safety levels with respect to human exposure to radio frequency electromagnetic fields, 3 kHz to 300 GHz.

IEEE Std C95.6. 2002. Standard for safety levels with respect to human exposure to electromagnetic fields, 0–3 kHz.

ISO 7010. 2003. Graphical symbols, safety colours and safety signs: Safety signs used in workplaces and public areas.

Karpowicz, J. 2007. Electromagnetic fields. In *Occupational Risk—The Base on Assessment Methods*, ed. W. M. Zawieska, 227–258. Warsaw: CIOP-PIB.

Karpowicz, J., A. Bortkiewicz, K. Gryz, R. Kubacki and R. Wiaderkiewicz. 2008. Electromagnetic fields and radiation of frequency 0 Hz–300 GHz. Rationale documentation for revision harmonizing workers' permissible exposure level with the Directive 2004/40/EC. *Principles and Methods of Assessing the Working Environment* 4(58):7–45.

Karpowicz, J., and K. Gryz. 2007a. Practical aspects of occupational EMF exposure assessment. *Environmentalist* 27(4):535–553.

Karpowicz, J., and K. Gryz. 2007b. *Electromagnetic Fields in Office and Non-industrial Spaces.* Warsaw: CIOP-PIB.

Karpowicz, J., M. Hietanen, and K. Gryz. 2006. EU Directive, ICNIRP guidelines and Polish legislation on electromagnetic fields. *Intern J Occup Saf Ergon* 12(2):125–136.

Karpowicz, J., M. Hietanen, and K. Gryz. 2007. Occupational risk from static magnetic fields of MRI scanners. *Environmentalist* 27(4):533–538.

Koradecka, D., M. Pośniak, E. Jankowska, J. Skowroń, and J. Karpowicz. 2006. Chemical, dust, biological and electromagnetic radiation hazards. In *Handbook of Human Factors and Ergonomics*, 3rd ed., ed. G. Salvendy, 945–964. Hoboken: John Wiley & Sons.

Reilly, P. J. 1998. *Applied Bioelectricity: From Electrical Stimulation to Electropathology.* New York: Springer-Verlag.

WHO. 1993. *Environmental Health Criteria 137: Electromagnetic Fields (300 Hz–300 GHz).* Geneva: World Health Organization. Also available at http://www.inchem.org/documents/ehc/ehc/ehc137.htm.

WHO. 2006. *Environmental Health Criteria 232*: *Static Fields*. Geneva: World Health Organization. Also available at http://www.who.int/peh-emf/publications/reports/ehcstatic/en/index.html.

WHO. 2007. *Environmental Health Criteria 238: Extremely Low Frequency Fields (ELF).* Geneva: World Health Organization. Also available at http://www.who.int/peh-emf/publications/elf_ehc/en/index.html.

10 Static Electricity

Zygmunt J. Grabarczyk

CONTENTS

10.1 INTRODUCTION

Static electricity is the formation, accumulation, and disappearance of an unbalanced electric charge Q. The unbalanced charge is usually formed on the surface of solid or liquid objects, which can be compact or broken up (e.g. dust and haze particles), of low electrical conductivity (dielectrics, semiconductors), and can conduct objects insulated from the ground. Also, the external electric field induces the electric charge on the surface of conducting object. In the electrified object, there is at least a local lack of equilibrium between the number of elementary positive and negative charges.

Gas particles can be electrified or ionised only by ionising radiation or electrical discharges. Gases (e.g. atmospheric air) carry aerosol particles, such as dust, smoke, ice crystals, or haze electrified by other mechanisms.

The change in the charge is equivalent to the electric current flow. The intensity of the electric current flow is proportional to the speed of the charge change. The current pulse intensity during the electrostatic discharge (ESD) may be up to a few dozen amperes.

The separation of positive and negative charges requires working against the coulomb (C) force. This work is converted into the potential energy of the electric field. When the unbalanced charge disappears, the energy of the field is converted into other forms, such as thermal (above 90% of the local energy), mechanical (e.g. acoustic), and electromagnetic (Gajewski 1987).

The unbalanced charge can decay slowly and continuously (taking from milliseconds to many hours and according to an exponential dependence on time) or rapidly (from tens of nanoseconds to a few hundred microseconds). In the first case, the neutralisation of excess charge via the conduction of a material is called the *dissipation* of the charge. In the second case, ESD, the discharge can be in the form of an electric current pulse in a gas, especially in the air, between the objects (which have enough electric potential difference to ionise the air), or in the form of a direct current pulse when the objects touch (*contact discharge*), if the potential difference is not enough to ionise the air.

10.2 TYPES OF ELECTROSTATIC DISCHARGES

ESDs are the main sources of electrostatic hazards. They occur when the intensity of the electric field locally achieves a strength of about $E_s = 3$ MV/m, the strength that is necessary to cause an avalanche ionisation of the air. ESDs can be classified as:

1. *Corona discharge*: Occurs around the edge and point of conducting blades or thin conductors when their radii of curvature are smaller than 5 mm and the electrical potential is high in relation to the surroundings. The onset voltage U_s for a corona discharge depends on the radius of the curvature (e.g. for $r = 0.1$ mm, U_s is about 2 kV and for $r = 1$ mm, U_s is about 6 kV). The current intensity of this discharge can be from picoamperes to a few hundred microamperes. The ignition potential of this discharge is generally very low, but it can ignite sensitive media for which the minimum ignition energy (MIE; see Section 10.3) is around several microjoules (e.g. hydrogen, carbon disulfide, acetylene). At a current intensity higher than 200 μA, a corona discharge can ignite the vapours of some hydrocarbons (Britton 1999). They are also used as an ion source to neutralise the electrostatic charge.
2. *Spark (capacitive) discharge*: Occurs between the conducting electrodes when the radii of curvature are above 50 mm and the voltage between them exceeds 320–350 V (Jones and King 1991; Kaiser 2006). The conducting liquid can also act as an electrode. The duration of a spark discharge is from the tens up to hundreds of nanoseconds, depending on the resistivity of the electrodes. It can ignite vapours, flammable gases, explosives, most of the combustible dusts, and about 90% of the dusts caused by ESDs (Jones and King 1991). Their energy is close to the energy accumulated in the electric capacity between objects:

$$W = \frac{CU^2}{2} = \frac{QU}{2} \tag{10.1}$$

where C is the electric capacity of both the conducting objects or one conducting object in relation to the ground, U is the potential difference between the conducting objects or between the conducting object and the ground, and $Q = UC$ is the charge accumulated in capacity C. Theoretically, this energy is unlimited, but in practice, it is between a few dozen up to a few hundred millijoules (mJ). This discharge can be prevented by bonding and earthing all of the conducted objects.

TABLE 10.1
Approximate Values of the Electric Capacities of Some Unearthed Metallic Objects Commonly Used in Industry

Metallic Objects	Capacitance, pF (Order of Magnitude)
Small screw	1
Coin	2
Shovel	10
Small hand tools	10–20
Small container (up to 0.05 m^3), hopper	10–100
Medium container, drum, vessel (up to 0.5 m^3)	50–500
Flange connection (unbonded)	10–500
Some elements of plant equipment	100–1000
Human body	70–300
Filter elements	10–100
Large machine, large container or vessel	100–1000
Vehicle	300–1000
Road tanker	~1000
Large silo with dielectric lining, rail tanker	up to 100,000
Sphere with radius R, far from the ground	~$110 \times R$ (m)

Spark discharge risk assessment is relatively simple and can be made by comparing the energy accumulated in the capacity C with the MIE. The risk is negligible if the stored energy (Equation 10.1) does not exceed 0.1 MIE.

The electric capacities of commonly used metallic objects, measured to the ground, are shown in Table 10.1.

3. *Brush discharge*: Occurs between conductors when the radius of the curvature of the smaller electrode is between 5 and 50 mm and the electrostatic field intensity exceeds 0.5 MV/m. It more often appears between the conductor and the electrified dielectric, solid, or liquid. This discharge ignites most flammable gases and vapours, including hydrocarbons (MIE about 0.2 mJ). The minimum electric potential differences that initiate brush discharge is around a dozen kilovolts (for a liquid, between 20 and 60 kV [Cross 1987]). The charge neutralisation during a single discharge occurs only on a limited surface of the dielectric (up to a few dozen square centimetres). The energy of the discharge is as follows (Cross 1987 after Masuda):

$$W = K(U_1 q - q^2/2C_d) \quad (10.2)$$

where K is the experimentally determined coefficient (approximately 0.08), U_1 is the initial potential of the dielectric surface, q is the charge transferred by the discharge, and C_d is the capacity equal to the transferred charge q

divided by the potential drop at the surface during discharge. Applying a charge neutraliser (particularly passive ionisers) or reducing the speed of the technological processes that cause the electrification can prevent this discharge.

4. *Propagating brush discharge (PBD)*: A multichannel discharge (with plasma channels radially coinciding) occurring along the surface of the highly electrified, thin dielectric layer and backed by the conducting surface (the flat capacitor) if the conducting object (e.g. the worker's finger) approaches the dielectric surface or as a result of the electric breakdown of the dielectric layer. This is the strongest ESD, in which energy can exceed 1 J, and it can ignite all flammable vapours, gases and hazes, and most combustible dusts, and can be dangerous for humans. PBDs occur when the dielectric layer thickness is between 0.1 and 8 mm, the surface density of accumulated energy is more than 10 J/m^2 and the surface density σ of the accumulated charge is more than 0.25 mC/m^2. PBDs can be prevented by avoiding placing the dielectric layers (or paint covers) on the metal surfaces, increasing the thickness of these dielectrics above 10 mm, reducing this thickness below 0.5 mm, or applying a dielectric for which the breakdown voltage is smaller than 4 kV. Some authors (Jones and King 1991; Cross 1987, after Heidelberg) recommend the following dielectric thicknesses, depending on the accumulated charge density σ:

   - If $\sigma \leq 0.25$ mC/m^2, the dielectric should be much thinner than 1 mm.
   - If $\sigma \geq 0.25$ mC/m^2, the dielectric should be much thicker than 1 mm.

5. *Cone discharge (Maurer's discharge, bulking brush discharge)*: Occurs along the conical surface of the electrified heap of the dielectric material towards the walls of the vessel or silo. According to Glor (1988), this type of discharge is possible during the intensive inflow of a coarse (particle diameter more than 1 mm), highly insulating (volume resistivity ρ_v more than 10^{10} Ωm), and charged loose material into the vessel or silo. Its maximum effective energies are probably not more than 20 mJ. It can ignite flammable gases, vapours and some sensitive dusts.
6. *Lightning-like discharge*: This type has not been observed in industrial practice, but occurs from the dust clouds during tornadoes, dust storms, and volcanic eruptions (Britton 1999). It cannot occur in silos smaller than 60 m^3 or of diameter less than 3 m (Boschung et al. 1977). Examples of possible discharges in silos are shown in Figure 10.1.

The accidental introduction of a conducting and grounded object into the interior of a filled silo can cause a spark discharge between the objects and the silo. The introduction of thin (diameter smaller than 3 mm) and grounded wires or rods into the silo before starting its filling can neutralise the electric charge of the poured material, because of corona discharges from the wires.

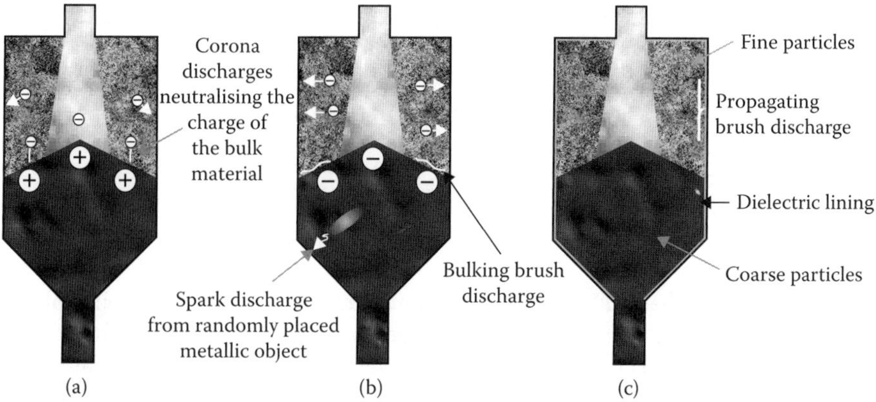

FIGURE 10.1 The potential electrostatic discharges that can occur while filling up the silo with dielectric bulk material: (a) coarse and fine particles are charged in reverse—corona discharges occur and neutralise the charge of heap; (b) coarse and fine particles have the same charge polarity—cone discharges occur, not grounded piece of metal inside the heap can cause the spark discharge; and (c) dielectric lining at the conducting walls can cause propagating brush discharge.

10.3 ELECTROSTATIC HAZARDS

If the dissipation rate of the electrostatic charge is insufficient, the accumulation of the charge leads to a strong electric field and an incendiary discharge; these can occur due to the widespread processing and application of nonconducting synthetic or natural materials and flammable dielectric liquids and dusts, which can be electrified easily and can store the accumulated charge for a long time. The three groups of hazards caused by the ESDs are as follows:

1. Initiation of deflagration or explosion of explosive atmospheres and materials
2. Shocks to humans
3. Damage to the semiconductor electronic devices

The ESD energy, usually in the range of several microjoules to a few hundred millijoules (seldom a few joules), sometimes exceeds the ignition energy of the explosive atmospheres and explosive materials. ESDs cause about 9% of dust explosions. According to the Polish National Headquarters of the State Fire Service 105 fires were caused by ESDs from 1999 to 2000.

Explosive atmospheres are categorised by their MIE, which is defined as the minimum energy that can ignite a mixture of specified flammable material with air or oxygen, and is measured according to the standard procedure.

The typical MIE ranges of flammable and explosive materials are as follows:

- Primary explosives: 0.001–0.1 mJ
- Mixture of gases and vapours with the air: 0.0004–1.0 mJ
- Sensitive dust clouds: 1–10 mJ
- Dust clouds: 10–5000 mJ

MIE values in atmospheres with increased concentrations of oxygen can be of a smaller magnitude.

ESDs from the charged human body do not create direct health or life physiological risk, but they can cause involuntary responses resulting in falls or injuries. The human response to ESD is dependent on the gender, age, physical characteristics, and skin moisture and sensitivity. The electrical capacity of the body is also significant; it is usually in the range of 70–300 pF (picofarad). The electrostatic potential of the body, which is the perception threshold, varies from person to person from 0.6 up to 7 kV depending on the surface in contact with the conducting object and the method of contact. The perception threshold is commonly agreed to be about 2 kV. With respect to these potentials, the energy thresholds range from 0.05–0.8 mJ (Guderska 1981) up to 0.5–2 mJ (Britton 1999), which exceed the MIE of the majority of the mixtures of flammable gases and vapours with the air. ESDs are rated as unpleasant for the range of 1–15 mJ, 15–40 mJ as very unpleasant, 40–100 mJ as painful, and 100 mJ as a severe shock. According to Britton (1999), unconsciousness is possible at 1–10 J and cardiac arrest at above 10 J. Frequent ESDs occurring on a limited area of the skin surface can cause local irritation and dermatitis. Usually, the electric potential of an electrified human does not exceed 20 kV (more often 10 kV), and the energy of the ESD from the body does not exceed 100 mJ, which can be painful but not dangerous. Discharge through the body from different electrified objects of higher potentials and accumulated energies can be dangerous to life. Such energies sometimes have PBDs and spark discharges from objects, whose potential is of the order of dozens or more kilovolts and that have capacities above 1000 pF. The contact of the human body with the electrodes of charged capacitors or with the wires supplying high-voltage direct current (DC) is very dangerous, even some time after disconnecting the wire from the DC supplier.

Human exposure to an electrostatic field is not harmful if it does not cause discharges from the body due to electrostatic induction (see Section 10.4). The electrostatic field does not penetrate the body and is suppressed by the skin about 10^{12}-fold (Polk and Postow 1996).

Electronic semiconductor devices, especially those made by *metal oxide semiconductor* technology, magnetoresistive and thin-layer devices, or printed boards, are particularly sensitive to ESDs. ESD (through-unit or nearby) can cause thermal damages to the thin conducting lattices or voltage breakdowns of thin insulating layers inside the semiconductor devices. This may happen during production, transportation, service, and usage of ready-made devices. ESDs can disturb the function of electronic devices, so in some cases, such as medical electronic diagnostic apparatus and data centres, antielectrostatic equipment of rooms (especially the floors) is obligatory to prevent the formation and accumulation of electrostatic charge. ESDs from the human body cause damage to electronic devices and also ignition of explosive atmospheres, even if the discharge is invisible and imperceptible by humans.

10.4 FORMATION OF UNBALANCED ELECTROSTATIC CHARGE

The generation of an excessive electrostatic charge requires the delivery of additional energy to the system. The phenomena most crucial for the safety at work are discussed in this section.

The most widespread phenomenon is electrification by contact and friction (Figure 10.2). At the contact surface between two solid objects, solid and liquid, or two liquids, diffusion of elementary charge carriers (electrons, ions, and holes) occurs. The carriers tend to occupy the lowest energy level. The detachment and separation of two different objects always leads to their electrification. Mutual friction provides a better contact and increases the surface temperature, thereby increasing the mobility of the charge carriers and intensifying the charge. Friction often causes the exchange of materials and results in a more intensive electrification than electrification by contact.

The friction causes electrification in objects in common operations such as wiping, pouring, pneumatic and cochlear transportation, and electrifying conveyor belts; in humans while walking on a synthetic floor or rising from synthetic seats (e.g. getting out of a car when the moisture of the air is low); in vehicles moving on rubber tyres; in liquids during flow and mixing; and in ice crystals due to the rapid expansion of gases causing the freezing of water vapour. While processing loose materials (e.g. sieving, crumbling, pouring, milling, pneumatic, and cochlear transportation), the mass density of the generated charge is in the range of 10^{-5} to 10^{3} μC/kg. Single-phase dielectric liquids, while flowing with the velocity v (m/s) through the metal pipeline, electrify at approximately $5v$ (μC/m^{3}).

The conducting objects (e.g. the human body) placed in the external electrostatic field are electrified by *electric induction* (Figure 10.3), which is the result of drawing aside the positive and negative charges by Coulomb force. Attaching a conducting (especially earthed) object to another conducting object can cause a spark discharge, which is especially crucial during the production of paper, synthetic foils, tyres, and so on.

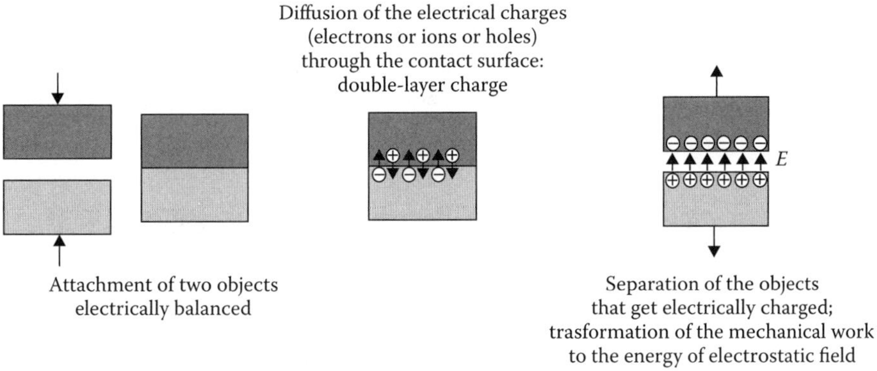

FIGURE 10.2 Electrification by contact.

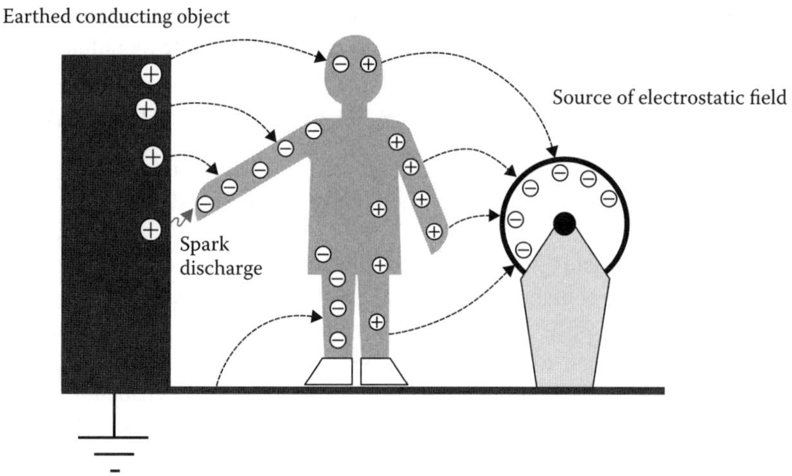

FIGURE 10.3 Electrification of a conducting object by induction.

If a conducting object touches the other conducting and electrified object, the object is then *electrified by conduction*; for example, an electric shock experienced by two persons during handshaking. This is also possible if a stream of ions or electrified particles drifts towards the object (dielectric or insulated conductor) in the external electrostatic field.

Electrification during the crumbling and atomisation of solid materials and liquids can be very intensive and dangerous. The fractions of coarse and fine particles in such operations usually have opposite polarity, and the gravitational separation of the fractions generates a very strong electric field between them. This occurs when silos are filled with bulk dielectric materials (powders, granules, or petals). It can also occur when a stream of liquid comes in contact with an obstacle and a highly electrified haze or spray is formed. This happened in 1969 when the oil vapours in three Shell tankers exploded while the interior tanks were being washed with streams of water (Chang et al. 1995). Streams of an oil and water mixture are electrified more intensively than the both liquids separately.

In polyphasic liquids, the settlement and separation of solid or liquid particles immiscible with the basic constituent and forced by gravitation or inertia can form a high-intensity electric field inside the liquid and cause ignition. This is called *electrification by sedimentation*.

10.5 METHODS OF PREVENTING ELECTROSTATIC DISCHARGE

Protection against ESDs is always necessary in the vicinity of explosive atmospheres and wherever electronic diagnostic apparatus are used. In the European Union, prevention of ESDs and electrostatic charge accumulation in explosive atmospheres are mandated by directives 94/9/EC (1994), 1999/92/EC (1999) and, to some extent, by 2006/42/EC (2006).

The resistivity of the materials is significant in the prevention of electrostatic hazards. Resistivity is divided into three groups (CLC/TR 50404, 2003):

- *Insulating*: Volume resistivity $\rho_v > 10^9\ \Omega\cdot m$ and/or the surface resistivity $\rho_s > 10^{10}\ \Omega$
- *Dissipative*: $10^4\ \Omega\cdot m < \rho_v \leq 10^9\ \Omega\cdot m$ and/or $\rho_s \leq 10^{10}\ \Omega$ (or surface resistance $R_s < 10^{10}\ \Omega$) measured at ambient temperature and 50% relative humidity
- *Conductive*: $\rho_v < 10^4\ \Omega\cdot m$

The resistivity limits specified in some other documents may differ slightly from these values.

According to the model shown in Figure 10.4, the chain of events leading to the accident or to damage of the sensitive devices is as follows:

1. Appearance of a sensitive medium (explosive atmosphere, explosive material, human body, or sensitive electronic unit or device)
2. Essential electrification
3. Excessive accumulation of charge and energy higher than that igniting, damaging, or causing shock
4. ESD through the sensitive medium

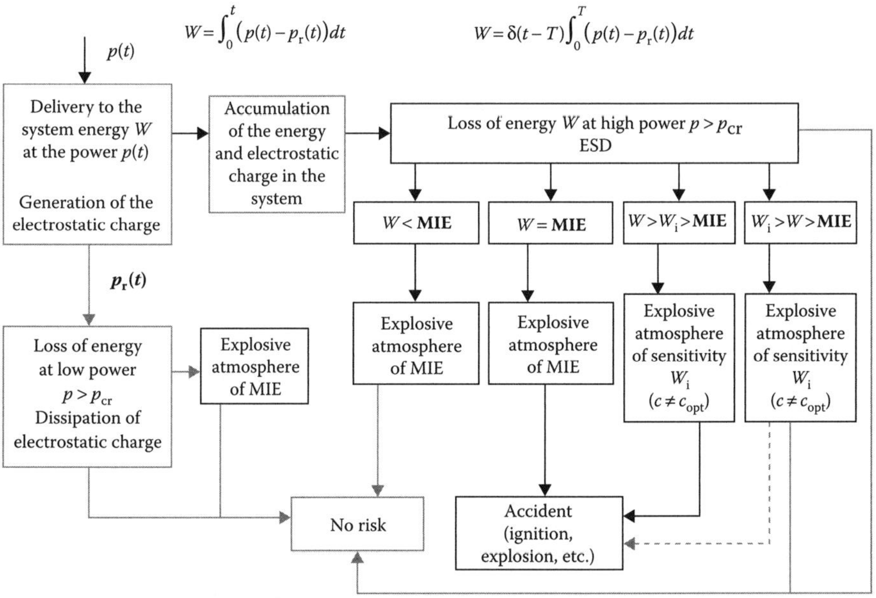

FIGURE 10.4 The physical model of hazard formation of the ignition of explosive atmosphere by the electrostatic discharges: W = the energy of discharge; W_i = the ignition energy in given conditions; MIE = the minimum ignition energy; p = the power of the discharge; p_r = the dissipated power; p_{cr} = the critical (igniting) value of the power; and c = the concentration of flammable factor in the atmospheric air.

Any interruption to this sequence prevents an accident. The general aim in preventing ESDs is to break this chain via points 2 through 4.

10.6 AVOIDING EXCESS ELECTRIFICATION

Selecting appropriate materials to be in contact during technological processes or decreasing the intensity of their mutual friction (e.g. decreasing of the linear speed of the flow or mixing of liquids with a conductivity smaller than 50 pS/m; pS is picosiemens) can limit electrification's intensity and reduce its accumulation.

The triboelectric series of different materials was introduced primarily by Cohen in 1898 and then developed by other researchers (Harper 1998) and organises commonly used materials according to their permittivity (ε). Cohen noticed that if two materials are brought into contact, the one with a higher permittivity receives the positive charge (usually its work function is also lower; Lüttgens and Wilson 1997). The more distant the contacting materials in the triboelectric series, the more electrified they are. The simplified triboelectric series is shown in Table 10.2 in descending order from the material receiving the most positive charge to the material receiving the most negative charge; however, this order is not absolute, and there are exceptions.

The electrification of nonconducting flammable liquids can be avoided by preventing splashing, which causes the formation of an electrified haze or spray. For

TABLE 10.2
Approximate Triboelectric Series for Commonly Used Materials

			Donor (+)		
1	Asbestos	16	Rock-salt (0)	31	Polyester (polyethylene terephthalate [PET])
2	Leather	17	Steel (0)	32	Celluloid
3	Glass	18	Wood (0)	33	Styrene
4	Quartz	19	Acrylic glass (PMMA)	34	Acrylic fibres
5	Mika	20	Amber	35	Polyvinylidene chloride (PVDC)
6	Human hair, rabbit fur	21	Sealing wax	36	Polyurethane (PU)
7	Polyoxyethylene (POE)	22	Resins	37	Polyethylene (PE)
8	Nylon	23	Polystyrene	38	Polypropylene (PP)
9	Wool	24	Rubber	39	Polyvinyl chloride (PVC)
10	Animal fur	25	Ni, Cu, Ag, C, Co, Sn, Sb, brass	40	Polyvinyl acetate (PVA)
11	Pb, Ca, Mg	26	Sulphur	41	Silicon
12	Silk	27	Pt, Au	42	Polytetrafluorethylene (PTFE)
13	Al	28	Synthetic rubber		Acceptor (−)
14	Paper	29	Acetate		
15	Cotton (0)	30	Rayon (regenerated cellulosic fibre)		

example, the outlet of the hose should be placed at the bottom of the tank while filling up tankers with fuel.

In explosion-hazard zones, rapid expansion of compressed gases should be prevented. The expansion of a gas is accompanied by cooling, which can result in freezing or condensation of some gases or vapours. The droplets or ice crystals striking at the obstacles can be strongly electrified and can cause ESDs. Cooling of the expanding, previously compressed carbon dioxide causes the formation of electrified ice. Therefore, carbon dioxide is not recommended for rapid curtains or fire extinguishers in explosion-hazard zones. A similar situation can occur when compressed air or water steam escapes from openings. If a flammable gas expands through an opening, the risk is obviously greater. However, LPG produces charged drops easily, and can escape through an opening.

Charge accumulation can be effectively limited by suitable and continuous neutralisation. This can be achieved in several ways:

- The resistivity of the electrifying material by the addition of antistatics (for which chemical formulas are seldom revealed) can be decreased. The dangerous electrification of flammable liquids can be eliminated if their conductivity is 250 μS/m or higher. In bulk materials, contact with earthed metallic installation parts, containers, and so on does not cause risk if the value of their resistivity ρ_v does not exceed 10^4 Ωm.
- The resistance of the path of charge leakage to the ground can be reduced, for example, by using metallic earthed pipes for the transport of liquids and powders, electrostatic conducting floors, and grounded containers, and assuring a relative air humidity higher than 60%.
- Leakage resistance to the ground of the conducting objects (e.g. the human body and the resistance of shoes, in series) not higher than 1 MΩ in explosion risk zones and not higher than 1 GΩ in data centres or electronic plants should be assured. The same conditions are required for conducting or dissipative floors where the leakage resistance should be measured by the space between a suitable electrode placed on the floor and the point of the earthing.
- The unbalanced charge with a stream of air ions, which are emitted by the corona ionisers or sources of radioactive radiation (especially the alpha; do not apply plutonium because of its toxicity) can be eliminated. The active corona ionisers can be supplied with high-voltage DC, alternating current (AC), or high-frequency voltage. The passive corona ionisers have earthed electrodes, and the source of the necessary electric field is an electrified object to be neutralised. The best solution is an arrangement of the electrodes and their supply, at which the electric field of the electrified object 'draws out' the ions of the necessary polarisation.
- In processes causing strong electrification of flammable liquids and powders, the material should be kept motionless during several relaxation times τ of the material ($\tau = \varepsilon\rho$; ε being permittivity and ρ the volume resistivity of the material) before undertaking any actions that could provoke discharge

from their surfaces (e.g. before the sampling of petrol just after mixing, with the earthed metal sampler on metal rod).

- Electrified workers are one of the dangerous sources of electrostatic hazards in explosive atmospheres, in the electronics industry, and so on; the electrostatic potential should be reduced to lower than 100 V (or even less for the electronic devices) by using the conducting or dissipative floors, wall and chair covers, antistatic clothes, and conducting footwear.

In the electronics industry, special electrostatic protection areas (EPAs) are used in which all conducting parts (including workers) are equipotential due to bonding and earthing and the trolleys and shelves are equipped with conducting wheels and frames. The workers have special bands on the wrist or under the knee, joined by the elastic conductor to the common connecting point, with the table covering, floor, antistatic clothes, and so on. This point is usually earthed (joint with the protective earth). These precautions should generally assure the effective resistance of the worker's body to the ground less than 1 GΩ (but not less than 100 kΩ if electric shock is possible). To avoid the risk of electric shock, the wrist- or under-knee band and elastic conductor should not be metallic. Electronic devices and components should be transferred and kept in many-layered antistatic packaging to protect them against electrification, ESDs, and electrostatic field. The intensity of the electrostatic field in the surroundings of sensitive devices should not exceed 10 kV/m. Detailed values of the resistance of the mentioned units and the rules of EPA arrangement are specified in international (IEC, CENELEC) and national standards and technical reports with:

- Spark discharges can be prevented in explosion-hazard zones by equalising the electric potential of all conducting objects (machines, installations, apparatus, tools, and so on) with firm connections and earthing. This does not concern just very small objects whose capacitance is less than about 3 pF in the vicinity of explosive gases and vapours or 10 pF in the vicinity of explosive dusts. The effective resistance of the earthing installation should be smaller than 10 Ω, and the effective resistance of the passage between the earthed objects or between the object and the ground should be less than 100 Ω. The earthing installations have to be continuous (e.g. chains are inadmissible). The ground leakage resistance of the rotating shafts cannot exceed 100 Ω. The earthing is used independently of the other precautions.
- The bonding of potentially insulated items in explosive atmospheres should be applied (e.g. the metallic flanges of the connected pipes for flammable liquids or bulk materials, if they can potentially be insulated by seal, paint, or corrosion, should be bonded with independent wires). In hazard zones, trolleys should have conducting wheels, which should be grounded through a conducting floor. The transmission conveyor belts should be made of dissipative materials, and rolls should be made of the metal and should be grounded. The speed of the belt in the hazard zones 0, 1, and 2 should not exceed 0.5 m/s and in the zones 20, 21, and 22 should not exceed 5 m/s.

- In explosion-hazard zones, workers should wear antistatic clothes and footwear. Their underwear should be made of cotton and they should not take off or put on pieces of clothing while in this zone.
- Low speeds should be used in processes in which the materials are being electrified, such as pouring, sieving, mixing, and grinding, especially while creating mixtures. Pouring powders from insulating bags (like paper or polyethylene) or from containers with an inner lining into metallic vessels with flammable liquids should be avoided. Brush discharges from bags to vessel walls can occur and ignite the vapours of the solvent. The workers should stand on a conducting (or at least concrete) floor and use conducting bags or containers.
- Nonconducting pipes should not be used to transport the powders and flammable liquids. If possible, the speed of those liquids should not exceed 1 m/s. The inner lining of the metallic pipes, large containers, and silos can cause the development of PBDs, if the breakdown strength of the lining is higher than 4 kV.
- Depending on the kind of bulk materials and the explosive atmospheres, the use of flexible intermediate bulk containers (FIBCs, also called 'big bags') should be preceded by the proper choice of the FIBC (there are four types: A, B, C, and D [CLC/TR 50404 2003]).

Detailed principles for preventing electrostatic hazards can be found in many standards and guides (e.g. CLC/TR 50404 2003; NFPA77 2007; BS5958 1991; JNIOSH–TR–No. 42 2007).

10.7 MEANING OF THE PROPER BEHAVIOUR OF WORKERS

Workers and management follow safety rules and procedures for the prevention of ESD hazards, particularly in the explosive atmospheres, in accordance with adequate standards and regulations. Most ignitions and explosions caused by ESDs are caused by deviation from these rules and procedures, by lack of knowledge and training, or even by carelessness. The use of improper or bad-quality clothes and footwear, use of unearthed metal items and tools, pouring flammable liquids from/into insulated or unearthed metallic vessels by unearthed persons, and standing on nonconducting floors are dangerous. The floor of the work area should be clean, as flammable dust can settle on the floor and can create an explosive atmosphere from a blast of air (accidental or caused by explosion).

An increase in the concentration of oxygen in the atmosphere is particularly dangerous, as this reduces the MIE by a few orders of magnitude. Oxygen should not be used for scavenging (cleaning) pipes because of the risk of explosion of the dust and oxygen mixture.

10.8 SUMMARY

ESDs can cause ignition and explosion of mixtures of air, gas, or liquid oxygen (and other oxidants, such as chlorine, nitric oxide, ozone, nitrite, and nitrate) with flammable gases, vapours, and hazes of flammable liquids or dusts. The mixtures of flammable gases or vapours and dusts are particularly dangerous because dust can be the source of ESDs. ESDs can also cause electric shock and are destructive to electronic devices; they should be prevented by reducing the intensity of electrification, increasing the conductivity of the objects, equalising the electric potential of metallic objects in hazardous areas to zero, and avoiding large surfaces on dielectric objects. Safety procedures should be established and followed by both workers and management. Proper equipment and training for workers are also necessary.

REFERENCES

Boschung, P., W. Hilgner, G. Luttgens, B. Maurer, and A. Widmer. 1977. An experimental contribution to the question of existence of lightning-like discharges in dust clouds. *J Electrostat* 3:303–310.

British Standard BS5958. 1991. Part 2—Code of practice for control of undesirable static electricity.

Britton, L. G. 1999. *Avoiding Static Ignition in Chemical Operations*. New York: Center for Chemical Engineers of the American Institute of Chemical Engineers.

Chang, J.-Sh., A. J. Kelly, and J. M. Crowley, eds. 1995. *Handbook of Electrostatic Processes*. New York: Marcel Dekker.

CLC/TR 50404. June 2003. Electrostatics—Code of practice for the avoidance of hazards due to static electricity. Technical Report. Europe: European Committee for Electrotechnical Standardization.

Cross, J. 1987. *Electrostatics: Principles, Problems and Applications*. Bristol: Adam Higler.

Directive 1999/92/EC of the European Parliament and of the Council of 16 December 1999 on minimum requirements for improving the safety and health protection of workers potentially at risk from explosive atmospheres (15th individual Directive within the meaning of Article 16(1) of Directive 89/391/EEC). OJ EU, L 23/5728.1.2000.

Directive 2006/42/EC of the European Parliament and of the Council of 17 May 2006 on machinery, and amending Directive 95/16/EC (recast). OJ EU, L 157/24, 9.6.2006.

Directive 94/9/EC of the European Parliament and the Council of 23 March 1994 on the approximation of the laws of the Member States concerning equipment and protective systems intended for use in potentially explosive atmospheres. OJ EU, L 100/119.4.94.

Gajewski, A. S. 1987. *Static Electricity—The Recognizing, Measurement, Prevention, Elimination*. Warsaw: IWZZ.

Glor, M. 1988. *Electrostatic Hazards in Powder Handling*. Letchworth: Research Studies Press, Ltd.

Guderska, H. 1981. The influence of electrostatic discharges on the physiological responses of the human body. Ph.D. diss. Wroclaw: Wroclaw University of Technology.

Harper, W. R. 1998. *Contact and Frictional Electrification*. Morgan Hill: Laplacian Press.

JNIOSH–TR–No. 42. 2007. Recommendation for requirements for avoiding electrostatic hazards in industry. Technical Recommendation of National Institute of Occupational Safety and Health, Japan.

Jones, T. B., and J. L. King. 1991. *Powder Handling and Electrostatics: Understanding and Prevention*. Chelsea, MI: Lewis Publishers.

Kaiser, K. L. 2006. *Electrostatic Discharge*. Boca Raton, FL: CRC Press.

Lüttgens, G., and N. Wilson. 1997. *Electrostatic Hazards*. Oxford: Butterworth-Heinemann.

NFPA 77. 2007. *Recommended Practices on Static Electricity*. Quincy, MA: International Codes and Standards Organization.

Polk, Ch., and E. Postow. 1996. *Handbook of Biological Effects of Electromagnetic Fields*. Boca Raton, FL: CRC Press.

11 Electric Current

Marek Dźwiarek

CONTENTS

11.1 INTRODUCTION

Human beings have been exposed to electrical hazards since the dawn of time. Initially, electrical hazards were associated with natural phenomena, like atmospheric or electrostatic discharges. The first effective protective device was the lightning conductor invented by Benjamin Franklin. However, only when humans developed industrial applications of electric current at the end of nineteenth century did people realise the dangers of electrical energy. The first standardisation body—the International Electrotechnical Commission—was then established, aimed at developing principles of prevention from electrical hazards. The commission has since developed a few thousand standards covering different issues of prevention, such as the effects of electric current upon humans and the environment.

Today, in power supply systems of receivers in industry, civil engineering, agriculture, and other branches of industry, electrical energy is usually transferred by applying a voltage of 400/230 alternating current (AC) with a frequency of 50 Hz (hertz). However, the risk of electric shock is inherent, and is the most important hazard posed, in such cases (Karski 2003). Electric shock is the most frequent cause of accidents (Fordyce et al. 2007), and is fourth in the list of most fatal accidents, after falling down, losing control over a car, or being hit with a falling object. The consequences of electric shock are usually tragic, often resulting in death. Protection against electric shock is therefore crucial. To ensure effective protection against electric shock, safety principles should be followed when using electrically powered systems and only the devices that have been designed according to the requirements given in the standards and rules should be used.

11.2 DEFINITIONS

Live part: A conductor or conductive part intended to be energised in normal use and that conducts electric current; for example, socket pins, cable phase wires (also neutral conductor N), switchgear conductor bars, motor windings, electrical equipment, and so on.

Exposed conductive part: A conductive part of electrical equipment, but not a part of the working equipment; for example, electrical machine enclosures, switchgear enclosures, motor enclosures, luminaires, and so on.

Extraneous conductive part: A conductive part, not a part of electrical equipment and liable to introduce a potential, generally the earth potential; for example, all pipelines, building structure elements, belt conveyors, and so on.

Direct contact: The contact of persons or livestock with live parts that may result in electric shock.

Indirect contact: The contact of persons with exposed or extraneous conductive parts, which could become live under fault conditions or other disturbances.

11.3 GENERAL CHARACTERISTICS OF ELECTRIC SHOCK HAZARDS

Electric shock is a shocking current flow through a human body when it comes into contact with two points of different electric potentials. The difference between potentials may be a touch voltage, which appears when a shocking current flows through the following paths: upper extremity–upper extremity (hand–hand), upper extremity–lower extremity (hand–leg), hand–torso, and leg–leg.

Direct contact can cause electric shock to a human; that is, touching a live part, which, by definition, operates normally under a voltage higher than the safe one relative to the reference potential (Figure 11.1). Direct contact usually happens when work safety requirements are not satisfied, for example if one does not follow the instructions, if the design or use of the device is improper, if the device has not been adapted to environmental hazards, if a damaged or incomplete device is used, and so on.

Electric shock can also be caused by indirect contact (Figure 11.2), that is, contact with the exposed conductive elements (the elements of an electrical system or electrical equipment that are not live parts) or extraneous conductive parts (i.e. metal objects not connected to the electrical circuits). When the voltage in such an element or between the elements, relative to the earth, exceeds the safe voltage, the person touching the element will receive an electric shock. This usually happens when a fault appears in the insulating system, and is accompanied by an earth fault. Some extraneous causes of electric shock include the propagation of currents after

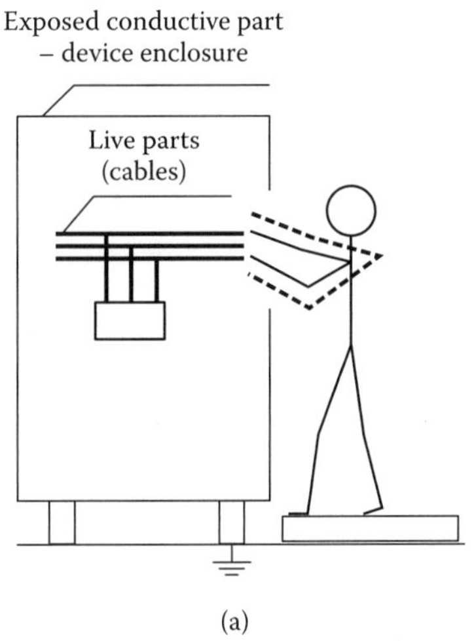

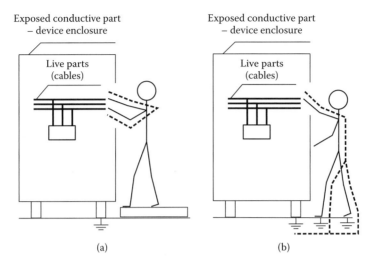

FIGURE 11.1 Electric shock due to direct contact: (a) shock current flowing along the hand–hand path, (b) shock current flowing along the hand–leg path.

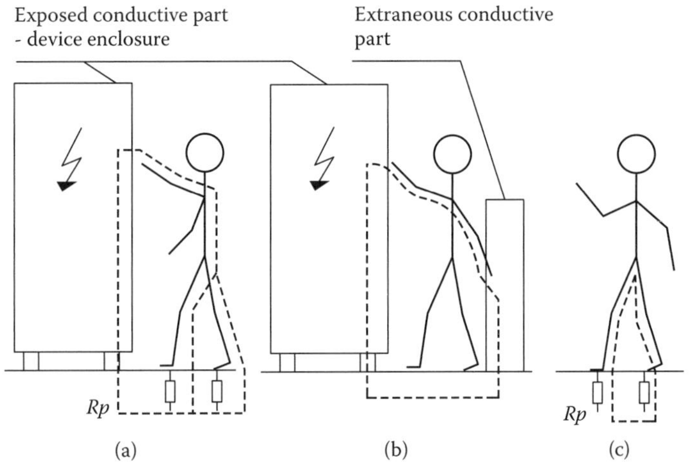

FIGURE 11.2 Electric shock due to indirect contact: (a) shock current flowing along the hand–leg path, (b) shock current flowing along the hand–hand path, and (c) shock current flowing along the leg–leg path.

an atmospheric discharge, stray currents approaching other objects, or a step voltage accumulating between the feet of a person standing on a conductive floor.

11.4 PARAMETERS

The electric current can be transmitted as a direct current (DC) or an alternating current (AC). Industrial and household systems usually have AC systems. DC is used in battery-supplied systems or internal parts that are usually unexposed electrical circuits.

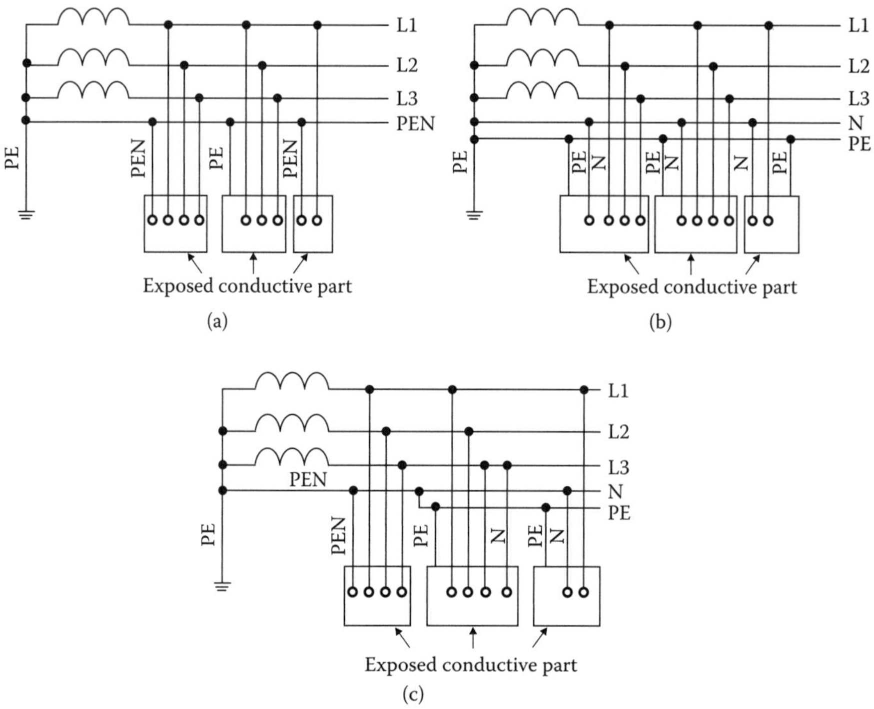

FIGURE 11.3 Possible network arrangements in the TN system: (a) TN-C, (b) TN-S, and (c) TN-C-S.

The electrical systems can be divided into circuits as follows, depending on the hazard level:

- Extra-low voltage: Up to 25 V for the DC and 60 V for the AC
- Low voltage: Up to 1000 V for the DC and 1500 V for the AC
- High voltage: Over 1000 V for the DC and 1500 V for the AC

High-voltage systems are used mainly for electrical energy transfer and power supply of high energy–consumption machines and devices. Special rules and safety requirements apply to such systems because they pose hazards. Only authorised persons can access the system. Unauthorised people cannot get close to this type of system; therefore, this chapter will discuss only commonly used low-voltage systems.

Depending on the connection of neutral conductor and earthing, power supply systems using AC networks can be divided into three basic systems: TN (TN-C, TN-S, TN-C-S), TT, and IT, which are described in the IEC 60364-3:1993 standard. Figures 11.3 through 11.5 show possible system network arrangements.

In the TN network system, each point is directly earthed and the exposed conductive parts are connected to it by protective conductors. Such a system can be arranged in three ways (Figure 11.3):

- *TN-S system*: In the whole electrical network, separate protective and neutral conductors are used.
- *TN-C system*: In the whole electrical network, one conductor plays the roles of protective and neutral conductors, respectively.
- *TN-C-S system*: In part of the electrical network, one conductor plays the roles of protective and neutral conductors, respectively.

In a TT network system, one point is earthed directly and the exposed conductive parts are connected with an earth electrode, independent of the network system earth electrode (Figure 11.4). In an IT network system, all live parts are insulated from the earth, one point is connected to the earth through impedance and the exposed conductive parts of the electrical system are earthed, independent of each other or together, or are connected to the system earthing (Figure 11.5).

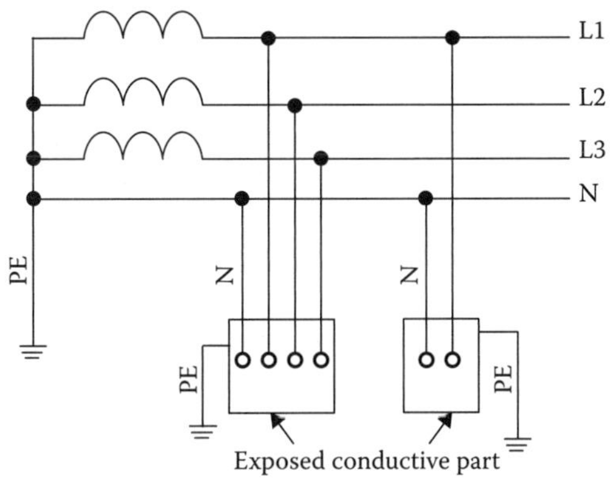

FIGURE 11.4 System of the TT network.

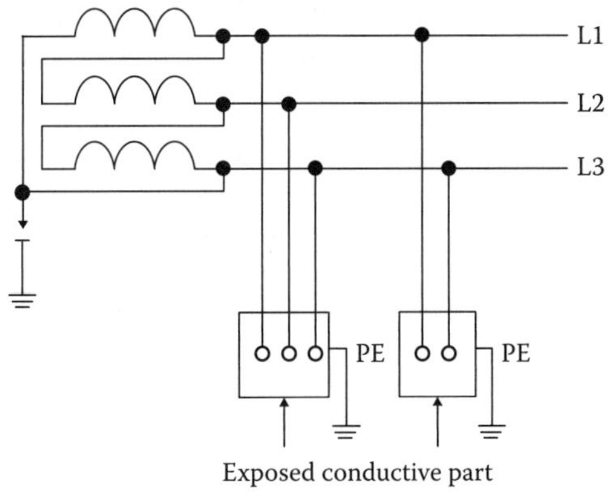

FIGURE 11.5 System of the IT network.

11.5 MEASUREMENT TECHNIQUES

The voltage and intensity of the current are basic network parameters. Commonly available analogue or digital measuring devices can be applied when measuring the parameters of an electrical network. Depending on the determined needs, very simple devices or very complex multifunctional measuring systems can be used.

To ensure effective protection against electric shock, the parameters of the protective circuit must be checked on a regular basis for its entire life time. The check is comprised of the following:

- Testing the insulation resistance of power supply and protective cables
- Testing the electric strength of insulation
- Checking the effectiveness of protection against electric shock, which is determined based on the measurement results of the continuity of the protective bonding circuit

Only persons who have relevant certificates should measure the parameters. A variety of measurement devices are available that were designed to measure the parameters of electrical systems.

The frequency the electrical system parameters must be checked depends on the environmental conditions under which the system is used. Strzałka (1999) has provided sample frequencies for a periodic check of electrical system parameters, depending on environmental conditions of usage (Table 11.1).

11.6 MECHANISMS BY WHICH ELECTRIC CURRENT AFFECTS THE HUMAN BODY

Electric current affects humans in both direct and indirect ways. In *direct contact*, current flows through the body and causes physiological changes, that is, changes

TABLE 11.1
Sample Frequencies for a Periodic Check of Electrical System Parameters

Exploitation Environment	Frequency of Checking the System Parameters	
	Insulation Resistance	Effectiveness of Protection against Electric Shock
Caustics	At least once a year	At least once a year
Explosion hazard	At least once a year	At least once a year
Open outdoor space	At least once every 5 years	At least once every 5 years
Very high (about 100%) and high (75%–100%) humidity	At least once every 5 years	At least once a year
High temperature of the air (over 35°C)	At least once every 5 years	At least once a year
Fire	At least once a year	At least once every 5 years
Dust	At least once every 5 years	At least once every 5 years
Others	At least once every 5 years	At least once every 5 years

in biological, chemical, and physical characteristics. This can cause the following symptoms:

- Pain due to contractions of transverse muscles through which the current flows. The muscle contractions make it impossible to get out of the electric shock, and may cause falling or mechanical injuries (e.g. damage of joints).
- Failure of blood circulation, breathing, sight, heat regulation, and equilibrioception
- Burns to the epidermal tissue and internal organs due to the large amount of heat supplied to the body; this may cause necrosis or carbonisation of tissues
- Syncope
- Ventricular fibrillation, usually causing death

After the current ceases, the following health distortions can occur: pallor, tremor, perspiration, and mental disorders such as apathy or euphoria. These disorders can occur at different times, from minutes to months after the electric shock.

Indirect contact, which is not always accompanied by a current flow through a human body, may cause mechanical injuries due to falling from a height or dropping an object or burns due to contact with objects of a high temperature or with fire. Electric arc phenomena are also relatively dangerous; they may be spontaneous or caused by human activity. *Combined electric shock* is when a discharge current flows through a human body and the arc breaks out very close to a human with no current flow. As a result, the following symptoms may appear:

- Extensive burns, even of the extremities or other parts of the human body, often causing death
- Increase in the temperature of vitreous bodies and injury of the retina due to infrared radiation
- Injuries to the cornea due to ultraviolet radiation
- Melting of metal parts and insulation carried by a hot flowing gas on unprotected surfaces of the body and the cornea
- Mechanical injuries due to the scattering of the damaged electrical device elements or falling due to a shock wave

Pathophysiological consequences of electric shock depend on a variety of factors:

- The type of current (current pulse, DC, and AC of higher frequencies [15–100 Hz] can cause most dangerous consequences to humans)
- Voltage and intensity of the electric shock current and the shock duration
- The path through which the current flows into the body
- Psychophysical condition of the injured person

These factors indicate the important role the total impedance of the human body (both the epidermis and internal organs) plays in an electric shock.

The impedance of the epidermis varies depending on its physical condition (especially temperature and humidity), the area of contact with the electric shock circuit electrode, the contact voltage, the intensity and frequency of the electric shock current, and the shock duration.

Internal resistance depends mainly on the path through which the current flows into the body and where it reaches its highest values along the hand–hand and hand–leg paths.

According to the recommendations given in IEC/TS 479-1:2005 for protection against electric shock, when the contact voltage is about 230 V at a frequency of 50 Hz, the body impedance for 90% of the adults is as follows, depending on the environmental conditions:

- *Under normal conditions*: $R \geq 1000\ \Omega$ relative to earth; the term 'normal conditions' applies to the premises of houses, offices, schools, and so on.
- *Under special conditions*: $R < 1000\ \Omega$ relative to earth; the term 'special conditions' applies to the areas where the skin can be wet, such as open spaces, bathrooms, and showers; farm buildings; industrial halls of humidity higher than 75% at a temperature higher than 35°C or lower than −5°C; rooms with low-resistance floors; and some special environmental conditions such as swimming pools and interiors of metal containers, where the human body can be wet or in contact with a conductive floor over a large area.

The perception limit is 0.5 mA (milliamperes) for the AC at 50 Hz flowing along the paths from the left hand to the feet, independent of the flow duration.

The subject of the electric shock can escape when a current of 10 mA flows for up to 2000 milliseconds; for higher currents of longer durations, muscle contractions keep the person from easing his or her grip. Higher values of electric shock current cause pathophysiological consequences, such as pain, temporary disorders of the cardiovascular system such as blood pressure increase, reversible breathing distortions due to contractions of the chest muscles, and the possibility of ventricular fibrillation; their severity depends on the duration of the shock.

If the duration of the electric shock current is long, ventricular fibrillation will begin, accompanied by respiratory tract paralysis, which may lead to death, at a current intensity of 30 mA. For protection against electric shock, the limiting value of the shock duration should not be longer than 1 second. A higher electric shock current may also cause injuries to the skeletal system (breaks due to muscle contractions or damage to joints) as well as extensive burns to the skin and internal organs, causing necrosis or carbonisation of tissues.

The consequences of electric shock, especially neurological and neurophysiological effects, are felt for a long time after the accident (Bailey et al. 2008; Kharbouch et al. 2008).

11.7 HAZARDOUS SITUATIONS WHEN EXECUTING DIFFERENT TASKS

The level of electric shock hazard depends on the types of both the task and the operated device. The persons maintaining and repairing electrical networks are particularly exposed, as they usually come very close to live parts. As a rule, the power supply to the electrical installation should be disconnected during maintenance and repair works. However, accident analyses presented by Karski (2003) indicate that cutting off the power supply is not a complete solution due to improper design or negligence of safe work rules. Despite cutting off the power supply, part of the circuit can sometimes still be energised. One should therefore strictly follow the safety and health instructions when executing such tasks at work and make sure that conductive parts are not energised before starting the work.

Electric shock accidents often happen when operating office equipment due to faults in the protective circuits and electrical equipment that were not detected during periodic measurements. The probability of accident substantially increases when the measurements are not properly performed and when maintenance and repair works are neglected. The inspection of electrical equipment is often neglected in office buildings, artisan workshops, small enterprises, and the construction industry.

Electrical equipment at construction sites is often in poor condition due to its exposure to atmospheric hazards and mechanical damages caused by machines and its adaptation to the work. The hazard posed by overhead power transmission lines is very important, as machines at very high buildings may contact such lines and pose electric hazards to both the operators and other people.

Electric shock accidents are rare in the manufacturing industry. The results of research conducted at the Central Institute for Labour Protection–National Research Institute (CIOP-PIB) and presented by Karski (2003) prove that such accidents practically do not happen in large- and medium-size enterprises because they employ special staff to inspect and keep the electrical equipment in good condition. Applied technologies such as electrolytic coating or technologies also cause hazards that result in electrostatic charge accumulation. Electric shock accidents may then happen due to electrostatic discharge (see Chapter 10).

11.8 PROTECTION AGAINST ELECTRIC SHOCK

Protection against electric shock requires the following three elements:

- Protection against direct contact
- Protection against indirect contact
- Additional safety measures

11.8.1 Technical Protection against Direct Contact

Protection against direct contact involves the prevention of contact between the human body and the live part of the device. Based on the designs of electrically powered devices and their environmental hazards, protection against electric shock can be executed using the following:

- *Insulation of the live parts*: Application of basic insulation protects against both intended and unintended contacts. Replacement of the element with a new one is required if the insulation capacity is lost or if there are mechanical damages.
- *Application of guards, enclosures, and barriers*: This includes placing live parts inside casings or fences that allow for at least IP2X protection level, or the use of guards that prevent contact with the energised elements. The casing and fences should be fixed as firmly as possible and made up of materials that ensure their functionality within the lifetime specified by the manufacturer. The disassembly of casings, fences, and guards should be difficult (e.g. require special tools) or be possible only after eliminating the hazards against which they protect.
- *Application of obstacles, portable barriers, and placing live parts outside the reach of the operator*: This reduces the risk of electric shock only when contact with energised elements results from deliberate action. The means of protection can be disassembled relatively easily. It is very rare in solid designs, which are not moveable, but is used very often in maintenance and repair works executed in electrical installations.
- *Application of residual current protective devices (additional safety measures)*: Using switches with a low residual current response ($\Delta I \leq 30$ mA) improves efficiency against direct contact if the protective means are ineffective (Figure 11.6). This method cannot prevent the risk of electric shock due to direct contact, but can reduce its consequences; therefore, this should not be the only protective measure.

11.8.2 Technical Measures of Protection against Indirect Contact

Protection against indirect contact is the prevention of electric shock when a person touches exposed conductive parts, which are energised due to insulation damage. The following measures can be applied:

- *Automatic disconnection of supply*: Short circuits are created through the protective conductors, connecting the exposed conductive parts with a neutral point or the earth (depending on the network structure), combined with protective means ensuring that the supply is disconnected. This task may be executed using overcurrent protection means or residual current protective devices.
- *Protection by provision of class II equipment or by equivalent insulation*: This protection prevents dangerous energising of the exposed conductive parts in case there is basic insulation damage. Double insulation, reinforced insulation, or equivalent insulation is used. Class II equipment is commonly used as an additional safety measure, especially when dealing with moving equipment or hand tools such as power tools (Figure 11.7).

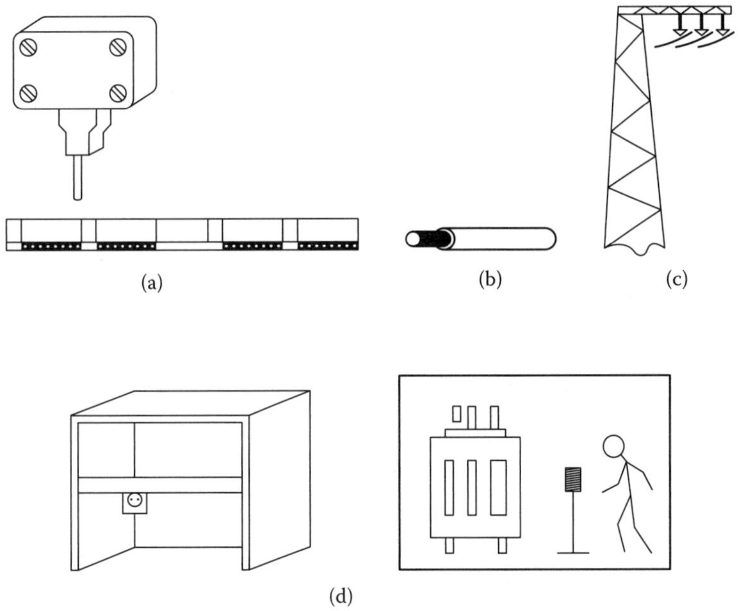

FIGURE 11.6 Sample protection measures against electric shock: (a) guard, (b) insulation, (c) placing live parts outside the reach, (d) fence.

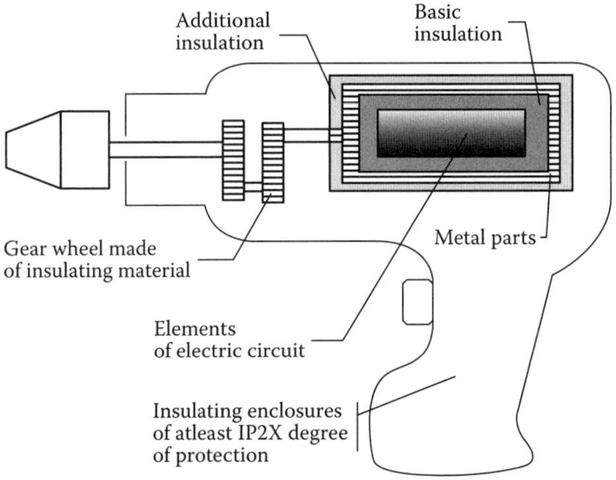

FIGURE 11.7 Sample of class II equipment.

- *Insulation of a workstation*: Preventing simultaneous contact with parts that may have different potentials. The surroundings of a given device should be completely insulated and at the same time no earthed objects or protective conductors should be connected.
- *Application of electric separation in power supply*: A receiver is actuated through a separating transformer or separating converter for a DC. When

using this type of protective means, no electric shock can happen because there is a one-point contact with the live part. However, its application is strictly limited.

- *Application of equipotential bonding not being earthed*: This prevents dangerous touch voltage. All exposed and extraneous conductive parts are connected using unearthed local equipotential bonding.

11.8.3 Organisational Safety Measures for Protection against Electric Shock

Besides technical safety means, organisational safety measures also play an important role in reducing hazards posed by electrical energy, electrostatic energy, or the exploitation of machines and devices. The following organisational safety measures may be followed:

- Ensure that the safety requirements have been satisfied.
- Ensure that the persons operating or maintaining the specified machines and devices have the relevant permissions.
- Ensure that the relevant instructions on the machine and device operation are available.
- Spread knowledge of the safe operation of electrical machines and devices and focus attention on initial and periodic employee training.
- Perform a periodic medical examination of employees.

11.9 GOOD PRACTICES

Good practices in the operation of electrically powered equipment include keeping the equipment in a good technical condition and following the schedules for periodic measurement of its protection against electric shock. The following aspects of work organisation should also be emphasised:

- Periodic training of employees
- Appointing only the persons with relevant permissions to perform all operations of electrical installations and devices
- Ensuring that the equipment is used for what it was intended
- Displaying tables and warning signs about the hazards of electric shock
- Performing periodic inspection of nonmobile electrical devices at least once a month and in the following cases:
    - Before starting the device after changes have been made to its electrical and mechanical parts or after any repairs have been done
    - Before starting the device after a one-month (or longer) break in its operation
    - Before starting the device after it has been transferred
    - While reporting all the repairs and inspections in the device maintenance book

When operating electrically-powered devices, the following rules should be followed:

- Use the plugs that fit into the sockets on the receivers.
- Avoid direct connection of the conductors with the socket (without a plug).
- Use only the devices for which the enclosures reveal at least an IP44 protection level.
- Apply the separating transformers when using the devices of category I.
- Carefully follow the manufacturer's recommendations concerning the use of the devices.
- Try to avoid exposing the electrical equipment to rain or using it under explosive, humid, or damp environmental conditions.
- Use special tools designed for industrial applications when a long-lasting continuous task needs to be executed.
- Use the conductor only for what it was intended.
- Do not move the device by grabbing its conductors or disconnect the plug-socket combination by pulling the conductor.
- Protect the conductors against high temperatures, oils, lubricants, and sharp edges.
- Cut off the power supply to a device that is not operating or on which replaceable equipment is mounted.
- Make sure that the device is in the 'off' state before connecting it with a power supply.
- Do not make any changes in the electrical system or in the design of electrical devices without professional help.

Obeying the aforementioned rules will considerably reduce the possibility of dangerous events when operating electric devices.

11.10 ACTIONS TO BE TAKEN IN CASE OF ELECTRIC SHOCK

Special protection equipment is typically not used in the operation of electrically powered devices. Protection against electric shock can be provided by using other methods. Contact with a source of electric current may cause skin burns, syncope, cardiac standstill, or even death. When a person is subject to an electric shock of a voltage lower than 1 kV for a short time, there is no harm to the human body; however, when the time of exposure is longer, it might be dangerous. The injured person should be freed from the electric current as soon as possible, however, the person under electric shock is as dangerous as the source of current itself. To free the injured person from an electric current of voltage lower than 1 kV, the following should be done:

1. Disconnect the proper electric circuit.
2. Open the proper switching devices.
3. Take the cutouts from the power supply circuit.
4. Cut out or tear the conductors from the power supply side using special tools.

5. Make a short circuit of conductors from the power supply side.
6. Take out the cutouts using special holders.

When the aforementioned actions cannot be taken, the following should be done:

1. Drag the injured person out of the energised device using special tools or objects made of dry wood or plastic.
2. Prevent the injured person from falling down, which can result from cutting off the power supply.

References

Bailey, B., P. Gaudreault, and R. Thivierge. 2008. Neurologic and neuropsychological symptoms during the first year after an electric shock: Results of a prospective multicenter study. *Am J Emerg Med* 26(4):413–418.

Fordyce, T., M. Kelsh, E. T. Lu, J. D. Sahl, and J. W. Yager. 2007. Thermal burn and electrical injuries among electric utility workers, 1995–2004. *Burns* 33(2):209–220.

IEC 60364-3. 1993. Electrical installations of buildings, Part 3: Assessment of general characteristics.

IEC/TS 479-1. 2005. Effects of current on human beings and livestock, Part 1: General aspects.

Karski, H. 2003. Technical means of protection against electric shock. In *Principles of Accident Prevention.*, ed. Z. Pawowska. Warsaw: CIOP-PIB.

Kharbouch, H., et al. 2008. Cataract after electric shock. *J Fr Ophtalmol* 31(Suppl. 1):133.

Strzałka, J. 1999. *Guidance for Periodical Inspections of Technical Devices, Electrical Installations and Lighting Rods*. Warsaw: COBR.

12 Electric Lighting for Indoor Workplaces and Workstations

Agnieszka Wolska

CONTENTS

12.1 INTRODUCTION

Lighting is the use of light (visual radiation) to make places, objects, and their environments visible. Lighting directly influences the speed and reliability of vision and determines how forms, figures, colours, and features of the surface (e.g. texture) of objects are seen. Lighting is very important to maximising visual performance and visual work efficiency and, at the same time, minimising visual fatigue and the risk of perception mistakes. Lighting has three main functions:

- Ensure safety for people in interior environments.
- Ensure proper conditions for performing visual tasks.
- Assist in creating a proper luminous environment.

12.2 HISTORY OF LIGHTING

12.2.1 The Sun and Fire

The history of lighting is as old as the history of man. The most important light source, and for a long time the only one known to man, is the sun. The first artificial light source was fire, which was discovered accidentally. Humans believed fire was a 'gift from the heavens', discovered when a bolt of lightning struck a tree or a bush, setting it on fire (http://www.mts.net/~william5/history/hol.htm). Fire, like the sun, is not only a source of lighting but also plays an important role in human life and the development of culture. Humans could not kindle fire; they could obtain it only after a thunderbolt struck or from the lava expelled from an active volcano. Fire was therefore guarded day and night, without extinguishing. Light was an object of desire and filled humans with awe. According to Greek mythology, Prometheus stole fire from the heavens (Olympian gods) and gave it to the human race with instructions how to use it. Thus, he became the benefactor of humankind. In Roman mythology, the Vestal Virgins (the virgin priestesses of Vesta, the Roman goddess of the hearth) maintained Vesta's sacred fire. When a sacred fire went out, the Vestal Virgin on duty was flogged. The fire of the hearths of Roman homes symbolised stability. It was said, 'Keep the home fires burning, and keep the home thriving' (Fowler 2006). The Slavs worshiped many gods, and among them Swarog and his son Swarozyc were identified with light: the father with the sun, his son with fire. The Slavs considered the sun a heavenly body that was kind and beneficial and called it the face or the eye of God. To point a finger at the sun or to stand with one's back to the sun during harvest time were prohibited. Swarozyc was the god of both sacred and hearth fire. The Slavs' relationship with fire consisted of worship, fear, and affection. Glowing embers were kept burning at each house and were given to the son when he moved to his own house. This tradition existed in most cultures and was so deep-rooted that at the beginning of the twentieth century in some regions villagers still did not allow a burning house to be extinguished after it was struck with a thunderbolt, because it was 'fire from heaven' sent by God.

12.2.2 Evolution of Artificial Light Sources

The first portable lamp was a resinous chip, that is, a piece of tarry wood, which was later improved to form a torch (the resinous chip with one side muffled up with linen or cotton fibres soaked in animal or vegetable fats as fuel). The torch was used in ancient times for lighting castle rooms (e.g. in Syria in the sixth century BC). In about the tenth century BC, oil lamps were used in Egypt. Those lamps were at first made of naturally occurring materials, such as drilled rocks and stones filled with animal fat.

Dry moss, later replaced by tied plant fibres, was used as a candlewick. The ancient Egyptians, Etruscans, Greeks, and Romans used either standing or hanging iron and bronze oil lamps. The oldest glass lamp came from ninth century AD Persia.

The Etruscans used candles in the second century BC, the Chinese used them in the fourth century BC, and the Europeans first used wax and later, in the eighteenth century AD, the oil lamp. Aime Argand constructed a new oil burner with a circular wick mounted between two cylindrical metal tubes so that the air was channelled through the centre of the wick, as well as outside it, thus steadying the flame and improving the flow of air. The propagation of gas lamps then began. In 1820, stearin candles and in 1860 paraffin candles were invented. One of the most critical events in the history of the development of light sources was the discovery of petroleum distillate and the construction of the first paraffin (kerosene) lamp in 1853 by a Polish constructor, Ignacy Lukasiewicz. Fuels used previously, such as whale oil, paraffin, rape, linen, and hemp oils, were replaced by cheaper, lighter fuels that produced more light. In a very short time, petroleum distillate completely replaced other fuels used for lighting purposes. Kerosene lamps were produced as standing, hanging, wall, or technical lamps used in mining, cars, and railways.

Gas lighting began to be used at the same time as kerosene lamps. Gas lamps did not require storing fuel, and burning them did not result in an unpleasant odour. However, special installation was needed to connect a gas lamp with the place where the gas was produced; this installation was only profitable in dense urban settlements. These lamps were used in the nineteenth century in Europe (first in London in 1809, later in Paris in 1819, and in Warsaw in 1856; Williams, 2005) and North America (in Philadelphia in 1916) to illuminate public places, but were rarely used in residential houses. Today these lamps are mainly used as tourist attractions as they require liquid gas, propane, methane, or butane. Another type of gas lamp was called a carbide lamp and produced and burned acetylene. However, these lamps were quickly replaced with electric lighting.

In the second half of the nineteenth century, electric lamps, which were much more luminous, became common. They had a good spectrum of emitted light, and they could generate a highly luminous flux. Arc lamps such as carbon arc lamps were the first electric lamps. In these lamps, light is produced by heating electrodes and ionised gases in the arc. This kind of light source was later replaced with incandescent lamps, in which light is emitted by an electric filament as a result of a flow of current. Thomas Edison constructed the first practical and useful incandescent lamp in 1879. The next light source was the gas discharge lamp, which produced light by forming an electric arc between tungsten electrodes placed inside an arc tube filled with both gas and metal salts. The first discharge lamp, a high-voltage neon-filled tube, was a neon lamp constructed in 1910 that emitted a red light. Later, mercury vapour lamps (1927), sodium vapour lamps (1935), and fluorescent lamps (1938) were constructed. Dynamic development of light sources occurred in the second half of the twentieth century, when the following types of lamps were constructed: halogen lamps, metal halide lamps, three-band phosphor fluorescent lamps of high luminous efficacy, long-life, compact fluorescent lamps with a high colour-rendering index (integrated and nonintegrated), electrodeless lamps (fluorescent induction and sulphur lamps), and light-emitting diodes (LEDs, which are already being used for illumination).

12.3 BASIC TERMS IN LIGHTING TECHNOLOGY

Experimental studies show that red, violet, and blue lights stimulate the visual system significantly less than green and yellow lights of the same radiant power. In addition, the same monochromatic radiation differently stimulates the photoreceptors of the retina, the cones and rods. The efficiency of radiation in stimulating visual sensation depends on the radiation spectrum and spectral radiant power (the radiant power of monochromatic radiation for specific wavelengths) in a given spectrum. Therefore, physical attributes that describe radiation (e.g. radiant flux, radiant power, and radiant power density) are insufficient to describe the *efficiency* of that radiation in stimulating visual sensations (Bąk and Pabjańczyk 1994). The term 'spectral luminous efficiency' describes the efficiency of monochromatic radiation in stimulating visual sensation; other related terms are 'relative luminous efficiency function' or 'relative spectral sensitivity'.

12.3.1 CIE Standard Photometric Observer

Light measurements are most useful in evaluating lighting conditions if they are sensitive to the spectral power distribution of radiant energy (as described by spectral luminous efficiency) in the same way as the human eye (Rea 1993). Because there are significant differences in the vision of individual observers, the International Commission on Illumination (CIE) has determined standard observer response curves (*CIE 1931 Standard Observer*) for two borderline cases: when the eyes adapt to light (photopic vision—vision mediated essentially by the cones) and to dark (scotopic vision—vision mediated essentially by the rods). Thus, there are two standardised relative spectral luminous efficiency functions (called *CIE standard photometric observers*) that describe the average sensitivity of the human eye to light of different wavelengths at two opposite adaptive states of the photoreceptors of the retina: photopic $V(\lambda)$ and scotopic $V'(\lambda)$. The maximum photopic weighting function is 555 nm (green-yellow light), and the maximum value of scotopic function is 507 nm (green light; Figure 12.1).

The standardisation of the eye sensitivity weighting function is key to photometry. *Photometric quantities* are the quantities that describe optical radiation with regard to the visual effects of that radiation and are weighted using the spectral luminous efficiency function. Photometry is the measurement of visible radiation (light) with a sensor with a spectral responsivity curve equal to the CIE standard photometric observer. The often-measured photometric quantities are luminous flux, luminous intensity, illuminance, and luminance.

12.3.2 Photometric Quantities

Luminous flux: The basis of the photometric quantities system. It is derived from the radiant flux by evaluating the radiation according to its action upon the CIE standard photometric observer. Photopic spectral responsivity $V(\lambda)$ is commonly used (unit: lumen [lm]) to evaluate lighting conditions and lighting equipment.

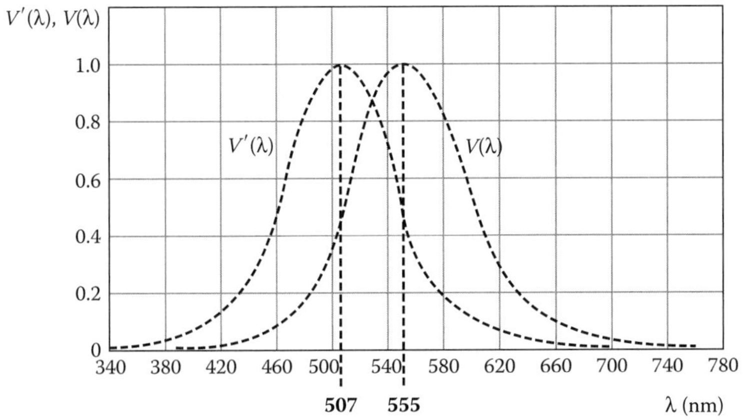

FIGURE 12.1 The relative spectral luminous efficiency for photopic vision $V(\lambda)$ and scotopic vision $V'(\lambda)$.

Luminous intensity: The (solid) angular density of the luminous flux in a given direction in a unit solid angle containing that direction (unit: candela [cd]).

Illuminance: At a point on the surface is the areal density of the luminous flux incident at a point on a surface (unit: lux [lx]).

Luminance: The physical measure of brightness. It refers to the magnitude of the sensation that results from viewing surfaces from which light comes to the eye. The intensity of the sensation is greater if brightness is greater, which means that the source seems to emit a greater luminous flux per unit area. Therefore, brightness is treated as the equivalent to visual sensation intensity measure. A relationship between brightness and the retinal illuminance level, which is the basis of luminance determination (unit: candela per square metre [cd/m^2]), can be assumed.

12.4 LIGHTING PRINCIPLES

To perform visual tasks in optimal conditions, the workstation should be illuminated in a way that ensures visual comfort. Three conditions are essential for visual comfort (Bąk 1981):

1. Full ability to distinguish details
2. Efficient perception, without any risk for the observer
3. Perception without any discomfort or excessive fatigue

There are three main groups of lighting principles (Bąk 1981):

1. Physiological
2. Aesthetic
3. Economic

Physiological principles are the most important for the visual system (Bąk 1981):

- Sufficient luminance in relation to the visual task
- Sufficient contrast between the visual task and its surroundings
- Avoidance of details with too-small angular dimensions and a too-short time for perception
- Uniform luminance of the surroundings

Excessive deviation from these rules for lengthy visual work leads to visual fatigue and a decrease in work efficiency.

Aesthetic principles are derived from analysis of the influence of light on the mental state of man. There are two main aesthetic principles:

1. Increasing the attractiveness of the picture
2. Creating a mood

The selection of luminaires with similar lighting characteristics but different prices and the selection of energy-saving light sources and ballasts depends on the employer's economic principles, which often focus on reducing excessive costs.

All of the above-mentioned principles are linked with one another and, depending on the visual work, the difficulty of the visual task, and the purpose of the lit interior, each has different groups of principles that are most important. Physiological principles are the most significant in designing indoor lighting. These principles must be followed, especially for interiors where visual tasks are difficult and lengthy. Aesthetic principles are also most significant in interiors, where the main purpose is to create a special mood or impression with lighting (e.g. in the case of lighting shop windows or a stage). Therefore, physiological principles that do not contradict the aesthetic ones can be used. Economic principles also must be considered in evaluating and selecting lighting equipment. Today, special attention is paid to the use of energy-saving solutions.

12.5 BASIC LIGHTING PARAMETERS

Lighting quality should be ensured through the use of specific lighting equipment. It can be described with the following lighting parameters:

- Illuminance
- Illuminance uniformity
- Luminance distribution
- Control of glare
- Colour appearance and colour rendering
- Flicker

12.5.1 ILLUMINANCE

The most important aspect of interiors not intended for work is the general impression created by the lighting. The ability to discern the features of a human face is the minimum criterion of a suitable illuminance level; for this, the luminance should be about 1 cd/m^2. This value can be obtained in normal lighting conditions, when the illuminance level on the horizontal plane is about 20 lx, which is the lowest acceptable value of illuminance for interiors not intended for work (EN 12464-1 2004). Seeing the features of a human face without effort is possible when luminance is about 10–20 cd/m^2 and the background is not black or very dark. This means that illuminance on the vertical plane must be at least 100 lx and on the horizontal plane about twofold greater. Thus, an illuminance of at least 200 lx is required in continuously occupied areas, regardless of the visual tasks performed (EN 12464-1 2004; Pawlak and Wolska 2005).

The illuminance values (i.e. the value below which the average illuminance on a specific surface is not allowed to fall; EN 12464-1 2004) for various types of interiors, tasks, and activities are presented in the lighting standard EN12464-1 (2004; in European Union countries).

12.5.2 Illuminance Uniformity

Illuminance uniformity is the ratio of minimum measured illuminance to the average illuminance of a given work plane (task area, immediate surrounding area; Wolska 2005). The recommended illuminance uniformity should not be lower than (EN 12464-1 2004)

- 0.7 in the task area
- 0.5 in the immediate surrounding area

The task area is the partial area in the work plane in which the visual task is carried out, and the immediate surrounding area is a band at least 0.5 m wide around the task area within the visual field.

12.5.3 Luminance Distribution

Differences in surface reflectances cause differences in luminances at a given illuminance level. The illuminance can therefore be appropriate for a visual task while the luminance balance in the whole interior is unacceptable. A well-balanced luminance distribution should be provided in the visual field to improve visual performance, visual acuity, contrast sensitivity, and efficiency of the visual functions such as accommodation, convergence, papillary contraction, and eye movements.

The luminance distribution can be described either by luminance ratios (luminance contrasts) of characteristic surfaces in adjacent and distant surroundings of a visual task or by the reflectance of major interior surfaces. European standard EN 12464-1 (2004) contains criteria for evaluating luminance distribution, based on the recommended ranges of reflectances for major interior surfaces, that is, for ceilings it is 0.6–0.9, for walls 0.3–0.8, for work planes 0.2–0.6, and for floors 0.1–0.5.

12.5.4 Glare Control

Glare is a vision state in which there is a sensation of discomfort or a reduction in the ability to perceive details, objects, or both. It is caused by an inappropriate luminance range or extreme contrasts in space or time. The following types of glare are distinguished based on certain conditions (Figure 12.2):

- *Direct glare* is caused by a bright object in or close to the direction of the object that is being looked at.
- *Reflected glare* is caused by specular reflections of bright objects in glossy surfaces.

Three types of glare can be distinguished by their effects:

1. *Discomfort glare* causes discomfort, annoyance, and irritation without necessarily impairing the vision of details or objects. It increases with time but disappears immediately after the source of glare is eliminated (e.g. a lamp has been switched off; Wolska 2005).
2. *Disability glare* impairs visual ability for a short but noticeable time without necessarily causing discomfort.
3. *Blinding glare* is a strong, short, but noticeable glare, during which nothing can be seen. This is an extreme kind of disability glare.

The control of discomfort glare is one of the major aspects of lighting. When discomfort glare is properly limited, disability glare and blinding glare do not exist. The unified glare rating (UGR) tabular method is recommended for evaluating discomfort glare (CIE 1995 and EN 12464-1 2004). In this method, the UGR is

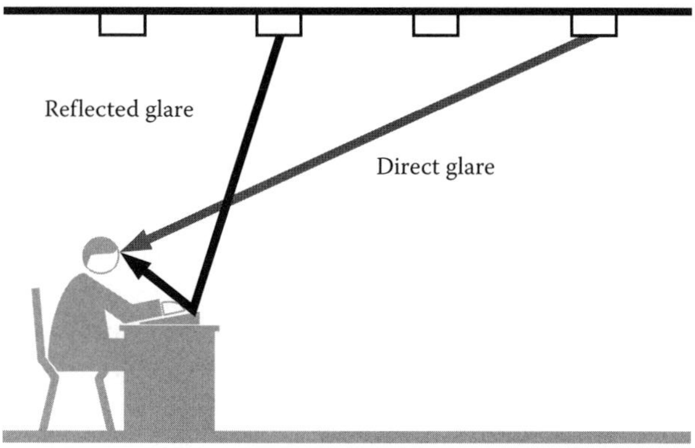

FIGURE 12.2 Direct and reflected origin of glare. (Adapted with permission from Pawlak, A., and A. Wolska. 2005. *Lighting of Indoor Work Places and Work Stations*. 4th edition, vol. 12. Warsaw: CIOP-PIB.)

calculated when the lighting system is being designed, using the following formula (EN 12464-1 2004):

$$\mathrm{UGR} = 8\log\left(\frac{0.25}{L_b}\sum\frac{L^2\omega}{p^2}\right)$$

where L is the luminance of the bright parts of each luminaire in the direction of the observer's eye, L_b is the background luminance (calculated from the formula: $E_{ind}\,\pi^{-1}$, where E_{ind} is indirect illuminance at the eye of the observer), ω is the solid angle of the bright parts of each luminaire at the observer's eye, and p is the Guth position index for each luminaire.

The UGR limits for specific interiors, tasks, or activities are presented in the lighting requirement tables of standard EN 12464-1 (2004). The UGR values calculated for the lighting installation being designed should not exceed the values given in the standard for the specific interior, activity, or task. The UGR is not measured at the workplace because it is only a calculation parameter. Photometric data of luminaires, the geometry of their displacement (arrangement), and the position of the observer's eye are considered when determining the UGR.

The degree of discomfort glare depends not only on luminance in the visual field but also on the type of activity being performed. More demanding visual tasks will result in stronger feelings of discomfort and will require stricter glare control (a lower value of UGR). When a worker must move in order to perform particular activities, the degree of discomfort is smaller than when the worker stays in one place. Therefore, the UGR limits are different for different activities and tasks.

12.5.5 Colour Appearance of Light and Colour Rendering

The colour of light refers to the apparent colour (chromacity) of the light emitted by a given source and is described by the correlated colour temperature (T_c) in kelvin (K). White light is grouped according to its correlated colour temperature: warm-white (warm), intermediate (neutral), and daylight (cold-white). The colour temperature should increase along with an increase in the level of illuminance, as shown in Table 12.1.

TABLE 12.1
Recommended Ranges of Correlated Colour Temperatures Depending on the Level of Illuminance

Illuminance (lx)	Recommended Correlated Colour Temperatures of Light Sources (K)
<300	<3300
300–500	3300–5000
>750	>5000

TABLE 12.2
Light Sources and Groups of Colour Rendering

Light Source	Group of Colour Rendering				
	$R_a \geq 90$	$90 > R_a \geq 80$	$80 > R_a \geq 60$	$60 > R_a \geq 40$	$40 > R_a \geq 20$
Incandescent	•	•	–	–	–
Tubular fluorescent	•	•	•	•	–
High-pressure mercury	–	–	–	•	•
Metal halide	•	•	•	–	–
Self-balasted mercury lamp	–	•	•	–	
High-pressure sodium	–	•	•	•	•
Low-pressure sodium[a]	–	–	–	–	–
Fluorescent induction	•	•	–	–	–

[a] Low-pressure sodium lamps emit monochromatic yellow light, and R_a is not determined for that source.

The appearance of an object can change when it is illuminated with different light sources. It is therefore important to select proper colour rendering for specific visual tasks. Colour rendering describes the appearance of colours under a given light source as compared with their appearance under a reference source. A general colour-rendering index R_a has been introduced so that colour-rendering properties can be objectively identified. The maximum value of 100 represents excellent colour rendering. This is the value of solar light and most incandescent lamps. Examples of rendering colour groups are presented in Table 12.2.

In interiors where people work or stay for long periods, lamps should have an R_a lower than 80 (EN 12464-1 2004). For visual tasks that demand colour recognition, such as printing or textile work, the value of R_a should be close to 100.

12.5.6 Flicker

An unsteadiness of the visual sensation induced by a light stimulus whose luminance or spectral distribution fluctuates with time (EN 12665 2003) is known as *flicker*. Flicker can be caused by a 50-Hz power supply of high-pressure lamps with magnetic control gears, defects of control gears or lamps, or decreases in the power supply voltage. The most frequent kind of flicker is the pulsation of discharge lamps (fluorescent, high-pressure mercury, or sodium) equipped with magnetic control gear and supplied from 50-Hz mains. Flicker can be perceived by the visual system and cause fatigue, however, if those kinds of sources are used to illuminate machinery with rotary or reciprocate parts, then there could be a stroboscopic effect: the moving parts of machinery can appear to be stationary, operating at a reduced speed, or even in reverse rotation. This is dangerous and can cause an accident.

The use of high-frequency electronic control gear in discharge lamps increases the frequency of the power supply of those sources from several to several dozen kilohertz. This gear can reduce flicker and stroboscopic effects and reduce power consumption by several percent, which contributes to higher energy savings.

12.6 IMPROPER LIGHTING AS A HAZARDOUS FACTOR IN WORK ENVIRONMENT

12.6.1 Lighting and Visual Fatigue

Factors that can cause visual fatigue can be divided into external (environmental) and internal (visual system) factors.

External factors include the following (Wolska 1998):

- Too-low or too-high illuminance
- Glare
- High luminance contrasts and colour contrasts in the visual field or no contrast
- Flicker
- Simultaneous use of natural and artificial light in a spectrum that is not properly adjusted to daylight
- Lengthy visual effort
- Incorrect placement of the visual task area in relation to the location of the luminaires

Internal factors include the following:

- Uncorrected or inadequately corrected refractive errors
- Accommodation and convergence disorders
- Binocular vision disorder
- Eye diseases
- Other diseases

Visual fatigue occurs when the visual system is strained and can be manifested as the following:

- Ocular symptoms (sore, tired, itchy, dry, burning eyes)
- Visual symptoms (blurred, double vision)

- Systemic symptoms (headaches, tiredness)
- Transient changes in visual functions, such as a decrease in the amplitude of accommodation and a longer distance of near point of convergence

People with any of the above internal factors experience greater visual fatigue more frequently compared to people without internal factors. Working in incorrect lighting conditions for a long time weakens the visual system and refractive disorders can emerge or become more severe.

12.6.2 Poor Visibility as a Result of Incorrect Lighting and Occupational Accidents

An analysis of occupational accidents indirectly caused by incorrect lighting showed that the main causes were poor visibility resulting from insufficient lighting, lack of lighting, and glare. Poor visibility could also be the result of flicker, stroboscopic effects, or poor colour rendering, which impair the discrimination of safety colours. However, no information is available on accidents related to these lighting parameters.

Poor visibility as an indirect cause of the analysed accidents was caused by the following:

- Insufficient level of illuminance and/or illuminance uniformity in the work area, making it difficult to detect the hazard (accident group I)
- Lighting equipment in poor condition or lack of equipment, so that people undertake activities to improve work conditions (accident group II)
- Incorrect lighting in an explosive atmosphere (accident group III)
- Incorrect lighting in work spaces and where there is vehicle traffic (accident group IV)

Table 12.3 lists the numbers and kinds of accidents from the various groups for 197 analysed accidents.

Poor visibility caused by unsuitable lighting is often ignored so that work is done in poor lighting conditions. Work in poor lighting conditions and activities to improve lighting (installation, repair, or maintenance of lighting equipment) that

TABLE 12.3
Number and Kinds of Accidents in Different Groups of Accidents

Kind of Accident	Number of Cases				
	Group I	Group II	Group III	Group IV	Total
Incident (no injury)	13	6			19
Injury (of different severity)	29	38	7	4	78
Fatal	20	66	5	9	100
Total	62	110	12	13	197

does not conform to safety procedures and is not done by authorised personnel can cause accidents.

12.7 LIGHTING COMPUTER WORKSTATIONS—EXAMPLES

Regardless of the type of computer screen, the following lighting systems can be used to light computer workstations (Wolska et al. 2003):

- Direct lighting with 'dark-light' luminaires, which are equipped with fluorescent tubes covered with an optical system built of aluminised mirror parabolic-shaped louver. Figure 12.3 is an illustration of a ceiling-mounted dark-light luminaire. Figure 12.4 illustrates the typical polar curves of the luminous intensity of this type of luminaire and the correct arrangement of luminaires in relation to the workstations in the room.
- Indirect lighting with free-standing or pendant-mounted luminaires. Figure 12.5 illustrates the polar curves of luminous intensity of this type of luminaire and the correct arrangement of luminaires in relation to the workstations in the room.
- Direct-indirect lighting
  - With pendant-mounted luminaires: The polar curves of luminous intensity of light emitted downward follow the requirements for dark-light luminaires, and the ceiling luminance follows the requirements for indirect luminaires. Figure 12.6 illustrates the polar curves of luminous intensity of this luminaire type and the correct arrangement of luminaires in relation to the workstations in the room.
  - The polar curves of luminous intensity of light emitted downward have a diffuse distribution and the luminance of optical elements transmitting and reflecting light is below 500 cd/m^2.

FIGURE 12.3 A ceiling-mounted dark-light luminaire.

- With ceiling-mounted 'soft-light' luminaires: The polar curves of luminous intensity of light emitted downward have a diffused distribution and the luminance of optical elements transmitting and reflecting light is below 500 cd/m^2. Figure 12.7 illustrates a surface-mounted soft-light luminaire and its polar curves.
- Compound lighting with one of the above general lighting systems (usually indirect) and the low-luminance task-lighting luminaire (Figure 12.8).

Due to their luminous intensity distribution, dark-light luminaires have the following advantages:

- At workstations located between two lines of luminaires (Figure 12.4), the illuminance values are higher than under luminaires in circulation areas.
- The emphasised directions of light emission are perpendicular to the line of sight, limiting the influence of the specular portion of luminous flux reflection from the work plane, which impairs recognition of details.
- The relatively small number of luminaires enables the required illumination of the interior while using only a small amount of power, ensuring less power consumption.

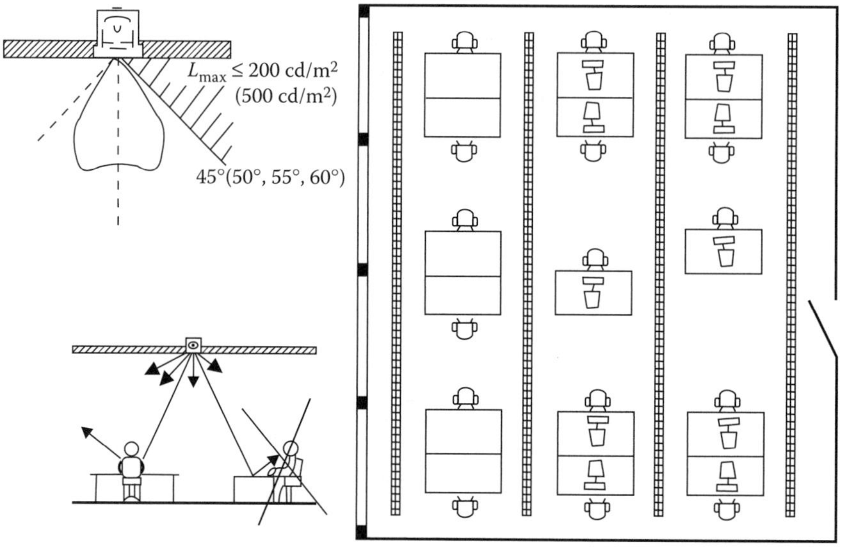

FIGURE 12.4 Examples of the polar curves of luminous intensity of a dark-light luminaire and the correct arrangement of dark-light luminaires in relation to workstations in the room. (Adapted with permission from Wolska, A., A. Gedliczka, and J. Bugajska. 2003. Ergonomic requirements. In *Computer Work Stations: Health and Ergonomic Aspects*, ed. J. Bugajska, 95–142. Warsaw: CIOP-PIB.)

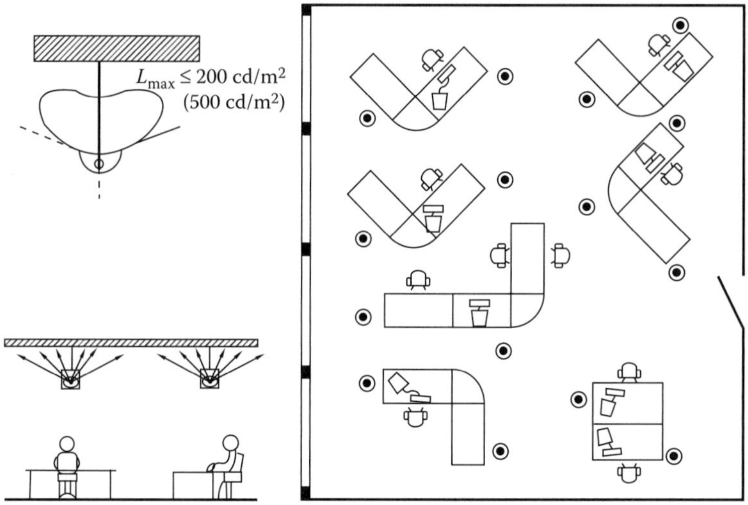

FIGURE 12.5 Examples of the polar curves of luminous intensity of an indirect luminaire and the correct arrangement of dark-light luminaires in relation to workstations in the room. (Adapted with permission from Wolska, A., A. Gedliczka, and J. Bugajska. 2003. Ergonomic requirements. In *Computer Work Stations: Health and Ergonomic Aspects*, ed. J. Bugajska, 95–142. Warsaw: CIOP-PIB.)

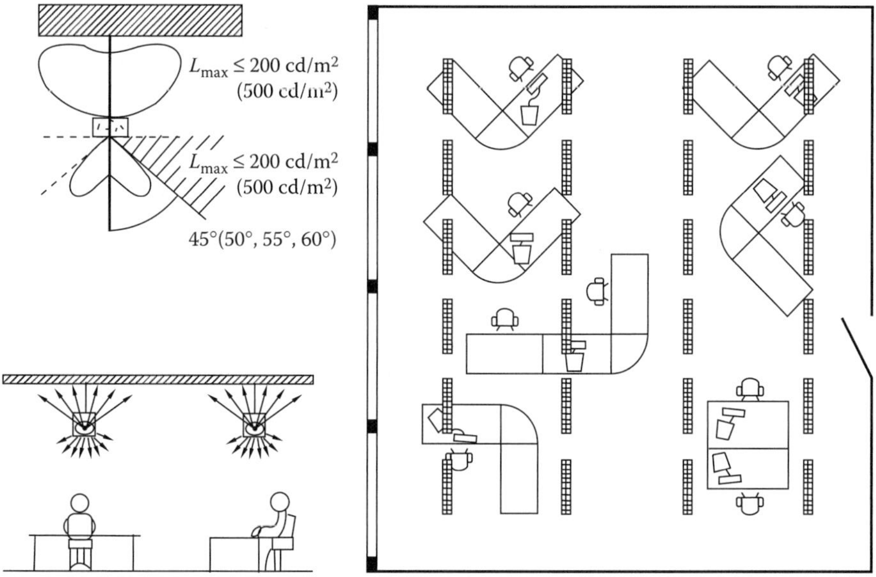

FIGURE 12.6 Examples of the polar curves of luminous intensity of a direct-indirect luminaire and the correct arrangement of dark-light luminaires in relation to the workstations in the room. (Adapted with permission from Wolska, A., A. Gedliczka, and J. Bugajska. 2003. Ergonomic requirements. In *Computer Work Stations: Health and Ergonomic Aspects*, ed. J. Bugajska, 95–142. Warsaw: CIOP-PIB.)

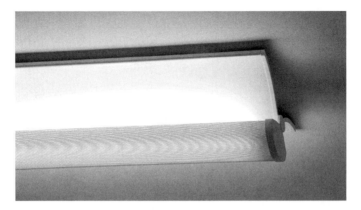

FIGURE 12.7 A view of a ceiling-mounted 'soft-light' luminaire.

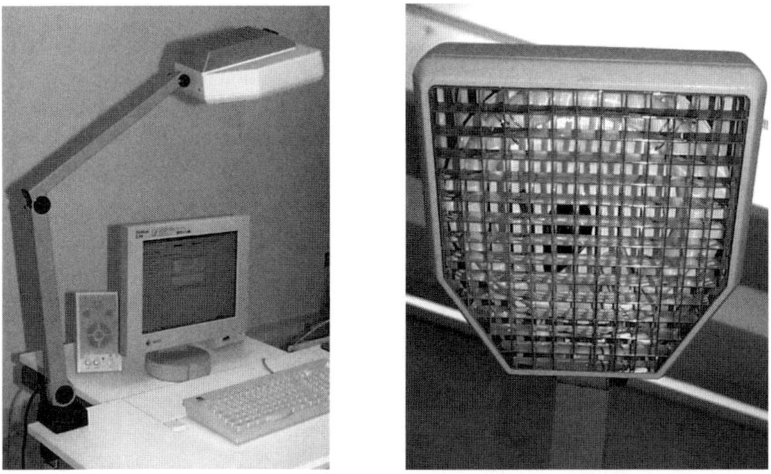

FIGURE 12.8 An example of a task lighting luminaire suitable for illuminating a computer workstation and a view of a parabolic mirror louver that provides antiglare shielding.

Indirect lighting can completely eliminate the reflection of the luminaires on the display screen. However, this lighting system is more energy-consuming than a direct lighting system as it requires more luminaires to provide the same required illuminance level, and the lack of luminance contrasts in the room can cause monotony.

A direct-indirect system has the advantages of both of the above systems. Users often prefer this system if the conditions in Figure 12.6 are fulfilled.

The use of a compound lighting system, in which workstations are additionally illuminated with task lighting luminaires, is advisable only when there is no direct or reflected glare from that luminaire. Task lighting luminaires are often a source of glare; only luminaires specially designed for computer work and with a suitable profile of reflector and louver (which reduces glare and reflections; see Figure 12.8)

FIGURE 12.9 An example of a task lighting luminaire of an asymmetric luminous intensity curve.

should be selected. If this kind of task lighting luminaire cannot be provided, luminaires without louver can be used, but they should be equipped with a special reflector, preferably asymmetric (Figure 12.9).

If dark-light luminaires are installed, computer workstations should be located so that the line of sight is parallel to the line of luminaires and windows. The workstations should be located between the lines of luminaires and not directly under them to avoid reflections from the work plane and the keyboard (Figure 12.10). This kind of restriction does not apply to indirect lighting and soft-light luminaires.

Computer workstations should not be located near windows, and workers should not sit in front of a window during daylight, because of the nonuniform luminance distribution in the field of view. On the other hand, if a workstation is located in such a way that the window is behind the worker, daylight is reflected on the screen. Therefore, computer workstations should be placed perpendicular to the window and at least 1 m away from it.

In summary, suitable lighting systems for computer workstations should be selected based on the following assumptions:

- The type and intensity of computer work performed
- The users' preferences, as each lighting system creates a different luminous environment
- Economic possibilities (energy savings, costs of a new lighting system)

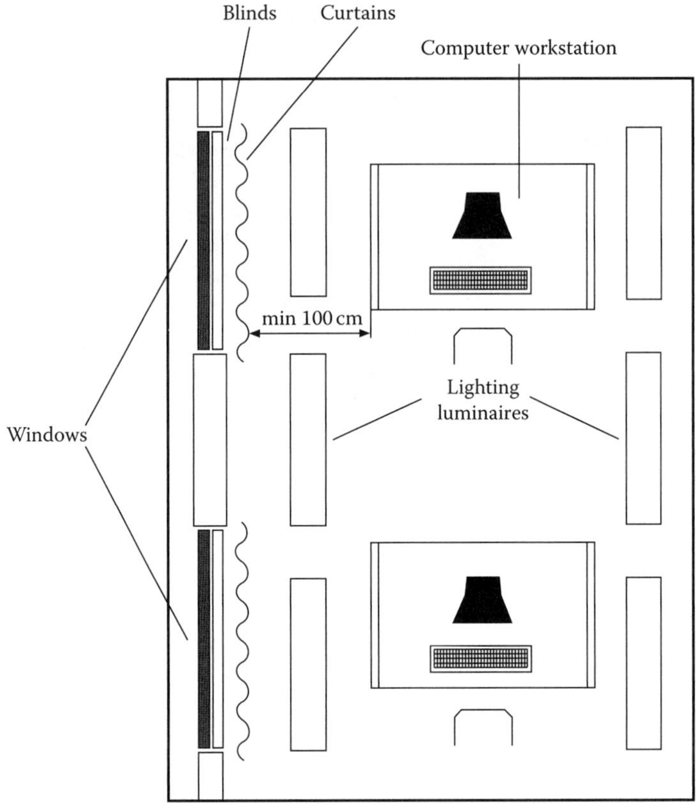

FIGURE 12.10 Location of a computer workstation in relation to dark-light luminaires and windows. (Adapted with permission from Wolska, A., A. Gedliczka, and J. Bugajska. 2003. Ergonomic requirements. In *Computer Work Stations: Health and Ergonomic Aspects*, ed. J. Bugajska, 95–142. Warsaw: CIOP-PIB.)

REFERENCES

Bąk, J. 1981. *Lighting Engineering*. Warsaw: PWN.

Bąk, J., and W. Pabjańczyk. 1994. *Fundamentals of Lighting Engineering*. Lodz: Technical University of Lodz.

CIE. 1995. Discomfort glare in interior lighting. Technical Report CIE no 117. Vienna: International Commission on Illumination.

EN 12464-1. 2004. Light and lighting. Lighting of workplaces, Part 1: Indoor workplaces.

EN 12665. 2003. Light and lighting. Basic terms and criteria for specifying lighting requirements.

Fowler, R. 2006. Vestal virgins of Rome. Privileged keepers of Rome's home fires. http://ancienthistory.suite101.com/article.cfm/vestalvirginsrome (accessed August 2008).

Pawlak, A., and A. Wolska. 2005. *Lighting of Indoor Work Places and Work Stations*. 4th edition, vol. 12. Warsaw: CIOP-PIB.

PN-84/E-02033. 1984. Electric lighting of indoor workplaces.

Rea, M. S., ed. 1993. Measurements of light and other radiant energy. In *Lighting Handbook. Reference & Application*. 8th ed., ed. M. S. Rea, 27–64. New York: IES of North America.

Williams, B. 2005. *A History of Light and Lighting. Edition 2.3, 1990–2005*. http://www.mts.net/~william5/history/hol.htm (accessed August 2008).

Wolska, A. 1998. Psychophysiology of vision. In *Lighting Engineering '98. Handbook*. eds. W. Dybczyński and J. Grzonkowski, 135–189. Warsaw: PKOś.

Wolska, A. 2005. Human aspects of lighting in working interiors. In *International Encyclopedia of Ergonomics and Human Factors*. Vol. 2. ed. W. Karwowski, 1793–1799. Boca Raton, FL: Taylor & Francis.

Wolska, A., A. Gedliczka, and J. Bugajska. 2003. Ergonomic requirements. In *Computer Work Stations: Health and Ergonomic Aspects*, ed. J. Bugajska, 95–142. Warsaw: CIOP-PIB.

13 Noncoherent Optical Radiation

Agnieszka Wolska and Władysław Dybczyński

CONTENTS

13.1 INTRODUCTION

Optical radiation is an important environmental factor essential to normal human development and activity. Both insufficient and excessive exposure to optical radiation can cause many negative biological effects. Optical radiation is a natural component of solar radiation and can be produced artificially by electrical and technological sources. Electrical sources are used in many industrial processes, medical treatments, and in the cosmetic industry. Optical radiation is also emitted into the environment as a by-product of processes like welding and plasma cutting and by hot objects like blast furnaces (Wolska et al. 2001).

Visual radiation, or light, was the first part of the spectrum to be discovered and investigated. In 1666, Isaac Newton investigated the solar radiation spectrum using a lens–prism system, in which the prism was placed just before the focus of the lens. This created a picture of the visual radiation spectrum—a rainbow. It was only in 1752, however, that Thomas Melvill noticed the spectral lines of the individual colours of light. In 1800, William Herschel discovered infrared radiation while measuring the temperature of the individual colours of light that were dispersed by a prism when white light was passed through it. The highest temperature of the spectrum occurred outside the red light, where there was no visible radiation. This experiment proved for the first time the existence of radiation invisible to the human eye.

In 1801, Johann Wilhelm Ritter, a German scientist, was experimenting with light dispersion in a prism. He discovered that the radiation of wavelengths shorter than that of violet light darkened silver chloride the most. He called this invisible radiation *chemical rays*; later, they came to be called *ultraviolet radiation*. With Ritter's discovery, the range of solar radiation was extended by two subranges: infrared and ultraviolet, and the complete spectral range was called *optical radiation*.

13.2 GENERAL CHARACTERISTICS OF OPTICAL RADIATION

Optical radiation is a portion of the electromagnetic spectrum whose wavelength ranges between 100 nm and 1 mm. The optical radiation spectrum is made up of ultraviolet radiation (UV), visible radiation (VIS), and infrared radiation (IR). Ultraviolet and infrared radiation ranges consist of bands A (near), B (medium), and C (far). According to CIE No. 17.4 (1987) directive 2006/25/EC (2006), the following wavelengths of ultraviolet radiation make up these bands:

- UV-A: 315–400 nm
- UV-B: 280–315 nm
- UV-C: 100–280 nm

However, radiation of wavelengths shorter than 200 nm does not occur in workplaces, because it is almost wholly absorbed by the air.

The thresholds of visible radiation are not strictly defined, and it is generally assumed that the minimum is 360–400 nm and the maximum is 760–830 nm. Most often, the visual radiation range is assumed to be 380–780 nm.

Individual bands of infrared radiation consist of the following wavelengths (CIE No. 17.4 1987 Publication and Directive 2006/25/EC, 2006):

- IR-A: 780–1400 nm
- IR-B: 1400–3000 nm
- IR-C: 3000–1 mm

There are two groups of optical radiation parameters: those that define the intensity of radiation and those used to evaluate the hazards to human health caused by such radiation.

13.2.1 Radiometric Quantities—Parameters That Define the Intensity of Radiation or Its Sources

The following radiometric terms are used to describe the intensity of optical radiation:

- Irradiance
- Radiant exposure
- Radiance

Irradiance (*at a point on a surface*; E) is the ratio of radiant power ($d\Phi_e$) incident on a specific point located on an elementary surface to the area of the elementary surface (dS); in other words, it is the surface density of radiant power (power density). Irradiance at a point on a surface (expressed in watts per square metre) is expressed by the following equation:

$$E = \frac{d\Phi_e}{dS} \tag{13.1}$$

Radiant exposure (N) is the product of irradiance and time and is expressed in joules per square metre; in other words, it is the surface density of the radiant energy received:

$$N = E \cdot t \tag{13.2}$$

Radiance (L) in a given direction, at a given point on a surface, is the radiant power ($d\Phi_e$) emitted from a unit area of a surface containing the specific point (dA) at a given solid angle ($d\omega$) in a specified direction (θ). It is expressed in watts per square metre per steradian by the following equation:

$$L_e = \frac{d^2\Phi_e}{dA \cdot \cos\theta \cdot d\omega}$$

where θ is the angle between the normal surface and the specified direction of radiation.

13.2.2 Radiometric Quantities Used in Photobiology—Parameters Used in Evaluating Health Hazards Caused by Optical Radiation

The following terms are also used in the evaluation of health hazards caused by optical radiation:

- Effective irradiance
- Effective radiant exposure
- Effective radiance

- Total exposure duration
- Single exposure duration

Effective irradiance (E_{eff}) or *effective radiant exposure* (N_{eff}) is the calculated irradiance or radiant exposure in a given range of wavelengths between λ_1 and λ_2, spectrally weighted with a specific biological spectral weighting function $X(\lambda)$ and expressed by the following equations:

$$E_s = \sum_{\lambda_1}^{\lambda_2} E(\lambda) \cdot X(\lambda) \cdot \Delta\lambda \qquad (13.3)$$

$$N_s = E_s \cdot t \qquad (13.4)$$

where $E(\lambda)$ is the spectral irradiance from the measurements of irradiance for wavelength λ expressed in watts per square metre per nanometre, t is the exposure duration in seconds (total or single, depending on the type of hazard considered), $X(\lambda)$ is the relative spectral weighting function (relative spectral effectiveness of radiation for inducing a specific biological effect) and is unitless, $S(\lambda)$ is the actinic relative spectral effectiveness of ultraviolet radiation (UVR) that causes adverse health effects on the skin and eye (conjunctivitis, keratoconjunctivitis, erythema, skin burns, skin cancer, etc.), and $\Delta\lambda$ is the bandwidth in nanometres of the calculated or measured intervals.

Effective radiance (L_{eff}) is the calculated radiance of the source in a given range of wavelengths between λ_1 and λ_2, spectrally weighted with a specific biological spectral weighting function, $X(\lambda)$, and expressed by the following equation:

$$L_s = \sum_{\lambda_1}^{\lambda_2} L(\lambda) \cdot X(\lambda) \cdot \Delta\lambda \qquad (13.5)$$

where $L(\lambda)$ is the spectral radiance from the measurements of radiance for a specific wavelength λ, expressed in watts per square metre per steradian per nanometre ($Wm^{-2}\ sr^{-1}\ nm^{-1}$).

Single exposure duration (t_i) is the duration of eye or skin exposure to optical radiation during a specific single action (task) and is expressed in seconds.

Total exposure duration (t) is the total duration of a series of emissions or of a continuous radiation emission incident on the skin or eye during a whole work shift and is expressed by the following equation:

$$t = \sum_{i=1}^{n} t_i \qquad (13.6)$$

where n is the number of exposures.

13.3 SOURCES OF OPTICAL RADIATION

The sources of noncoherent optical radiation are divided into natural, electrical, and technological (by-products of technological processes). The sun is the most important natural source of optical radiation; it sustains life on Earth. Solar radiation reaching the Earth's surface under a clear sky contains about 7% ultraviolet, 43% visual, and 50% infrared radiation. These values can change depending on the geographical location, such as the latitude (various incident angles of radiation on the surface of the ground), altitude (different thicknesses of the atmosphere transmitting the radiation), the time of year (affecting the distance from the sun), the time of the day (affecting the elevation of the sun), cloud cover, and the transmission properties of the atmosphere (ozone, aerosols, air pollutants, and haze). All of these factors influence the spectral characteristics of solar radiation, irradiance, radiance, and other parameters describing the radiation.

Figure 13.1 shows the spectral distribution of solar radiation in the following conditions: beyond the Earth's atmosphere (AM0), on the surface of the Earth, on the equator at noon at an equinox (AM1), and when the sun's rays pass through the atmosphere at a 60° angle (AM2). The graphs show the absorption of radiation by ozone (O_3), water (H_2O), and carbon dioxide (CO_2). Absorption increases with the distance of the optical path traversed by the radiation (at the higher latitudes), and at times other than midday, such as in the morning and in the evening.

The moon, stars, planets, and so on are also natural sources of optical radiation. The colour temperature of solar radiation is variable, changing from about 6000 to 5250 K (kelvin; measured in Poland) to about 4800 K or below (after sunrise and before sunset). The maximum irradiance of solar radiation on the Earth's surface is

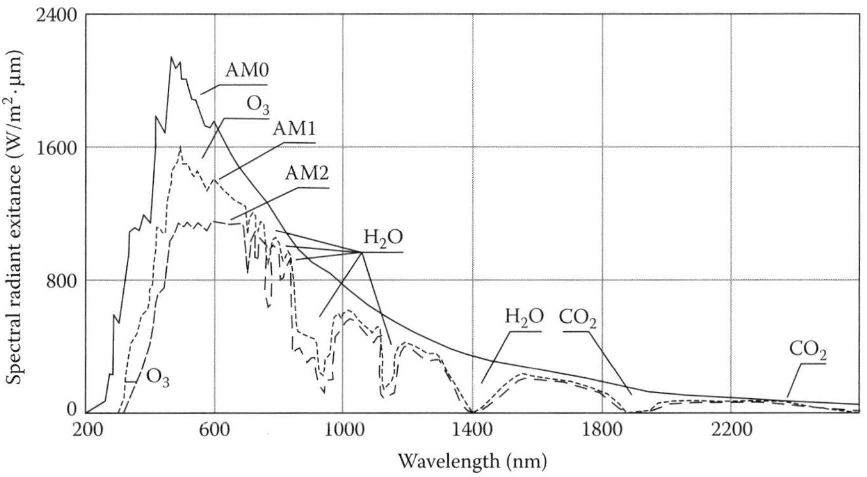

FIGURE 13.1 Spectral distribution of solar radiation: AM0 is beyond the Earth's atmosphere, AM1 is on the surface of the Earth on the equator at noon at an equinox, and AM2 is the sun's rays passing through the atmosphere at a 60° angle (Reprinted with permission from Dybczyński, W. 1999. Optical radiation. In *Occupational Safety and Ergonomics*, ed. D. Koradecka, 1:461–486. Warsaw: CIOP.)

about 925 W/m^2, and in Poland it is about 800 W/m^2 (Podogrocki 1998). Real total irradiance is higher because of the scattered radiation from the sky. AM1 radiation is the total irradiance and consists of irradiances from specific spectral ranges: about 65 W/m^2 of ultraviolet radiation, about 400 W/m^2 of visual radiation, and about 460 W/m^2 of infrared radiation. Irradiance for a fully overcast sky is many times lower than that for direct solar radiation on a sunny day, and the ultraviolet and infrared radiations decrease significantly. Luminance distribution is uniform in an overcast sky. The luminance is highest at the zenith and tapers off toward the horizon, regardless of the temporary position of the sun. The optical radiation most often encountered in daily life is emitted by electrical sources. The two main ways electrical sources generate radiation are incandescence (temperature) and luminescence.

Incandescence or *temperature radiation* is an optical radiation generation that emits radiant energy through the thermal excitation of atoms or molecules. Molecules of both gases and solids are constantly in motion. This movement is a function of temperature; an increase in temperature causes a more rapid movement of the molecules. On heating, atoms and molecules start to vibrate and rotate, raising their energy to higher levels. At higher temperatures, electronic transitions in atoms and molecules result in the release of visible and infrared radiation along with heat (Rea 1993). Since the vibration frequencies of molecules vary, the radiation they emit consists of different wavelengths and the spectrum is continuous. Radiation is most frequently generated in this manner.

Luminescence is the generation of optical radiation by matter, whose intensity at specific wavelengths is greater than the intensity of temperature radiation at a specific temperature of the matter. This radiation is characteristic of specific matter. In the luminescence phenomenon, the single electrons are excited and removed, after receiving additional energy, from the outer shell of an atom (valence electrons) to a farther orbit with a higher level of energy. These electrons remain at this higher energy level for a short time, and on their return to an intermediate or previous (normal) level they release the excess energy as a quantum of radiation. Energy so emitted is strictly related to the difference in energy levels and is directly proportional to the frequency of the electromagnetic wave; radiation, therefore, consists of a specific number of spectral lines corresponding to specific frequencies. This phenomenon contrasts with incandescence, where the irregular excitation of the free electrons of innumerable atoms at a high temperature gives rise to radiation of all wavelengths to form a continuous spectrum (Rea 1993).

Recombination also generates optical radiation. If a large amount of energy is supplied to an atom (e.g. in a collision with an electron), the atom can lose one or more electrons and undergo ionisation. If a free electron is close enough to the ionised atom, it can be attracted to it and take the place of the lost electron. It would then release part or all of its energy in the form of a quantum of optical radiation. The amount of that energy could be different, so the frequency of obtained quantum is different. Radiation generated by recombination could form a continuous spectrum. The energy needed to excite atoms could be delivered in different forms: electrical, radiant, chemical, thermal, and so on. The following terms have been introduced to describe such forms of radiation energy: electroluminescence (used in high discharge lamps), photoluminescence (used in fluorescent tubes), chemiluminescence (used in light sticks), or bioluminescence (present in some living organisms such as fireflies that glow in the dark).

An *open electric arc* is used for welding and cutting metals. Carbon arc lamps with special applications operate on a similar principle. Emission of radiation from an electric arc depends mainly on the electrode material and the intensity of the electric current. When special welding electrodes are used (rutile, wolfram, or alkali electrodes), the sources of emission are small, so luminance is extremely high. The arc emits ultraviolet, visible, and infrared radiation during welding, and immediately after the arc disappears, the welded material emits a thermal infrared radiation that depends on its temperature. Similar phenomena occur in arc furnaces used for melting metals. Depending on the power of the furnace, the intensity of the radiation emitted by the inner parts of the heating chamber can reach very high levels, especially for infrared radiation.

Infrared lamps (Hering 1992) are special types of incandescent lamps that emit mostly infrared radiation from the IR-A band. Visual radiation, if produced, has secondary significance. These incandescent lamps are equipped with a glass bulb in the form of a paraboloid with a mirror coating; the light centre of the filament is placed in the focus of the paraboloid. The electrical parameters needed to operate the filament are chosen so as not to exceed a temperature of 2000 K. The electric power delivered to the lamp is then converted mostly to infrared radiation and is transmitted by the frontal cupola of the lamp. The cupola is usually painted in highly-saturated red to reduce visible radiation. The cupola could have some elements that diffuse radiation at a specific solid angle. The power of infrared lamps varies from 100 to 250 W and their lifetime is between 300 and 5000 hours. These lamps are commonly used in drying processes in dye works, varnish works, heating processes in the food industry, for breeding animals, and for therapeutic medical treatments.

There is another type of infrared lamp that is constructed differently and resembles linear halogen lamps in appearance. These lamps are made of silicon glass tubes with coiled tungsten filaments that work in a halogen atmosphere (the filament autoregenerates because of the halogen regenerative cycle). Their output power ranges from 300 to 3000 W and their average lifetime is about 5000 hours. In some products, a part of the tube's circumference is coated with a thin gold layer, which reflects radiation and increases the lamp's intensity at a specific solid angle, thereby increasing the lamp's operational efficacy.

Ultraviolet lamps emit mostly ultraviolet radiation; if there is visible radiation present in its emission spectrum, it has secondary significance. Fluorescent lamps resemble fluorescent tubes and most commonly generate UV radiation. A fluorescent lamp is a low-pressure gas discharge source that generates a different spectrum of UV radiation, predominantly with fluorescent powders, commonly called phosphors, which are activated by UV-C energy (at a wavelength of 253.7 nm) generated by a mercury arc (Rea 1993). Fluorescent ultraviolet lamps are used in technological processes like optical copying, drying glues and varnishes (for example silicon), irradiating printed-circuit boards, and photopolymerisation. They are also commonly used in insect traps, banknote testers and phototherapy. Choosing an appropriate phosphor enables the adaptation of the spectral distribution of these lamps to the requirements of a specific process or application. Germicidal lamps are fluorescent lamps without any phosphor coating on the inner walls. They emit ultraviolet radiation mainly in the UV-C band, with the main spectral line at 253.7 nm, and also some radiation in the UV-A, UV-B, and visible bands, with only six spectral lines of low radiation efficiency. The output power of these lamps can range from 6 to 36 W, and they are commonly used for disinfecting surfaces, air, and water, for medicinal applications, and in the food and pharmaceutical industries.

Ultraviolet lamps are also designed for use as high-intensity discharge (HID) lamps. These include mercury lamps or lamps with some elements added to the arc tube (like metal halides) to increase the emission of radiation in the UV-A, UV-B, and UV-C regions. In high-pressure mercury lamps, radiation is produced by an electric current passing through mercury vapour. The amount of mercury in the lamp determines the final operating pressure, which is in the 200–400 kPa (kilopascals) range in most lamps (Rea 1993). The operating pressure of the lamp also determines the spectral power distribution of radiation. The higher the operating pressure, the greater the proportion of longer-wavelength radiation. For example, the output power of lamps used in tanning beds and booths is 400–2000 W, and the power of the UV-A radiation is 450 W. Metal halide lamps are more effective than mercury lamps, because the chemical composition of the filler is optimally matched to the intended application of the specific lamp. Metal halide lamps of 150–4000 W are used in industrial applications. In these lamps, about 22.5% of the electric power is used in the following bands of ultraviolet radiation: 17% in UV-A and about 4.5% in UV-B and UV-C.

Like fluorescent lamps, high-pressure mercury lamps are produced with a filter that blocks the emission of visual radiation. They then emit an almost monochromatic radiation of 365 nm. A 125-W lamp produces irradiance of 0.15 W/m^2 at a distance of 1 m.

13.4 BIOLOGICAL EFFECTS OF OPTICAL RADIATION

Only absorbed radiation can cause biological effects. The two types of reactions produced in biological tissues by optical radiation are photochemical and thermal. The effects of exposure depend on the physical parameters of the radiation (e.g. the wavelength and spectral power distribution), the magnitude of the absorbed dose of radiation, and the optical and biological features of the exposed tissue (e.g. the skin type).

Some optical radiation is good for humans. Ultraviolet radiation is essential for the production of vitamin D_3, assists in reducing cholesterol levels, and accelerates the healing of wounds, infections, and some skin diseases. Infrared radiation helps heal inflammations and accelerates the healing of injuries of the joints and the soft parts of the body's extremities. However, excessive exposure to optical radiation can have many negative effects on the skin and eyes; it also can affect the body's immune system (Tables 13.1 and 13.2). In the ultraviolet range, especially below 320 nm, injuries to the photochemical mechanisms of tissue are predominant. At higher wavelengths, photon energy will most likely be high enough to cause a fall in photochemical reaction, virtually to zero if the wavelengths are over 800 nm. When wavelengths are over 380 nm, thermal reactions begin to appear; the chances of

TABLE 13.1
Relationship between Eye Injuries and the Range of Optical Radiation

Range of Radiation	Eye Injury
UV-A	Lens injuries, for example, cataracts, occurring after many years of chronic exposure
UV-B and UV-C	Cornea and conjunctiva injuries, for example, keratoconjunctivitis (welder's flash), photokeratoconjunctivitis
VIS and IR-A	Photochemical and thermal retinal injuries, for example, photoretinitis, retinal burns, scotoma (blind spot in the fovea)
IR-B and IR-C	Thermal injuries of the cornea and lens, for example, corneal burns and lesions, cataracts, occurring after many years of chronic exposure

TABLE 13.2
Relationship between Skin Injuries and the Range of Optical Radiation

Range of Radiation	Skin Injury
UV	Erythema, sunburn, skin aging (long-term), skin cancer (long-term)
VIS	Thermal damage
IR	Thermal damage

their occurrence increase with an increase in wavelengths. Wavelengths over 800 nm induce only thermal reactions in biological tissues (Wolska et al. 2001).

13.4.1 Eye Injuries

Different parts of the eye, depending on the wavelength of optical radiation, can suffer hazardous effects. The cornea absorbs ultraviolet radiation of wavelengths 200–215 nm (UV-C) and infrared of wavelengths 1400 nm or greater (IR-B and IR-C). The lens absorbs near ultraviolet radiation (UV-A) and partially absorbs wavelengths of 780–3000 nm (IR-A and IR-B). Visible radiation (400–780 nm) and near infrared radiation (IR-A) are transmitted to the retina. Thus, the parts of the eye most affected by optical radiation are the cornea, retina, and lens. Table 13.1 lists the adverse effects of eye exposure to different wave bands of optical radiation.

13.4.2 Skin Injuries

The skin is the largest organ of the body and has the greatest risk of contact with optical radiation. The parts of the skin most likely to be exposed are uncovered parts of the body: the hands, head, neck, and arms. The skin has four major components: the stratum corneum (the dead layer), epidermis, dermis corium, and subcutaneous tissue. The outermost skin component—the dead layer—protects the living tissue beneath it from water loss, injury, and radiant energy. The epidermis is the outermost layer of living tissue; colour changes like tanning take place there. The dermis gives the skin elasticity and supportive strength. The subcutaneous tissue is made up mostly of fatty tissues that insulate and absorb shock.

The effect of optical radiation on skin tissue varies with the irradiance of the incident radiation, absorption of tissues at the incident wavelength, duration of exposure and the effects of blood circulation, and heat conduction in the affected area. The short-term effects of exposure to optical radiation over the exposure limit values are skin burns or erythema. Thermal injury usually follows temperature elevation in skin tissues or photochemical injury from excessive levels of UV. Long-term adverse effects of repeated or chronic exposure to optical radiation can be delayed. Only ultraviolet radiation is believed to cause long-term effects like skin aging and skin cancer. Table 13.2 lists the adverse effects of exposure of the skin to different ranges of optical radiation.

13.5 CRITERIA FOR EVALUATING HEALTH HAZARDS CAUSED BY OPTICAL RADIATION

Studies on the spectral effectiveness of optical radiation for specific harmful health effects determined the criteria for health hazard evaluation and exposure limit values. The relative spectral effectiveness (action spectrum) of optical radiation is different for different specified biological responses. A normalised action spectrum is the wavelength dependence of the dose of monochromatic radiation required to induce a certain biological response, and is commonly normalised to '1' at the wavelength of the 'maximum action', that is, where the smallest dose is sufficient to induce the required effect (Sliney 2007). The following are examples of relative spectral effectiveness:

- *The erythema action spectrum*: The relative spectral effectiveness of ultraviolet radiation to produce erythema in human skin (i.e. reddening of the skin as in sunburn; CIE 125 1997; PN-T-06588 1979)
- *Photokeratitis action spectra*: The relative spectral responsivity for an inflammation of the cornea following overexposure to ultraviolet radiation with a peak at 288 nm (CIE 106 1993)
- *Photoconjunctivitis action spectra*: The relative spectral responsivity for an inflammation of the conjunctiva following overexposure to ultraviolet radiation with a peak at 260 nm (CIE 106 1993)
- *Action spectrum for photocarcinogenesis (nonmelanoma skin cancer)*: The relative spectral effectiveness of ultraviolet radiation to cause nonmelanoma skin cancer with a peak at 299 nm (CIE 138 2000)
- *Germicidal action spectrum*: The effectiveness of ultraviolet radiation in causing the inactivation of microorganisms in water, air, and on surfaces, with a peak at 257 nm (PN-T-06588 1979)
- *Vitamin D_3 action spectra*: The effectiveness of ultraviolet radiation to produce vitamin D_3 in human skin, with a peak at 296 nm (PN-T-06588 1979) or at 298 nm (CIE 174 2006)
- *Actinic action spectra*: The actinic relative spectral effectiveness of ultraviolet radiation to cause adverse effects of the skin and eye like conjunctivitis,

keratoconjunctivitis, erythema, skin burns, and skin cancer, with a peak at 270 nm (directive 2006/25/EC; ICNIRP 2004)

- *Blue light photochemical retinal hazard*: The spectral weighting function for retinal photochemical injury by radiation in the range 300–700 nm, with a peak at 440 nm (directive 2006/25/EC 2006; ICNIRP 1997; PN-T-05687 2002; PN-T-06704 2003)
- *Retinal thermal hazard*: The spectral weighting function for retinal thermal injury function by visible and IRA radiation in the range of 380–1400 nm, with a peak at 440 nm (directive 2006/25/EC 2006; ICNIRP 1997; PN-T-06704 2003; PN-T-05687 2002)

Some of the action spectra listed above, such as that for erythema, differ from one another because of the differing results of various scientific studies (CIE 125 1997; PN-T-06588 1979). Because some countries use different exposure limit values, calculated or measured for different action spectra for the same biological response, it is impossible to compare evaluated hazards; some countries are less restrictive than others. Directive 2006/25/EC (2006) was an attempt to standardise the criteria used to evaluate hazards related to optical radiation. The directive includes criteria and exposure limit values for optical radiation, which member states must implement by 27 April 2010. The criteria and exposure limit values for noncoherent optical radiation that are obligatory in Poland are the same as those in that directive.

13.5.1 Criteria for Evaluating Health Hazards Caused by Artificial Noncoherent Optical Radiation in the EC

Exposure limit values represent the conditions under which nearly all individuals can be expected to be repeatedly exposed without acute adverse effects and, based upon the best available evidence, without a risk of delayed effects (ICNIRP 2004).

13.5.1.1 Ultraviolet Radiation

The fundamental goal of evaluating the health hazards arising from ultraviolet radiation is preventing cornea and conjunctiva inflammation, cataracts, erythema, photoaging of the skin, and skin cancers. The following parameters have been determined for 8-hour periods (for the calculations, total duration of exposure is 8 hours):

- Effective radiant exposure in the spectral region of 180–400 nm (spectrally weighted by $S(\lambda)$ or the actinic relative spectral effectiveness of UVR; Figure 13.2) incident upon unprotected skin and eye
- Radiant exposure (unweighted) in the spectral region of 315–400 nm incident upon the unprotected eye

The values obtained are compared with the relevant exposure limit values prescribed in directive 2006/25/EC (2006).

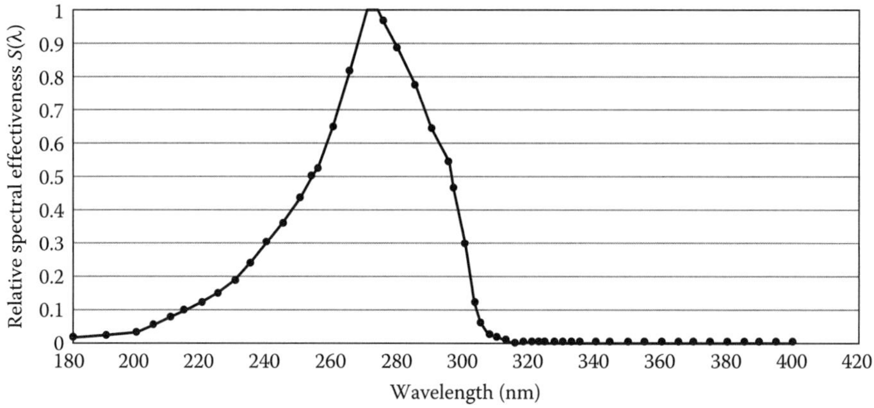

FIGURE 13.2 Actinic relative spectral effectiveness S(λ) of the UVR for inducing adverse effects on the skin and eyes.

13.5.1.2 Visual Radiation

The fundamental goal of evaluating the health hazards arising from visual radiation is preventing photochemical and thermal injuries to the retina. Intensive visual radiation, especially in the range of blue light between 400 and 500 nm, can cause retinal damage or diseases. Exposure time of less than 10 seconds mainly causes thermal injuries, whereas longer exposure can have photochemical effects.

Photochemical retinal hazards have been determined in the range of 300–700 nm by calculation or measurement of the effective radiance (spectrally weighted by the $B(\lambda)$ spectral weighting function for retinal photochemical injury; Figure 13.3) and the total time of exposure.

Thermal retinal hazards have been determined in the range of 380–1400 nm and for sources of high luminance over 10,000 cd/m^2. They are evaluated by calculating or measuring the effective radiance (spectrally weighted by the $R(\lambda)$ spectral weighting function for retinal thermal injury; Figure 13.3) and the duration of a single exposure.

The values of the above-mentioned parameters obtained by calculations or measurements are then compared to the relevant exposure limit values listed in Directive 2006/25/EC (2006).

13.5.1.3 Infrared Radiation

The fundamental goal of evaluating the health hazards caused by infrared radiation is preventing thermal injuries of the cornea, lens, retina, and skin. Hazards are determined separately for the retina, the cornea and lens, and the skin.

Intensive infrared radiation, especially in the IR-A range, can cause retinal thermal injuries. If a single exposure time is less than 10 seconds, retinal thermal hazard should be evaluated in the same way as visual radiation. However, if a single exposure time is over 10 seconds, thermal hazard evaluation should be determined in the same way as the range 780–1400 nm, by calculating or measuring effective radiance

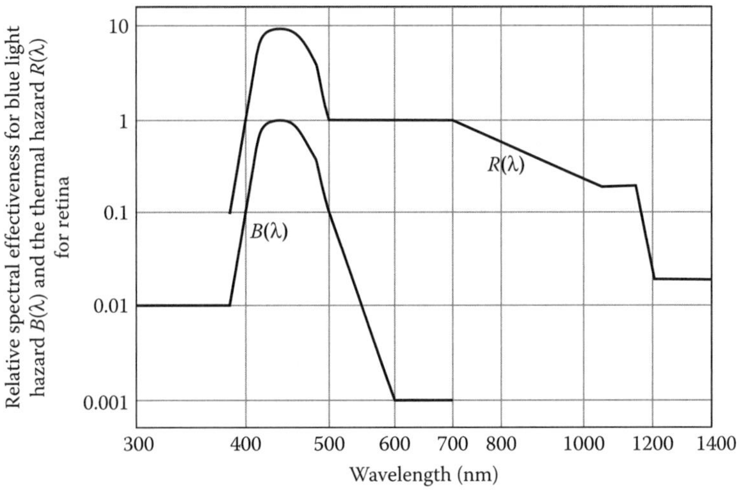

FIGURE 13.3 Retinal thermal hazard function $R(\lambda)$ and retinal photochemical hazard function $B(\lambda)$. (Reprinted with permission from PN-T-06704. 2003. Compilation of maximum permissible exposure to non-coherent (non-laser) optical radiation.)

(spectrally weighted by the $R(\lambda)$ spectral weighting function) and the duration of a single exposure.

Thermal hazards to the cornea and lens are evaluated by calculating or measuring irradiance (unweighted) in the wavelengths of 780–3000 nm and the duration of a single exposure.

Thermal hazards to the skin are evaluated by calculating or measuring irradiance (unweighted) within the wavelengths of 380–3000 nm and the duration of single exposure no longer than 10 seconds. For longer exposure, the system's thermal strain should be evaluated.

The values of these parameters obtained from the calculations or measurements are then compared with the relevant exposure limit values listed in directive 2006/25/EC (2006).

13.6 MEASUREMENTS OF OPTICAL RADIATION PARAMETERS USED FOR EVALUATING HEALTH HAZARDS CAUSED BY OPTICAL RADIATION

Excessive exposure to optical radiation can be harmful to human health; therefore, the occupational risk related to environmental factors present at workstations with sources of optical radiation must be evaluated. This is also the reason for periodically measuring the relevant parameters of optical radiation and comparing them with exposure limit values. Measurement methods and commonly used radiation measurement devices are described in standards EN 14255-1 (2005) and EN 14255-2 (2005).

Three types of radiometry were used in our survey of the physical quantities of optical radiation: visual, physical, and photographic. In visual photometry, because the eye detects the optical radiation, the measurement is valid only in the visual radiation range and is classified as subjective (Dybczyński 1996). In photographic radiometry, photosensitive photographic material detects the radiation. After photochemical treatment, it turns dark, and the degree of its darkening depends on the dose of radiant exposure. The optical density of the material is the measure of the physical quantity under consideration. Both subjective assessment and photographic radiometry are used rather rarely. In physical radiometry, the photoelectric converter plays the role of an optical radiation detector, classified as objective (rather than subjective, described above) because measurement equipment is used for detection of radiation. Photoelectric converters, or photodetectors, are chosen or spectrally corrected in a way that its spectral sensitivity range

- Covers the whole wavelength range of the measured radiation.
- Measures the entire range with the same sensitivity at each wavelength or with a specified spectral weighting function according to the biological effect considered.

Two measurement geometries are used to survey physical quantities of optical radiation. According to the first, the radiant power incident on the active measurement head comes from the hemisphere and irradiance is measured accordingly. The other geometry measures a part of the radiant power from a specified solid angle (usually narrow), and thus the entire radiance is measured.

The two major types of equipment for measuring optical radiation are radiometers and spectroradiometers. A *radiometer* is a broadband device designed to measure physical optical radiation quantities, and consists of a measurement head (the detector), a converter with an amplifier, and an analogue or digital display. The converter often transforms the photoelectric current into a voltage signal, which is then conveyed to the voltmeter system. Because radiometers are usually supplied with an assortment of detectors, changing the detector and programming the radiometer (e.g. by choosing a relevant factor and measurement range for the detector) enables measurement of different magnitudes of irradiance and radiance in different wavelength ranges, both unweighted (having constant spectral sensitivity) and weighted (having spectral sensitivity according to specified weighting function).

Spectroradiometers measure the radiation spectrum (spectral power distribution), which can be used to calculate unweighted (nonselective) and weighted (effective) parameters.

Spectroradiometers are of two major types: one equipped with a scanning monochromator and the other with an array detector. The former is usually a system incorporating a monochromator (single or double), an input device (e.g. integrating sphere, reflex telescope), a detector, a controller, and a computer with a special computer program for operating the whole system.

The monochromator is the most important component of the spectroradiometric system. It transmits a mechanically selected narrow band of radiation wavelengths chosen from a wider range of wavelengths available at the input. The spectral bandwidth is chosen according to need. Diffraction gratings or prisms are the optical elements used for spatial separation of the wavelengths. Depending on the optical radiation range, the prisms are made of glass, melted quartz, salt chloride, or potassium bromide. Photomultipliers, photoelements (silicon, indium arsenide, lead telluride, indium antimonide, lead selenide), photoresistors, thermopile, and so on are used as photoelectric converters (detectors of radiation). Spectroradiometric measurements of the whole range of optical radiation usually require more than one detector. For measurements of spectral irradiance an integrating sphere is used as input device, and for spectral radiance a reflex telescope is used.

This device scans the whole spectrum systematically, a procedure which takes some time (depending on the range of the spectrum and the measurement interval), and, therefore, reliable results can be obtained for continuous radiation only when the scanning process is underway (EN 14255-1 2005).

A spectroradiometer with an array detector measures the whole range of wavelengths instantly (within seconds or milliseconds), because the measured spectrum is projected onto the photodiode array. Reliable results can thus be obtained both for constant radiation and radiation that varies over time (EN 14255-1 2005).

The result of spectroradiometric measurement is a spectrum, which can be used to calculate weighted and unweighted quantities. This method of quantifying the optical parameters values is far more precise than using broadband radiometers, however, the equipment is much more expensive, often large, and impossible to use outside the laboratory. Dosimeters are also used to assess occupational risk at workstations; they measure radiant exposure during a work shift.

13.7 EXAMPLES OF HAZARDS CAUSED BY OPTICAL RADIATION AT WORKSTATIONS

Tables 13.3 through 13.5 present the optical radiation parameters used for evaluating hazards at electric welding, gas welding, and glassblowing workstations. Values that exceed the exposure limit values indicate the presence of high occupational risk (i.e., a health hazard).

Measurements of ultraviolet radiation emitted during electric and gas welding (Tables 13.3 and 13.4) show that for a total exposure time of 3 hours (10,080 seconds) during the work shift there are

- Hazards to the eyes, face, and skin on the hands during electric welding
- Hazards to the skin on the hands during gas welding

Measurements of visible radiation show that for the same exposure time during the work shift and angular subtense of a source lower than 11 mrad (millirad) there are

- Photochemical retinal hazards during electric welding
- No retinal hazard during gas welding

TABLE 13.3
Sample Results of Measurements of UV, VIS and IR Radiation for MAG Electric Welding of Steel St3 (Protected with $ArCO_2$) Rutile Electrode ER 146 BAILDON, Dimension: 3.25 mm; Welding Current: 123 A

Range of Radiation	Exposed Part of the Body	Distance from the Source (m)	Exposure Time(s)		Angular Subtense (mrad)	Irradiance	Effective Radiant Exposure (J/m^2)	Exposure Limit Values (Directive 2006/25/EC 2006)	
			Single	Total				Irradiance	Radiant Exposure (J/m^2)
Ultraviolet	Eyes	0.50	–	10,080	–	0.75 W/m^2 (effective)	7560	–	30 (effective)
	Face (skin)	0.50				1.09 W/m^2 (effective)	10,987		
	Hands (skin)	0.08				29.3 W/m^2 (effective)	295,344		
Visible	Eyes	0.50	–	10,080	10	64 $\mu W/cm^2$ (effective)	–	1 $\mu W/cm^2$ (effective)	–
Infrared	Eyes	0.50	19	–	–	22.7 W/m^2	–	1978 W/m^2	–

TABLE 13.4
Sample Results of Measurements of UV, VIS and IR Radiation for Gas Welding of Steel St3 (Acetylene: Oxygen = 4:1)

Range of Radiation	Exposed Part of the Body	Distance from the Source (m)	Exposure Time, (s)		Angular Subtense (mrad)	Irradiance	Effective Radiant Exposure (J/m^2)	Exposure Limit Values (Directive 2006/25/EC 2006)	
			Single	Total				Irradiance	Radiant Exposure
Ultraviolet	Eyes	0.46	–	10,080	–	0.00139 W/m^2 (effective)	14.1	–	30 (effective)
	Face (skin)	0.46				0.00219 W/m^2 (effective)	22.1		
	Hands (skin)	0.30				0.0033 W/m^2 (effective)	33.3		
Visible	Eyes	0.46	–	10,080	10.8	0.51 $\mu W/cm^2$ (effective)	–	1 $\mu W/cm^2$ (effective)	–
Infrared	Eyes	0.46	35	–	–	191 W/m^2	–	1251 W/m^2	–

TABLE 13.5
Sample Results of Measurements of IR Radiation for a Glassblower, during Different Activities at an Entrance to the Glass Furnace (Sodium Silicate Glass Mass at Temperature of 1180°C)

Activity	Single Time of Exposure (s)	Exposed Part of the Body	Distance from the Source (m)	Irradiance (W/m²)	Radiant Exposure (J/m²)	Exposure Limit Values (Directive 2006/25/EC 2006)	
						Thermal Hazard for Cornea and Lens (W/m²)	Thermal Hazard for Skin N_C (J/m²)
Preliminary glass mass taking	9	Eyes	0.85	3377	–	3464	–
	9	Face	0.85	3377	30,393	–	34,641
	9	Hands	0.45	10,616	95,544		34,641
Preliminary forming of nipple	27	Eyes	0.5	383	–	1520	–
Preliminary exhaust of nipple	17	Eyes	1.4	137	–	2150	–
Additionally taking the glass mass	8	Eyes	0.85	3377	–	3784	
	8	Face	0.85	3377	27,016	–	33,636
	8	Hands	0.45	10,616	84,928	–	33,636
Fundamental forming	36	Eyes	0.5	1043	–	1225	–
Fundamental exhaust	79	Eyes	1.4	280	–	679	–

FIGURE 13.4 Electric arc welding.

In both types of welding, infrared radiation did not pose a thermal hazard to the eyes and skin (Figure 13.4).

Measurements of infrared radiation emitted during a glass blower's activities at the glass furnace show the presence of thermal hazards to the uncovered skin on the hands during preliminary and additional glass mass taken from the entrance to the furnace (Figure 13.5). There was no thermal hazard to the eyes (cornea and lens) or to the skin on the face.

13.8 PROTECTION AGAINST OPTICAL RADIATION

The principal rule for protection is to avoid exposure. If this is impossible, a combination of three preventive measures should be taken:

- Time of exposure should be minimised.
- Distance from the source should be maximised.
- The body should be shielded against radiation.

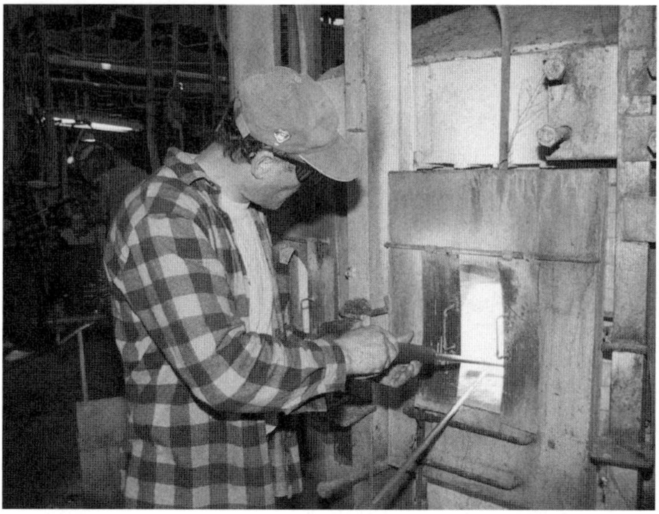

(a)

(b)

FIGURE 13.5 Glass blower work activities during (a) glass mass taking, (b) preliminary exhaust of nipple.

The hierarchy of control measures varies depending on the source of UVR and the risk assessment performed. When the source of optical radiation constitutes a hazard according to directive 2006/25/EC (2006), the employer must provide protection from exposure through a combination of engineering controls, administrative measures, and personal protective equipment (PPE).

Engineering controls include redesigning equipment or work processes to eliminate optical radiation in the space where workers perform work tasks, or withdrawing the workers from processes that can cause optical radiation hazards. Examples of this kind of control are installing shielding, enclosing the source in a container or dedicated room, using interlocks, and covering materials with appropriate nonreflective optical radiation materials and built-in radiation detectors and alarms (visual or auditory).

Administrative measures might include disseminating information about hazards and safety aspects and training workers, limiting access to areas where equipment emits optical radiation, limiting the time of exposure and increasing the distance of the user from the source, introducing hazard signs or warning lights, and monitoring the hazard by periodic measurements of optical radiation at workstations.

PPE is used if engineering and administrative controls do not provide appropriate protection. Protective equipment for the skin and eyes is imperative. Exposed areas of skin, especially the face, backs of the hands, head, neck, and wrist areas should be covered with working or protective clothing and hands with protective gloves.

When there is intensive exposure to UVR, work clothes should be made of special fabric for protection against UVR with a minimum clothing protection factor (CPF) value of 30 to guarantee high protection against natural UVR. Working clothes that protect against infrared radiation should be made of a fabric with high heat resistance. The basic criterion for work clothes that protect against infrared radiation is the irradiance value of workers' exposure at a specific workstation. Such protective clothing is used in steel works or for extinguishing fire.

The face can be protected with a face shield, helmet, or mask, which also protects the eyes. Suitable headwear protects the head and neck. Goggles, spectacles, visors, or face shields that absorb or reflect radiation can also protect the eyes.

The methods against for reduction of optical radiation and related hazards discussed in this chapter should not be treated as a one-time activity needed only when launching new workstations. Existing hazards and health effects require continuous supervision to be sure that workers follow the safety procedures, especially wearing the appropriate PPE and controlling the technical conditions.

REFERENCES

CIE Publ. no 17.4. 1987. *International Lighting Vocabulary*. Wien: International Commission on Illumination (CIE).

CIE Publ. no 106. 1993. *CIE Collection in Photobiology and Photochemistry*. Wien: International Commission on Illumination (CIE).

CIE Publ. no 125. 1997. *Technical Report. Standard Erythema Dose: A Review*. Wien: International Commission on Illumination (CIE).

CIE Publ. no 138. 2000. *CIE Collection in Photobiology and Photochemistry 2000*. Wien: International Commission on Illumination (CIE).

CIE Publ. no 174. 2006. *Technical Report. Action Spectrum for the Production of Previatmin D3 in Human Skin*. Wien: International Commission on Illumination (CIE).

Directive 2006/25/EC of the European parliament and of the Council of 5 April 2006 on the minimum health and safety requirements regarding the exposure of workers to risks arising from physical agents (artificial optical radiation) (19 individual Directive within the meaning of Article 16(1) of directive 89/391/EEC), OJ L 114/38, 2006.

Dybczyński, W. 1996. *Optical Radiation Measurements*. Bialystok: Technical University of Bialystok.
Dybczyński, W. 1999. Optical radiation. In *Occupational Safety and Ergonomics*, ed. D. Koradecka, 461–486. Warsaw: CIOP.
EN 14255-1. 2005. Measurement and assessment of personal exposures to incoherent optical radiation, Part 1: Ultraviolet radiation emitted by artificial sources in the workplace.
EN 14255-2. 2005. Measurement and assessment of personal exposures to incoherent optical radiation, Part 2: Visible and infrared radiation emitted by artificial sources in the workplace.
Hering, M. 1992. *Fundamentals of Electrothermics*. Warsaw: WNT.
International Commission on Non-Ionizing Radiation Protection (ICNIRP). 1997. Guidelines on limits of exposure to broad-band incoherent optical radiation (0.38 to 3 μm). *Health Phys* 73(3):539–554.
International Commission on Non-Ionizing Radiation Protection (ICNIRP). 2004. Guidelines on limits of exposure to ultraviolet radiation of wavelengths between 180 nm and 400 nm (incoherent optical radiation). *Health Phys* 87(2):171–186.
PN-T-06588. 1979. Protection against optical radiation. Ultraviolet radiation. Terms, definitions and units.
PN-T-05687. 2002. Protection against optical radiation. Measurement methods of visible and infrared radiation at the workstations.
PN-T-06704. 2003. Compilation of maximum permissible exposure to non-coherent (non laser) optical radiation.
Podogrodzki, J. 1998. *Natural Optical Radiation Handbook*. Warsaw: PKOś SEP.
Rea, M. S., ed. 1993. *Lighting Handbook. Reference & Application*. 8th ed. New York: IES of North America.
Sliney, D. H. 2007. Radiometric quantities and units used in photobiology and photochemistry: Recommendations of the International Commission on Illumination. *Photochem Photobiol* 83:425–432.
Wolska, A., S. Marzec, and G. Owczarek. 2001. *Principles of Hygienic Assessment of Non-Laser Optical Radiation*. Warsaw: CIOP.

14 Laser Radiation

Grzegorz Owczarek and Agnieszka Wolska

CONTENTS

14.1 INTRODUCTION

Laser radiation is not a natural phenomenon, but is produced by a specialised device known as a laser (an acronym for light amplification by stimulated radiation). Lasers generate electromagnetic radiation with optical radiation wavelengths ranging from 100 nm to 1 mm. Lasers use stimulated emission of radiation, and their radiation differs significantly in its physical properties from the classical optical radiation emitted by conventional sources such as ultraviolet, infrared radiation, or light sources used for lighting purposes. Many techniques are used to produce laser beams, so lasers can vary greatly.

The history of lasers is relatively short. In 1960, American physicist Theodore Maiman and his coworkers built the first ruby laser, beginning the dynamic development of lasers and associated technology. Laser radiation is used in various technological processes in industry (laser cutting, welding), medicine (laser surgery), science, military applications (laser tracking), and cosmetics (laser biostimulation). Lasers are also used in electronics, film, and the music industry (laser printers, CD/DVD players and recorders). Telecommunication devices increasingly use fibre-optic cables to transfer information. Laser radiation is also used to project images and create spectacular visual effects.

14.2 CHARACTERISTICS OF LASER RADIATION

The basic action of a laser is to stimulate the emission of energy quanta in an active medium, also called laser substance or active optical medium. The laser function stimulates an active optical medium and then releases energy as a quantum of coherent radiation. The basic elements of a laser are an active medium, an optical resonator, and a pumping system. The active medium is an appropriately selected solid, gas, or liquid. A supply of energy to this solid, gas, or liquid via pumping initiates the laser action. The resonator is an optical system consisting of two mirrors, one of which has a high reflectance. The other is an output coupler mirror; of the two, at least one is partially transmitting. Using a resonator concentrates the radiation in a very narrow spectrum range at the cost of the remaining range. The distinctive feature of a laser setting it apart from other optical radiation sources is its greater density of spectral radiation. The pumping mechanism transports as many electrons as possible in the active medium to an excited state. The arc of a flash lamp, another laser radiation, an electric current, chemical reactions, atomic impact, or electrons injected into a substance can pump a laser.

Laser radiation has a high level of monochromatic directional coherence, with the angle of beam divergence usually not exceeding a few meridians. Concentrating the entire energy of laser radiation into an incredibly small spectrum range and small solid angle could also be achieved over time. Another significant feature of lasers is that they emit a polarised beam. A laser usually emits one or a few wavelengths of radiation in a specific power range and can be tailored for a particular use. Table 14.1 shows examples of the use of some laser types.

Lasers are classified according to different features such as the type of resonator, pumping mechanism, active medium, or operational system. There are lasers with stable, nonstable, linear, and ring resonators. The excitation mechanisms used are electric current and photon exposure chemical reactions. There are lasers of continuous and pulse work. Lasers are classified as three- or four-level according to the methods and types of transfer of electrons occurring between levels of the active medium. However, the general classification standard most often used is the state of the active medium. There are solid-state lasers (solid-state rods or glass as a cover), semipermeable lasers (coupling lasers), liquid lasers (for dye), and gas lasers (atom, ion, molecular lasers). Those most widely found in technological applications are CO_2, Nd:YAG and excimer lasers.

14.3 DANGERS ASSOCIATED WITH LASER RADIATION

Laser radiation is dangerous to human health, especially to the eyes and skin as this type of radiation can transport vast amounts of energy and their absorption can injure these tissues (Sliney and Wolbrasht 1980). The eyes are the most vulnerable (Wolska and Konieczny 2006). The dangers posed to the different parts of the eye depend on the laser's wavelength. The cornea can absorb the far ultraviolet (UV) wavelength range from 200 to 215 nm and infrared (IR) of over 1400 nm. The lens absorbs near UV and IR, and the retina absorbs visible radiation or near IR. Retina injuries, which are extremely dangerous, are caused by radiation of 400–1400 nm range. The laser beam, whose diameter is a few millimetres, can be concentrated on the retina into a small spot of 10 μm diameter; that is, an irradiance of 1 mW/cm² (milliwatt per square centimetre) at the eye cornea enlarges to 100 W/cm² on the retina and therefore can injure it severely. The degree of damage caused differs depending on the location of its focus. Injury around the fovea centralis can cause permanent blindness. Figure 14.1 is a schematic diagram of an eye injury (Owczarek 2002).

The risk of skin injury caused by a laser beam is very high. The most vulnerable skin areas are the head, hands, and skin on the shoulders. However, radiation must

TABLE 14.1
Examples of Use of Lasers

Type of Laser	Wavelength (nm)	Type of Work, Time of Impulse	Ability	Examples of Use
Ruby Al_aO_3:Cr^{3+}	694.3	Pulse, from a few to over 10 microseconds	0.1%–0.5%	Welding, melting, drilling, dentistry, pulse holography, biology, distance measurement
Neodymium Nd^{3+}: YAG	1064.6 1300, 1400	Stable or pulse from few picoseconds to over 10 milliseconds	0.1%–10% (depending on pump type—higher with diode pump)	Telecommunications, laser tracking, controlled nuclear reactions, surgery, microprocessing, cutting, distance measurement
Neodymium on glass Nd: glass	1050–1060	Stable or pulse	1%–5% (with lamp pump)	Optical amplifier to create a gigantic pulse (GW), initiation of controlled nuclear reactions, cutting, microfusion
Semiconductor GaINAsP, GaAs, ALGaAS	800–1600	Stable or pulse	60%–75%	Fibre-optic cable telecommunications, geodesy, printing (indirectly as a pump to cut matrix), recording and playing CDs/DVDs
Titanium Al_2O_3: Ti^{3+}	Tunable: 665–1130	Stable or pulse from few femtoseconds	0.01%–0.1% (depending on pump type)	Determining the level of atmospheric contamination (LIDAR systems), isotope separation, biomedical research
He-Ne	632.8	Stable	0.1%	Metrology, holography, interferometry
Ne-Cu (Cu–vapour)	510.6 and 578.2	Pulse	Do 3%	Precise material processing, dermatology
Nitrogen (N_2)	337.1	Pulse 10 nanoseconds	20%	Spectroscopy, photochemical reactions
CO_2	More often 10,600	Stable or pulse	30%	Material processing, cutting, welding, surgery, dentistry, laser tracking, controlled nuclear reactions, isotopes separation
Alexandrite	Tunable: 710–820	Stable or pulse	0.3%	Determining the level of atmospheric contamination (LIDAR systems), medicine, spectroscopy
Excimer KrCl, ArF, KrF, XeCl, XeF	157, 193, 248, 308, 351	Pulse	1%–2%	Surgery (ophthalmology, cardiosurgery), precision engineering, marking, hole-making
Erb on glass Er: glass	1540	Pulse	0.2%	Safety for eyes in distance measurement
Erb Er: YAG	2940	Pulse	1.5%	Medicine, biomedical research

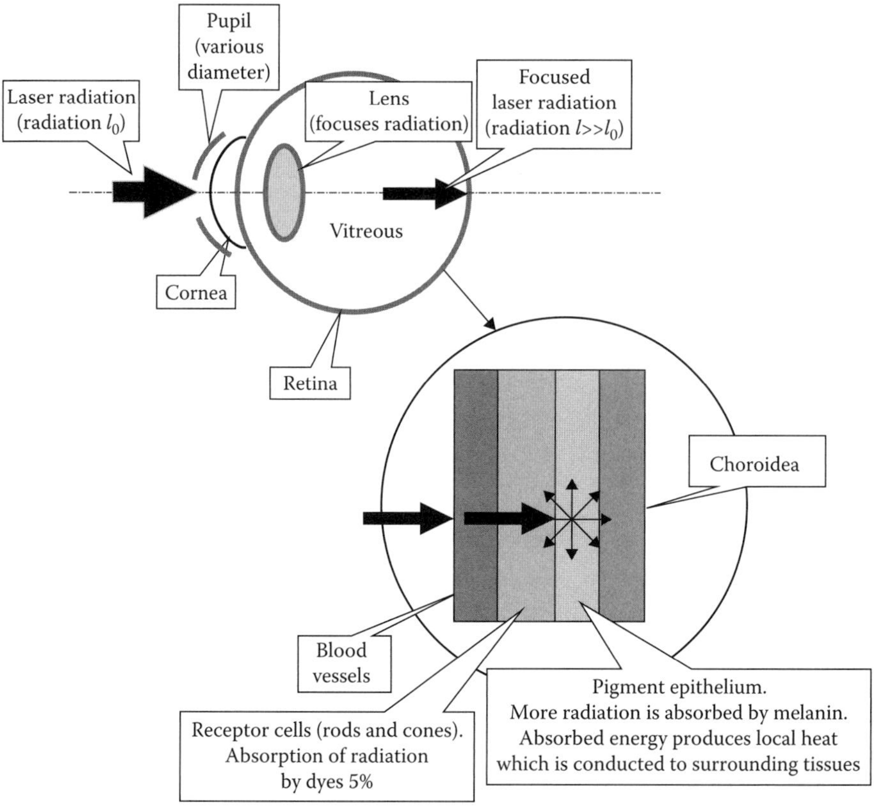

FIGURE 14.1 Diagram of laser-induced eye hazard. (Reprinted with permission from Owczarek, G. 2002. *Protection Against Laser Radiation*. Warsaw: CIOP-PIB.)

be far more powerful to injure skin than it does to injure the eye. Visible or IR laser radiation can cause mild erythema; if powerful, it can also cause burns. Short pulses of high peak-power laser radiation can carbonise tissues.

14.4 CLASSES OF SAFETY FOR LASERS

Even though laser devices have special protective covers and come with instruction manuals, accidents can occur during their operation. Laser radiation accounts for 44% of these (http://technologialaserowa.republika.pl/bhp.html). Hence, knowledge of laser safety classifications, which indicate the danger level of a laser, is important.

The severity of injuries caused in the tissue depends on the laser's wavelength and radiation power. Accordingly, lasers are categorised into seven classes, namely 1, 1M, 2, 2M, 3R, 3B, and 4 (PN-EN 60825-1 2005). Previously, lasers were only of five classes (1, 2, 3A, 3B, and 4). Producers of lasers must affix labels containing information on the classification to enable users to prevent the dangers related to laser usage. Table 14.2 lists the characteristics of the various classes of lasers.

TABLE 14.2
Classification of Lasers and Laser Devices

Class	Characteristics
1	Lasers that are safe under reasonably foreseeable conditions of operation, including the use of optical instruments for intrabeam viewing.
1M	Lasers that emit radiation in the wavelength range of 302.5–4000 nm and which are safe under foreseeable conditions of operations, but may be hazardous if the user uses optics in the beam.
2	Lasers that emit visible radiation in the wavelength range of 400–700 nm, where eye protection is normally afforded by aversion responses, including the blink reflex.
2M	Lasers that emit visible radiation in the wavelength range of 400–700 nm, where eye protection is normally afforded by aversion responses, including the blink reflex. However, viewing the output may be more hazardous if the user uses optics in the beam.
3R	Lasers that emit radiation in the wavelength range of 302.5–10^6 nm, where direct intra-beam viewing is potentially hazardous.
3B	Lasers that are normally hazardous when direct intra-beam exposure occurs. Viewing diffuse reflections is normally safe.
4	Lasers that are capable of producing hazardous diffuse reflections. They may cause skin injuries and could also constitute a fire hazard. Their use requires extreme caution.

Source: PN-EN 60825-1. 2005. Laser safety products. Part 1: Equipment classification, requirements, and user's guide.

Except for the devices in class 1, lasers are dangerous because of the harm the radiation may cause to the eyes or skin. The most dangerous are devices in class 4, which are used in cutting, welding, marking, and sometimes in medicine (e.g. laser lances). Great care is called for in using these kinds of lasers. Any laser used must be properly classified and precautions for preventing the dangers of direct radiation must be listed. Protection from reflected and scattered laser radiation could also be required for the eyes and skin of employees who are in the vicinity of class 4 lasers. Reflected or scattered radiation is also dangerous to humans, which must be remembered when selecting a safe place to operate such lasers.

The highest level of laser radiation that does not cause eye or skin injury is defined legally in the EU directive 2006/25/ EU. These defined extreme values apply to accidental, short-term human exposure to such radiation but do not apply to exposures for medical, rehabilitation, or optical computer tomography purposes. Exposure to laser radiation that exceeds the defined values of maximum permissible exposure (MPE) is indicative of occupational risk potentially hazardous to health.

14.5 ASPECTS OF SAFE LASER USAGE

Organising a proper and safe workplace for laser usage requires a detailed evaluation of the dangers arising from laser operation. The three basic elements of danger evaluation are (1) the potential danger of the laser itself, (2) the environment where it is installed, and (3) user awareness.

Extremely caution is necessary when working with lasers. Emitted radiation is characterised by a significantly larger power than that from classical sources. The dangers of laser devices relate not only to laser beam radiation, but also to construction and functioning factors such as

- Electric current
- Steam and gases (e.g. in laser surgery, cutting tissue produces gas)
- Fire or explosion (e.g. flammable materials that may burn because of high-power laser radiation)
- Those that accompany other types of radiation (not laser radiation; e.g. danger of high-frequency radiation or X-ray radiation emitted by the laser)

Each factor noted above is a potential hazard arising from the laser device. The location of the laser is crucial, as it relates not only to the laser's efficiency but also to safety. Essential elements of safe laser use include the application of proper safety interlocks and using personal and collective protective equipment. Employees also should be educated about lasers, especially the following aspects:

- Laser operation procedure
- Use of control procedures, warning labels, and so on
- Accident reporting procedure
- The biological effects of radiation on the eyes and skin

Rooms where the laser might emit an uncovered radiation beam should be equipped with the following:

- Proper lighting, because pupils expand less in proper lighting in comparison with dark or badly lit places; the smaller the expansion of the pupil the less laser radiation can enter the eye, minimising potential damage
- Matt walls to prevent accidental and dangerous mirror-like reflections from smooth walls
- Proper window coverings to prevent escape of laser radiation
- Proper labelling of the entrance to the room where the laser device is used to warn of potential danger (Figure 14.2 displays an example of a warning label)

Table 14.3 lists the basic needs of and recommendations for laser users.

14.6 PERSONAL AND COLLECTIVE PROTECTORS

The most important rule for the use of personal protective equipment is that it should always be used if other methods of eliminating or preventing danger are exhausted. This rule also applies to laser radiation. Covering the laser beam or isolating it is the best protection, but can only be carried out if the construction of the laser device allows it. Laser covers, also called screens, can be categorised into two types: those that prevent dangerous primary laser radiation and those that prevent secondary

FIGURE 14.2 Laser radiation warning label. Signs and edges should be black and background should be yellow. (Adapted with permission from PN-EN 60825-1. 2005. Laser safety products. Part 1: Equipment classification, requirements, and user's guide.)

TABLE 14.3
Basic Needs of and Recommendations for Laser Users

Requirements and Recommendations	Class of Laser						
	Class 1	Class 1M	Class 2	Class 2M	Class 3R	Class 3B	Class 4
Laser safety officer					+[a]	+	+
Use of access panel and safety interlock						+	+
Key control						+	+
Use of laser beam stop or attenuator						+	+
Laser radiation emission warming					+[a]	+	+
Use of warning labels						+	+
Covers for laser beams					+	+	+
Avoidance of beam reflection					+	+	+
Personal eye protectors						+[b]	+[b]
Protective clothing						+[c]	+[c]
Training in the field of safety work with lasers					+	+	+

[a] Required only during emission of radiation beyond the visible range.
[b] Required if exceeding MDE values in the area of laser radiation.
[c] Required if laser radiation creates a potential hazard.

radiation produced during laser processing of materials; such covers are integral elements of laser device.

Safety screens are designed to control laser radiation; they are generally constructed like curtains and are installed on special stands or attachment points. Glasses, goggles, and face shields with special protective filters protect the eyes from dangerous laser radiation of wavelengths from 180 to 1000 μm.

The monochromatic nature of laser radiation necessitates protective filters that are designed for each specific type of laser. Eye covers are used to weaken dangerous laser radiation in the visible spectrum range from 400 to 700 nm to the definite values of class 2 lasers ($P \leq 1$ mW for stable lasers—in this case, physiological reactions such as blinking that protect the eye), and are known as personal eye protectors for alignment of laser systems.

REFERENCES

Directive 2006/25/EU of the European Parliament and of the Council of 5 April 2006 on the minimum health and safety requirements regarding the exposure of workers to risks arising from physical agents (artificial optical radiation) (19th individual Directive within the meaning of Article 16(1) of Directive 89/391/EEC). OJ L 114/38, 27.4.2006.

http://technologialaserowa.republika.pl/bhp.html/ (accessed August 2008).

ICRP 2007. Publication 103. The 2007 Recommendations of the International Commission on Radiological Protection, 37(2–4).

International Commission on Non-Ionizing Radiation Protection (ICNIRP). 2000. Revision of guidelines on limits of exposure to laser radiation of wavelengths between 400 nm and 1.4 μm. *Health Phys* 79(4):431–440.

Owczarek, G. 2002. *Protection Against Laser Radiation*. Warsaw: CIOP-PIB.

PN-EN 60825-1. 2005. Laser safety products, Part 1: Equipment classification, requirements and user's guide.

Sliney, D. H., and M. L. Wolbarsht. 1980. *Safety with Lasers and Other Optical Sources*. New York: Plenum Publishing Corp.

Wolska, A., and P. Konieczny. 2006. Laser radiation—Health effects and safety aspects. In *Proceedings of Electrotechnical Institute*, 228, 283–296.

15 Ionising Radiation

Krzysztof A. Pachocki

CONTENTS

15.1 INTRODUCTION

Radiation protection refers to all the actions and activities necessary to prevent the exposure of humans and the environment to ionising radiation, and if prevention of such exposure is not possible, these actions should endeavour to reduce the harmful influence of this radiation on the health of future generations in the form of genetic consequences. With this in mind, the International Commission on Radiological Protection (ICRP) has formulated the following recommendation: No practice involving exposure to radiation should be adopted unless it produces sufficient benefit to the exposed individuals or to society to offset the radiation detriment it causes.

Radiation protection methods are based on scientific data and hypotheses formulated on the basis of that data. However, these data are not the sole basis for the rules that are generally formulated to protect humans from ionising radiation. Judgement, based on social criteria and the value system respected by most civilised societies, also plays a significant role in this process. The universal belief is that compliance with conditions of optimum radiation protection also assures sufficient protection of other species, although this may not be true for some specimens (Hrynkiewicz 2001).

The rules of radiation protection, composed by the ICRP, form a substantive base for the development of national legislation, although their integration into a nation's legal system is not mandatory. The International Radiology Congress founded the ICRP in 1928. The ICRP's task was to compose rules and practices to follow that would protect against X-rays used in medicine and radiation emitted by radium applicators. After World War II, the scope of the ICRP expanded significantly to cover all applications of ionising radiation, including those that resulted from the rapid development and widespread use of artificial radionuclides in nuclear power engineering, industry, medicine, science, agriculture, and so on.

In 1977, the concept of a *dose limit* replaced the *maximum permissible dose* and *effective dose equivalent* was introduced. The first comprehensive system of radiation protection was developed (ICRP-26). This system adopted a dose limit of 50 mSv/year (effective dose equivalent) for occupational exposure.

In 1980, the International Commission on Radiation Units and Measurements (ICRU) approved new doses for ionising radiation under the international system of units (SI system): for absorbed doses, the gray (Gy) replaced the earlier rad (radiation absorbed dose [rd, rad]); for dose equivalent, the sievert (Sv) replaced the former rem (roentgen equivalent man [rem]).

In 1990, changes in the recommendations (ICRP-60) raised the dose limits to a stricter level of 20 mSv/year for occupational exposure (effective dose; ICRP 1990).

The ICRP's recommendations formed the basis for the *International Basic Safety Standards for Protection against Ionizing Radiation and for the Safety of Radiation Sources* (IAEA 1996) formulated and published collectively by the organisations affiliated with the United Nations, namely the International Atomic Energy Agency (IAEA), the World Health Organization (WHO), the International Labor Organization (ILO), and the Food and Agriculture Organization (FAO), as well as the Organisation for Economic Cooperation and Development (OECD), the Nuclear Energy Agency (NEA/OECD), and the Pan American Health Organization (PAHO). National regulations on radiation protection in most countries are based on the ICRP recommendations of 1990 (ICRP 1990). The directives of the European Union are also based on these documents (OJ EC 1996; OJ EC 1997).

The ICRP uses all available sources of scientific information in its work, especially the periodic reports of the United Nations Scientific Committee on the Effects of Atomic Radiation (UNSCEAR 2000, UNSCEAR 2006), issued every 3 to 5 years.

Exposure to ionising radiation is of three basic types (ICRP 1977; ICRP 1990; ICRP 2007):

1. *Occupational exposure*: This refers to hazards relating to work and applies solely to persons employed in conditions where they face a threat of exposure to ionising radiation.
2. *Public exposure*: This refers to the presence of ionising radiation sources in the environment, for example, in potable water, air, or soil, and in objects of everyday use. It also refers to sources other than those naturally encountered in the given environment that are responsible for the radiation level in the natural environment.
3. *Medical exposure*: This includes all medical uses of radiation for diagnostic tests and in radiation therapy.

15.2 BASIC CONCEPTS, VALUES, AND UNITS OF MEASURE APPLIED IN RADIATION PROTECTION

15.2.1 Physical Basis of Radiation

Ionising radiation is electromagnetic radiation (gamma, γ; roentgen, X) or molecular radiation (corpuscular, e.g. alpha, α; beta, β), which, when passing through matter directly or indirectly produces ions (other than photons of ultraviolet radiation). Radiation is related to the emission and transfer of energy. When a body is said to radiate, this means that it emits energy. Sources of ionising radiation include the following:

- Radioactive substances (elements or their chemical compounds) such as radium (^{226}Ra)
- Equipment, for example, X-ray apparatus

Radioactivity is the independent disintegration of the atomic nuclei of certain isotopes associated with the emission of the rays α, β and γ. Radioactive nuclides are often referred to as radionuclides. The nucleus of the newly created element can be stable or radioactive. In some cases, a whole chain of radionuclides forms, one born of the other. This is called a radioactive family or decay chain. When a radionuclide disintegrates with a defined energy, α and β particles and γ quantum are ejected, and the distribution of this energy is called the energy spectrum. Roentgen rays (X-rays), similar to γ rays, are electromagnetic radiation. The origins of these radiation types vary. Gamma rays are produced by stimulated atomic nuclei, while X-rays appear outside the nucleus because of electron retardation (retardation radiation; Gorczyca et al. 1997).

An important feature of ionising radiation is its hardness, that is, the degree to which it is absorbed by matter. The hardness depends on the type of radiation and its energy and rises as energy increases.

The penetration of α radiation (a stream of fast-moving helium nuclei) is very small. In air, its maximum range does not exceed 10 cm, and in tissue only fractions of millimetres. It can hardly penetrate through a single sheet of ordinary paper.

The ionisation caused by β radiation (a stream of fast-moving electrons or positrons), which consists of particles smaller than α particles and of a lower electrical charge, is small; hence, its range is much greater (in air, almost 60 times greater than α radiation of the same energy), and can reach distances of several metres. This radiation can even penetrate a metal shield of a several-millimetre thickness.

Gamma or X radiation (electromagnetic radiation) is very powerful and can penetrate very thick layers of concrete or steel. It is very difficult to define its range in matter. Consequently, the thickness of this layer of matter is usually expressed in terms of the thickness that is needed to weaken the intensity of this radiation by half (the so-called half-value layer).

15.2.2 Activity and Half-Life

The *activity* (*A*) of an amount of a radionuclide in a particular energy state at a given time is the quotient of *dN* divided by *dt*, where *dN* is the expectation value of the number of spontaneous transitions from that energy state in the time interval *dt*:

$$A = \frac{dN}{dt} \tag{15.1}$$

The SI unit of activity is the reciprocal of second (s^{-1}), called the becquerel (Bq).

$$1\ \mathrm{Bq} = 1\mathrm{s}^{-1}$$

The becquerel is a very small unit, hence in practice its multiple is used:

$$1\ \text{kBq (kilobecquerel)} = 10^3\ \text{Bq}$$
$$1\ \text{MBq (megabecquerel)} = 10^6\ \text{Bq}$$
$$1\ \text{GBq (gigabecquerel)} = 10^9\ \text{Bq}$$
$$1\ \text{TBq (terabecquerel)} = 10^{12}\ \text{Bq}$$

Before the introduction of the SI system, the unit in use was the curie (Ci), which denoted the equivalent activity of 1 g of radium and amounted to $3.7 \times 10^{10}\ s^{-1}$. The relationship between the curie and the becquerel is as follows:

$$1\ \text{Bq} = 27\ \text{pCi} = 27 \times 10^{-12}\ \text{Ci}$$
$$1\ \text{Ci} = 37\ \text{GBq} = 3.7 \times 10^{10}\ \text{Bq}$$

Since the curie, unlike the becquerel, is a large unit, in practice its submultiples are used:

$$1\ \text{mCi (millicurie)} = 10^{-3}\ \text{Ci}$$
$$1\ \mu\text{Ci (microcurie)} = 10^{-6}\ \text{Ci}$$

1 nCi (nanocurie) = 10^{-9} Ci
1 pCi (picocurie) = 10^{-12} Ci

The degree of radioactive contamination of various substances (including food) is defined by its activity concentration. The *activity concentration* (A_V; also called *volumic activity* or *activity divided by volume* or *activity per volume*) of a specified radionuclide in a volume is the activity A of the radionuclide in the volume divided by the volume V.

$$A_V = \frac{A}{V} (\text{Bq/kg}) \tag{15.2}$$

The *specific activity* (A_m; also called *massic activity, activity divided by mass,* or *activity per mass*) of a specified radionuclide in a sample is the activity A of the radionuclide in the sample divided by the total mass m of the sample.

The *surface activity concentration* (A_F; also called *areal activity concentration*) of a specified radionuclide on a surface is the activity A of the radionuclide on the surface area F divided by the area.

The time taken for the number of nuclei of the radionuclide to decrease by half is called the *half-life* $T_{1/2}$ of its radioactivity. Each radionuclide has a unique half-life, for instance, 5730 years for carbon-14, 24,000 years for plutonium-239, 4.47×10^9 years for uranium-238, 8 days for iodine-131, about 30 years for cesium-137, and 1600 years for radium-226. The half-life of radon-222, a radioactive gas, is 3.8 days, and for potassium-40 is 1.28×10^9 years. For various radionuclides, these values range from fractions of seconds to millions of years. In the subsequent half-life periods, the activity of the radionuclide progressively declines to 1/2, 1/4, 1/8, and 1/16 of its original value (IAEA 2004). Therefore, a radionuclide's activity can be computed at any time if its activity at that moment is known.

During radioactive disintegration, the number of atoms of the isotope declines with time.

$$N(t) = N(0)e^{-\lambda t} \tag{15.3}$$

where $N(0)$ is the starting number of nuclei at time $t = 0$, λ is the so-called disintegration constant, and is typical for the given radioactive isotope, e is the basis of natural logarithms and e = 2.71828. The correlation described in Equation 15.3 is graphically depicted in Figure 15.1.

15.2.3 Radiation Doses

A *dose* is the volume of radiation energy absorbed by the matter. The expression dose or *radiation dose* is usually used in a general sense; however, in a discussion on radiation protection its usage should be precise, for example, absorbed dose, equivalent dose, and so on.

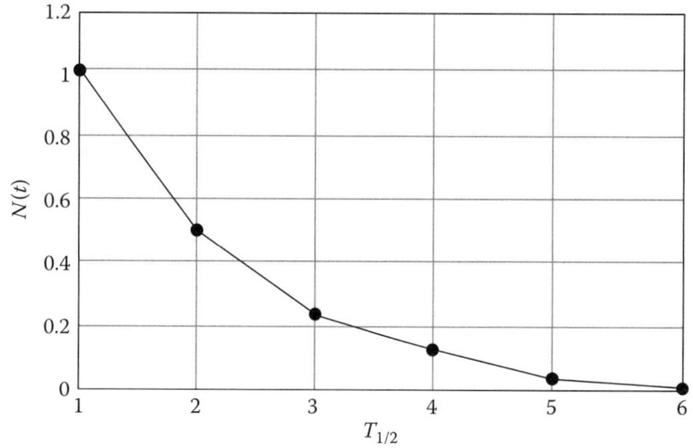

FIGURE 15.1 Correlation of the sample activity as a function of time.

15.2.4 Exposure Dose

The *exposure dose* is the amount of air ionised by X-ray or γ radiation and is defined by the equation:

$$X = \frac{dQ}{dm} \tag{15.4}$$

where dQ is the absolute value of the total charge of the ions of one sign produced in the air when all the electrons and positrons liberated or created by photons in the air of mass dm are completely stopped. Exposure dose is expressed in coulombs per kilogram (C/kg).

Until recently, exposure dose was expressed in roentgens (R). It was equal to 1 R if 1 cm^3 of dry air contained in 2.08×10^9 pairs of ions. The relationship between the roentgen and the coulomb is as follows:

$$1\ \text{R} = 2.58 \times 10^{-4}\ \text{C/kg} = 258\ \text{mC/kg}$$
$$1\ \text{C/kg} = 3876\ \text{R} = 3.876\ \text{kR}$$

The increment in exposure for a suitable small interval of time divided by the interval of time is the exposure rate $\dot{X}$:

$$\dot{X} = \frac{dX}{dt} \tag{15.5}$$

The exposure rate unit in the SI system is ampere per kilogram (1 C/kg·s = 1 A/kg). In the earlier system, the exposure rate was measured in roentgens per hour (R/h), or more often, in milliroentgens per hour (mR/h). The relationship between the old and new units is as follows:

$$1\ \text{pA/kg} \approx 0.014\ \text{mR/h}$$

The concept of exposure dose is currently used less frequently and is being replaced by the more universal concept of absorbed dose.

15.2.5 Absorbed Dose

Ionising radiation loses energy as it passes through a medium. Absorbed dose (D) is the measure of absorption. Absorbed dose is the ionising radiation energy transferred to the matter contained in a volume element divided by the mass of this element. It is expressed by the following formula:

$$D = \frac{dE}{dm} \tag{15.6}$$

where dE is the mean value of transferred energy and dm is the mass of matter contained in the volume element.

The legal unit for absorbed dose is joules per kilogram (J/kg) and its special name is gray. The term 'absorbed dose' is a fundamental dosimetric quantity. In radiation biology, clinical radiology, and radiation protection the absorbed dose, D, is the basic physical dose quantity and is used to describe ionising radiation of all types and all irradiation geometries.

The absorbed dose is 1 Gy when 1 kg of the medium through which the radiation passes absorbs an energy of 1 J. It is usually expressed in centigrays (cGy = 10^{-2} Gy). This unit is not included in the SI system, and the rad is used as the measure of absorbed dose. A dose is said to be 1 rad when 1 g of the medium through which the radiation passes absorbs an energy of 100 ergs.

$$1 \text{ rad (rd)} = 1 \text{ cGy} = 0.01 \text{ Gy}$$
$$1 \text{ Gy} = 100 \text{ rad}$$

Absorbed dose is a universal concept. It applies to every type of ionising radiation and every medium through which this radiation passes. It is used to define absorption of ionising radiation for a wall, for a concrete or lead shield, and for all living organisms, including human beings.

15.2.6 Equivalent Dose

The absorbed dose is the absorption of radiation by various materials, including the human body. However, solely stating the amount of absorbed energy is insufficient for live organisms, because, aside from physical phenomena, the tissues are home to numerous biological processes depending on the type of radiation. Therefore, the concept of *equivalent dose* (H_T) was introduced to determine the degree of energy absorbed by live organisms while taking into account the biological consequences for various types of radiation. H_T is the dose absorbed in the tissue or organ T, weighted for the type and energy of the ionising radiation R and expressed by the formula:

$$H_T = \sum_R w_R D_{T,R} \tag{15.7}$$

where $D_{T,R}$ is the absorbed dose from ionising radiation R, averaged over the tissue or organ T and w_R is the weighting factor for radiation R. Values for the weighting factor are set forth in the recommendations of the ICRP (ICRP 1990; ICRP 2007). These values are based on experimental data for the relative biological effectiveness (RBE) of various types of radiations at low doses, on biophysical considerations, and on judgements.

The unit of equivalent dose is joules per kilogram and has the special name sievert. Under the old units, the equivalent dose was expressed in rem:

$$1\ \text{Sv} = 100\ \text{rem}$$
$$1\ \text{rem} = 0.01\ \text{Sv} = 1\ \text{cSv} = 10\ \text{mSv}$$

15.2.7 Effective Dose

The harmful consequences of exposure to ionising radiation depend on whether the whole body is irradiated or only its individual organs. Much of the research and most experiments show that irradiation of various tissues with the same equivalent doses results in different consequences for the body as whole. The conclusion is that radiation damage to various tissues has varying influences on the effectiveness and functioning of the whole body. The concept of *effective dose* (E) was introduced to illustrate the overall hazard for the body when exposure of the body, organs, or tissues is uneven. E is the sum of the weighted equivalent doses from external and internal irradiation of tissues and organs, expressed by the formula:

$$E = \sum_{T} w_T \cdot H_T = \sum_{T} w_T \sum_{R} w_R D_{T,R} \tag{15.8}$$

where $D_{T,R}$ is the absorbed dose from ionising radiation R, averaged over the tissue or organ T, w_R is the weighting factor for radiation R, w_T is the weighting factor for the tissue or organ T and $\sum_T w_T = 1$. The sum is calculated for all organs and tissues of the human body considered sensitive to stochastic effects. The w_T values are chosen to represent the contributions of individual organs and tissues to the overall radiation detriment of stochastic effects.

The unit for effective dose is the sievert ($Sv = J\ kg^{-1}$), as it is for equivalent dose and effective dose as well as for some operational dose quantities.

When discussing the irradiation of a single organ or tissue, the concept of equivalent dose H_T is used; when discussing exposure of the whole body, several organs or tissues the concept of effective dose E is used.

The values of weighting factors for tissues, w_T, take into account not only the incidence and mortality caused by malignant carcinomas and the risk of hereditary anomalies, but also the shortening of life expectancy caused by carcinomas of the various organs. Effective dose is used to assess the risk of uneven irradiation of the body, and its values, calculated according to Equation 15.8, correspond in numerical terms—from the standpoint of the risk of stochastic consequences—to the values of equivalent dose in the case of even irradiation.

The effective dose E received for a specified time is determined as the sum of the effective dose E_z from external exposure during that time and the committed doses resulting from the radioactive nuclide intake during that same time. This is evaluated for a period of 50 years from the intake time (for children, from the intake time until the age of 70 years). For the purpose of dose limit determination, 'specified time' denotes either 1 or 5 years depending on the criterion adopted. Effective dose E (in sieverts) for an individual from the age group g is defined by the formula:

$$E = E_Z + \sum_j e(g)_{j,p} J_{j,p} + \sum_j e(g)_{j,o} J_{j,o} \tag{15.9}$$

where $e(g)_{j,p}$ and $e(g)_{j,o}$ denote unit committed effective doses for individuals from age group g, that is, the committed effective doses (in sieverts) received by these individuals from the intake of unit activity (i.e. 1 Bq) of nuclide j by ingestion (index p) or by inhalation (index o). These doses, depending on the manner of nuclide transfer to the alimentary system and from the digestive tract into the body fluids (described by the value of f_1 factor), and depending on the absorption rate in lungs (which may be fast, moderate or slow), are given in tables in recommendations and laws (IAEA 1996; Council Directive 96/29/Euratom..., 1996).

$J_{j,p}$ and $J_{j,o}$ denote activities in becquerels of the nuclide j for intake by ingestion (index p) or by inhalation (index o).

If the equivalent dose from internal contamination in a tissue or organ received in unit time, that is, time derivative H_T of this dose, is known, then the committed effective dose received in time τ is described by the formula:

$$H_T(\tau) = \int_{t_0}^{t_0+\tau} \dot{H}_T \, dt \tag{15.10}$$

where t_0 denotes the moment of nuclide intake. If the τ value is undefined, then the integration time should be given as 50 or 70 years.

A dose limit includes the sum total of the doses from external and internal contamination.

Weighting factor values for tissues and various types of radiation are presented in Tables 15.1 and 15.2. In the new ICRP recommendations (ICRP 2007), the limits required for radiological protection remain unchanged but the radiation and tissue weighting factors used in the calculations of these quantities have been revised. The revised w_R values are given in Table 15.3.

Tissue weighting factors have again been changed and extended to a wider range of organs. A set of w_T values was chosen for the new ICRP 2007 recommendations based on epidemiological studies on cancer induction in exposed populations and risk assessments for heritable effects. The respective values of relative radiation detriment are shown in Table 15.4. They represent the mean values for humans averaged over both sexes and all ages and thus are not related to the characteristics of particular individuals.

TABLE 15.1
Values of Tissues (Organs) Weighting Factor w_T[a]

Tissue (Organ), T	Tissue (Organ) Weighting Factor w_T
Gonads	0.20
Bone marrow (red)	0.12
Colon	0.12
Lungs	0.12
Stomach	0.12
Bladder	0.05
Breasts	0.05
Liver	0.05
Oesophagus	0.05
Thyroid	0.05
Skin	0.01
Bone surface	0.01
Remainder[b,c]	0.05

[a] Values are determined for a representative group of people, with equal numbers representing both genders and a broad age spectrum; they may be utilised to define effective doses independent of gender, for both exposed workers and for members of the public.

[b] For calculation purposes, the 'remainder' is comprised of the following tissues (organs): adrenals, brain, upper large intestine, small intestine, kidneys, muscles, pancreas, thymus, uterus, or others, which may be selectively irradiated.

[c] In exceptional cases, when a single tissue (organ) listed under the 'remainder' receives an equivalent dose exceeding the largest dose for any of the 12 organs listed in the table for which w_T values have been determined, then the weighting factor 0.025 for this tissue (organ) should be used, and the factor of 0.025 for the average dose for other tissues (organs) listed under the 'remainder'.

TABLE 15.2
Values of the Radiation Weighting Factor w_R

Radiation Type and Energy Range, R	Radiation Weighting Factor, w_R
Photons, all energies	1
Electrons and muons, all energies	1
Neutrons, energy <10 keV	5
10–100 keV	10
>100 keV–2 MeV	20
>2–20 MeV	10
>20 MeV	5
Protons, excluding the recoil protons, energy >2 MeV	5
α particles, fission fragments, heavy nuclei	20

TABLE 15.3
Radiation Weighting Factors[a]

Radiation Type, R	Radiation Weighting Factor, w_R
Photons	1
Electrons and muons	1
Protons and charged pions	2
α particles, fission fragments, heavy ions	20
Neutrons	A continuous function of neutron energy

[a] All values relate to the radiation incident on the body or, for internal sources, emitted from the source.
Source: Data from ICRP. 2007. *Ann ICRP* 37(2–4).

TABLE 15.4
Tissue Weighting Factors w_T

Organ or Tissue	ICRP Tissue Weighting Factors, w_T	
	1990	2007
Gonads	0.20	0.08
Bone marrow	0.12	0.12
Lower large intestine	0.12	0.12
Lung	0.12	0.12
Stomach	0.12	0.12
Bladder	0.05	0.04
Breasts	0.05	0.12
Liver	0.05	0.04
Oesophagus	0.05	0.04
Thyroid	0.05	0.04
Bone surface	0.01	0.01
Skin	0.01	0.01
Brain	In remainder organs	0.01
Salivary glands	0	0.01
Remainder organs[a]	0.05	0.12

[a] Remainder organs under 1990 include adrenals, brain, kidney, muscle, pancreas, small intestine, spleen, thymus, upper large intestine, and uterus; remainder organs under 2007 include adrenals, extrathoracic region, gall bladder, heart, kidneys, lymphatic nodes, muscle, oral muscosa, pancreas, prostate, small intestine, spleen, thymus, and uterus.
Source: Data from ICRP. 1990. *Ann ICRP* 21(1–3); and ICRP. 2007. *Ann ICRP* 37(2–4).

15.3 BIOLOGICAL EFFECTS OF IONISING RADIATION

Radiation-induced changes in biological matter begin when live tissue absorbs radiation energy, which causes, among other things, the ionisation or excitation of atoms or particles, setting off a chain of secondary biological reactions. The ionisation and excitation of the atoms that constitute live matter is the first link in a chain of changes that lead to the biological effects of radiation. The interaction of ionising radiation with a live organism over time can be broken down into several phases occurring subsequently, for example, *physical interactions* (energy deposition, excitation or ionisation, initial particle tracks; time: 10^{-15}–1^{-12} seconds), *physico-chemical interactions* (radical formation, diffusion, chemical reactions, initial DNA damage, DNA breaks or base damage; time: 10^{-9}–10^{-3} seconds), *biological responses* (repair processes, damage fixation, cell killing, mutations, transformations, aberrations; time: 1 second^{-1} day), and *medical effects* (proliferation of 'damaged' cells, promotion or completion, teratogenesis, cancer, hereditary defects; time: 10 days–100 years).

A cell's genetic material, the DNA, is most sensitive to radiation. Damage to DNA, if not flawlessly repaired, can lead to cancerous transformations or death of the cell.

The effect of radiation on live tissue depends on many factors and is therefore very complicated. After irradiation, the body's reaction depends on two major parameters—radiation hardness and the RBE. Other parameters include dose size and intensity, the type of exposure (single or fractioned, that is, spread over time), and the properties of the irradiated object, such as exposed body area, age and sex, individual and species sensitivity, temperature, metabolic activity and hormonal balance, hydration, and oxygenation of the exposed biological material. Cells' sensitivity to radiation is greater if their proliferation activity (ability to multiply) is higher and the tissue differentiation is smaller.

The irradiated organism sometimes does not demonstrate any identifiable symptoms for a prolonged period following exposure. The consequences of such irradiation could exist in latent form and develop gradually, manifesting themselves even after dozens of years.

There are two categories of biological consequences of ionising radiation:

1. *Deterministic effects (nonstochastic)*: The radiation effects for which a threshold dose generally exists, above which the severity of the effect is greater. These include all of the well-known complications in radiation therapy.
2. *Stochastic effects*: Radiation effects that generally occur without a threshold dose level, whose probability is proportional to the dose and whose severity is independent of the dose. These include malignant carcinomas.

Thus, one of the basic differences between stochastic and nonstochastic (deterministic) consequences is that a specific dose, called the threshold dose, must be exceeded in order to cause the latter. No specific dose threshold has been defined for stochastic consequences. Preventing nonstochastic consequences is relatively simple, especially if the dose rate is small or fractioned, because the threshold doses for consequences that are significant for pathophysiology are large, ranging from a few to several dozen grays. The situation is different for stochastic effects. No threshold means that every dose, even a very small one, carries an increased probability of effects. The appearance of exposure consequences can then be analysed only in probability terms, that is, risk assessment.

The only credible ways of obtaining information on the correlation between radiation dose and the risk of carcinoma in humans are observation and epidemiology tests of persons subjected and not subjected to radiation. Malignant carcinomas caused by ionising radiation are not different in clinical or morphological terms from carcinomas observed in populations not subjected to radiation. There is no proof that radiation induces many malignant carcinomas (e.g. cancer of the cervix and uterus, cancer of the prostate, malignant lymphomas, or chronic lymphocytic leukaemia). The shortest latency period for leukaemia does not exceed 2 years, and for cancer (solid tumours), 5–10 years. The average risk coefficient of a stochastic effect caused by ionising radiation as the result of occupational exposure between the ages of 18 and 65 amounts to about 4.7×10^{-2} Sv^{-1} (Table 15.5).

Assessment of the risk of hereditary changes, which in humans can be caused by irradiation of the reproductive cells leading to dominant and recessive mutations and hereditary mutations relating to gender, is currently based on the extrapolation of experimental data obtained through animal research. The risk coefficient of serious genetic damage caused by radiation is estimated to be about six times smaller than that of causing radiation-induced carcinomas (mortal and treatable), and ranges from 0.8 to 10^{-2} Sv^{-1}.

TABLE 15.5
Nominal Probability Coefficients of Stochastic Consequences for Individual Tissues and Organs

Tissue or Organ	Aggregated Detriment (10^{-2} Sv^{-1})	
	Whole Population	Workers
Bladder	0.29	0.24
Bone marrow	1.04	0.83
Bone surface	0.07	0.06
Breast	0.36	0.29
Colon	1.03	0.82
Liver	0.16	0.13
Lung	0.80	0.64
Oesophagus	0.24	0.19
Ovary	0.15	0.12
Skin	0.04	0.03
Stomach	1.00	0.80
Thyroid	0.15	0.12
Remainder	0.59	0.47
Total	5.92	4.74
Probability of severe hereditary disorders		
Gonads	1.33	0.80
Grand total (rounded)	7.3	5.6

TABLE 15.6
Detriment-Adjusted Nominal Risk Coefficients (10^{-2} Sv^{-1}) for Cancer and Heritable Effects

Exposed Population	Cancer		Heritable Effects		Total	
	ICRP 2007	ICRP 1990	ICRP 2007	ICRP 1990	ICRP 2007	ICRP 1990
Whole population	5.5	6.0	0.2	1.3	5.7	7.3
Workers	4.1	4.8	0.1	0.8	4.2	5.6

The stochastic risk figures for the 2007 ICRP Publication 103 (2007) are given in Table 15.6 along with the previous values from ICRP Publication 60 (1990).

Based upon cancer incidence data, detriment-adjusted nominal risk coefficients for cancer are 5.5×10^{-2} Sv^{-1} for the whole population and 4.1×10^{-2} Sv^{-1} for adult workers. The corresponding values specified in ICRP Publication 60 are 6.0×10^{-2} Sv^{-1} and 4.8×10^{-2} Sv^{-1}, respectively.

The ICRP states that the overall approximated risk coefficient of 5×10^{-2} Sv^{-1}, on which current international radiation safety standards are based, continues to be appropriate for radiation protection because the decreases in total stochastic risk (cancer and genetic effects) are small and do not justify any change in the values of dose limits. A key assumption in the ICRP philosophy is that stochastic effects exhibit a linear-no-threshold (LNT) dose response. The LNT hypothesis remains fundamental for averaging and summing up the doses used for effective dose and for the system of dose record keeping. The ICRP believes that LNT remains a pragmatic, realistic, and conservative tool for radiological protection.

15.4 SOURCES OF IONISING RADIATION AND THEIR APPLICATIONS

The source of ionising radiation is any radiation source or device that produces ionising radiation. In industry, meters employing radiation isotopes are used to measure various values, control production processes, and so on. These include meters that measure material thickness and density, the levels of solids or liquids in tanks, or the thickness of a top layer on a base, and meters that measure the concentrations of acids, dustiness of the air, and so on. These devices employ mainly closed sources of γ or β radiation, and their functioning is usually based on the absorption of radiation passing through matter or of radiation dispersion.

A very popular method of ‘nondestructive’ testing is *industrial radiography*, which uses γ- or X-rays to inspect items such as welded elements. Aside from radiation hardness and the correlation of its absorption with the thickness of material, these tests also use another physical phenomenon—the blackening (density increase) of photographic film caused by radiation. The tested element is placed between the radiation source and the detector containing the film. If the element is homogenous, its whole mass absorbs radiation equally and the photographic film is blackened evenly (i.e. its density increases evenly). If it is not homogenous, radiation absorption and the blackening of film are uneven. Apparatus used in industrial radiography are called flaw detectors; if γ radiation is used, they are frequently called *gammagraphic apparatuses*. Aside from classic radiophotography used to identify leaks and flaws in important construction elements (e.g. pipelines), which use γ and X radiation, in recent years several special radiography methods have gained importance, namely *neutronography* (for analysis of light materials), *proton radiography*, *microradiography* (to test very small objects), and *dynamic radiography*.

Radiation techniques are used in various areas of industry. They are used to sterilise disposable medical equipment, to modify polymers, materials, and semiconductor-based devices, and to colour fabrics, glass, and artificial—and sometimes even natural—stones. There are millions of tons of products manufactured or modified with radiation worldwide, and this is constantly on the rise. Radiation techniques are based on irradiation of raw materials and finished products with an electron beam or γ radiation. An interesting example of these techniques is thermoshrink tubes and tapes, which provide perfect electrical insulation. They are used wherever durable and hermetic joints of construction elements are required, for example, in assembly of ventilation ducts, conduits, and electrical cables. Radiation techniques are also employed in cleaning exhaust fumes from installations that burn fuels such as coal. Irradiation of the gas with an electron beam causes reduction of the sulphur dioxide (SO_2) emission by 95%, and nitrogen oxides (NO_x) by 80%.

Radioactive tracer methods are used most frequently for analysis of objects and various states of matter. They generally entail the addition of isotopes to materials sent over distances, subjected to mixing processes, or that change their state of matter during technological processing (dissolving, vaporisation, etc.). This allows changes in the intensity of radiation at various places in the flow or movement of the analysed material to be tracked. Such tests are conducted with very small volumes of low-activity isotopes so that their application does not disturb the analysed processes. Trading methods are used to establish the flow of materials and determine their speed and trajectory dispersion. These methods are also used to analyse the processes and phases of mixing and separating components and hence are used in the glass, paper, chemical, and metallurgy industries. Radioactive tracers are also used to analyse the degree of wear on materials and tools, diagnose corrosion processes, analyse greases and lubricants, and locate and measure leaks in tanks and pipelines.

One of the most important ionising radiation methods is *activation analysis*—nuclear analysis of the composition of materials. This method enables identification of contaminants in semiconductors, luminophores, and other materials of high purity. Activation analysis is also used to determine the residual amount of heavy metals in waste (e.g. in ashes) and nitrogen in grain, fertilisers, and the like. The significant advantage of this method is its discrete, simultaneous sampling of many elements. Activation analysis, as its name suggests, is based on the phenomenon of activation—usually neutron activation. The analysed sample is radiated with neutron flux from a reactor or from another source (e.g. radium–beryllium) so that it becomes radioactive. Under the given activation conditions, the volume of the radioactive isotope is proportional to the volume of the stable 'parent' isotope, which enables identification and marking of the very small volumes present in the analysed sample. The energy spectrum of the radioactive isotope so obtained is then analysed with a special device known as β or γ spectrometer to determine the type and energy of the radiation emitted by the isotope. The energy spectra of various isotopes are known, so analysis of the sample spectrum enables determination of the isotopes—the elements that comprise this sample as well as their volumes.

In the food and pharmaceutical industries, large doses of radiation are used for food preservation and sterilisation of medical equipment, syringes, and dressings. The source of the radiation used for these purposes is primarily cobalt (^{60}Co), with an activity of dozens thousands terabecquerels (TBq), in doses of several dozens of kilograys, or electron accelerators. In geological research, neutron-emitting sources are used to search for mineral deposits and to measure the humidity (or density) of soil.

The ionising properties of radiation are also used in devices known as *electric charge eliminators*. In the production of paper, foil, fabric, and photographic film—wherever nonconducting material comes into contact with and moves along various elements of machinery—the resulting friction produces electric charges. Buildup of a large charge presents very serious fire and explosion hazards in technological processes that employ flammable and explosive substances. Use of radiation to strongly ionise the air in the vicinity of the charged material can prevent such mishaps. Smoke detectors also often use radiation sources (Am-241, Pu-239).

Radioactive decay properties are also used to measure time. Various radioisotopes are used to determine the age of historical findings, the most popular being the carbon method, which measures the contents of radioactive carbon 14 in the remains of organisms and enables determination of their age.

The use of radiology and radioactive isotope techniques in medical diagnostics and radiotherapy (X-ray tests, computer tomography, nuclear medicine, radiotherapy) is a great success of science. These techniques enable the identification of flawed structures or activity of the organs. The test results are often crucial to forming the right diagnosis. Radiotherapy used to fight carcinomas often proves decisive in patients' survival. Ionising radiation currently dominates scientific applications in various fields, especially physics, chemistry, and biology.

Listing all the applications of ionising radiation is impossible here. It is used in agriculture, food preservation, water source location, medical diagnostics and therapy, sterilisation of medical equipment, and detection of pollution in and cleansing of the natural environment. This radiation is used also to alter the chemical structure of materials, construct high-sensitivity smoke detectors, and analyse the pollution of rivers, water reservoirs, and underground aquifers. Nuclear (radioactive isotope) techniques are used in mining, geology, and archaeology. They enable precise determination of the age of rocks or minerals, as well as of the remains of organisms.

15.5 DETECTION OF IONISING RADIATION

Radiometers are devices used to directly evaluate hazards that may arise when working with sources of ionising radiation. Depending on their application, radiometers can be used to measure radioactive contamination, dose, and absorber dose rate.

Radiation indicators are nongraduated devices and are not fitted with measuring devices. They are used only to determine the presence of radiation in a given area and to provide a rough indication of its intensity.

A radiometer consists of the following:

- A detector to detect ionising radiation
- A meter circuit
- Power supply

The fundamental element of these devices is the detector, which detects the ionisation occurring in gases, liquids, or solids under radiation. It transforms the radiation it detects into electrical signals. Detectors are placed in special compartment that is linked to the main body of the apparatus and is called the probe for measurements. In universal radiation meters, this set is comprised of several interchangeable stalks containing detectors that are sensitive to various types of radiation and different ranges of energy. Radiometer are broken down into meters that measure α, β, γ, X-ray, and neutron radiation. These meters can be stationary or portable, with power supplied from mains or batteries.

The instrument used most frequently in radiation protection is the gaseous detector. Gaseous detectors include the following:

- Ionisation chambers
- Geiger–Müller counters
- Proportional counters

The ionisation chamber consists of two metal electrodes placed on very strong insulators in a gas-filled enclosure. These electrodes are connected to a power source, which produces an electrical field that causes a flow of ionisation current between the electrodes (i.e. a flow of ions created by radiation). When the voltage between the electrodes attains a certain level, called the saturation voltage, almost all of the ions produced by radiation reach the electrodes. Because the voltage of the ionisation current is a function of the intensity of the radiation reaching the ionisation chamber, this method enables determination of the exposure. The ionisation current referred to above is very weak, and needs to be amplified if it is to be recorded.

Because the γ and X radiation interact with matter similarly, identical ionisation chambers are used to record measurements for both types. However, both types of radiation are of similar energy levels and are within the same metering range. Ionisation chambers are typically made of metal (steel, aluminium) or plastics such as bakelite, organic glass, polystyrene, and so on, covered on the inside with electrically conducting material. Homogenous ionisation chambers are constructed in a manner such that the effective atomic number of the material from which the walls of the enclosure are made corresponds to the effective atomic number of air or human tissue. These chambers and electrodes come in different shapes: cylindrical, spherical, ellipsoidal, flat, and reticulated. The power supply voltage to these electrodes ranges from 300 to 800 V. Ionisation chambers are used to measure X-ray, α, β, and γ radiation.

Proportional counters, similar to those used in the ionisation chamber, consist of two electrodes enclosed in a cylinder filled with air or a special mixture of gases. The voltage employed is higher (up to 2000 V) than that in the ionisation chamber. Ions and electrons appearing as a result of the ionising radiation's action in the stronger electric field increase in energy and produce additional positive ions and electrons (secondary ionisation) during collisions with gas particles. Secondary ionisation is when the number of charges reaching the electrodes is greater than the number of charges originally produced. The detector's operating range (from several hundred to 2000 V) is called the range of proportionality, because the number of ions that collect on the electrodes—that is, the strength of the electric signal emitted by the detector—is proportional to the number of parent ions the radiation particle produces. Detectors that work in this range are called proportional counters. They are also used to measure the quality of the radiation—that is, to measure the distribution of energy within the spectrum of radiation, for example, α or β radiation, because the number of parent ions produced depends on the energy of the particle.

Proportional counters are used primarily to detect particle α and especially β radiation. With the use of appropriate converters similar to those in the ionisation chambers, they can be used to detect neutron radiation. Proportional counters are less frequently used for γ and X radiation.

A further increase in the voltage between electrodes of the detector causes a sudden increase in gaseous amplification. After passing through the limited proportionality area, radiation reaches the range of the Geiger–Müller counters. In this range, gaseous amplification is very strong. The proportionality between the number of pairs of parent ions and the number of ions that collect on the electrodes no longer exists. This phenomenon is known as cumulative ionisation. In the early phase, discharges occur precisely as they do in the proportional counter, but near the electrode, due to collisions of electrons with atoms of noble gases filling the detector's chambers, the latter emit photons. These photons propagate immediately to fill the whole counter and those that reach the cathode or the particles of the multiatom gases can cause emission of secondary electrons through the photoeffect phenomenon. These photoelectrons support further ionisation inside the counter, collide with atoms of noble gases, and cause further emission of photons, creating even more photoelectrons, and so on.

Above a certain level of voltage, referred to as the threshold voltage, impulses obtained from the counter are almost identical, regardless of the energy of the particles that initiate the process. Cumulative ionisation differentiates the Geiger–Müller counter from the ionisation chamber and the proportional counter. A Geiger–Müller counter built in the same manner could register only the first particle, or the first quantum of radiation. To prevent this, cumulative ionisation is stopped within the meter by reducing the voltage between electrodes (reducing energy of the accelerated electrons), or by adding another gas to the gas that is within the meter, called quenching. The second gas is usually multiatom, such as alcohol or ether vapour. A Geiger–Müller counter is usually cylindrical and can have a window; its cathode is made from metallised glass or plain metal. The design of the meter enables detection of all types of radiation, but they are used most often to detect β rays because they are 100% effective for this purpose.

Scintillation counters operate on the luminescence phenomenon, which appears in some substances under the influence of ionising radiation. The luminescence phenomenon is found in gases, liquids and solids. Radiation causes flare-ups (scintillations) in the scintillator that are registered by photomultipliers of very high light sensitivity, which transform them into electrical impulses.

The superiority of scintillation counters over gas-based counters is attributable to the high density of the applied scintillators. This allows for high absorption of primary radiation and therefore a high effectiveness. Its other important features are a high counting speed and proportionality between the energy of recorded particles and the amplitude of the impulse produced. Applications of scintillation counters are numerous; they can be used to measure almost all types of radiation.

Semiconductor detectors have many advantages, such as low power demands, a very small background count and direct transformation of radiation energy into electron or hole pairs without having to use luminous energy. These factors have led to an increased interest, and features such as high resolution, linearity, fast growth of impulses, and a small size have led to their growth in popularity. Except for one important difference, the functioning of semiconductor detectors is similar to the operations of the ionisation chamber. The density of the materials used—usually silicon or germanium—is many times greater than the density of gas (air) used in the conventional ionisation chamber. Ionising radiation raises the conductivity of semiconductors (crystals). The intensity of the current in the electrical circuit changes with the strength of the radiation dose.

Semiconductor detectors are used mainly for spectrometry measurements of γ and X radiation. They are usually built with single crystals of germanium. Semiconductor detectors with single silicon crystals are usually used to measure α and β radiation and neutrons.

Photographic dosimeters (referred to as the film badge), used frequently in isotopic laboratories and X-ray rooms, consist of a plastic cassette containing photographic film that detects radiation. The extent of the film's blackening depends on the radiation dose as well as the volume of absorbed radiation. This detector is used to define the type of absorbed radiation and its energy (Figure 15.2).

The cassette housing has a window through which radiation not absorbed by the cassette's material reaches the film. Filters—thin sheets of metal—are placed on the inside of the cassette, partly covering the film's surface and absorbing various kinds of radiation. Identifying the type of radiation that produced a certain absorbed dose and further establishing whether it was the result of a single dose or the aggregate of many small doses calls for comparing the blackening of the film in various places with different filters.

Photometric dosimeters are used to measure doses of β radiation with energy greater than 0.5 MeV, γ radiation, and X radiation, and, with the use of an additional filter and a nuclear emulsion, also doses of thermal and fast neutron radiation. These dosimeters cannot be used to measure α radiation.

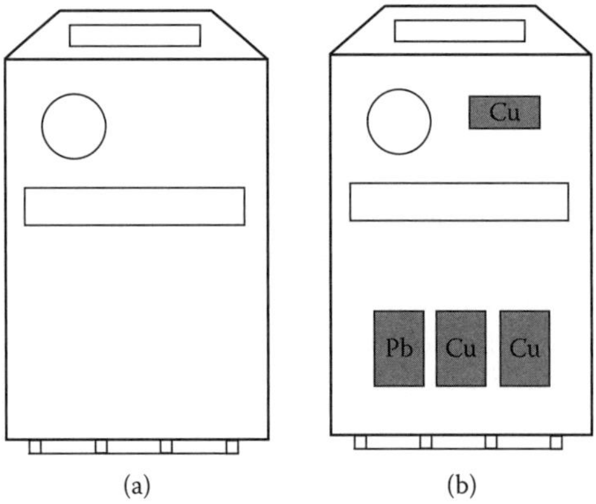

FIGURE 15.2 A film badge for X radiation: (a) external appearance and (b) system of filters.

15.6 RULES OF RADIATION PROTECTION—CRITERIA FOR ASSESSMENT OF HAZARDS

Regulations for protection from ionising radiation cover only the types of sources and situations that submit to measurement. Radiation from sources such as natural radionuclides within the human body (e.g. potassium—^{40}K, radium—^{226}Ra) or cosmic radiation cannot be limited. Radiation protection is currently based on two systems (Hrynkiewicz 2001): licensing and supervision and dose limitation.

The licensing system requires that the purchase of radioactive substances (sources of ionising radiation), and their possession, usage, and removal be allowed only for specific purposes, in appropriately prepared locations, and solely by authorised persons.

Regulations for radiation protection are formed based on three fundamental rules: justification, optimisation, and limitation. These principles are defined (ICRP 2007) as follows:

- *The principle of justification*: Any decision that alters the radiation exposure situation should do more good than harm.
- *The principle of optimisation of protection*: The likelihood of incurring exposure, the number of people exposed, and the magnitude of their individual doses should all be kept as low as reasonably achievable, taking into account economic and societal factors.
- *The principle of application of dose limits*: The total dose to any individual from regulated sources in planned exposure situations other than medical exposure of patients should not exceed the appropriate limit.

These rules enable the achievement of the fundamental goals of radiation protection, which are to prevent deterministic consequences under conditions of intended exposures performed for peaceful purposes, with the exception of radiotherapy, and minimising the probability of stochastic exposure. The justification of these activities entails the analysis of profits and losses relating to the utilisation of new ionising radiation sources and demonstrating that their benefits would surpass all costs and losses, including those relating to health. Optimisation of radiation protection means keeping doses as low as is reasonably achievable—the ALARA rule.

Ionising radiation poses certain hazards to human health. Therefore, appropriate caution must be exercised in using ionising radiation. Adherence to the following rules can reduce absorbed doses of ionising radiation:

- The shorter the time spent near an ionising radiation source, the smaller is the radiation dose absorbed.
- The farther from an ionising radiation source one is, the safer it is.
- Use of shields is beneficial; they weaken radiation.

Radiation types vary in hardness and in the ways they interact with matter. Shields are therefore made of different materials. For example, shields that afford protection from X and γ radiation are made of heavy materials (lead and depleted uranium), and those that afford protection from β radiation are made of light materials (aluminium and plastics). In practice, shields for protection from α radiation are not needed because its hardness is negligible.

Radiation shields can be *permanent* (e.g. a wall), *movable* (e.g. screens, containers for storage of sources), and *personal* (e.g. protective goggles, rubber gloves, aprons of lead rubber).

In the case of small doses, a simple proportionality between the risk of stochastic consequences and the collective dose is expected, but complete avoidance of such consequences is not possible due to a lack of threshold dose. However, the costs and inconveniences tied to even lower values usually rise rapidly and are disproportionately high in relation to the achieved reduction of effective dose. They become less and less justified in the context of social utility. Thus, the avoidance of very small risks at a huge cost cannot be deemed socially desirable.

Optimisation of radiation protection almost always refers to a specific source and must be applied in the design of equipment which poses hazards for exposed persons (IAEA 2002). To help achieve this goal, industrial standards for structural optimisation are used, especially for equipment and apparatus in general as well as for design and construction of buildings, rooms, and shields to protect workers from radiation. Optimisation is also required in everyday operational activities in technological processes and should be the most important part of the safety culture in the workplace. The quantitative methodology tools used for protection optimisation originate from the essential dimensions, namely the effective collective dose, which is the product of average dose and the number of exposed persons.

In summary:

- *Justification*: Prior to commencing activities that utilise new types of ionising radiation, the head of an organisation prepares justification to establish that the scientific, economic, and social benefits expected from such activity will prevail over any possible human health detriment and damage to the environment that may result from it.

- *Optimisation*: The head of an organisation ensures that the principle of optimisation is implemented in all activities carried out—after accounting reasonably for economic and social factors—so that the number of workers and members of the public exposed and the ionising radiation doses shall be as low as possible.
- *Limitation (dose limits)*: The sum of ionising radiation doses to the workers and general public, received from all types of activities, shall not exceed established dose limits.

15.6.1 Dose Limits

The head of any organisation whose employees work in conditions where they may be exposed to ionising radiation must ensure the following:

- Medical care for the employees is always available and necessary personal protective equipment and dosimetric equipment, appropriate to the hazardous conditions, are provided.
- Individual absorbed doses are measured, and data on dosimetric measurements in the work environment are collected and recorded.

Dose limits are established to define the scope of exposure under normal conditions, that is, the radiation sources for which surpassing the limit must be deemed unacceptable. The established dose limits apply to persons employed under hazardous ionising radiation conditions, including pregnant and breastfeeding women, interns, and students exposed to ionising radiation in the course of employment or education, and to members of the general public. Dose limits also apply to long-term exposure caused by the removal or containment of a past radiation accident or past activity as well as the exposure due to a rise in natural ionising radiation resulting from human activity (e.g. in mines, in caves and other underground locations, and in aviation, other than work performed by ground crews).

Dose limits do not include exposure to natural radiation if this radiation did not increase because of human activity. They specifically do not include exposure resulting from natural radioactive nuclides in the human body, cosmic radiation at ground level, and exposure above the ground surface caused by radioactive nuclides present in the intact earth crust. They do not apply to persons subjected to ionising radiation for medical purposes.

Doses for persons employed in conditions where the probability of exposure to ionising radiation exists are determined as follows:

- Based on measurements of individual doses or dosimetric measurements conducted in the work environment, taking into account the types of radiation and energy, in a situation in which there is external exposure
- Based on radioactive contamination measured in the work environment, or after determining the content of radioactive substances in the body of the exposed person, in a situation in which there is internal contamination
- Using biological dosimetry methods, specifically when there is a possibility of uncontrolled exposure of the whole body or its parts

Doses for the general public are determined based on the following:

- Results of radiation monitoring of the environment, including measurement of the dose rate and the contents of radioactive substances in the elements of the environment
- Measurements of radioactive substances in potable water and food
- Data on the activity and type of radioactive substances discharged into the environment
- Results obtained after the application of biological dosimetry methods, especially where there is a possibility of hazard to these persons as the result of a radiation accident

Exposure of employees and the general public to ionising radiation is evaluated based on the effective doses absorbed by these persons, and the equivalent doses, which are determined based on indexes.

15.6.1.1 Workers

For persons working in conditions of ionising radiation hazards, the dose limit, expressed as the effective dose, amounts to 20 mSv per calendar year. This dose can rise to the level of 50 mSv in the course of a given calendar year if within the subsequent five calendar years its cumulative level does not exceed 100 mSv. The values of dose limit, expressed as equivalent dose, within a calendar year when the aforementioned condition for effective dose is met, are as follows (Table 15.7):

- 150 mSv for the lens of the eye
- 500 mSv for the skin (averaged over any 1 cm^2 area of exposed part of skin)
- 500 mSv for the hands, forearms, feet, and ankles

15.6.1.2 Pregnant and Breastfeeding Women

From the time she informs the head of the organisation of her pregnancy, a woman may not be employed in conditions under which the foetus could receive an effective dose higher than 1 mSv. A breastfeeding woman may not be employed in conditions exposing her to internal or external contamination.

15.6.1.3 Minors, Apprentices, and Students

For apprentices and students ages 18 years and over, the same dose limits apply as for persons employed under conditions of ionising radiation hazard. Persons less than 18 years old may be employed under conditions of ionising radiation hazard solely for the purpose of training or apprenticeship.

For apprentices and students 16 to 18 years old, the dose limit, expressed as the effective dose, is 6 mSv per calendar year. The specific dose limit, expressed as the equivalent dose, amounts to the following values in the course of a year:

- 50 mSv for lens of the eye
- 150 mSv for the skin (averaged for any 1 cm^2 area of exposed part of skin)
- 150 mSv for the hands, forearms, feet, and ankles

TABLE 15.7
Dose Limits of Ionising Radiation

Persons	Effective Dose[a] (mSv)		Equivalent Dose[b] (mSv)	
	In a Year	In a Year, Conditional	For the Lens of the Eye	For the Skin (Averaged for Any 1 cm^2 Area of Exposed Skin) and for the Hands, Forearms, Feet, and Ankles
Workers and apprentices aged 18 years and over	20	50 (limited to 100 mSv during the subsequent 5 years)	150	500
Apprentices aged 16–18 years	6	–	50	150
Members of the general public and apprentices under 16 years old	1	The dose of 1 mSv may be exceeded on the condition that the average for the subsequent 5 years does not exceed 5 mSv	15	50 (average value defined for 1 cm^2 of skin surface), no limit for limbs
Pregnant woman Foetus	A pregnant woman may not be employed under conditions that could cause the foetus to absorb an effective dose exceeding 1 mSv. A breastfeeding woman may not be exposed to internal and external contamination.			

[a] Effective dose: The sum of equivalent doses absorbed from external and internal contamination, determined with the use of appropriate organ or tissue weighting factors, which illustrates exposure of the whole body.

[b] Equivalent dose: Dose absorbed by a tissue or organ, determined on the basis of ionising radiation type and its energy.

The same dose limit that applies to members of the general public applies to apprentices and students under 16 years.

15.6.1.4 Members of the General Public

The dose limit for members of the general public, expressed as the effective dose, is 1 mSv per calendar year. Specific dose limits, expressed as the equivalent dose, amount to the following values in the course of a calendar year:

- 15 mSv for the lens of the eye
- 50 mSv for the skin (averaged for any 1 cm^2 area of exposed part of skin)

An effective dose equal to 1 mSv can be exceeded in the course of a given calendar year, if within the subsequent five calendar years its cumulative level does not exceed 5 mSv.

15.6.2 Exposure Categories

To match the methods of exposure assessment with the anticipated exposure level for workers employed in organisations, two categories of workers, depending on the magnitude of exposure, are established:

1. *Category A*: Workers who may be exposed to an effective dose exceeding 6 mSv in 1 year or to an equivalent dose exceeding three-tenths of the dose limit values for the eye lens, skin, and limbs
2. *Category B*: Workers who may be exposed to an effective dose exceeding 1 mSv in 1 year or to an equivalent dose exceeding one-tenth of the dose limit values for the eye lens, skin, and limbs

To adapt the actions and measures used for radiological protection of workers to the magnitude and type of potential exposure, the head of the organisation must use the following classifications of workplace sites:

- *Controlled areas*: Where there is a possibility of receiving doses established for category A workers, a possibility of radioactive contamination spreading, or a possibility of occurrence of large variations in the ionising radiation dose rates
- *Supervised areas*: Where there is a possibility of receiving doses established for category B workers and which have not been classified as controlled areas

Assessment of employee exposure should be based on controlled measurements of individual doses or dosimetric measurements in the work environment. Category A employees are subjected to exposure assessment based on regular individual dose measurements; if they may be exposed to internal contamination that influences the level of effective dose, they are also subject to internal contamination measurement. If individual dose measurement is impossible or insufficient, assessment of the individual dose absorbed by a category A employee can be performed based on measurements of individual doses conducted for other exposed employees of the same category, or based on dosimetric measurements in the work environment. Category B employees are subject to assessment of exposure based on dosimetric measurements in the work environment in a manner that establishes whether they are properly included in this category. The permit can contain a condition for assessing the exposure hazard for category B employees performing work specified in that permit, based on individual dose measurement.

15.6.3 Exceptional Situations

In special cases, radiation accidents excluded, category A employees who give consent may receive doses which exceed the dose limit if it is necessary to perform a specific task. Such exposure is not allowed for apprentices, students, and pregnant and breastfeeding women if it is likely to cause radioactive contamination of the body. The head of the organisation must justify the need for such excess exposure and discuss in detail beforehand the related issues with the interested employees (volunteers) or their representatives as well as with the authorised doctor and radiation protection inspector. This behaviour must be documented in writing. Doses absorbed by the employee in such situations should be recorded separately. The employer may not remove the employee from regular tasks or to move him or her to another position without his or her consent.

15.6.4 Assessment of Occupational Risk

Based on the radiation protection system that defines admissible exposure conditions at work stations, an assessment of the risk resulting from occupational exposure to ionising radiation consists of:

- *Identification of the source of ionising radiation*: This is based on technical documentation of equipment containing radioactive sources or producing ionising radiation, as well as on data specified in the equipment's usage permit. The review should apply also to the maintained records and to control of radioactive sources.
- *Analysis of work conditions (workstation, premises)*: This should be conducted to ensure fulfilment of the technical and radiation protection requirements that apply to laboratories using sources of ionising radiation and the conditions in which these sources are used.
- *Exposure conditions*: These are the actions performed and the specific procedures and rules for working with the ionising radiation source that influence the size of exposure. The analysis must define the type of radiation; whether exposure was constant (during one shift), sporadic, or fractioned; whether exposure level was constant or variable; and whether external and/or internal contamination was possible.
- *Personnel and equipment*: The following should be determined: (1) the number of persons exposed to radiation; (2) whether they were employees,

students, apprentices, or pregnant or breastfeeding women; and (3) whether they were of legal age or minors. It is also necessary to establish whether they received medical care and that no contraindications were found for their employment in conditions exposing them to ionising radiation; that they were fully capable of work, or capable under certain conditions; whether they received appropriate radiation protection training; and whether individual protective equipment was used.

- *Measurement of the exposure level*: Employees are subject to obligatory systematic assessment of the level of ionising radiation exposure. The frequency and type of dosimetry have to be defined, and that the employer conducted individual dosimetry or dosimetric measurements in the work environment and the type of dosimetric equipment the employer has must be established. The levels of measured effective and equivalent doses are also important.
- *Evaluation of occupational risk*: For ionising radiation, a *three-degree risk evaluation scale* can be used. The exposure classification described above may also be applied to ionising radiation. Employees must be informed of the results of the occupational risk evaluation. The following criteria for a three-degree risk assessment for employees can be applied:
  - *High risk* arises where the permissible exposure or dose limits are exceeded.
  - *Medium risk* arises when category A employees absorb doses and such employees include those who may be exposed to effective dose higher than 6 mSv within the period of 1 year, or an equivalent dose exceeding three-tenths of the dose limits for the lenses of the eye, skin, and limbs.
  - *Low risk* arises when category B employees absorb doses and such employees include those who may be exposed to effective dose higher than 1 mSv within the period of 1 year, or an equivalent dose exceeding one-tenth of the dose limits for the lenses of the eye, skin, and limbs.
- *Corrective and/or preventive actions*: The head of the organisation must have the employees evaluated for exposure. If the analysis suggests a need for corrective actions, he or she determines further exposure limits for the employees.

If a high risk is established, actions that reduce the risk level by limiting exposure should be undertaken. Such options should also be considered for those exposed to medium risk.

Regardless of the level of exposure, the ALARA principle should be applied so that the number of workers and members of the public exposed is as low as possible and the ionising radiation doses absorbed are as small as possible. Monitoring the effectiveness of actions undertaken to reduce doses absorbed by employees is also necessary.

15.6.5 Medical Supervision

A person can be employed under threat of exposure conditions if a properly qualified doctor issues a certificate that there are no counterindications for such employment.

Medical supervision of category A employees is the responsibility of the head of the organisation and the doctor with access to the information necessary to issue certificates for those employees attesting to their ability to perform the given work and to provide information on environmental conditions at the worksite. Medical supervision covers preliminary medical examinations before employment to determine whether the employee can be employed in a category A position. Regular, periodic checkups should occur to establish whether the employee can continue to perform his or her duties.

If tests show that any of the dose limits have been exceeded, the head of organisation must require that the employee undergo medical tests. Further work under conditions of exposure then requires the approval of an authorised doctor.

15.7 CONCLUSIONS

Ionising radiation is an element of the natural environment, which means that humans have always been and will always be subjected to this radiation from natural sources. The development of civilisation gave humans the ability to produce artificial sources of ionising radiation. These offer many benefits, such as clean energy, electricity, and heat necessary for further development; the ability to diagnose and treat diseases; applications in the fields of industry, agriculture, science, and so on. The use of ionising radiation offers concrete benefits to individuals and mankind as a whole, but at the same time poses a great threat when it is used carelessly. The principal goal of radiation protection is to derive the greatest possible benefit from ionising radiation while also minimising its undesirable consequences. This can be achieved through reasonable action, including the development, implementation, and constant application of the provisions instituted to afford protection to all from unnecessary risk, while reducing losses, in particular health losses, to a level considered safe and allowing for reasonable use of all technical capacities in this field.

REFERENCES

Council Directive 96/29/Euratom of 13 May 1996 establishing basic safety standards for health protection of workers and general public against the dangers arising from ionizing radiation. OJ EC, 1996, No. L 159, Vol. 39.

Council Directive 97/43/Euratom of 30 June 1997 on health protection of individuals against the dangers of ionizing radiation in relation to medical exposure. OJ EC, 1997, No. L 180, Vol. 22.

Gorczyca, R., and K. Pachocki, et al. 1997. *Radiation Protection in an X-ray Room: Guidebook for Radiation Protection Inspector*. Warsaw: EX-POLON.

Hrynkiewicz, A. Z., ed. 2001. *Man and Ionizing Radiation*. Warsaw: Scientific Publishers PWN.

IAEA. 1996. *International Basic Safety Standards for Protection Against Ionizing Radiation and for the Safety of Radiation Sources*. Vienna: IAEA.

IAEA. 2002. Optimization of radiation protection in the control of occupational exposure. Safety Reports Series No. 21. Vienna: IAEA.

IAEA. 2004. *Radiation, People and the Environment*. Vienna: IAEA.

ICRP. 1977. Recommendations of the International Commission on Radiological Protection. ICRP Publication 26. *Ann ICRP* 1(3).

ICRP. 1990. Recommendations of the International Commission on Radiological Protection. ICRP Publication 60. *Ann ICRP* 21(1–3).
ICRP. 2007. Recommendations of the International Commission on Radiological Protection. ICRP Publication 103. *Ann ICRP* 37(2–4).
UNSCEAR. 2000. *Sources and Effects of Ionizing Radiation (Report to the General Assembly).* New York: United Nations Scientific Committee on the Effects of Atomic Radiation.
UNSCEAR. 2006. *Effects of Ionizing Radiation (Report to the General Assembly).* Vol. 1. New York: United Nations Scientific Committee on the Effects of Atomic Radiation.

16 Thermal Loads at Workstations

Anna Bogdan and Iwona Sudoł-Szopińska

CONTENTS

16.1 INTRODUCTION

Designers of workstations and employers alike have made considerable efforts to provide optimal work conditions suited to workers' capabilities. Nevertheless, there are still environments that subject individuals to thermal loads of various types. A hot environment is particularly strenuous if heat sources are large and/or when a work activity requires a considerable expenditure of energy, that is, one that is performed mainly in protective clothing. A cold environment, on the other hand, is strenuous when work tasks are performed in the open air, usually in winter and in windy conditions. Such an environment can also be present when work activities are carried out indoors where the ambient temperature is below the values established for the 'thermal comfort zone', that is, 17°C–21°C (Gwóźdź 2004).

The impact of thermal environments on workers' health and, consequently, on their productivity and physical and intellectual capabilities is determined by adaptation abilities of an individual and by the solutions the employer applies to reduce the harmful influence of the thermal environment.

Numerous European and international standards address the impact of workplace thermal environments on the individual. These assessment methods first determine specific indices of thermal load and then compare them with limit values. If the values are exceeded, the standards suggest specific solutions aimed at reducing the risk.

16.2 GENERAL CHARACTERISTICS

The *predicted mean vote* (PMV) index classifies the thermal environment into three categories: hot, moderate, and cold. A PMV index ranging between –2 and +2 is the 'comfort zone'. A hot environment is one in which the PMV index is higher than +2. A cold environment is one in which the PMV index is lower than –2.

Determinations about the thermal environment's influence on a human body are based on the heat balance equation. This formula provides the parameters that influence the thermal equilibrium between a human and the environment, that is, between endogenic heat and exogenic heat. The body's metabolic processes generate the former, while the latter is derived externally, that is, from food intake (Fanger 1970; ASHRAE 2005; Krause 2000; Figure 16.1). Energy loss resulting from the intensity with which one performs a work task is an example of the endogenic factor. Exogenic factors include heat transfer resistance through clothing and air parameters such as air temperature, mean radiant temperature, water vapour pressure, and air velocity. All of these factors influence the rate at which the body emits heat into the environment in order to maintain homothermy (i.e. maintaining the body's core temperature at a relatively stable level of 37 ± 0.3°C; Grether 1973).

Fanger developed the thermal balance equation, and after numerous modifications, its form is as follows (ASHRAE 2005):

$$M - W = (C + R + E_{sk}) + (C_{res} + E_{res}) + (S_{sk} + S_{cr}) \qquad (16.1)$$

where M is the rate of metabolic heat production (in watts per square metre [W/m^2]), W is the rate at which mechanical work is accomplished (W/m^2), $C + R$ is the rate of convective and radiation heat loss from skin (sensible heat loss; W/m^2), E_{sk} is the rate

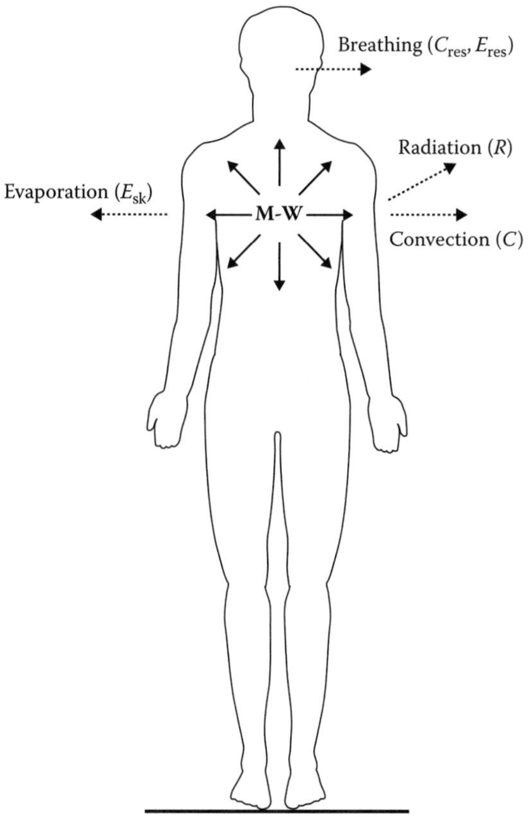

FIGURE 16.1 Human–environment heat exchange.

of evaporative heat loss from skin (latent heat loss; W/m^2), C_{res} is the rate of convective heat loss from respiration (W/m^2), E_{res} is the rate of evaporative heat loss from respiration (W/m^2), S_{sk} is the rate of heat storage in the skin compartment (W/m^2), and S_{cr} is the rate of heat storage in the core compartment (W/m^2).

In addition to the variables presented in the thermal balance equation, many other factors affect thermal response and comfort at the workplace. They are divided into two major categories:

1. Environmental parameters:
   - Air temperature
   - Air velocity
   - Relative humidity
   - Radiant temperature asymmetry
2. Individual (personal) parameters:
   - Thermal insulation of clothing
   - Metabolic rate
   - Age
   - Acclimatisation

16.2.1 Environmental Parameters

The temperature of the air greatly influences thermal comfort. High temperatures can cause severe psychophysiological dysfunctions. When the temperature rises above 25°C, the following symptoms can appear: dissatisfaction, lack of concentration on work activities, an increase in the number of mistakes made, and a decline in dexterity in work tasks requiring precision. When the temperature rises above 30°C, physiological dysfunctions occur, including fluid-electrolyte imbalance, an excessive load on the heart and the circulatory system, and considerable fatigue that could lead to exhaustion (Gwóźdź 2004).

Studies by Traczyk and Trzebski (2004) revealed that exposure to an air temperature higher than 21°C reduces psychophysical perception by about 60% in relation to the neutral temperature (16°C–21°C). Furthermore, they noted that when the 26°C threshold is considerably exceeded, attention level, perception, and reflexes are impaired to a great extent, in particular during short exposures, that is, those lasting not longer than 120 minutes. The fall in perception was about 16%. Prolonged exposure to temperatures below the comfort level causes vasoconstriction. Not only does blood flow in the limbs decrease, but also the temperature of the legs decreases by as much as 10°C. The lowest permissible skin temperature is 20°C and the lowest core temperature is 36°C. A further decrease in the core temperature leads to many dysfunctions: when the core temperature falls below 36°C, the metabolic rate rises; when it falls below 35°C, severe shivering occurs; below 33°C, severe hypothermia occurs; at 32°C, consciousness is clouded, shivering ceases, and blood pressure becomes difficult to measure; and at 30°C, muscle rigidity increases and pulse and blood pressure are difficult to measure (Nims 1999).

16.2.1.1 Air Velocity

Air velocity and the difference between the skin temperature (t_{sk}) and the ambient air (t_a) determine the thermal sensation temperature (ASHRAE 2005; Śliwowski 2000), which increases with an increase in the difference ($t_{sk} - t_a$). Furthermore, an increase in air temperature causes an increase in the amount of heat transferred to the environment by convection. Hence, maintaining the air temperature of the room at one level results in a thermal sensation of cold when compared to thermal effects in an environment without perceptible air flow. For instance, air velocity below 0.1 m/s most often occurs in naturally ventilated buildings; the threshold for strong sensations is 0.2 m/s; for perceptible movement, 0.25–0.5; for air flow causing hair to wave, 0.5–1.0; and for normal walking speed, 1.0–2.0 (Śliwowski 2000).

16.2.1.2 Relative Humidity of the Air

The relative humidity of the air affects the comfort of the workers. Its influence is greater in higher air temperatures, that is, at values higher than the set point for thermal comfort, it produces intensive sweating episodes (ASHRAE 2005; Śliwowski 2000). High relative humidity (about 70%) hinders heat dissipation from the skin surface and creates conditions favourable to the growth of indoor bacteria and mould. Low relative humidity in heated rooms can cause dryness of skin and nasal mucous membranes. It can also cause throat aches and headaches and contribute to air pollution. Hence, for health reasons, indoor relative humidity should be between 40% and 70%.

16.2.1.3 Asymmetry in Air Temperature

The presence of a vertical temperature gradient in a room may cause different thermal sensations in the parts of the body, such as a thermal sensation of cold at the feet or head and a comfortable thermal sensation in other parts of the body. Thermal radiation asymmetry should not exceed 10°C, and the ground temperature should not exceed 24°C. Overheating of the feet may lead to vasodilation or even to swelling (Śliwowski 2000).

16.2.2 Individual Factors

16.2.2.1 Clothing Thermal Insulation

In a cold environment, the body's natural thermoregulatory mechanisms find it difficult to maintain thermal balance. For their health and safety, workers must therefore wear protective clothing whose thermal insulation is properly adjusted to the surroundings and to the activity of the concerned worker (Śliwińska 1988). The amount of heat transferred through clothing ensembles is determined by clothing surface area and the temperature gradient between the skin and external surface of the clothing. The unit used for measuring clothing thermal insulation is the *clo,* a term introduced in 1941 by Gagge instead of $m^2 \cdot K/W$ (square meters times kelvin per watt; Parson 2003). One clo represents the thermal insulation necessary to maintain a state of thermal comfort in an environment where the air temperature is 21°C, humidity is 50%, and air flow velocity reaches 0.01 m/s. In SI units, 1 clo is equivalent to a heat transfer resistance of $0.155\ m^2 \cdot K/W$. Eskimo clothing has the highest insulation value (4 clo), while the value of summer clothing is about 0.6 clo, and winter clothing 1 clo.

16.2.2.2 Metabolic Rate

The intensity of the work performed influences the metabolic rate of heat produced by the human body. Metabolism is a key element in determining the thermal insulation of clothing and in designing indoor air parameters for thermal comfort. The metabolic rate value depends on the activity of an individual, his or her age, and genetic and health conditions, as well as the work environment conditions. The unit used to express the metabolic rate is a *met* (1 met = $58.2\ W/m^2$ and is the metabolic rate of a sedentary person). Table 16.1 shows the metabolic rate for typical activities when performed continuously.

16.2.2.3 Age

Ageing brings about changes in the efficiency of tissues and regulatory mechanisms. These phenomena reduce the human body's ability to adapt to hot and cold environments. Research studies have determined that women and men above the age of 45 find it more difficult to work in a hot environment. The acclimatisation process

TABLE 16.1
Examples of Metabolic Heat Generated for Various Activities

Activity	Metabolic Rate (met)
Resting in a reclining position	0.8
Resting	1.0
Resting in a standing position	1.2
Moderate activity in a standing position (light industry)	1.6
Mean activity in a standing position (house cleaning)	2.0
Walking at a speed of 5 km/h	3.0
Heavy work in a standing position	3.4
Running at a speed of 15 km/h	9.5

requires a higher physiological effort in this age group as compared to younger persons (Marszałek 1997). Studies have also revealed that women and men above the age of 45 react to a hot environment with a lower perspiration rate than younger persons. This phenomenon is regarded as symptomatic of a decreased function in sweat glands rather than a diminution in their number (Marszałek et al. 2007).

16.2.2.4 Acclimatisation to Hot and Cold

Acclimatisation to a hot environment is the process of adapting to permanent or repetitive exposure to heat. It leads to favourable physiological changes that raise an individual's tolerance of thermal environments. These changes include decreased skin blood volume and activation of a sweating reaction at a lower ambient temperature when compared to nonacclimatised persons, that is, individuals not subjected to everyday exposure to heat during the week preceding work. Sweat composition also changes in order to maintain the sodium ion level (Kociuba-Uściłko 2004). Furthermore, acclimatised persons experience a smaller increase in their core temperature (0.5°C–0.8°C) and in heart contractions (10%–15%) when compared to nonacclimatised persons (Gwóźdź 2004).

Persons employed in hot surroundings and resuming work after a long absence must adapt to the heat. This process can be conducted either in a climatic chamber or in the hot work environment. The specified process that should be followed (Sołtyński and Marszałek 2004) is as follows:

- On the first working day, the time spent in hot surroundings should not exceed 50% of the duration of an entire shift. The exposure time should gradually be extended by 10% every day until the sixth day, after which working in the hot environment can be continued throughout the whole shift.
- On the first and second day, the working time in a hot environment should be restricted to 35% of the shift, on the third and fourth to 50%, on the fifth and sixth to 65%, and on the seventh day, it can be extended to cover the whole shift.

16.3 IMPACT OF THERMAL LOAD EFFECT ON THE HUMAN BODY

16.3.1 Changes in the Circulatory System

The basic response of the human body to hot environments is vasodilation, which is the increase of blood flow to the skin from 5% in conditions of thermal comfort to as much as 20% in a very hot environment. The body can then transfer more heat to its surface, the skin. Simultaneously, the body stimulates the sweat glands and secretes sweat onto the surface of the skin (Kociuba-Uściłko 2004; Krause 2000). Further exposure to hot surroundings exceeds the adaptive capabilities of the human body and may lead to unfavourable reactions such as a rise in the heart rate (to above 140 beats per minute), lowered blood pressure following an increase in the circular resistance of blood vessels, and finally to ischaemia of internal organs and circulatory collapse (Gwóźdź 2004).

In a cold environment, stress results in vasoconstriction that reduces cutaneous and subcutaneous blood flow to some areas of the body, in particular to the peripheral regions (Krause 2000). Blood is directed to the blood vessels located deeper in the body, which increases central blood volume and, at times, causes a rise in blood pressure.

16.3.2 Sweat Glands

When the body temperature or the ambient temperature rises above 28°C–32°C, sweat is secreted all over to the body to increase cooling by evaporation. However, the intracellular and extracellular volumes of body fluids are reduced with heat loss. Due to the body's need to maintain fluid-electrolyte balance, secreted sweat should not exceed the limit value of 5% of the body mass.

16.3.3 Skeletal Muscles

The human body reacts to a cold environment through shivering and nonshivering thermogenesis. Stimulation of the reflex skeletal muscles provokes an increase in the metabolic rate, which is the source of heat for the human body. Consequently, tension in resting muscles increases, causing shivering, a source of heat. The amount of heat generated by this reaction depends on the ambient temperature and the duration of its influence. In extreme conditions, the basic metabolic rate can rise by 4 or even 5.5 times.

16.3.4 Heat Disorders

Extreme ambient temperature conditions and impairment of the body's thermoregulatory mechanism can cause excessive heat loss. This results in a cooling of the body (hypothermia) or insufficient heat loss leading to overheating (hyperthermia; Kociuba-Uściłko 2004).

In extremely cold conditions, the body's core temperature falls below 35°C and leads to a decrease in performance, reduction in muscle contraction power, and physical and behavioural disorders such as a prolonged reaction time. Lowering the muscle temperature by 10°C decreases precision of movements and makes maintaining balance difficult. Individuals working in such environments gradually become apathetic and mechanisms of their self-preservation instincts are disturbed (Gwóźdź 2004). Further hypothermia results in decreased heart output and stroke volume. In turn, vascular systemic resistance increases. When the body temperature falls to 28°C, the risk of ventricular fibrillation with myocardial irritability is magnified. A further decrease in the core temperature to 23°C–25°C may cause dysfunction of the circulatory and respiratory systems, clouding of consciousness, liver and kidney failure, and body fluid–electrolyte imbalance (Kociuba-Uściłko 2004).

In a hot environment, where the amount of heat absorbed by or created in the body exceeds the amount of heat lost, hyperthermia occurs. Dehydrated persons suffering from sweating dysfunctions may experience heat stroke, which can cause death due to thermal shock at a core temperature of 42°C–43°C (Kociuba-Uściłko 2004). Failure to balance a loss of electrolytes leads to dysfunctions of the alimentary canal and contracture of skeletal muscles. Hyperthermia may occur even with sufficient sweat secretion. It causes apathy, fatigue, behavioural and physiological changes, and, in extreme cases, death due to cardiovascular collapse (Kociuba-Uściłko 2004). Studies show that in 40% of white-collar workers, an increase in core temperature by only 1°C caused as much as a 60% increase in the number of mistakes they made. In another case, the number of occupational accidents occurring in hot surroundings rose by 18%–24% when compared to thermoneutral environments. Workers experienced reduced concentration, critical abilities, dexterity, and coordination when performing tasks requiring precision (Gwóźdź 2004).

16.4 ASSESSMENT OF WORKERS EXPOSURE TO THERMAL LOADS

Numerous national, European, and international standards have addressed the problems workers experience due to workplace exposure to hot or cold environments. These documents examine and determine thermal loads based on indices characteristic of a given work environment, that is, wet bulb globe temperature (WBGT) for hot environments and t_{wc} and required clothing insulation (IREQ) for cold environments. The standards also include reference values (the 'permissible' values of the aforementioned indices), which are the levels of hot or cold conditions to which almost all individuals could be exposed without suffering any adverse health effects. The indices are based on the typical effects of a given environment on an individual over prolonged periods, excluding peak loads, that is, temporary extremes of cold or hot or work that is more or less intense than usual.

When the permissible threshold of temperature is exceeded in a hot environment, the thermal load should be reduced by modifying the work environment or the level of load. Reflective shields between radiant sources and workers (see Section 16.8), limiting exposure or work time and using heat reflective clothing or reducing thermal insulation of clothing (see also Section 16.9) are effective methods of reducing thermal load. A detailed analysis of the thermal load should also be carried out by measuring core temperature and the volume of evaporated sweat (Parson 2003). The ISO 7933 standard (2004) describes the method for assessing these values.

When the t_{wc} exceeds permissible values, exposure time must be reduced (see Sections 16.7 and 16.8) and workers must wear protective clothing with higher thermal insulation (see Section 16.9).

16.4.1 Hot Stress Index

Assessment of exposure to a hot environment uses the WBGT index, which was developed by the US Navy Marine Corps Recruitment Depot at Parris Island in Southern Carolina, in 1957 (Parson 2003). The term WBGT comes from the names of the measurement devices used, that is, wet bulb globe temperature.

The WBGT index has the advantage of recording measurements on-site without causing any major disturbance in the performance of work; the measurements last only an hour. Another advantage of the WBGT index is that actual thermal loads are assessed at a given workstation and these measurements take into account durations representative of the whole shift. The method's limitations lie in its reference values, that is, the WBGT heat exposure limits (ISO 7243 1989; ASHRAE 2005). The reference values were developed based on tests with young men wearing normal permeable clothing (0.6 clo). Moreover, compliance with ISO 7243 (1989) prevents only an increase in core temperature above 38°C. It does not guarantee that other physiological criteria such as the heart rate or the volume of the evaporated sweat will be maintained within healthy limits. These reference values are included in many documents on standards and norms, for example, in ISO 7933 (2004).

16.4.2 Cold Stress Indices

Two parameters are used to assess a cold environment and its effects on the human body (Parson 2003):

1. t_{wc} (wind chill temperature) is an indicator of the local impact of cold wind on the human body.
2. IREQ (required clothing insulation) enables assessment of the overall impact of a cold environment. It is determined by specifying the thermal insulation values recommended for a given environment to limit the risk of body cooling (ASHRAE 2005; Parson 2003).

The t_{wc} defines the local impact of wind on the human body when the ambient temperature is below 10°C. Wind chill temperature causes a sensation of ambient temperature when a wind velocity of 4.2 km/h producces the same sensation of cold as perceived by a worker in actual environmental conditions.

IREQ is a way to calculate the thermal insulation required of protective clothing used for work performed with a specified intensity and in specified surroundings. Such clothing prevents a decrease in the core temperature to below 36°C (Figure 16.2). The IREQ index developed by Holmer (1984) was based on the thermal balance concept proposed by Langer. It is expressed in clo units.

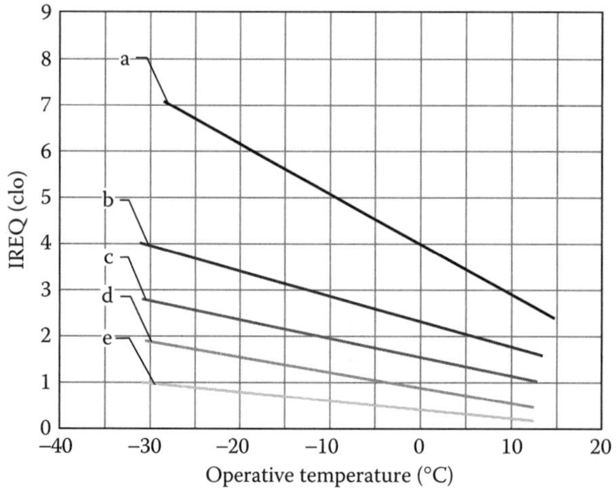

FIGURE 16.2 Required thermal insulation as a function of the metabolic rate (activity) of a human: (a) rest, (b) light work, (c) moderate work, (d) intensive work and (e) very intensive work.

TABLE 16.2
Principles of Clothing Insulation Selection

$I_{clr} < IREQ_{min}$	Insulation is insufficient, exposure time must be shortened.
$IREQ_{min} < I_{clr} < IREQ_{neutral}$	Insulation is sufficient to protect against cooling of the body, thermal conditions are described by a worker as 'acceptable' and surroundings as 'cold' or 'neutral'.
$I_{clr} > IREQ_{neutral}$	Insulation values are adequate, a risk of overheating exists with higher physical activity.

The standard EN ISO 11079 (2007) describes the methodology for calculating the t_{wc} and IREQ indices and their reference values (i.e. threshold values of air velocity). The standard describes the following levels for the IREQ index:

- $IREQ_{min}$ is the minimal thermal insulation required to maintain the body's thermal equilibrium at a subnormal level of mean body temperature, that is, equal to 30°C and at the ratio of the required dissipation of metabolic heat to the maximum value of 0.06. $IREQ_{min}$ is the highest limit value of behavioural and physical stress related to cooling of the body during a shift.
- $IREQ_{neutral}$ is the thermal insulation required to ensure conditions of thermal neutrality in the body, that is, maintain thermal equilibrium at the normal level of mean body temperature. This value requires no or minimal cooling efforts.

The criteria for selecting clothing that provides the thermal insulation sufficient to establish the defined level of heat balance (on the assumption that I_{clr} is the insulation value used at the workstation) are given in Table 16.2.

16.5 METHODS OF THERMAL LOAD ASSESSMENT IN THE WORKPLACE

The methods commonly used to assess workers' exposure to a hot or cold environment are based on the indices mentioned earlier, that is, WBGT, t_{wc} and IREQ. Their use is widespread because they are easily determined and applied in real-life conditions at workstations. The reference values are included in the recommended standards.

16.5.1 Measurements and Calculations of the WBGT Index

There are three phases to the methodology for measuring the WBGT permissible values:

Phase I: Measure the wet bulb temperature, t_{wb}, temperature of the black globe, t_{gb}, and the air temperature when tested outside the building, t_a (Kabza and Kostyrko 2003/2004).

The measurements should be recorded when the thermal load is at the maximum, for example, in the middle of the day during the summer. In a homogenous environment, measurement devices should be placed on a single level—the level of the stomach (Figure 16.3). In a thermally heterogeneous environment they should be placed on three levels—the levels of the head, stomach, and ankles (Figure 16.4).

The values measured are entered into the appropriate formula, depending on the site where the measurements were recorded:

- For indoor conditions or conditions outside the building with no solar radiation:

$$\mathrm{WBGT} = 0.7t_{nw} + 0.3t_g \tag{16.2}$$

FIGURE 16.3 Sets of sensors for measuring air parameters: (a) wet globe thermometer, (b) dry air thermometer, and (c) wet bulb.

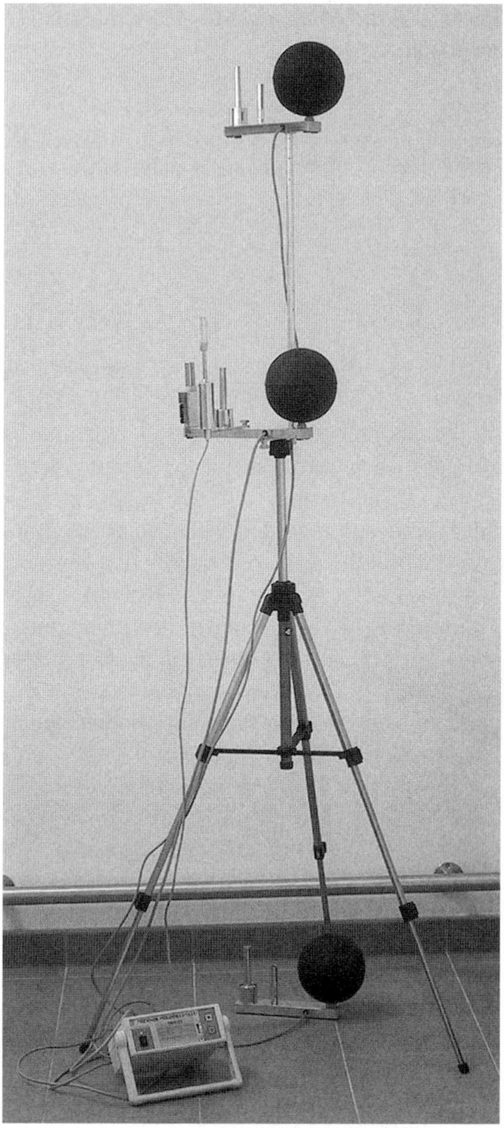

FIGURE 16.4 Set of three sensors for measuring air values in heterogeneous thermal conditions.

- For outside conditions with solar radiation:

$$WBGT = 0.7t_{nw} + 0.2t_g + 0.1t_a \tag{16.3}$$

If the worker is in a standing position, measurement sensors are placed at the height of 0.1, 1.1, and 1.7 m from the ground level. If the worker is in a sitting position, the sensors are placed at the height of 0.1, 0.6, and 1.1 m. The WBGT weighted-average value is then calculated:

$$\text{WBGT} = \frac{\text{WBGT}_{\text{head}} + 2\text{WBGT}_{\text{stomach}} + \text{WBGT}_{\text{ankles}}}{4} \tag{16.4}$$

In an environment where the physical parameters and the intensity of the work change frequently during a work shift, a representative value for a given parameter must be set. This is calculated based on constant measurements of this parameter (e.g. temperature, energy expended) and its weighted-average determined from Equation 16.5 below. Measurements of each parameter are carried out within one hour. Depending on variable environmental conditions and the intensity of the work, the measurements should include work periods wholly representative of the given workstation. Each period covers a phase of work and a phase of rest, assuming that the worker's rest conditions are the same as those in which he or she works.

$$\bar{p} = \frac{(p_1 \cdot t_1) + (p_2 \cdot t_2) + \cdots + (p_n \cdot t_n)}{t_1 + t_2 + \cdots + t_n} \tag{16.5}$$

where $p_1, p_2, \cdots p_n$ are the parameter values obtained at time points $t_1, t_2, \cdots t_n$, $t_1 + t_2 + \cdots + t_n = 1$ h.

Phase II: Determine the energy expended for each analysed work cycle. Measurements are based on tables presenting the metabolic values for various activities (ASHRAE 2005; Fanger 1970) or on the method for measuring the metabolic level at a workstation included in the standard ISO 8996 (1990).

Phase III: Compare the WBGT weighted-average value calculated from Equations 16.2 through 16.5, with reference to the appropriate values for energy expended. These reference values are given in the standard ISO 7243 (1989) and other documents (ASHRAE 2005).

When determining a reference value, the following aspects have to be taken into consideration: information concerning acclimatisation at the workstation or lack thereof, air flow, and clothing thermal insulation (for steam and water-resistant clothing, WBGT reference values will be lowered). Comparing the WBGT calculated value to the reference value determines whether the value measured at a given workstation (phase I) with reference to a given individual working within a given intensity (phase II) is within the safe range or exceeds the WBGT permissible value.

16.5.2 Measurements and Calculations of t_{wc} and IREQ Indices

Measuring the local impact of the cold environment using the t_{wc} index is a two-phase process. Measurements are recorded with the devices shown in Figures 16.3 and 16.4.

1. *Phase I*: Measurements of temperature and air velocity in a zone occupied by workers are recorded at least once every 4 hours in a period of extreme negative thermal loads. The mean values are then calculated, which take into account changes in the parameters that occur during work.

   Wind chill temperature (t_{wc}) is defined by the following equation:

$$t_{wc} = 13.12 + 0.6215 t_a - 11.37 v_{10}^{0.16} + 0.3965 t_a v_{10}^{0.16} [°C] \tag{16.6}$$

where v_{10} is the standard meteorological value of air velocity measured at a height of 10 m above ground level.

2. *Phase II*: The t_{wc} index-obtained value is compared with the reference values according to ISO 11079 (2007). The required clothing insulation IREQ is analogous to the calculation of the WCI (wind chill index). It is carried out in three phases:
   - *Phase I*: Measurements of the air temperature, air velocity and black globe temperature are recorded. Their mean values are then calculated giving due consideration to changes in the parameters of time and space. The worker's energy expended is determined by means of tables or direct measurements at the workstation.
   - *Phase II*: The $IREQ_{neutral}$ and $IREQ_{min}$ reference values (permissible values) are determined based on formulas and the programme *Calculation of Required Clothing Insulation (ireq), Duration Limited Exposure (dlim), Required Recovery Time (rt), and Wind Chill Temperature (twc)* detailed in the standard ISO 11079 (2007).
   - *Phase III*: The required clothing insulation reference values given in the standard are compared with the thermal insulation of the clothing used in the workplace. Clothing ensembles with insulation levels higher than the $IREQ_{min}$ should immediately replace clothing with lower thermal insulation values.

16.6 WORKING IN COLD AND HOT ENVIRONMENTS—STATE OF RISK

Despite a systematic reduction in the number of workers exposed to adverse climatic conditions (Table 16.3) in Poland, there are still numerous workstations where work is performed in cold (e.g. in cold stores, slaughter houses) or hot environments (e.g. in steelworks, coal mines).

TABLE 16.3
Numbers of Workers Exposed to Hot and Cold Environments in Poland in the Years 2001–2005

Type of Environment	Numbers of Persons (in Thousands) in a Given Calendar Year				
	Years				
	2001	2002	2003	2004	2005
Hot	26.4	21.3	19.3	18.4	18.3
Cold	21.3	21.0	17.0	15.5	17.4

Source: Data from GUS. 2005. *Working Conditions in 2004*. Warsaw: Central Statistical Office of Poland; and Ministry of Labour and Social Policy. 2006. *Assessment of Safety and Hygiene of Work Conditions in 2005*. Warsaw: Ministry of Labour and Social Policy.

16.7 EXAMPLES OF GOOD PRACTICE IN PREVENTING THE ADVERSE EFFECTS OF A HOT OR COLD ENVIRONMENT ON THE HUMAN BODY

Differences in perception of heat, thermoregulation processes, and considerable differentiation in the time and space of thermal conditions at workstations mean that each worker's load or risk must be individually assessed at each workstation (Krause 2000). Proposed solutions to reduce thermal load can be categorised by hot and cold environments and are described in the next section.

16.7.1 Solutions Falling within the Scope of Health Care Prevention

- Solutions for hot environments include:
  - Ensure workers are acclimated to heat and that water loss is replaced. Provide access to fluids supplemented with salt (where the diet is low in sodium) and balance electrolyte shortages caused by intensive sweating.
  - Frequently replace the protective clothing of workers to eliminate the risk of skin infections, as fabrics stiffened with salt secreted with sweat can damage the epidermis (Krause 2000).
  - Assess physical fitness in the preliminary medical examination of workers required to wear protective clothing. Measuring the ability to maximise oxygen use per minute (Krause 2000), enables the selection of individuals fit to work in this type of clothing. Procedures should also include annual medical examinations and entail measurements of physiological parameters that assess the efficiency of the thermoregulatory system.
- Solutions for cold environments include:
  - Provide access to hot drinks (except caffeinated drinks) and hot, high-calorie meals.
  - Replace wet innerwear or wet clothes due to their lowered thermal insulation parameters.

16.7.2 Organisational Solutions

- Automate working activities if permissible limits of expended energy are exceeded.
- Work in teams of at least two persons to ensure mutual supervision.
- Ensure compliance with working time limits and work-rest cycles as recommended in the standards. For instance, for an activity calling for expending high or moderate energy and performed in a 4-hour working shift, in conditions where the temperature of air ranges from –35°C to –37°C and the air velocity is 2.6 m/s, working time cannot exceed a maximum of 40 minutes with at least four breaks. Rest can be provided to workers in a hot environment by assigning tasks with a lower intensity or load than the work tasks

performed at the usual workstation, for example, stipulating the working and resting periods during a shift in percentages as 25%/75%, 50%/50% or 75%/25% in accordance with the patterns proposed in the standard ISO 7243 (1989).

- Provide training on health hazards caused by thermal loads. Training should include presentations on the symptoms of thermal stress (heatstroke, frostbite, hypothermia, and so on), information on the potential influence of medicaments on thermal sensations and the efficiency of thermoregulatory reactions, as well as instructions on the proper use of personal protective equipment.
- Introduce systems of supervision and control to ensure that workers fulfil the appropriate recommendations.

16.8 COLLECTIVE PROTECTION

An appropriately designed workstation effectively reduces the hazards present in cold or hot surroundings. The following aspects should be considered at the design stage:

- Ensuring control of hot or cold temperature sources.
- Ensuring control of air parameters in the working areas.

Solutions appropriate to a hot environment include:

- Using production processes, machinery, and equipment that do not generate heat or generate only a small amount of heat.
- Automating technological processes.
- Isolating equipment that are sources of heat by providing separate rooms or situating them outside the building, by cooling and shielding with water shields or shields made of insulation materials, aluminium sheets, aluminium foil, or absorptive glass, or by shielding with heat exchangers, where the shield is cooled internally with air or water.
- Installing local ventilation or air conditioning and suspended or ventilated ceilings.
- Eliminating ventilation leaks or installing local steam-shaft elements.
- Increasing the coefficient of partition heat resistance if work is performed in rooms where the cause of high temperature is excessive heat gain through nontransparent partitions (walls, ceilings). To this end, the following solutions can be attempted: adding more layers of insulation, installing suspended ceilings, installing shields outside buildings, or coating external surfaces of the walls with solar reflective material.
- Using double- or triple-glazed windows filled with noble gas (most frequently argon) to reduce the heat transfer coefficient in buildings where large wall areas are occupied by transparent partitions (windows).

In rooms where the above-mentioned solutions cannot be implemented, changes in the organisation of the working area and time should be implemented. Generally, these changes include minimising exposure time to heat, providing protective equipment with appropriate ventilated or cooled thermoinsulation properties, and creating facilities for rest in air-conditioned or heated rooms.

Working in a cold environment requires appropriately designed machines that allow handling without removing protective gloves. Metal seats, all handles, and the control panels must be covered with thermoinsulation materials.

16.9 PROTECTIVE CLOTHING FOR INDIVIDUALS

Protective clothing worn by workers in a hot or cold environment should meet some specified criteria. In a hot environment, clothing must have low thermal insulation. Additionally, protective and barrier clothing should be used to offset high ambient heat, especially by individuals performing tasks requiring physical exertion (e.g. firefighters, steelworkers). Workstations and clothing can be equipped with various cooling systems:

- *Water*: The clothing ensemble depends on the purpose it is meant to serve. It can protect only the head, or can be a vest cooling the head and trunk, or a set of underwear cooling the whole body. No clothing ensembles can cool the hands or feet. The temperature of the water supplied to the cooling system should be 20°C.
- *Air*: Cooling by means of air between the worker's skin and his or her clothing, at a temperature of 10°C and relative humidity of 20%. Air can be supplied on a constant basis or periodically where activities require changes of location.
- *Ice or phase change materials (PCM)*: Vests with ice and/or PCM packs can be supplied for short and intense work activities (these can be used for a maximum of 2–3 hours).

Thermal insulation in protective clothing used in a cold environment has parameters equal to the value of the IREQ index. The latter is determined for the actual work conditions, that is, considering the ambient thermal conditions and the intensity and energy expense of activities performed. If work tasks were performed in a moderate or hot environment prior to commencing activities in cold surroundings, clothes and underwear that become dampened need to be replaced.

When the air temperature falls below –1°C, workers should wear protective gloves that can prevent frostbite and at the same time ensure adequate manual dexterity. The worker must avoid glove contact with volatile liquids, for example petrol (Sołtyński and Marszałek 2004). The combined impact of cold and mechanical vibrations transmitted through the handles of a machine poses an unusual hazard. At a temperature of +14°C, there is a risk of spasms of the blood vessels in the fingers induced by a thermoregulatory dysfunction of cutaneous blood circulation (Koradecka 1980). Individuals working outside buildings should wear glasses to protect themselves against snow and hail.

16.10 SUMMARY

Numerous studies have shown that physical fitness and intellectual capabilities peak in conditions of thermal comfort, whereas hot or cold surroundings have a negative effect on health and the quality of work (Grether 1973; Fanger 1970; Nadler and Busch 2002).

Despite the advancements of modern technology, there are still many workstations where workers are exposed to hot and cold environments. A challenging task faced by designers, employers, and occupational safety and health services is to design workstations that conform to ergonomic rules and preserve work conditions that were created under close supervision.

Relevant normative documents contain guidelines on the design of such workstations and on risk assessment methodology. A systematic approach to risk assessment of workers exposed to unfavourable environments is necessary so that work conditions can be improved and workers' health protected.

REFERENCES

American Society of Heating, Refrigerating and Air Conditioning Engineers. *HVAC Fundamentals Handbook*. Atlanta, GA: Academic Press.

Fanger, P. O. 1970. *Thermal Comfort*. New York: McGraw-Hill.

Goodfellow, H. D., and E. Tahti. 2001. *Industrial Ventilation Design Guidebook*. Atlanta, GA: Academic Press.

Grether, W. F. 1973. Human performance at elevated environmental temperature. *Aerosp Med* (44):747–755.

GUS. 2005. *Working Conditions in 2004*. Warsaw: Central Statistical Office of Poland.

Gwóźdź, B. 2004. Human in an inustrial environment with elements of ergonomics. In *Human Physiology with Elements of Applied and Clinical Physiology*, eds. W. Traczyk and A. Trzebski, 266–277. Warsaw: PZWL.

Holmer, I. 1984. Required clothing insulation (IREQ) as an analytical index of cold stress. *ASHRAE Trans* (90):116–128.

ISO 7243. 1989. Hot environments. Estimation of the heat stress on working man, based on the WBGT-index (wet bulb globe temperature).

ISO 8996. 1990. Ergonomics. Determination of metabolic heat production.

ISO 7933. 2004. Ergonomics of the thermal environment. Analytical determination and interpretation of heat stress using calculation of the predicted heat strain.

ISO 11079. 2007. Ergonomics of the thermal environment. Determination and interpretation of cold stress when using required clothing insulation (IREQ) and local cooling effects.

Kabza, Z., and K. Kostyrko. 2003/2004. *Metrology of Indoor Thermal Environment and Environmental Physical Quantities*. Opole: Technical University of Opole.

Kociuba-Uściłko, H. 2004. Thermoregulation. In *Human Physiology with Elements of Applied and Clinical Physiology*, eds. W. Z. Traczyk and A. Trzebski. Warsaw: PZWL.

Koradecka, D. 1980. *Influence of Factors Connected with a Use of Typical Vibration Tool at Work on the Peripheral Blood Circulation*. Warsaw: CIOP.

Krause, M. 2000. Human thermoregulation and thermal loads. In *Science of Work—Safety, Hygiene, Ergonomics*, Vol. 4, ed. D. Koradecka, 107–122. Warsaw: CIOP.

Marszałek, A. 1997. Age and heat tolerance. *Ergonomia* 20(1):41–47.

Marszałek, A., G. Bartkowiak, W. Kamieńska, and A. Stefko. 2007. *Ageing Workers in a Hot Environment*. Warsaw: CIOP-PIB.

Ministry of Labour and Social Policy. 2006. *Assessment of Safety and Hygiene of Work Conditions in 2005*. Warsaw: Ministry of Labour and Social Policy.

Nadler, E., and C. Busch. 2002. Effects of hot and cold temperature exposure on performance: A meta-analytic review. *Ergonomics* (45):682–698.

Nims, D. K. 1999. *Basics of Industrial Hygiene*. Canada: John Wiley & Sons.

Parson, K. 2003. *Human Thermal Environments: The Effects of Hot, Moderate and Cold Environments on Human Health, Comfort and Performance*. London: Taylor & Francis.

Śliwińska, E. 1988. *Thermal Comfort in Cold and Thermally Neutral Workstations*. Wroclaw: Technical University of Wroclaw.

Śliwowski, L. 2000. *Indoor Thermal Environment of Interiors and Thermal Comfort of Humans in Rooms*. Wroclaw: Technical University of Wroclaw.

Sołtyński, K., and A. Marszałek. 2004. Thermal loads. In *Occupational Risk Assessment. I. Methodological Basis*, ed. W. M. Zawieska, 3:253–268. Warsaw: CIOP-PIB.

Traczyk, W. Z., and A. Trzebski. 2004. *Human Physiology with Elements of Clinical and Applied Physiology*. Warsaw: PZWL.

17 Atmospheric Pressure (Increase and Decrease)

Wiesław G. Kowalski

CONTENTS

17.1 INCREASED ATMOSPHERIC PRESSURE

17.1.1 Introduction

A hyperbaric environment is an environment with a high atmospheric pressure, whose characteristic factors significantly affect the human body during occupational activities (working in caissons, diving), military underwater operations, and recreational scuba* diving. For untrained individuals, those who do not observe adequate precautions and safety measures, elderly persons, and those with health problems, exposure to such an environment could prove dangerous with a risk of death. However, hyperbaric procedures are employed in the treatment of some medical conditions. Knowledge of this field covers a range of issues significant in the context of occupational safety, the safety of qualified underwater recreation participants, and appropriate medical practices.

* SCUBA = self-contained underwater breathing apparatus.

17.1.2 Selection of Candidates for Scuba Diving

According to the present regulations, any physician may sign fitness certificates for amateur scuba diving. It is therefore necessary to determine basic certification principles including a list of possible contraindications in order to minimise the risk of accidents due to the divers' health. Those organising training for amateur divers frequently rely on the participants' self-assessment or on the opinions of physicians who are not well acquainted with problems related to hyperbaria.

Appendix B to European standard EN 14153 (2003) contains the scope of medical history of recreational scuba divers, including an adequate preparticipation form to be completed by the examining physician. The following factors require a physician's consultation: if the person is over age 45, smokes tobacco, has a high cholesterol level, has a history of diseases, for example, asthma, all possible conditions of the lungs and heart, chest surgery, arterial hypertension, ear diseases, thrombosis, epileptic seizures, uses of certain medicines, and so on. Adequately trained physicians should take advantage of the certification provisions to ensure divers' safety. Scuba divers should undergo examinations every year, and more frequently on specific occasions such as participation in competitions. Determining the probability of the risk of sudden disability or loss of consciousness of intending participants is essential. If there are any particular health related limitations, the diver has to be made aware of his or her condition and of the acceptable exertion level, which is dependent on the type of dive. Factors such as the depth, duration of the dive to the defined depth, water temperature, and so on should be considered (Krzyżak 2006).

Adequate training is essential for divers' safety, and should include learning how to follow certain diving principles and avoid the risk of injuries due to lung distension. The risk of such injuries is greatest during ascent through the water or due to sudden dilation of the gas drawn from the breathing apparatus while diving under a pressure greater than atmospheric pressure. An increase in lung pressure of this nature can result in vesicular damage if the air is not exhaled during emersion. A change in depth of 10 m in fresh water and 9.75 m in seawater causes a change in pressure equal to 1 atm, and is equivalent to the difference between atmospheric pressure at sea level and in aerospace. Improper ascent from depths of 2–3 m can result in lung tissue damage. Individuals with symptoms of airway obstruction or emphysematous spaces, patients with a history of chest surgery or proneness to spontaneous pneumothorax, and those with asthma or abnormal lung functions revealed by a spirometric examination, should not dive using a diving apparatus (Dolatkowski and Ulewicz 1973; Krzyżak 2006; Olszański and Siermontowski 2002).

Increased mucous secretion in the airways, encountered in patients with acute conditions such as bronchitis, pneumonia, and bronchiectasia, is also a definite contraindication for scuba diving and for working under hyperbaric conditions. A medical examination including a chest radiograph of persons should precede their return to work in a hyperbaric environment after a recovery period.

Diseases of the cardiovascular system also disqualify candidates if the pathological changes (like organic heart diseases, septum defects, or valve stenosis) significantly impair systemic function. Asymptomatic valvular insufficiency without changes revealed by ECG record and echocardiography does not disqualify the participant. Partial slowing down of the heart rate in elderly nontraining individuals may indicate a heart condition or the use of cardiotonic agents. A blockage of the left bundle branch is usually a contraindication for diving, particularly in elderly individuals. Exercise ECG examination is recommended for patients with a blockage of the right bundle branch. Persons diagnosed with Wolff–Parkinson–White (WPW) syndrome, supraventricular tachycardia (which can cause loss of consciousness), and artioventricular blocks of the second or third degree should not engage in diving activity (Krzyżak 2006).

Ischaemic heart disease and the resulting heart rhythm disorders or a history of myocardial infarction are also definite contraindications. Individuals over 40 years of age require an evaluation for the risk factors for ischaemic heart disease and should undergo exercise ECG examination. Effectively controlled arterial hypertension is not a definite contraindication for diving. Elderly persons over 50 years of age should be made aware of the risk of sudden death involved in diving; such incidents may also prove dangerous for other persons diving in their close vicinity due to the risk of panic attacks and a need for immediate ascent.

Voluntary dehydration in hot climates is an additional risk factor for amateur divers. After a sudden change from cold to hot weather conditions in the course of travel from one climate zone to another, thirst is no longer indicative of a person's water requirements. The body may also experience sudden cooling (hypothermia), even in warm water. Under such conditions, a substantial physical effort could also contribute to a critical disturbance of the body's functional equilibrium (Klukowski 2005).

The pressure in the middle ear and paranasal sinuses should be equal to the ambient pressure; therefore, individuals with chronic inflammation of the ear and paranasal sinuses or tube and acoustic meatus should refrain from diving. Persons with a history of ear surgery are at risk of internal ear damage, while those with internal ear diseases may experience dizziness and emesis while diving and are at risk of drowning. Persons who experience barotrauma of the internal or middle ear due to diving require adequate laryngological evaluation and treatment (Dolatkowski and Ulewicz 1973).

Neurological disorders with concomitant convulsions, loss of consciousness, and stress-related syncopes and mental disorders such as psychoses, addictions, personality disorders, mental defects, inclination to panic attacks, and claustrophobia are also disqualifying, as they may interfere with diving safety.

Digestive system disorders such as gastroesophageal reflux are a contraindication for diving. Patients diagnosed with hiatus hernia should not dive until the condition is surgically corrected, as changes in pressure can contribute to the formation of so-called air traps and result in severe pain.

Hypoglycaemia from diabetes mellitus is a definite contraindication for diving. It is not only the health conditions of amateur divers that may be dangerous, but also the diving itself may exacerbate existing disorders and health may deteriorate (e.g., diving in the early period of pregnancy may contribute to foetus malformation; Dolatkowski and Ulewicz 1973; Krzyżak 2006; Dolmierski et al. 1981).

17.1.3 Hyperbaric Environment Disorders Due to Pressure Changes

17.1.3.1 Decompression Sickness

The physiological origin of disorders caused by pressure changes is the saturation of tissues with respiratory gases such as compressed air or other gas mixtures that are inhaled during immersion in water. The removal of gases dissolved in tissues and blood during a fast ascent through water is accompanied by gas bubble formation, resulting in decompression sickness (DCS). If the ratio between the partial pressure of a neutral gas in tissues and the ambient pressure does not exceed 2:1, the symptoms of DCS do not occur. Thus, rapid ascent through water from a depth of 7–8 m during recreational diving is safe, except for longer, multiple diving episodes within a single day. The speed values for safe ascent through water at larger depths are defined in special decompression tables. The theoretical fundamentals as well as applications of these values in different cases are very complicated and constitute a separate, more detailed area of knowledge (Dolatkowski and Ulewicz 1973; Konarski et al. 2005; Krzyżak 2006).

Although the greatest potential danger of DCS occurs due to errors committed during decompression (ascent through water), divers may experience symptoms of the disease even during a correct ascent through water if any pathological conditions are present. Joint pain is characteristic of a mild course of DCS; and neurological symptoms and symptoms of the heart and lungs are characteristic of the severe course. Treatment consists of decompression in pressure chambers. After DCS associated with neurological symptoms is treated, the changes in the central nervous system slowly recede and the diver may resume activity after 1–3 months. Persistent residual neurological symptoms should be confirmed by thorough psychoneurological examination including magnetic resonance imaging and computed tomography scans and are serious indications for permanent exclusion from diving. Any defects of the interatrial septum should be excluded in amateur divers affected by DCS with neurological symptoms. DCS may be chronic as well as acute. The chronic form is called aseptic necrosis, occurring in professional divers working in caissons due to multiple or lengthy exposures to high pressures. The prevalence of this form of DCS depends on the extent of decompression procedures observed by the diver (Dolatkowski and Ulewicz 1973; Krzyżak 2006; Olszański and Siermontowski 2002).

17.1.3.2 Pulmonary Barotrauma and Gas Embolism

The physiological effects of breathing under hyperbaric conditions differ and depend on the features of the given environment—mostly, the depth and the type of respiratory gas. The conditions of heat conduction also change, as do the optical and acoustic properties of the ambient environment. High hydrostatic pressure exerted on the incompressible human body is, according to Pascal's law, the same at each point and equal to external pressure. Thanks to this phenomenon, additional massive pressure (greater than atmospheric pressure) equal to 90 tons and exerted on the whole body at a depth of 50 m does not crush it. However, administration of respiratory gas or gases with a pressure equal to the ambient pressure (or several mmHg higher) enables balancing the pressure exerted on the human body whilst diving. Understanding the principles of safe diving requires knowledge of the laws of gas, which describe the phenomena connected with changes in pressure, temperature, and gas volume, as well as the changes that occur in respiratory gases under hyperbaric conditions. A sudden change in depth during shallow diving may be more dangerous than an analogous change when diving deeper. As the body cannot equalise the pressure between the environment and body's air cavities, such changes result in tissue damage—namely barotrauma of the lungs, paranasal sinuses, middle and internal ear, teeth, and gastrointestinal tract. The most dangerous conditions include pulmonary barotrauma and gas embolism. Pulmonary barotrauma may develop even during ascent from minor depths (5 m) and can result in severe conditions or even death. The air inspired at that depth is decompressed during ascent through the water, and the closed larynx results in ectasia and pulmonary tissue damage. Secondary complications include the development of air embolisms, of which arterial air embolisms are the most dangerous.

As with DCS, it is essential to determine whether the gas embolism is the result of an obvious error committed while diving or is due to pathological changes in the lungs. Specialists recommend that persons with a history of pulmonary tissue damage quit diving (Dolatkowski and Ulewicz 1973; Krzyżak 2006; Olszański and Siermontowski 2002).

17.1.4 Other Properties of a Hyperbaric Environment

Pressure breathing causes excessive breathing gas mixture to dissolve in body fluids. Each gas exerts its own pressure (partial pressure) regardless of the pressure of the other gases in the mixture. An increase in pressure alters the physiological effects of different gases on the human body, for example, high oxygen pressure is toxic and the effect of neutral nitrogen resembles that of narcotic substances. Hence, persons diving to large depths are given breathing mixtures consisting of oxygen and much lighter gases, namely helium and hydrogen in different proportions, to replace nitrogen (Krzyżak 2006; Konarski et al. 2005).

Diving safely requires the proper training of participants and observance of depth limits and time limits on diving with stopped respiration. Amateur diving at depths of over 10–20 m requires a thorough knowledge of respiratory and circulatory physiology under such conditions and the application of multiple technical precautions including safeguarding.

Adequate physical and mental preparation, including building tolerance to a decline in partial oxygen pressure in the body, a rise in carbon dioxide pressure, and the resulting respiratory muscle contractions, significantly improves diving safety as well as health and life protection (Krzyżak 2006; Olszański and Siermontowski 2002).

17.1.5 Diving in the Mountains and Flights after Diving

The generally accepted speed for ascent from the water is calculated for diving that starts at the sea level. The manometer indications in the diving equipment are calibrated to this pressure value. Diving at high altitudes, for example, in mountain lakes, requires proper calculation of the values as determined in appropriate tables, as well as the proper calibration of scuba diving equipment indicators.

At high altitudes, the risk of DCS is elevated due to the low atmospheric pressure, partial oxygen pressure, and low temperature of the water. At higher altitudes, earlier and greater changes are likely to occur. In a defined altitude range, the risk of DCS or pulmonary oedema diminishes after 2 weeks of acclimatisation. Breathing pure oxygen prior to diving flushes nitrogen from the body thereby affording protection from DCS. Thus, diving in the mountains requires a knowledge of physiology and some essential procedures that may need to be performed under such conditions.

Specific problems can occur during air travel to and from warm climatic zones for recreational diving. During the flight, the pressure inside the aircraft is reduced to correspond to the pressure at an altitude of about 2000 m above sea level. A person boarding 2–5 hours after diving is due for decompression, but is exposed to the reduced pressure during ascent of the airplane and subjected to further ambient pressure decrease (decompression) to the level set inside the aircraft. Partial oxygen pressure also decreases, and the microclimate inside the aircraft favours body dehydration. Such conditions can cause the formation of gas bubbles in the blood responsible for DCS. Joint pains, shortness of breath, and anxiety occurring after several hours are indicative of the onset of this condition. Most specialists recommend a 24-hour interval between the last dive and air travel (Krzyżak 2006; Kowalski 2002; Ernsting et al. 1999).

Diving shortly after air travel is also highly risky, particularly when several time zones have been traversed during the flight, especially in the eastern direction. Apart from the usual symptoms of fatigue and the perceptible effects of prolonged exposure to the low pressure and adverse microclimate inside the aircraft, an asynchrony of circadian rhythms can result, affecting multiple vital functions of the human body. Consequently, the body's functions do not attain the required level in the new time zone. Such conditions interfere with diving safety. The body resynchronises circadian rhythmicity and regains the requisite psychophysical condition at a pace of one time zone per day (24 hours). The body must also acclimatise to altered environmental conditions.

17.1.6 Treatment under Hyperbaric Conditions

Hyperbaric therapy is used for the treatment of DCS and air embolism in divers, as well as for the treatment of embolisms of different origins, carbon oxide poisoning, and anaerobic bacterial infections. It may also be used as complementary treatment for some diseases. In carbon oxide poisoning, oxygen physically dissolved in the plasma is supplied to the tissues under high pressure (Krzyżak 2006).

17.2 DECREASED ATMOSPHERIC PRESSURE

17.2.1 Introduction

Large populations can inhabit locations at relatively high altitudes, quite often over 5000 m above sea level (e.g., in the Andes or the Himalayas); however, for those not acclimatised, prolonged exposure to such conditions may result in death, even if the persons are not particularly physically active. Intensive or even moderate physical exertion can prove a substantial burden for the human body even at average altitudes.

An analysis of the effect of a low atmospheric pressure on the human body reveals three main types of disorders, namely:

1. Disorders due to partial reduction in oxygen pressure (high-altitude hypoxia)
2. Disorders due to the body's response to excessive pressure gradients between the pressure of gases dissolved in tissues and those present in the closed cavities of the body and the ambient pressure (DCS, altitude meteorism)
3. Disorders due to the limited opportunities available to the middle ear and paranasal sinuses for ventilation under conditions of rapid pressure changes

One may also experience disorders resulting from 'explosive decompression', which is caused by decompression inside an aircraft cabin. Such disorders are the focus of interest of a very small group of specialists. Pressure in an aircraft cabin falls rapidly after dehermetisation, for example, if a window breaks or a hole appears. Disorders occur because the pressure balance inside the aircraft takes several seconds to reach the ambient pressure level (usually corresponding to the altitude of 10,000–12,000 m above sea level). If everyone uses oxygen masks, the flight can continue safely. Without oxygen, the time of conscious activity (the 'reserve time') lasts no longer than several seconds. If the crew members remain conscious, they may be able to lower the altitude relatively quickly, and the passengers have a chance to survive even if they cannot wear their oxygen masks. As a rule, the parents or caregivers of children first put their masks on, and next help children and others wear theirs (DeHart and Davis 2002; Ernsting et al. 1999; Kowalski 2002).

17.2.2 High-Altitude Hypoxia

An important cause of high-altitude hypoxia is the fall in partial oxygen pressure in the inhaled air. Partial oxygen pressure decreases at higher altitudes. The values for atmospheric pressure and partial oxygen pressure at the sea level are 760 and 159 mmHg respectively, whereas the air pressure in lung vesicles is only 103 mmHg. Oxygen pressure in lung vesicles is lower than that of the ambient air, because water vapour saturates the air inhaled into the lungs, exerting a pressure of 47 mmHg at a temperature of 37°C, and mixes with the residual air present in the lungs.

Partial oxygen pressure in the ambient air decreases as altitude increases, and is 126 mmHg at an altitude of 2000 m, 110 mmHg at an altitude of 3000 m, and 79 mmHg at an altitude of 5500 m (i.e. half the value present at sea level). Thanks to certain characteristics of the haemoglobin dissociation curve, even a substantial initial fall in oxygen partial pressure to a level corresponding to the altitude of about 3000 m does not cause a significant decrease in oxygen-saturated haemoglobin. For example, at an altitude of 2000 m vesicular pressure decreases from the baseline value of 103 mmHg to a value of 76 mmHg, while the oxygen saturation of haemoglobin is still at 90%. This value significantly decreases the physical capacity of the human body during substantial aerobic exertion. Individuals with a low physical capacity, especially those over 50 years of age who plan to undertake mountain trekking or working at such altitudes should undergo an appropriate medical examination. The altitude accepted as the threshold of hypoxia tolerance is 1500 m. Hence, on board modern airliners, the pressure is usually reduced to the value corresponding to that altitude. A prolonged exposure to altitudes above 5000 m

TABLE 17.1
ICAO Standard Atmosphere Parameters (a Simplified Draft)

Altitude (m)	Air Pressure (mmHg)	Air Pressure (hPa)	Partial Oxygen Pressure (mmHg)	Temperature (°C)
0	760	1013	159	15
1500	634	845	133	5
2500	560	746	117	−1.25
3500	493	661	103	−8
4500	433	557	91	−14
5500	379	505	79	−21
6500	330	440	69	−27
7500	287	382	60	−34
8500	267	356	56	−37

is virtually impossible without and sometimes even after acclimatisation. Although acclimatised individuals can survive at altitudes above 8000 m without oxygen apparatus, their stay cannot be prolonged; nonacclimatised individuals risk loss of consciousness at such altitudes. The high-altitude exposure symptoms described here classify hypoxia into categories by altitude: moderate hypoxia occurs from 2400 to 3200 m; severe hypoxia, 3200 to 5200 m; and extremely severe hypoxia, above 5200 m. Table 17.1 lists standard atmospheric parameters as prescribed by the International Civil Aviation Organization (ICAO) standards* at basic altitudes (Barański 1977; Ernsting et al. 1999; Kowalski 2002).

Depending on the duration of exposure to decreased pressure, hypoxia is classified as follows:

- *Acute hypoxia*: Exposure lasts from 1 minute to 2 hours (aircraft cabin dehermetisation)
- *Chronic hypoxia*: Exposure lasts from hours to even many years
- *Constitutional hypoxia*: Occurs in inhabitants of high mountains

Four altitude zones have been determined based on the effects of acute high-altitude hypoxia:

1. *Neutral*: Altitudes up to 1500 m
2. *Complete compensation*: Altitudes from 1500 to 4000 m, at which compensatory responses of the circulatory and respiratory system develop
3. *Incomplete compensation*: Altitudes from 4000 to 5000 m, at which the compensatory mechanisms can no longer compensate for the oxygen deficit of the human body

* ICAO standard atmosphere is based on the relationship between the pressure and the altitude, occurring at the temperature of 45° north latitude.

4. *Borderline of humans durable exposure*: Altitudes above 5500 m for nonacclimatised individuals

In chronic altitude hypoxia, disorders of different intensities and different effects may develop depending on the speed of ascent to high altitudes, the amount of physical effort, and the exposure duration. At altitudes above 2500 m, there is a risk of acute mountain sickness (AMS). Its initial phase is characterised by headaches, nausea, vomiting, and sleep disorders, while more severe hypoxia (at higher altitudes) is characterised by acute pulmonary or cerebral oedema. This advanced stage of the disease requires prompt intervention using a portable pressure apparatus and quick transport to a hospital at a low altitude, which may prove life saving.

People knew of AMS even in ancient times. It most frequently appears in the mild form. Symptoms usually occur after 12 hours of exposure to high altitudes, are most intense during the second or third day of exposure and usually subside after two subsequent days. Continuation of climbing may cause symptoms to recur and develop into a more severe form of the disease. Evacuation to lowland areas results in the abatement of almost all symptoms. The symptoms of chronic mountain sickness (Monge's disease or CMS), a constitutional condition, include neurological and mental disorders apart from changes in the circulatory, respiratory, and haematopoietic systems (Ward et al. 1989).

Acclimatisation to conditions of altitude hypoxia consists of multiple adaptive responses that develop in a specified period.

Acclimatisation to average altitudes may be partially regarded as similar to the process of acclimatisation to high altitudes. Adaptation to an altitude of 2300 m requires two weeks of constant exposure; for an ascent to about 4600 m, the adaptation period for each successive 600 m level lasts one week.

Some compensatory responses develop very early. These include (1) increase in respiratory activity resulting in hyperventilation, (2) increase in blood flow at rest, and (3) increase in blood flow during submaximal efforts. Hyperventilation results from a progressive stimulation of aortic arch chemoreceptors and carotid bodies through decreased partial oxygen pressure in blood; however, it causes a rise in partial oxygen pressure in the lung vesicles. During the early acclimatisation period, submaximal effort causes heart rate and cardiac output to increase, while the ejection capacity remains unaltered and the blood flow rate also increases. In the mountains, physically active individuals often experience moderate dehydration with a sensation of dryness of the lips, oral cavity, and larynx. They must control their body mass and increase water intake. Later, an altered level of base-acid balance and haematological and cellular changes will develop in response to the change (Birch et al. 2005).

17.2.3 Decompression Sickness under Hypobaric Conditions

When working underwater, a rapid pressure decrease of 50% is relatively safe as compared to the pressure to which the diver was previously subjected. For DCS under hypobaric conditions, it is important that the decrease in pressure from the atmospheric to lower levels is safe, if the difference does not exceed 35–40%.

In rapid ascents to an altitude of about 8000 m, symptoms are to be expected. Such circumstances may be due to sudden dehermetisation of an aircraft at high altitudes, and in simulated conditions—such as during examination in a low pressure test chamber—when the practice of breathing pure oxygen for desaturation, that is, to flush nitrogen lavage from the body, was not followed. There are also rare cases of severe DCS at altitudes above 7000 m that can result in death. DCS is a rare occurrence in climbers because the ascent to altitudes above 7000 m is long and punctuated by rest periods. On the other hand, the burden of a significant effort effectively reduces the asymptomatic stage of DCS. The excellent solubility of nitrogen, the major constituent of the blood in the vesicles that is responsible for the condition, in fats means obese individuals are more susceptible to DCS. The symptoms of DCS under hypobaric conditions are the same as the symptoms of DCS in divers. They always occur in the same sequence and are indicative of the severity of the condition. However, the symptoms and further consequences of the disease are not as severe as they usually are in divers' DCS. The pilots examined under conditions of explosive decompression (a sudden pressure change associated with ascent to high altitudes, e.g. from 3000 to 8000 m) with DCS symptoms undergo treatment in a recompression chamber (DeHart and Davis 2002; Ernsting et al.1999).

17.2.4 Barofunction Disorders in the Middle Ear and Paranasal Sinuses

The auditory tube ventilates and drains the middle ear as its chondromembranous part opens during mandibular movements, yawning, and swallowing. This results in equalisation of the pressure inside the ear with the ambient pressure, which is essential at high altitudes, for example, whilst working at high altitudes, during airline travel, or driving through a region with significant altitude differences. A change of environment from one of a higher pressure to another of a lower pressure usually causes no problems. With a decrease of 15 mmHg in external pressure, the air in the middle ear opens up the osseous part of the auditory tube. Problems occur when the external pressure increases, for example, while descending mountains, or even in descent by lift in high-rise buildings. In such cases, pressure should be balanced, for example, by swallowing saliva or moving the mandibles to open and ventilate the auditory tube path. While this is usually sufficient, at times the pressure inside the oral cavity needs to be raised by attempting forceful exhalation with the mouth closed and the nose blocked. These ailments, which are more or less intensified after flights, are not rare, and result from changes of pressure inside the ear. Attempts to balance the pressure should be made as soon as the first symptoms appear. If the pressure inside the ear exceeds the ambient pressure by 80–90 mmHg, the auditory tube will close and pressure balance will be possible only after the next reduction of external pressure, namely after the pressure differences are reduced, which is impossible under flight conditions.

When there are rapid changes in pressure under emergency conditions, an adequately high pressure gradient and pathological changes in the auditory tube or the surrounding tissues (polyps, or even mild rhinitis), the pressure is not balanced, leading to tympanic membrane rupture and damage of the internal ear. These disturbances are accompanied by strong sensation of pain and sometimes vertigo. Similar problems occur when the paranasal sinus pressure and the ambient air pressure are not balanced. If, due to conditions such as rhinitis or pathological changes in the paranasal sinuses, the pressures are not balanced while moving from an environment of low pressure to another of high pressure, the suction from the negative pressure in the sinuses can cause mucosal congestion, transudation, or submucosal haematomas (DeHart and Davis 2002; Birch et al. 2005; Kowalski 2002; Rayman et al. 2001).

REFERENCES

Barański, S., ed. 1977. *Aerospace Medicine*. Warsaw: PZWL.

Birch, K., D. MacLaren, and K. Georg. 2005. *Instant Notes Sport Exercise Physiology*. Garland Science/BIOS Scientific Publishers.

DeHart, R. L., and J. R. Davis, eds. 2002. *Fundamentals of Aerospace Medicine*. Philadelphia, PA: Lippincott Williams & Wilkins.

Dolatkowski, A., and K. Ulewicz. 1973. *Essentials of Diving Physiology*. Warsaw: PZWL.

Dolmierski, R., S. R. Kwiatkowski, and J. Nitka. 1981. Neurological, psychiatric and psychological examination of divers in the light of their professional work. *Bulletin of the Institute of Maritime and Tropical Medicine* 32:141–152.

EN 14153. 2003. Recreational diving service. Safety related minimum requirements for the training scuba divers.

Ernsting, J., A. N. Nicholson, and D. J. Rainford, eds. 1999. *Aviation Medicine*. Oxford: Butterworth-Heinemann.

Klukowski, K., ed. 2005. *Medicine of Accidents in Transport*. Warsaw: PZWL.

Konarski, M., R. Kłos, and R. Olszański. 2005. Submarine crew rescue operations—Decompresion procedures. In *Medicine of Accidents in Transport*, ed. K. Klukowski. Warsaw: PZWL.

Kowalski, W., ed. 2002. *Aviation Medicine—Selected Problems*. Poznan: WLOP.

Krzyżak, J. 2006. *Diving Medicine*. Poznan: KOOPgraf.

Olszański, R., and P. Siermontowski. 2002. *ABC Diver's Health*. Gdynia: Drukarnia.

Rayman, R. B., J. D. Hastings, W. B. Kruyer, and R. A. Levy. 2001. *Clinical Aviation Medicine*. New York: Castle Connolly Graduate Medical Publishing, LLC.

Ward M. P., J. S. Milledge, and J. B. West. 1989. Chronic mountain sickness (CMS): Monge's disease. In *High Altitude Medicine and Physiology*, 405–412. London: Chapman and Hall Medical.

18 Mechanical Hazards

Krystyna Myrcha and Józef Gierasimiuk

CONTENTS

18.1 INTRODUCTION

Workers come in contact, directly or indirectly, with a variety of work-related objects that may be considered sources of potential mechanical hazards—mainly injuries—when in operation. Examples of such objects include tools, machines, equipment or instrumentation at workstations, raw materials, half-finished products, finished products, and so on. Mechanical hazards are a very important, unique group of physical hazards generated by the work process, whose consequences are instantaneous because of a combination of different aspects. Human beings have had to contend with hazards since the advent of modern industrial civilisation. The level of intensity of the negative consequences of hazards increases with the speed, mass, and other factors of the work equipment. Table 18.1 presents recent data on accidents at work in Poland (published by The Central Statistical Office [CSO]).

TABLE 18.1
Accidents at Work in Poland (2005–2008)

Accidents at Work	Year			
	2005	2006	2007	2008
Total number	84,402	95,465	99,171	104,402
Accidents caused by mechanical factors	72,508 (85.9%)	81,420 (85.3%)	84,328 (85.0%)	82,133 (78.6%)

Source: Central Statistical Office.

18.2 GENERAL CHARACTERISTICS OF MECHANICAL HAZARDS

Mechanical hazards are the physical attributes of various objects likely to cause injuries to human beings, for example, machines or their parts, tools, machined or treated objects, and forcefully ejected solid or liquid materials. The aforementioned elements pose basic mechanical hazards such as the possibility of crushing, smashing, cutting or cutting off, entanglement and friction or abrasion, being drawn in, trapped, impacted, punctured or stabbed, as well as high pressure fluid ejection, slipping, or tripping and falling. These hazards can occur during both normal, recommended machine operation as specified by the designer and/or manufacturer or due to the malfunction of the machine.

Examples of possible origins of mechanical hazards include the following:

- Mobile machinery and transported objects
- Moving parts
- Cutting parts, sharp edges, protruding parts
- Falling objects
- High-pressure fluids
- Rough, slippery surfaces
- Means of access (stairs, stepladders, ladders, walkways, access openings)
- Workstation positions relative to the ground (working at heights or depths)
- Other sources of hazards, for example, animals

Some examples of mechanical hazards and the parameters that influence the corresponding occupational risk are presented in Table 18.2.

18.3 TERMS RELATED TO MECHANICAL HAZARDS AND THEIR DEFINITIONS

Hazards: Potential sources of harm. The term ‘hazard’ can be qualified by its origin (e.g. mechanical hazard, electrical hazard) or the nature of its potential harm (e.g. electric shock hazard, cutting hazard, toxic hazard, fire hazard). By this definition, hazards are either permanently present

TABLE 18.2
Examples of Mechanical Hazards

Diagrams of Machine Components	Examples	Mechanical Hazards	Basic Parameters that Influence Occupational Risk
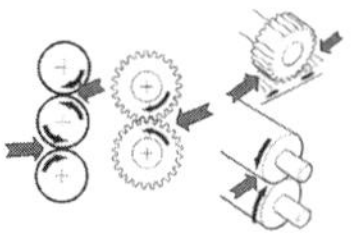	Toothed gears, rolling mills, roller conveyors	Crushing, smashing, drawing in	Torque, inertia (mass and speed), shape, dimensions, surface condition, accessibility
	Sawing machines: band saws, circular saws	Severing, cutting, ejection	Cutting and feeding speeds, tool mounting, strength
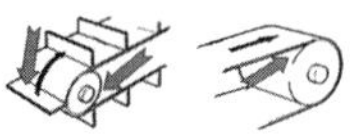	Conveyors: drags, belts	Trapping, impact, crushing, drawing in	Torque, dimensions, speed, shape
	Belt pulleys, screws	Drawing in, impact	Torque, inertia (mass and speed), shape, accessibility
	Spindles, drill, chucks	Drawing in	Torque, inertia (mass and speed), shape, surface condition, accessibility

Source: CIOP. 2002. *Machinery and Technical Equipment. Means of Protection Against Mechanical Hazards*. Warsaw: CIOP.

during the intended use of the machine (e.g. motion of moving components, the electric arc during welding, poor posture, noise emission, high temperatures) or could arise unexpectedly (e.g. explosions, accidental crushing resulting from an unintended or unexpected start up of machinery, ejection consequent to a breakage, or fall consequent to acceleration or deceleration).

Harm: Physical injury or damage to health.

Hazardous situations: Circumstances in which a person is exposed to at least one hazard. The exposure could result in immediate harm (e.g. mechanical hazard) or over an extended period (e.g. acoustic hazard).

Hazard zone (danger zone): Any space in and/or around machinery where a person is exposed to a hazard.

Risk: A combination of the probability of an occurrence of harm and its severity.

Adequate risk reduction: Risk reduction at least in accordance with legal requirements prescribed and consistent with the current state of the art.

Risk assessment: Overall process comprised of a risk analysis and a risk evaluation.

Protective measures: Measures intended to achieve risk reduction, implemented by:

- The designer of the machine or of its components (creating an inherently safe design, safeguarding and complementary protective measures, and information for use)
- The user (in organisations, by instituting safe working procedures, supervision, permit-to-work systems, additional safeguards, use of personal protective equipment, and/or training)

Information for use: Protective measures such as precautionary communication devices (e.g. texts, words, signs, signals, symbols, diagrams) used separately or in combination to convey information to users.

Safeguards: Guards or protective devices.

Guards: Physical barriers designed as part of a machine to provide protection to users and others in its vicinity. Depending on their construction, guards can be called casings, shields, covers, screens, doors, or enclosing guards. A guard may act

- Alone: it is then only effective when it is 'closed' for a movable type or 'securely held in place' for a fixed type.
- In conjunction with an interlocking device with or without a locking guard to ensure protection no matter the position of the guard.

Fixed guards: Guards affixed in a manner such that they can only be opened or removed with the use of tools or by destroying the fixing device (e.g. with screws, nuts, welding).

Movable guards: Guards that can be opened without the use of tools.

Adjustable guards: Fixed or movable guards that are adjustable as a unit or incorporate facilities for adjustment of part(s). The adjustment remains fixed for the duration of a particular operation.

Interlocking guards: Guards associated with interlocking devices; along with the control system of the machine, they perform the following functions:

- The hazardous functions of the machine 'covered' by the guard cannot operate until the guard is closed.
- If the guard is opened when the potentially hazardous functions of the machine are in operation, a stop command is given.
- When the guard is closed, the potentially hazardous functions of the machine 'covered' by the guard can operate, but the closure of the guard alone does not by itself start the hazardous machine.

Interlocking guards with guard locks: Guards associated with an interlocking device and a guard-locking device; along with the control system of the machine, they perform the following functions:

- The hazardous functions of the machine 'covered' by the guard cannot operate until the guard is closed and locked.
- The guard remains closed and locked until the risk due to the hazardous functions of the machine 'covered' by the guard ceases.
- When the guard is closed and locked, the hazardous machine functions 'covered' by the guard can still operate, but the closing and locking of the guard does not by itself start the hazardous machine functions.

Interlocking guard with start function (control guard): A special form of interlocking guard, which, once it has reached its closed position, gives a command to initiate the hazardous function of the machine without the use of a separate start control.

Protective devices: Safeguards other than guards.

Interlocking device, interlock: Mechanical, electrical, or other type of device designed to prevent the operation of hazardous functions of the machine under specified conditions (usually when the guard is not closed).

Safety function: A function of a machine whose failure could result in an immediate increase of risk(s).

Emergency stop: A function that is intended

- To avert arising or to reduce existing hazards to persons and damage to machinery or to work in progress.
- To be started by a single human action.

18.4 IDENTIFICATION OF MECHANICAL HAZARDS AND ESTIMATION OF RELATED OCCUPATIONAL RISKS

Mechanical hazards, together with the dangerous situations they can bring about, are identified based on an analysis of the operations being undertaken as well as the ways they are conducted. This includes both the time period that a person is present within the danger zone and the probability of contact with factors that pose mechanical hazards during normal operations as specified by the designer and/or manufacturer of the work equipment. This analysis is combined with an analysis of the probability that distortions will occur during the aforementioned operation, including potential consequences.

The analysis should cover the following:

- General features of the workstation; for example, location, equipment, and arrangement
- Types of operations and tasks performed by the worker(s), the methods they employ, and the time they take to perform
- Environmental conditions that could affect hazards at the particular workstation
- Information on the occurrence of accidents and potentially dangerous events
- Identification of possible sources of injuries and other types of damage to health
- Circumstances under which hazardous situations could occur
- Safety and health information that is available to workers

An analysis of general features of a workstation should specify its type and location in the plant (e.g. hall, umbrella roof, outdoor location) and its distance from other objects in the buildings is recorded. It should identify all the machines and devices, hand tools, installations, and other equipment used by the worker and their arrangement and location relative to the operator, the material and product storage rooms, and other items such as lubricants, coolers, or washing means.

Workers' methods of operation and their working habits at the workstation should be studied based on operation sheets, safety records, and health instructions. The results should be verified under real work conditions, taking a 'picture of the day' or using other records, including workers' output, details of their daily routines including handling of the equipment entrusted to each one's care, and the time allocated for and repetitiveness of each operation.

Environmental conditions that could elevate the occupational risk level related to mechanical hazards (e.g. improper illumination, dustiness) and/or pose other risks at the workstation (e.g. slippage due to unsealing of central heating installation, impact of objects ejected from a neighbouring workstation) should also be analysed.

Accidents and potentially dangerous events at a given workstation or at similar ones in the plant or other plants following similar manufacturing processes must be investigated to analyse the causes and circumstances of their occurrence.

The findings that emerge from such probes enable identification of conditions and possible conflicts between the operator and workstation elements (see Section 18.2, points 1–9), that is, the events that may lead to the worker's exposure to the mechanical hazard. Other effects that these elements may have upon a human and the influence of environmental conditions during normal operation, including possible disruptions, must also be considered.

The main factors that affect the occurrence of hazardous situations are as follows:

- Location of danger zones relative to the working area
- Type, shape, and roughness of the component's surfaces that a worker could encounter (cutting tools, sharp protruding edges, and so on) in the course of his or her work
- Relative position of objects and components which when in motion could produce crushing points (e.g. chains and belt transmissions, toothed gears)
- Likelihood of impact of the component or object in question upon the worker
- Kinematic energy of the elements in controlled or uncontrolled motion
- Potential energy of moving parts due to gravitational force, spring elements, or high-pressure and negative-pressure fluids

Risk is assessed based on the analysis of the hazardous situation for each mechanical hazard. It is estimated in terms of the probability of harm (injury or other health damage) and its severity. The probability of occurrence of harm depends on the following factors:

- Frequency and/or duration of exposure to the hazard
- Possibility of hazardous events occurring
- Possibility of avoidance or reducing the harm; this depends on how rapidly the hazardous event could lead to harm (suddenly, quickly, or slowly)
- Practical experience and knowledge of workers, especially their awareness of risk and ability to avoid or limit harm (e.g. reflexes, agility)

The following characteristics can be used to estimate and categorise the probability of occurrence of harm originating from mechanical hazards:

- Low
    - When frequency and exposure duration is low (occurrence less often than once in 6 months)
    - When the probability of the hazardous event occurring is low, that is, less than once a year
    - When the harm can be easily avoided
- High
    - When exposure occurs daily and the duration exceeds 2 hours
    - When the probability of the hazardous event occurring is very high, for example, a daily event
    - When the ability of harm avoidance is limited or the harm cannot be avoided at all
- Medium (all cases between high and low)

The following characteristics can be used to estimate and categorise the severity of harm:

- *Minor*: When the consequences do not involve any sick leave, for example, slight injuries requiring no more than first aid
- *Moderate*: A significant injury or illness requiring more than first aid, but the worker is able to return to the same job
- *Serious*: Death or permanent disabling injury or illness (e.g. amputation)

Based on the aforementioned factors and using the three-level scale given in Table 18.3, one can determine the level of risk associated with a given mechanical hazard. The basic criterion for acceptable risk is conformity to the requirements specified in the corresponding regulations and standards.

According to standard PN-N-18002 (2000), occupational risk is evaluated as follows:

- High risk is considered unacceptable; that is, for current work, immediate action should be taken to reduce the risk, while planned works should be started only after the reduction of occupational risk.
- Medium and low risks are considered acceptable, but medium risk still leaves room for improvement.

TABLE 18.3
Risk Matrix

Severity of Harm	Probability of Occurrence of Harm		
	High	Medium	Low
Serious	High risk	High risk	Medium risk
Moderate	High risk	Medium risk	Low risk
Minor	Medium risk	Low risk	Low risk

The results of risk assessment can be used to determine the actions necessary to eliminate mechanical hazards or reduce associated occupational risk.

18.5 MECHANICAL HAZARDS PRESENT IN VARIOUS TECHNOLOGICAL AND PRODUCTION PROCESSES

In practice, mechanical hazards are likely to occur in all manufacturing and production processes. Some hazards are industry-specific and occur in certain sectors and depend upon the machinery that is employed and the equipment that is used at workstations, as well as operational and work conditions. Mechanical hazards are important, especially in the building industry, metallurgical and metal products manufacturing, machine manufacturing and devices, transportation and storage, and trade and repair services.

18.6 REDUCTION OF RISKS RELATED TO MECHANICAL HAZARDS

Measures to eliminate or reduce risks related to mechanical hazards include the following:

- Suitable design (inherently safe design)
- Safeguards to protect persons from mechanical hazards that cannot be eliminated
- Additional measures to decrease the risks related to mechanical hazards

18.6.1 Elimination or Reduction of Risks by Design

Mechanical hazards can be minimised, if not eliminated, at the design stage of the machines and workstation equipment and at the stage of workstation organisation. At these stages, the designer decides the shapes and dimensions of objects as well as their layout and relative locations, kinematic parameters, and other factors affecting risk levels. Workers should also be given the opportunity to avoid dangerous situations or at least mitigate their consequences.

The shapes and dimensions of the design's elements influence the shape and dimensions of the danger zone. For example, where two rollers come into in contact with each other, the zone where material is drawn in and trapped assumes the form of a wedge whose acute angle depends on the roller diameters. The greater the acute angle of the wedge the higher is the risk of drawing in and trapping a body part such as an arm, leading to a crushing injury; the greater the acute angle, the lengthier the danger zone (Figure 18.1).

Flat, continuous, and smooth surfaces, small dimensions, and low speeds in machinery reduce the risk of trapping body parts or clothes and thereby reduce the probability of harm. Therefore, the designer should choose uniform wheels over wheels with arms, as well as choose lower speeds and mass of elements. This will reduce the kinematic energy of those elements and decrease the force, which is important in case those elements come in contact with a human body (Table 18.4 and Figure 18.2).

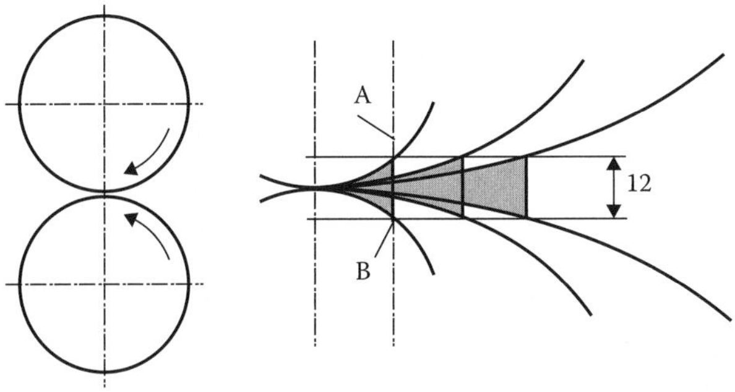

FIGURE 18.1 Mechanical hazard zone drawn for contacting rollers of different diameters. (Reprinted with permission from CIOP. 2002. *Machinery and Technical Equipment. Means of Protection Against Mechanical Hazards*. Warsaw: CIOP.)

TABLE 18.4
Maximal Values for Parameters of Objects in Contact with Parts of Human Body

Parameter	Maximal Values	
	Scenario 1	Scenario 2
Maximum force acting upon a part of a body	75 N	150 N
Maximum kinematic energy of a moving part	4 J	10 J
Maximum pressure a body is subject to	50 N/cm^2	50 N/cm^2

Source: CIOP. 2002. *Machinery and Technical Equipment. Means of Protection Against Mechanical Hazards*. Warsaw: CIOP.

The values given in the scenario 1 column in Table 18.4 should not be exceeded when the component acting upon the human body does not always return automatically (Figure 18.2a).

When the automatic return of the component is ensured, for example, after releasing a foot control or supplying the door with pressure sensitive edges or bars (Figure 18.2b), higher pressure may be applied; however, the pressure should not exceed the values given in the scenario 2 column in Table 18.4.

Eliminating mechanical hazards should focus on zones of possible contact between the moving objects and human body and should adhere to safety distances prescribed to prevent accidental severing, crushing, trapping, and so on. An example of a design made to eliminate the risk of severing a body part is shown in Figure 18.3. In the helical conveyor case, breaking the helical surface over a distance of S_1 (corresponding to the part of the body exposed to hazards) prevents it from crushing or severing body parts.

According to standard PN-EN 349+A1 (2008), the minimum gap relative to an arm should be at least 120 mm. However, the aforementioned procedure can be

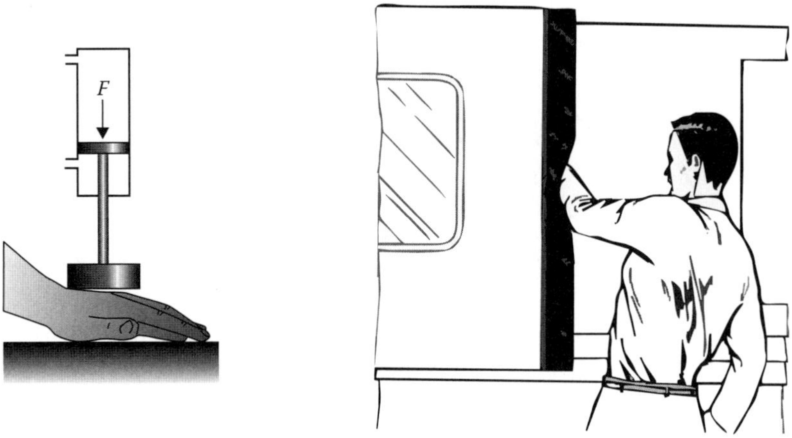

(a) (b)

FIGURE 18.2 Examples of protection from crushing by limiting force and energy. (Reprinted with permission from CIOP. 2002. *Machinery and Technical Equipment. Means of Protection Against Mechanical Hazards*. Warsaw: CIOP.)

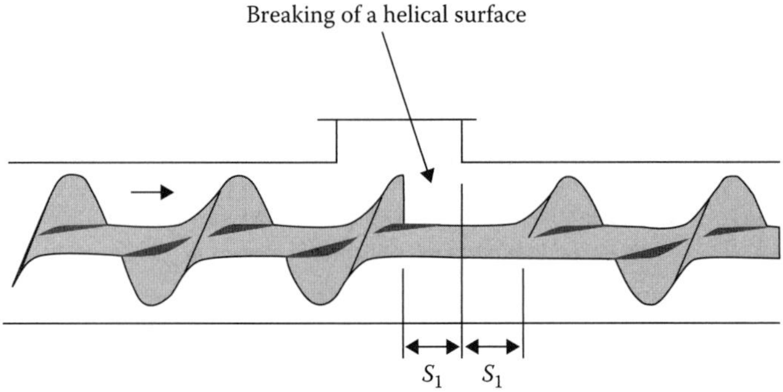

FIGURE 18.3 Design of a helical conveyor to eliminate crushing or severing hazards, for $S_1 \geq 120$ mm. (Reprinted with permission from CIOP. 2002. *Machinery and Technical Equipment. Means of Protection Against Mechanical Hazards*. Warsaw: CIOP.)

followed only in cases in which the properties of the transferred material allow it, that is, if the material does not block the conveyor—then the material itself would pose a hazard—or the material does not heat up, posing a new hazard.

Figure 18.4 illustrates the minimum gaps required between the fixed and moving part or between two moving parts to avoid crushing parts of the human body. These gaps are specified in the standard PN-EN 349+A1 and are based on anthropometrical measurements of the human body clothed in working attire. When the design process does not ensure that these gaps can be maintained, measures must be taken to prevent humans from entering the crushing zone, for example, by making the zone inaccessible.

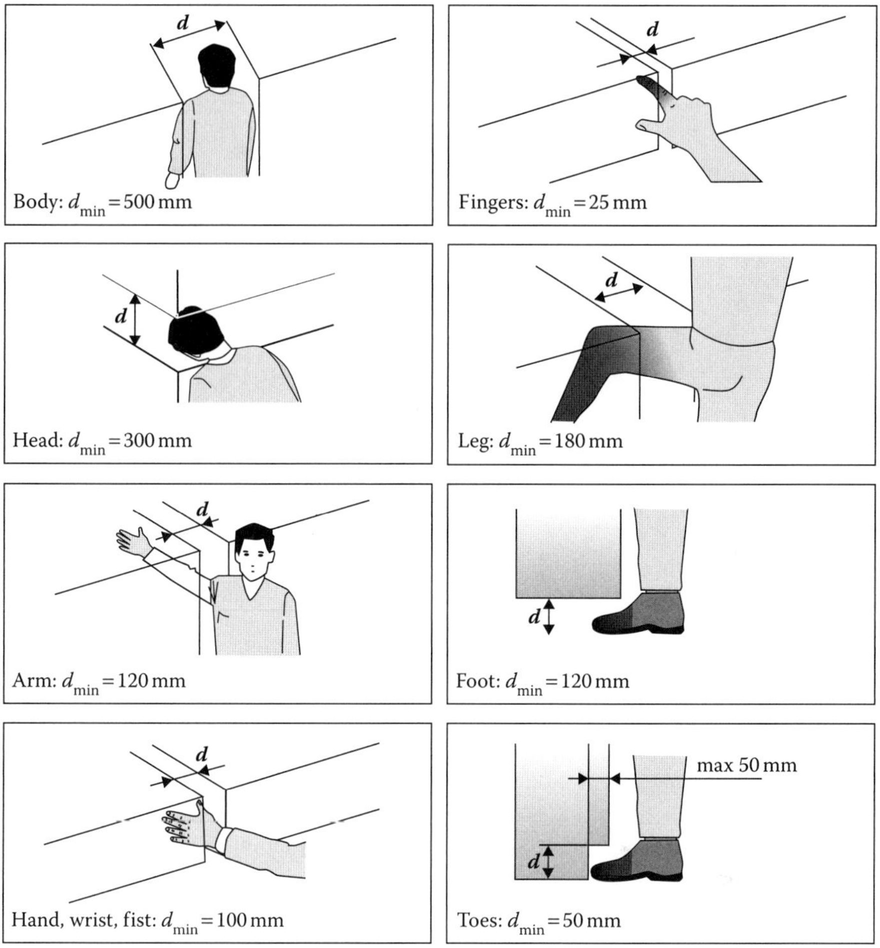

FIGURE 18.4 Minimum gaps required to prevent crushing. (Reprinted with permission from PN-EN 349+A1. 2009. Safety of machinery—Minimum gaps to avoid crushing of parts of the human body.)

Designs in which the mechanical hazard zone is located outside the area that can be reached by a human body part, for example, the arms (Figure 18.5), are most effective at eliminating the likelihood of injuries. Therefore, equipment manufacturers should aim to ensure that hazard zones do not overlap with areas of human activity.

Slippery, uneven, or wobbly surfaces most often are the cause of falls and the associated injuries and also raise the risk levels associated with movement by vehicle. Unevenness and roughness of a surface are caused mostly by the movement of vehicles, and often occur due to careless unloading and the use or misuse of chemicals. Therefore, the friction coefficient (preventing slippery conditions) and resistance to impact are the most important factors to be considered in flooring design and when choosing means of transportation.

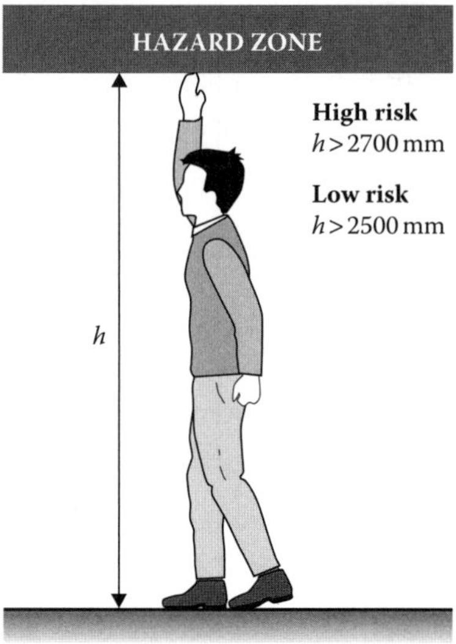

FIGURE 18.5 Location of the hazard zone outside the reach of a human arm. (Adapted with permission from PN-EN ISO 13857. 2008. Safety of machinery—Safety distances to prevent hazard zones being reached by upper and lower limbs.)

Eliminating mechanical hazards also requires activities that prevent disruption to normal machine operations and functions of workstation equipment. Disruptions can include structure damage, excessive loading, a rise in pressure or speed of rotation, and accidentally starting machines.

Damage to the structure or other reasons for disruption to the normal operation of work equipment can be prevented mainly by

- Not exceeding stress and strain limits; the strength of the components (including the coefficients of safety) should depend on the type and parameters of the load, taking into consideration the effects of the conditions of operation. If the element is crucial for safety, that is, slings, ropes, cabins, and roll-over or falling-object protective structures, the strength should be verified experimentally.
- Utilising devices that protect against the disruption of normal operations such as lifting capacity limiters, safety valves, shear pins, safety diaphragms, trip or lifting range limiters, and so on.

The consequences of disturbances to the normal operation of machines and other equipment include cracks, breaks, excessive deformation, or loosening or slackening of components. Such consequences can generate hazards that are difficult to identify, for example, overloading a crane may result in breaking a rope or crane jib or even crane roll over.

18.6.2 Safeguards

Safeguards can be classified into two main groups as follows:

- Physical (material) barriers separate the range of human reach in the working zones from the hazard zones, that is, the operational range of components that could generate mechanical hazards. These barriers are mainly guards that protect against contact with moving parts (especially in machines) and guard rails that protect personnel and objects from falls. Guards should also stop or capture materials, machined objects or energy, for example, chips, liquids, dust, vapours, gases, noise, or others that may be ejected, fall or be emitted from a machine. Mechanical hazards should first be considered when choosing a guard, also keeping in mind other hazards associated with the working process.
- The protective devices discussed in Chapter 28 can also be used for this purpose.

18.6.2.1 Guards

18.6.2.1.1 Requirements

Guards should meet the following requirements:

- Should be of robust construction
- Should not create additional hazards
- Should be situated at appropriate distances from the danger zones
- Should not be easy to defeat or to render nonoperational
- Should offer adequate viewing of the production process
- Should enable the performance of essential work for the installation and/or replacement of tools and for maintenance purposes by allowing access to the area where the work must be carried out, if possible without removing the guard or dismantling the protective device

These requirements also apply to other safeguards. These requirements must be maintained and checked periodically over the whole lifetime of the machines and other equipment.

18.6.2.1.1.1 Guard Strength The strength of the guards and their components is crucial; in many cases it determines their ability to protect. The guard should be able to withstand expected impacts from machine parts, machined objects, broken tools, ejected materials (both solid and liquid), unintended impacts, and so on. Guards, especially those affecting specified safety distances, should remain rigid and resist deformation and, if possible, withstand loads when used as ladders. The type C standards detail the strength requirements for guards and other protective devices. In the case of woodworking machines they apply to the various parts, for example, the riving knife or pawls.

In the case of a lack of the type C standard for the guard considered, its strength should be obtained from calculations including the foreseeable loads upon them. The strength should then be verified experimentally. The negative influences of factors such as thermal radiation, chemicals used in the current or proposed manufacturing processes, and environmental factors like ultraviolet radiation and corrosives, should be taken into consideration.

18.6.2.1.1.2 Elimination of Likely Hazards Posed by Guards The design of guards should ensure the following:

- There are no sharp edges, corners, or other protrusions.
- They do not cause injuries by pressure, impact of moving parts, and so on (pressures and force exerted should not exceed the values noted in Table 18.4).
- They do not create a crushing, shearing, or trapping environment.

18.6.2.1.1.3 Safety Distances Standard PN-EN ISO 13857 (2008) specifies the distances, called safety distances, required to prevent accessibility to the danger zone when reaching an arm through a regular opening (Table 18.5) and when reaching a lower limb through a regular opening (Table 18.6). Standard PN-EN 999+A1 (2008) is devoted to the proper positioning of protective equipment.

Figure 18.6 gives distances for particular cases in which the access of the operator's lower limbs is impeded when the operator remains in a standing position without additional support. The distances l relate to the height h from the ground or reference plane to the protective structure. These distances, however, are not safety distances, hence other safety measures might be required to prevent access.

To prevent a hazard zone from being reached over a guard (Figure 18.7 and Table 18.7), the relationship between guard height b, location of the danger zone (indicated by height a), and safety distance c, respectively, are specified in standard PN-EN ISO 13857 (2008) and should be maintained. The standard specifies safety distances for high- and low-risk levels, respectively (Table 18.7 gives distances for high risk). The table does not get additional values using the interpolation method. When dealing with intermediate values, those ensuring higher safety levels should be adopted.

18.6.2.1.1.4 Reducing Scope to Remove, Deactivate or Defeat Some parts of guards and the guards themselves should be fitted to the machine in a manner that allows their disassembly or removal only by use of tools. The ability to remove guards should be highly restricted, that is, discouraged by limiters, inverted hinges, and so on. The removal of guards can also be prevented by applying interlocking devices, which cannot be defeated using simple methods such as using adhesive tape or a moving connecting rod (pin) in the position of the closed guard, while in fact the guard remains open. The use of plane switches with a specified configuration or a unique key that corresponds to a given guard significantly reduces the possibility of defeating interlocking devices. Standard PN-EN

TABLE 18.5
Safety Distances To Be Maintained When Stretching the Arms through Regular Openings; Persons of 14 Years of Age and Above (Dimensions in Millimetres)

Part of Body	Illustration	Opening	Safe Distance, s_r		
			Slot	Square	Round
Fingertip	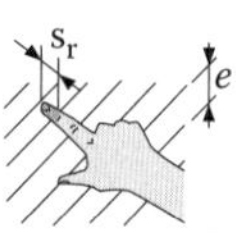	$e \leq 4$ $4 < e \leq 6$	≥2 ≥10	≥2 ≥5	≥2 ≥5
Finger up to knuckle join	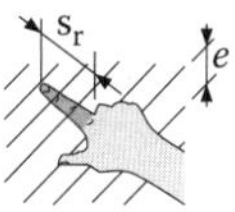	$6 < e \leq 8$ $8 < e \leq 10$	≥20 ≥80	≥15 ≥25	≥5 ≥20
Hand	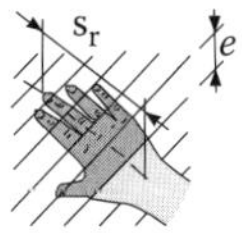	$10 < e \leq 12$ $12 < e \leq 20$ $20 < e \leq 30$	≥100 ≥120 ≥850[a]	≥80 ≥120 ≥120	≥80 ≥120 ≥120
Arm up to junction with shoulder	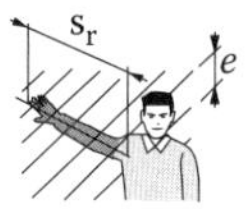	$30 < e \leq 40$ $40 < e \leq 120$	≥850 ≥850	≥200 ≥850	≥120 ≥850

Note: Bold lines in the table outline the part of the body restricted by the opening size.

[a] If the length of the slot opening is < 65 mm, the thumb will act as a stop and the safety distance can be reduced to 200 mm.

Source: Reprinted with permission from PN-EN ISO 13857. 2008. Safety of machinery—Safety distances to prevent hazard zones being reached by upper and lower limbs.

1088+A2 contains requirements for minimizing the possibilities for defeat of an interlocking device.

18.6.2.1.1.5 Visibility The guard should not limit visibility, especially of elements of which observation is required for carrying out technological processes. Where possible, the guards should be made of transparent materials that retain this property for lifetime, or an opening should be made in the guard to allow for observation. Obstructed visibility is a frequent cause of guard bypass.

18.6.2.1.1.6 Preserving Necessary Access without Disassembling Guards Guards should allow for the access necessary to change equipment or tools or to make adjustments to the machine, without a need for disassembling. Guard design should allow, for example, one to open the guard in a way that enables operations to continue while it remains connected to the machine.

TABLE 18.6

Safety Distances To Be Maintained When Stretching the Lower Limbs through Regular Openings; Persons of 14 Years of Age and Above (Dimensions in Millimetres)

Part of Body	Illustration	Opening	Safety Distance, s_r	
			Slot	Square or Round
Toe tip	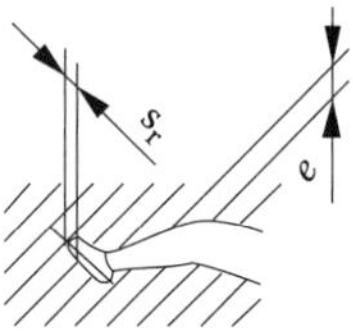	$e \leq 5$	0	0
		$5 < e \leq 15$	≥10	0
Toe		$15 < e \leq 35$	≥80[a]	≥25
Foot	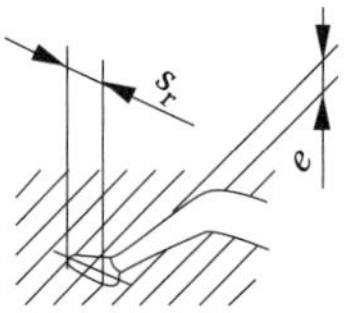	$35 < e \leq 60$	≥180	≥80
		$60 < e \leq 80$	≥650[b]	≥180
Leg (toe tip to knee)	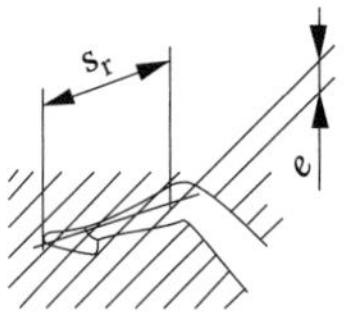	$80 < e \leq 95$	≥1100[c]	≥650[b]
Leg (toe tip to crotch)	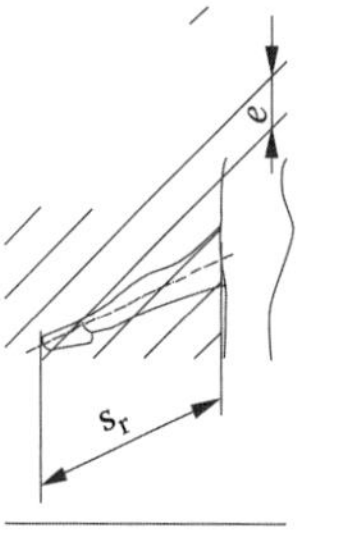	$95 < e \leq 180$	≥1100[c]	≥1100[c]
		$180 < e \leq 240$	Not allowed	≥1100[c]

Note: Slot openings with $e > 180$ mm and square or round openings with $e > 240$ mm will allow access for the whole body.

[a] If the length of the slot opening is ≤ 75 mm, the distance can be reduced to ≥50 mm.

[b] The value corresponds to a leg (tip of the toe to knee).

[c] The value corresponds to a leg (tip of the toe to crotch).

Source: Reprinted with permission from PN-EN ISO 13857. 2008. Safety of machinery—Safety distances to prevent hazard zones being reached by upper and lower limbs.

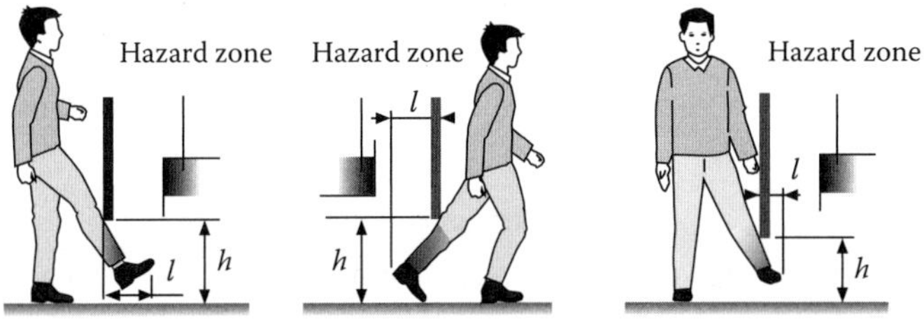

Height up to the protective structure h, in mm	Distance for impedance l, mm		
	Case 1	Case 2	Case 3
$h \leq 200$	≥ 340	≥ 665	≥ 290
$200 < h \leq 400$	≥ 550	≥ 765	≥ 615
$400 < h \leq 600$	≥ 850	≥ 950	≥ 800
$600 < h \leq 800$	≥ 950	≥ 950	≥ 900
$800 < h \leq 1000$	≥ 1125	≥ 1195	≥ 1015

FIGURE 18.6 Impeding free movement of lower limbs under protective structures. (Adapted with permission from PN-EN ISO 13857. 2008. Safety of machinery—Safety distances to prevent hazard zones being reached by upper and lower limbs.)

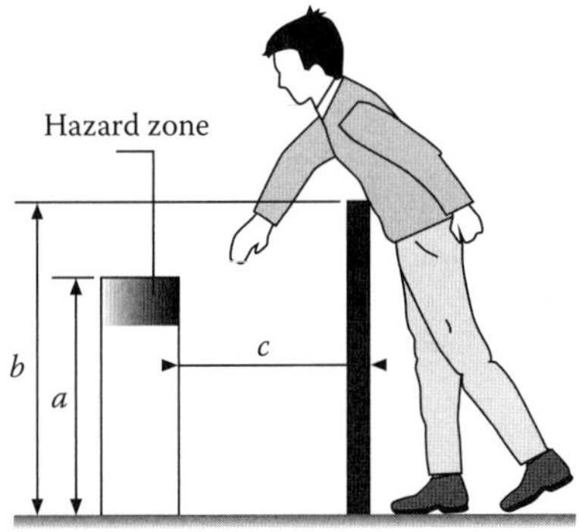

FIGURE 18.7 Safety distance c from the hazard zone of height a at which a protective structure of height b should be located. (Adapted with permission from PN-EN ISO 13857. 2008. Safety of machinery—Safety distances to prevent hazard zones being reached by upper and lower limbs.)

18.6.2.1.1.7 Classifications of Guards Guards can be divided into different types based on the following categories: the ways in which they can be mounted and operated, the possibility of adjustment, the type of material (solid, openwork), and the percentage of the danger zone that is protected. Schematic views of these types are presented in Figure 18.8. Figures 18.9 through 18.11 show examples of danger zone guard design.

The guard can perform its functions as follows:

- By itself (without an interlock): The guard is effective only in the 'closed' position; for a fixed guard 'closed' means 'connected with the mounting points'.

TABLE 18.7
Safety Distance *c* (in Millimetres) When Stretching Over Protective Structures—High Risk

Height of Hazard Zone 'a'	Height of Protective Structure b[a]									
	1000	1200	1400[b]	1600	1800	2000	2200	2400	2600	2700
	Horizontal Safety Distance to Hazard Zone 'c'									
2700	0	0	0	0	0	0	0	0	0	0
2600	900	800	700	600	600	500	400	300	100	0
2400	11,000	1000	900	800	700	600	400	300	100	0
2200	1300	1200	1000	900	800	600	400	300	0	0
2000	1400	1300	1100	900	800	600	400	0	0	0
1800	1500	1400	1100	900	800	600	0	0	0	0
1600	1500	1400	1100	900	800	500	0	0	0	0
1400	1500	1400	1100	900	800	0	0	0	0	0
1200	1500	1400	1100	900	700	0	0	0	0	0
1000	1500	1400	1000	800	0	0	0	0	0	0
800	1500	1300	900	600	0	0	0	0	0	0
600	1400	1300	800	0	0	0	0	0	0	0
400	1400	1200	400	0	0	0	0	0	0	0
200	1200	900	0	0	0	0	0	0	0	0
0	1100	500	0	0	0	0	0	0	0	0

[a] Protective structures less than 1000 mm in height are not included because they do not sufficiently restrict movement of the body.

[b] Protective structures lower than 1400 mm should not be used without additional safety measures.

Source: PN-EN ISO 13857. 2008. Safety of machinery—Safety distances to prevent hazard zones being reached by upper and lower limbs.

Associated with the interlocking device so that, together with the control system of the machine, all three requirements specified in the definition of an interlocking guard are satisfied. The interlocking guard can be also associated with a guard-locking device to ensure the requirements specified in the definition of an interlocking guard with locking are satisfied.

The interlocking device should be made either of reliable parts (for incidental access) or with redundant components and/or automatic monitoring (for frequent access). The interlocking devices can be used with both moveable guards and fixed guards, especially in cases when they are often disassembled when changing the equipment, for example, the side guards of the work zones of presses.

All contacts of interlocking and locking devices should open effectively. Effective opening occurs, for example, when

- The switch contacts separate as a direct result of a specified movement of the switch actuator through non-resilient element due to inelastic elements, for example, independent of springs (Figure 18.12a).

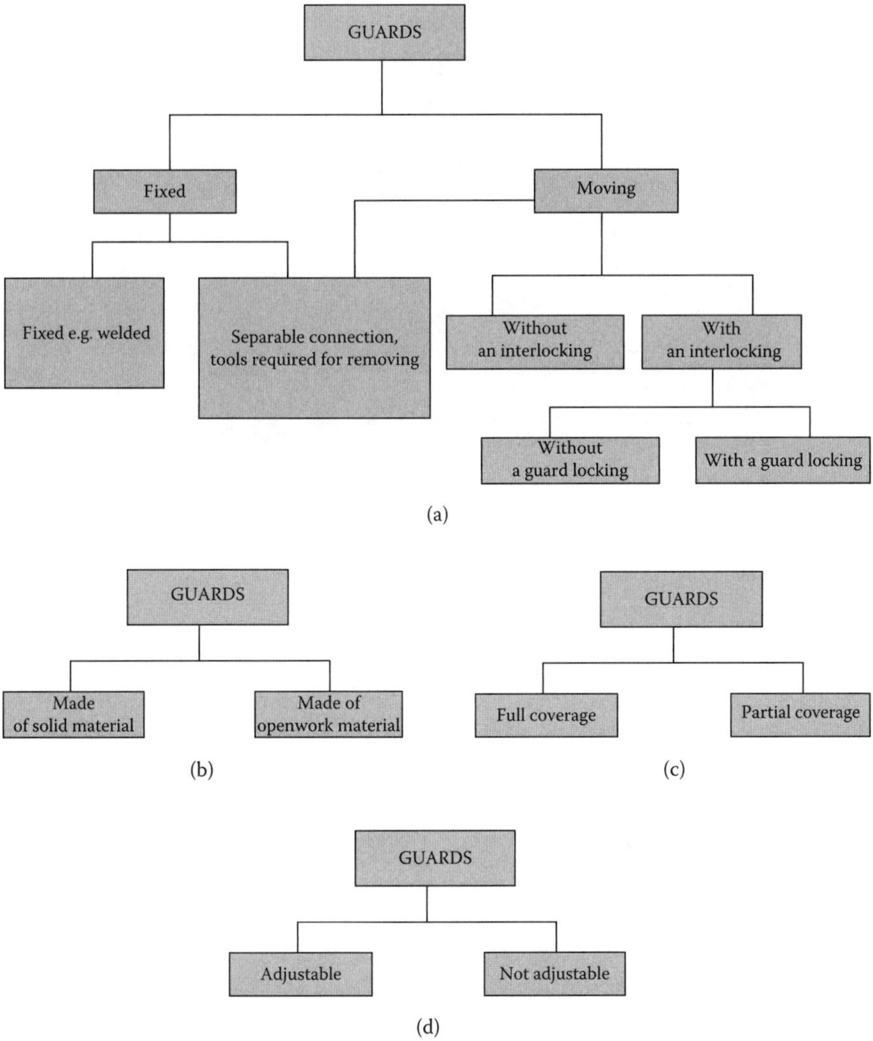

FIGURE 18.8 Types of guards classified by (a) mounting and operation, (b) the type of material they are made of, and (c) coverage of the dangerous zone and (d) adjustability. (Adapted with permission from Koradecka, D., ed. 1999. *Occupational Safety and Ergonomics*. Warsaw: CIOP.)

- Sensors monitoring the guard location (Figure 18.12b) are actuated in a positive mode, that is, actuated inevitably by other components in motion, either by direct contact or through rigid elements. The sensor switching opens the circuit and cuts off the power supply if a contact is stuck or broken, if a spring breaks or if the driving element is dirty. There is no such possibility in the nonpositive mode.

FIGURE 18.9 Movable guard for the working zone of a sawing machine and a riving knife for protection from ejected materials. (Reprinted with permission from CIOP. 2002. *Machinery and Technical Equipment. Means of Protection Against Mechanical Hazards.* Warsaw: CIOP.)

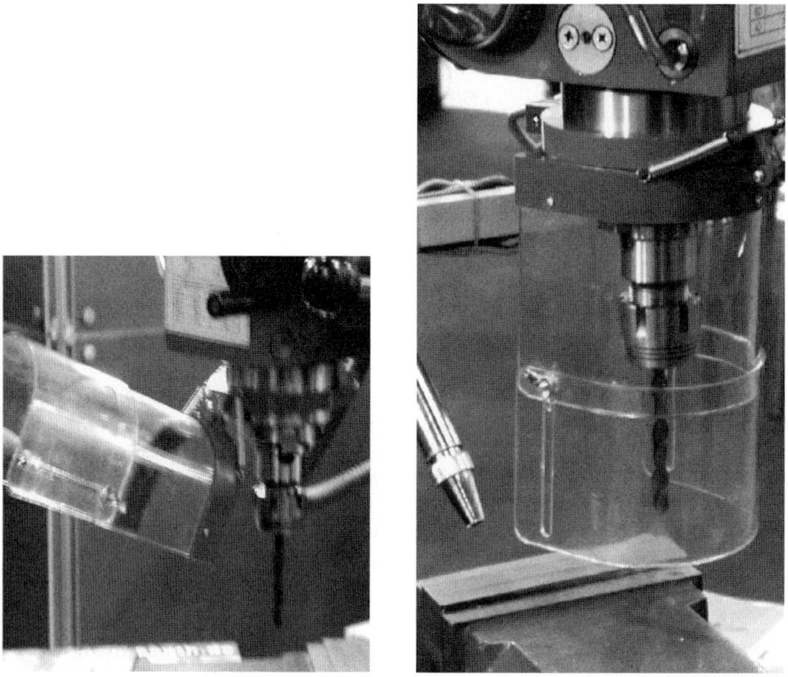

(a) (b)

FIGURE 18.10 Moving adjustable guard of the working zone of a drilling machine in two positions: (a) open and (b) closed.

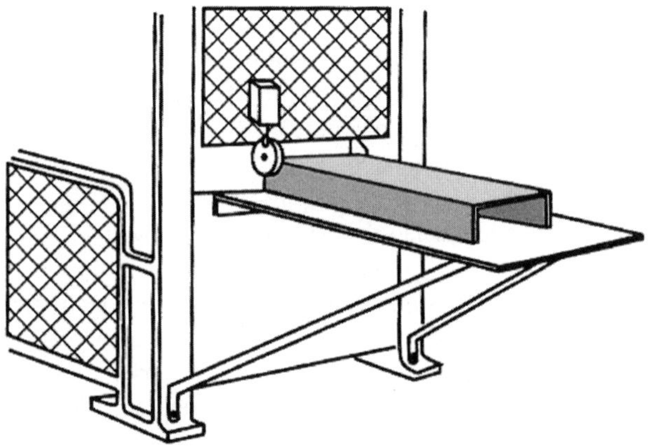

FIGURE 18.11 Movable guard with an interlock and fixed guards made of openwork material.

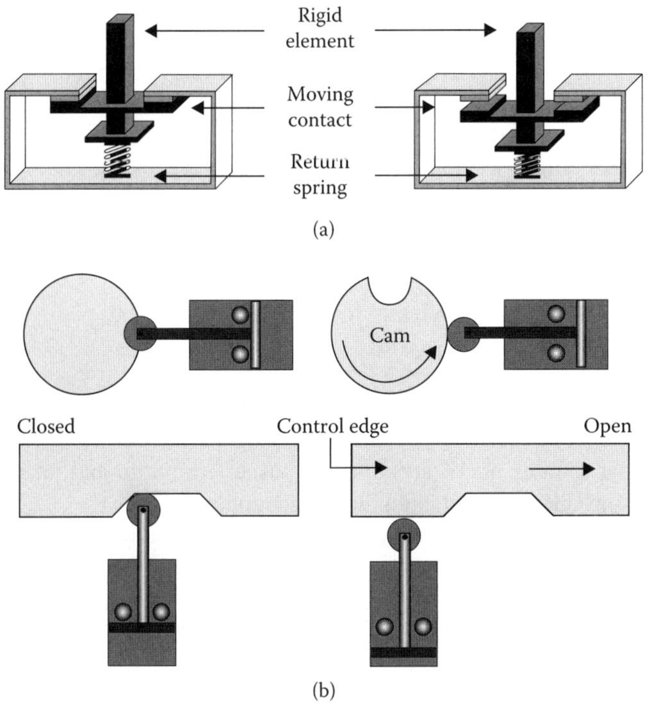

FIGURE 18.12 Examples of positive (a) opening of contacts and (b) mode of actuation of sensors. (Reprinted with permission from CIOP. 2002. *Machinery and Technical Equipment. Means of Protection Against Mechanical Hazards.* Warsaw: CIOP.)

18.6.2.1.2 Principles for Choice of Guards

The basic criteria for choosing a guard are as follows:

- The hazards generated by the machine together with the associated risk of operating the machine as intended, taking into consideration the possibility of misuse
- How frequently the equipment protected by the guard will need to be accessed

The tasks the machine performs significantly influences the type of guard that should be utilised. For example, if heat generated in the course of operation must be dissipated, the guards should be made of openwork material; if the manufacturing processes, however, generate dust or chemical hazards, solid guards allowing for mounting of extraction devices should be used. When the occupational hazard in the zones monitored by guards is high, access to those areas should be possible only after the risk has been reduced or eliminated. Interlocking guards with locking, or at least interlocking guards, are therefore be the best choice. If accessibility is essential, the recommendations given in Table 18.8 should be followed. Standard PN-EN 953+A1 (1999) provides guidance on choosing guards.

18.6.2.2 Other Physical Barriers

Guard rails are most often used for protection against falls off of something, while covers are for protection against persons falling into recesses, openings, or hollows. Where openings cannot be covered for technical reasons (e.g. in the course of building construction), or when the covers are opened or removed, danger zones should be fenced with guard rails. A guard rail is a hand rail located at a height of 1.1 m (except for scaffoldings, where the height should be 1 m), a knee rail located at half-height, and a stanchion at 0.15 m. Examples of guard rail applications are depicted in Figure 18.13.

18.6.3 Other Protective Means to Reduce Mechanical Hazard-Related Risks

18.6.3.1 Protective Means to Reduce Exposure to Hazard

If an operator should not directly penetrate a danger zone, he or she can use simple manual devices such as pliers, suckers, and pushers to put in and take out components and to introduce elements into the danger zone.

The need for a human presence in the working process within dangerous zones can be eliminated or limited by:

- Mechanisation and automation of operations
- Use of fault diagnosis systems
- Use of designs and methods of operation that prolong intervals between adjustments, lubrications, and other maintenance activities as well as between repairs

TABLE 18.8
Choice of Guards When There Is a Need for Access to Hazard Zones

	Type of Guard	Need for Access
Fixed	Without interlocking device	Guards being seldom removed (e.g. once per month) Guards of power transmission elements, for example, belt transmission and gear transmission
	With interlocking devices incorporating one switch with positive opening operation or two nonmechanically actuated position switches (inductive, magnetic)	Guards being removed and mounted more often (e.g. once a day) in the course of operations such as, adjustment, changing tools, when there arises the risk of unexpected start Side guards of presses, guards of drive wheels in a band sawing machine, and so on
Moveable	With interlocking devices incorporating one switch with positive mode actuation and with positive opening operation or two nonmechanically actuated position switches (inductive, magnetic)	Guard opened often (e.g. 10 times a day) Driving guards, guards of adjustment zones, and so on
	With interlocking devices incorporating two switches with positive opening operation and with positive mode actuation	Guards opened very often (e.g. a few times each hour) Guards for machines with manual loading and unloading performed within the hazard zone or close to it and so on
	With locking devices	Employed when the stoppage duration is longer than the access time Machines with rotating elements of high inertia, robotised devices
	Adjustable guards	Employed when the access to moving elements should be limited within the zones having openings, for example, for tools or products Guards of band sawing machines, circular saws, and so on

Source: CIOP. 2002. *Machinery and Technical Equipment. Means of Protection Against Mechanical Hazards*. Warsaw: CIOP.

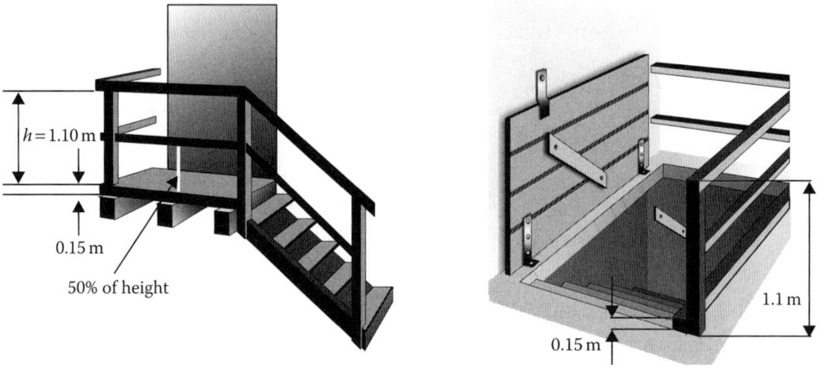

FIGURE 18.13 Applications of guard rails.

Automation may entail the application of robots, manipulators, conveyors, blowers, and so on. Mechanisation may entail the application of skids, feeders, unwinding drums, and so on. The application of those these devices should not present any new hazards to the humans who are present near these devices and machines

18.6.3.2 Other Means of Reducing Risks Associated with Mechanical Hazards

Other means of reducing risk associated with mechanical hazard are as follows:

- Means for safe access to machinery (working platforms and walkways, stairs, stepladders, and ladders according to PN-EN ISO 14122 parts 2, 3, and 4) for operation purposes and safe access openings to the maintenance points according to PN-EN 547+A1 parts 1, 2, and 3
- Attachment devices for easy and safe transfer of machines and their heavy parts (e.g. hooks, hook-type catches, eyebolts, guides for forks of forklifts)
- Means of isolation from all power supplies (e.g. locked isolating units in the isolating position) and for dissipating the accumulated energy, for example, in springs, pressure containers
- Means to mitigate the consequences of hazardous events, like emergency stop devices and devices to rescue persons trapped inside machines
- Means of warning about hazards using light and sound signalling devices, safety colours and signs, warning texts located directly on machines and other workstation equipment. Health and safety instructions should be made available to workers and made comprehensible for them; these also should include suggestions on safe practices in foreseeable atypical events that may occur in the course of work
- Use of protective clothes and shoes by workers to minimise mechanical hazards; for example, tight-fitting denim overalls, buttoned up sleeves, and suit legs reduce the possibility of trapping

Personal protective equipment should be supplied when these protective measures do not ensure reduction of risk to the acceptable level. It can protect against slips and falls, falling objects, and sharp edges (see Chapter 26).

REFERENCES

CIOP. 2002. *Machinery and Technical Equipment. Means of Protection Against Mechanical Hazards*. Warsaw: CIOP.

Koradecka, D., ed. 1999. *Occupational Safety and Ergonomics*. Warsaw: CIOP.

Myrcha, K., and Gierasimiuk, J. 2007. Mechanical hazards. In *Occupational Risk. Methodology of Assessment*, ed. W. M. Zawieska, s. 307–345. Warsaw: CIOP-PIB.

National Safety Council. 1988. *Accident Prevention Manual for Industrial Operations Engineering and Technology*. 9th ed. Chicago: National Safety Council.

Ordinance of the Minister of Labour and Social Policy of September 26th, 1997, on the general occupational health and safety provisions. Uniform text. DzU 2003 no. 169, item 1650; amended by DzU 2007, no. 49, item 330.

PN-EN 1088+A2. 2008. Safety of machinery—Interlocking devices associated with guards—Principles for design and selection.

PN-EN 999+A1. 2008. Safety of machinery—The positioning of protective equipment in respect of approach speeds of parts of the human body.

PN-EN 349+A1. 2009. Safety of machinery—Minimum gaps to avoid crushing of parts of the human body.

PN-EN 953+A1. 2009. Safety of machinery—Guards—General requirements for the design and construction of fixed and movable guards.

PN-EN 547+A1. 2009. Safety of machinery—Human body measurements— Parts 1, 2, and 3.

PN-EN ISO 14122. 2006. Safety of machinery—Permanent means of access to machinery—Parts 2, 3, and 4.

PN-EN ISO 13857. 2008. Safety of machinery—Safety distances to prevent hazard zones being reached by upper and lower limbs.

PN ISO 3864-1. Graphical symbols—Safety colours and safety signs—Part 1: Design principles for safety signs in workplaces and public area.

19 Biological Agents

Jacek Dutkiewicz

CONTENTS

19.1 INTRODUCTION

Biological risk agents are micro- and macro-organisms and the structures and substances they produce; they exert a harmful influence upon human organisms and lead to occupational diseases in the work environment (Dutkiewicz et al. 1988; Dutkiewicz et al. 2002; Dutkiewicz 2006). This comprehensive definition has been approved by experts in this field and encompasses not only infectious agents, but also allergens, biological toxins, and external parasites.

Biological agents in the work environment pose a significant and mounting problem to public health. On a global scale, at least several hundred million people are estimated to be exposed to these risks at work. In some occupational groups in Poland (farmers, foresters, health service workers), most diseases classified as occupational in nature are caused by biological agents of an infectious, allergic, or immunotoxic nature (Wilczyńska et al. 2006; Dutkiewicz 1998). Many diseases caused by biological agents, particularly in agriculture, are not diagnosed properly. This situation has shown gradual improvement with the development of the biomedical sciences, particularly molecular biology and immunology, which enable identification and characterisation of an increasing number of agents that often pose serious threats to the health of the working population.

19.2 PREVALENCE AND MEANS OF TRANSMISSION

Biological agents emerge from human, animal or plant organisms or their surfaces, as well as from various elements of the external environment such as soil, water, sewage, waste, manure, litter, oils, wood and dust, and air. Harmful biological agents in the work environment that are transmitted through dust and droplets in the air are of the greatest epidemiological significance. Germs, toxins, or allergens contained in the bioaerosol droplets or dusts and inhaled penetrate the bronchi and pulmonary alveoli, causing infectious, allergic, or immunotoxic diseases. They may also infiltrate the human body through the conjunctivae, the epithelium of the nasopharynx, or the skin (Cox and Wathes 1995; Lacey and Dutkiewicz 1994; Rylander and Jacobs 1994).

Harmful biological agents could also be transmitted through water, soil, infected objects (e.g. syringes and instruments used at health service facilities), infected animals (including haematophagous insects and arachnids), or through animal and plant products. Most often, they penetrate the human body through the epidermis of the limbs when there is direct contact (Dutkiewicz 2006).

Oral transmission and the occupational diseases caused by biological agents that penetrate the body via this route are of the least epidemiological significance and should be considered exceptional occurrences resulting from gross negligence of hygiene standards, such as lack of running water and clean eating areas at work, rather than from real exposure.

19.3 INFLUENCE ON THE HUMAN BODY

Under conditions of occupational exposure, biological agents may exert infectious, allergic, toxic, irritative, and carcinogenic actions. The most significant infectious and invasive diseases are those caused by viruses that are present among the health service workforce and diseases transmitted from animals to humans (also known as zoonoses), caused by bacteria, viruses, fungi, protozoa, and worms and common to farmers, foresters, employees of agriculture and the food industry, and related occupations (Dutkiewicz and Jabłoński 1989).

Allergic diseases caused by biological agents mostly occur among farmers and those engaged in other professions exposed to organic dust, plants, and animals. These are usually diseases of the respiratory system, which include hypersensitivity pneumonitis (also called extrinsic allergic alveolitis [EAA]), asthma and allergic rhinitis, and a specific variation caused by pollen known as pollinosis (Donham and Thelin 2006; Dutkiewicz 2006). Biological agents may also cause allergic skin diseases such as urticaria and contact eczema (also called allergic contact dermatitis), airborne dermatitis, and allergic conjunctivitis.

Many biological agents present in the workplace cause toxic reactions in humans, which most often manifests as skin inflammation, for example from toxic substances present in some plants or from venom injected by the bites of ticks or some small mites. Microorganisms that are inhaled with dust and the substances produced by them (endotoxin, peptidoglycan, glucans, mycotoxins) cause a specific toxic reaction in the lungs called an immunotoxic action (Burrell 1995). This may result in impairment of some components of the lungs' immune system, such as the alveolar macrophages, caused by mycotoxin action (Samson et al. 1994). However, a much more frequent response is excessive stimulation of the lungs' immune system caused by the activation of alveolar macrophages and release of cytokines and other inflammatory mediators (Burrell 1995); this was the case in a recently described disease called organic dust toxic syndrome (ODTS; Donham and Thelin 2006; Dutkiewicz 2006; Rylander and Jacobs 1994; Dutkiewicz 1998).

Irritation is a relatively mild form of reaction to biological agents, often mechanical in nature. Most often, it manifests as mucous membrane irritation (MMI) due to the biological components of dust.

Only a small number of biological agents in the work environment show a carcinogenic action. Prolonged inhalation of wood dust may lead to nasal adenocarcinoma (Maciejewska et al. 1993); mycotoxins also have carcinogenic action (Mikotoksyny 1984). A chronic hepatitis infection may lead to cancer and cirrhosis.

19.4 REVIEW OF BIOLOGICAL AGENTS IN THE WORK ENVIRONMENT

Biological agents are found among all living organisms, from the lowest (prions, viruses) to the highest (mammals and allergens produced by them). Biological hazards are most often divided into four groups based on the risk level for the employees exposed: group 1, no real hazard; group 2, moderate hazard; group 3, serious hazard; and group 4, very serious hazard, including death (Dutkiewicz et al. 2002; Directive 2000/54/EC; DzU 2005, no. 81, item 716).

19.4.1 Prions

Prions are infectious, mutated protein particles. They cause transmissible spongiform encephalopathy both in humans (Creutzfeldt–Jakob disease [CJD]) and in animals (bovine spongiform encephalopathy [BSE]). Farmers may possibly contract the disease from cattle, particularly the so-called 'variant' CJD, however, this hypothesis has not been confirmed by indisputable evidence (Dutkiewicz 2004a).

19.4.2 Viruses

Viruses of human origin, particularly hepatitis B (HBV; Figure 19.1) and C (HCV), pose a very serious epidemiological threat to the health service workforce and employees of laboratories (Beltrami et al. 2000; *Wirusowe zapalenie...* 1996). These viruses are most often transmitted through blood, as well as through serum and other bodily fluids. The late consequences of HBV infection may include primary carcinoma and cirrhosis. Protective vaccinations play a significant role in prevention of HBV infections; however, there is no effective vaccine to protect humans against HCV infection. Among the most frequently occurring occupational diseases caused by biological agents that cause infectious and parasitic diseases, viral hepatitis occupies the second place in Poland, after borreliosis; out of 151 cases recorded in 2005, 105 were HCV infections, and 46 were HBV infections (Wilczyńska et al. 2006). However,

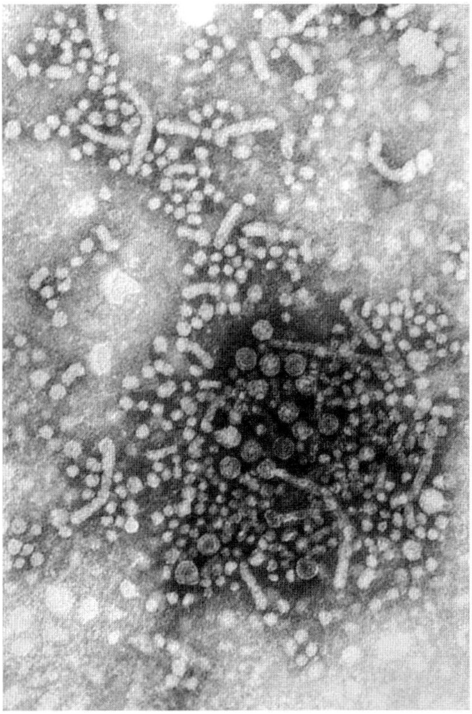

FIGURE 19.1 Transmission electron micrograph of hepatitis B virus (HBV; magnification × 150,000). (Photo by Betty Partin.)

the positive results of mass vaccinations of these vulnerable groups of people with a recombined yeast vaccine have generated some optimism, as has the increased availability of disposable syringes, which allows reduction of the viral reservoir.

The AIDS-causing human immunodeficiency virus (HIV) is another virus that is transmitted through blood. The virus dies very quickly in the external environment, so the probability of occupational infection by confirmed contact, such as by a scratch with a needle contaminated with blood of an infected patient, is estimated to be only around 0.3% (for HBV, it is 6%–30%, and HCV around 1.8%), and the number of documented cases of occupational infections throughout the world is low (Beltrami et al. 2000).

Viruses transmitted through the air and droplets cause another threat, to which the health service workforce, tutors, and teachers are exposed; these are adenoviruses, reoviruses, influenza viruses, respiratory syncytial virus (RSV), herpes simplex virus (HSV), and rubella virus. The virulent coronavirus (SCoV) poses a particularly serious threat to the health service workforce; it caused a rapid spread of the epidemic referred to as severe acute respiratory syndrome (SARS) in southeastern Asia in 2002–2003 (Dutkiewicz 2004a).

Diseases caused by viruses of animal origin usually following a mild course. The dominant group consists of pox-like skin diseases and influenza-like infections among farmers, caused by viruses found among ruminants (cattle, sheep). These viruses include contagious ecthyma, cowpox, pseudocowpox, bovine papular stomatitis, bovine vesicular stomatitis, and foot and mouth disease. The Newcastle virus causes acute conjunctivitis among poultry breeders and poultry-processing plant employees that is referred to as Newcastle conjunctivitis (Gliński and Buczek 1999).

Two groups of viruses found in wild rodents pose a serious threat for farmers: hantaviruses (Puumala, Hantaan, Seul, Sin Nombre) and arenaviruses (Junin, Guanarito, Machupo, Sabia, Lassa). People are infected by inhaling dust contaminated with the excrement of infected rodents or through broken skin; symptoms may include acute hemorrhagic fever (Dutkiewicz 2004a). Infection by viruses originating from African monkeys—Marburg, Ebola, and Simian B virus—could result in very serious consequences, including death; these diseases occur most often among the health service workforce and employees of laboratory facilities, vivaria, and zoos. The rabies virus poses a threat to veterinary medicine staff, employees of fox farms, vivaria, and zoos, and foresters. When the threat is particularly high, preventive vaccinations are administered (Gliński and Buczek 1999).

Among the viruses transmitted by haematophagous arthropods, the Central European tick-borne meningo-encephalitis virus poses the most significant threat to Polish foresters and farmers. The *Ixodes ricinus* and *Dermacentor reticulatus* ticks found in broadleaf and mixed forests transmit the infection. The most effective way to prevent the disease is protective vaccination administered to groups exposed to the threat (Skotarczak 2006; Cisak 2003; Dutkiewicz 1998).

The avian influenza virus, a subtype of influenza A virus (H5N1; Figure 19.2), is a new threat to poultry breeders, poultry-processing plant employees, and veterinary medicine personnel. This virus, which originated in southeastern Asia, causes

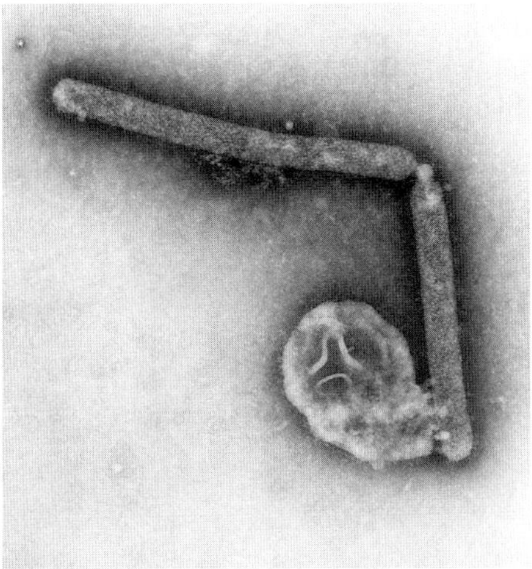

FIGURE 19.2 Transmission electron micrograph of avian influenza A virus (H5N1; magnification × 108,000). (Photo by Cynthia Goldsmith and Jackie Katz.)

serious, often lethal symptoms in humans. Other subtypes of avian influenza result in a moderately severe or mild infection.

19.4.3 Bacteria

Rickettsiae and chlamydiae are very small bacteria that also cause occupational diseases of animal origin. These include Q fever rickettsia (*Coxiella burnetii*), found mostly in sheep and also in cattle, rodents, and ticks. The bacteria may cause a polymorphic occupational disease known as Q fever. The psittacosis germ (*Chlamydia psittaci*) is found in many captive-bred and wild birds, including ducks, hens, pigeons, turkeys, and parrots. In professions requiring contact with birds, this bacterium causes a disease known as ornithosis, which takes the form of interstitial pneumonia and is accompanied by high fever (Gliński and Buczek 1999; Parnas 1960). In the 1990s, the list of biological agents posing risks to foresters and farmers included *Ehrlichia* and *Anaplasma* rickettsiae transmitted by ticks, which proliferate in the cytoplasmic vacuoles of leukocytes, usually granulocytes or monocytes. They cause ehrlichiosis (anaplasmosis), a multiple-system disease that is difficult to diagnose due to its nonspecific symptoms. In Poland, granulocytic anaplasmosis (ehrlichiosis) occurs and is caused by the bacterium *Anaplasma phagocytophilum* and transmitted by *Ixodes ricinus* ticks (Skotarczak 2006; Cisak 2003).

Borrelia burgdorferi (Figure 19.3), a spirochete also transmitted by *I. ricinus* ticks, causes Lyme borreliosis, a polymorphic disease; the initial phase usually takes the form of annular erythema migrans. Later, it may cause alterations of the joints, the central nervous system, the heart, and the vision organs (Skotarczak 2006; Cisak 2003). In Poland, borreliosis is a growing, serious health problem among foresters and farmers. It is among the most prevalent occupational diseases caused by infectious and parasitic biological agents; the number of cases in 2005 was 333, or 54.1% of all cases in this group of diseases (Wilczyńska et al. 2006). The most significant means of preventing borreliosis are appropriate protective clothing, tick repellents, and quick and skilled removal of ticks with tweezers or a special suction cup. This is a particularly important step, as the tick introduces germs into the blood circulation system as late as 24–48 hours after it attaches to a host.

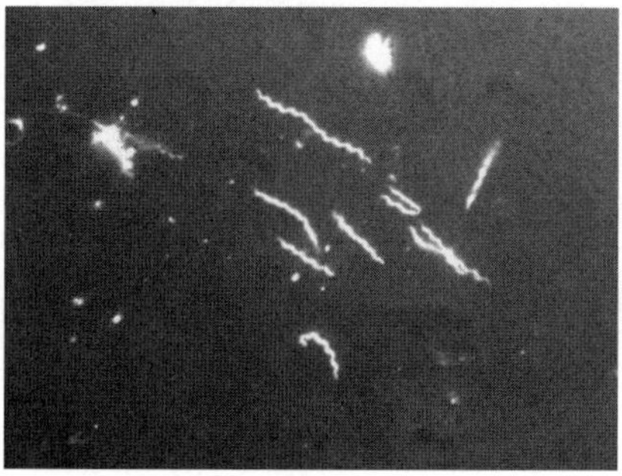

FIGURE 19.3 Light microscope of Borreliosis spirochete (*Borrelia burgdorferi;* magnification × 2000).

Leptospirae are aerobic spirochetes, now included in a single type species (*Leptospira interrogans*) of about 200 serovars. They are found in various wild and captive-bred animals and can cause acute diseases with a high fever (leptospiroses). The course of the disease depends upon the serovar that causes the infection (Parnas 1960).

Among gram-negative bacteria that cause occupational diseases of animal origin, the most significant are the *Brucella* species, which are responsible for the following animal diseases: cattle brucellosis (*Brucella abortus*), swine brucellosis (*Brucella suis*), and Malta fever (*Brucella melitensis*). In Poland, the most serious problem is *B. abortus*, which causes human brucellosis, a severe multiorgan disease typically occupational in nature and prevalent mainly in the western and northern parts of the country. The number of cases of brucellosis in Poland has fallen from 150–200 annually in the 1980s to 23–32 in 2000–2002 (Stojek 2003; Dutkiewicz 1998). The gram-negative vibrios *Campylobacter jejuni* and *Campylobacter foetus* may be the occupational risk factors in animal breeders as well as employees of slaughterhouses and meat-processing plants (Stojek 2003). In humans, these bacteria cause campylobacteriosis, which manifests as enteritis and gastritis.

The bacterium *Francisella tularensis* is found in many species of rodents; in Poland, mainly hares. It poses a threat to employees of game-processing plants and foresters, as well as to personnel of sugar factories; during the preliminary processing (rinsing, cutting) of beets originating from foci of tularaemia, workers could be exposed to inhalation of aerosols containing the excrement of infected field rodents (Dutkiewicz and Jabłoński 1989). In humans, the disease can take the form of ulceroglandular, glandular, oculoglandular, pulmonary, and gastrointestinal tularaemia (Stojek 2003; Dutkiewicz 1998).

Bacteria such as *Legionella pneumophila* (Figure 19.4) and other species of *Legionella* may cause an infectious disease not of animal origin in persons exposed to an occupational risk of inhalation of aerosol from warm water (20°C–50°C), sewage or moist soil—the typical biotopes of this bacterium. The disease may take the form of severe pneumonia or a milder Pontiac fever. *Salmonella* bacteria, often found in sewage, are a potential threat to employees of sewage treatment plants and municipal sewage facilities.

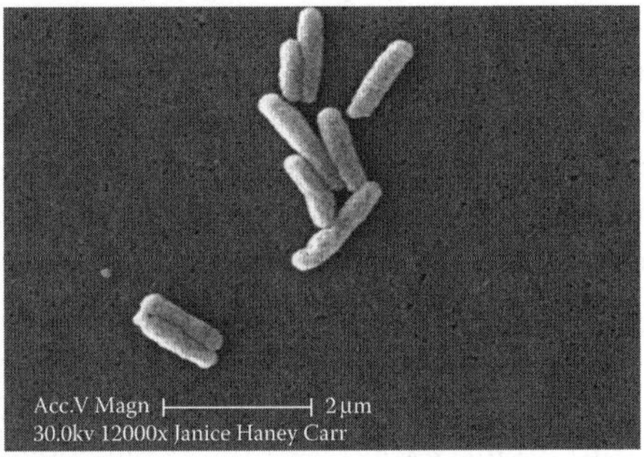

FIGURE 19.4 Scanning electron micrograph of *Legionella pneumophila* (magnification × 12,000). (Photo by Janice Haney Carr.)

Gram-negative rods of plant and animal origin are common in organic dusts and can cause allergic diseases. They also produce endotoxins, which cause immunotoxic diseases such as ODTS. The epiphytic species *Pantoea agglomerans* is a particularly significant pathogen (other names are *Erwinia herbicola, Enterobacter agglomerans*) commonly found on the surface of many plants, particularly cereal grains. These bacteria are invariably found in grain dust and cause EAA in farmers and other persons who may inhale this dust at their workplaces. They also produce a potent endotoxin (Dutkiewicz and Jabłoński 1989).

Endotoxins are biologically active lipopolysaccharides (LPS) that are integral components of the outer membrane of the cell walls of gram-negative bacteria. Structurally, they are macromolecules, resulting from the polymerisation of smaller LPS units with proteins and phospholipids of the cell wall. Fragmentation of the cell walls easily releases them into dust in the air in the form of microvesicles of 30–50 nm (Dutkiewicz 2004a). The concentration of endotoxins in air contaminated with organic dust is high and often exceeds the presumptive threshold level: 2×100 ng/m^3 (0.2 μg/m^3), which according to another scale of conversion, is equivalent to about 2000 endotoxin units (EU) in 1 m^3 of air (Dutkiewicz and Jabłoński 1989; Dutkiewicz 2006). Endotoxins inhaled by humans together with dust cause a nonspecific activation of lung macrophages, which release numerous substances known as inflammatory mediators. Symptoms include an inflammatory lung reaction, fever, gas exchange disorders, and bronchospasm (Burrell 1995; Rylander and Jacobs 1994).

Erysipelothrix rhusiopathiae is a gram-positive, rod-shaped pig pathogen. In humans, it causes skin wound infections, mainly in form of painful, red erythema. An occupational disease prevalent at quite high rates among butchers, slaughterers, and pig breeders, it has no severe effects. The rod-shaped bacterium of listeriosis (*Listeria monocytogenes*) can cause an occupational disease of animal origin, found among many species of captive-bred and wild animals as well as in silage and other fodders, soil, and water. It is a polymorphic disease whose most dangerous forms are encephalitis and meningitis. Streptococcus (*Streptococcus suis*), found in pigs, is a risk factor to which pig breeders and slaughterhouse employees are exposed. In humans, it can cause cerebrospinal meningitis, arthritis, pneumonia, endocarditis, and deafness.

The bacterium that causes anthrax poisoning, *Bacillus anthracis*, was once a frequent cause of occupational diseases among cattle and sheep breeders, as well as employees of meat-processing plants; it is now found mostly in developing countries (Dutkiewicz 1998). The causative agent of tetanus, the bacterium *Clostridium tetani*, can pose an occupational threat to farmers, gardeners, and other workers engaged in earthworks. A cut contaminated by soil, manure, or dust, causes infection. The characteristic symptom of tetanus is muscle spasms, caused by a strong protein toxin tetanospasmin and affecting the neuromuscular junctions of the striated muscles. Tetanus is a dangerous disease, and left untreated leads to death. Tetanus prevention includes active immunisation with a tetanus anatoxin vaccine, administered to groups exposed to the occupational threat of infection every 8 years. Anatoxin and antitoxin (antitetanus serum) should be administered to wounded persons.

The tubercle bacillus (*Mycobacterium tuberculosis*) poses a serious threat to the health service workforce, as shown by the outbreak of 91 cases of occupational disease in 2005 in Poland (Wilczyńska et al. 2006).

Thermophilic actinomycetes are filamentous, spore-forming bacteria that develop in damp fodders (mainly hay), contain a lot of water (30%–46%), and are able to self-heat to raise its temperature to the range of 55°C–70°C. They are considered the main cause of a well-known form of EAA referred to as 'farmer's lung'. The major sources of the pathogenic allergen of glycoprotein or glycopeptide structure are the following species: *Saccharopolyspora rectivirgula* (other names include *Micropolyspora faeni, Faenia rectivirgula*), *Thermoactinomyces vulgaris, Thermoactinomyces thalpophilus*, and *Saccharomonospora viridis* (Lacey and Dutkiewicz 1994). Mesophilic actinomycetes, belonging to the genus *Streptomyces* and often found in plant and soil dusts, may cause EAA in farmers and gardeners (Dutkiewicz and Jabłoński 1989).

19.4.4 Fungi

Many allergic and toxic occupational diseases are caused by filamentous fungi or moulds, particularly the so-called storage fungi, mostly belonging to the genera *Aspergillus* and *Penicillium*, which develop on the stored plant and animal material at elevated moisture and temperature levels. *Aspergillus fumigatus* (Figure 19.5) may cause not only EAA and occupational asthma, but also pulmonary aspergillosis (Di Salvo 1983). Many other species belonging to the *Aspergillus* and *Penicillium* genera can also cause EAA. Enzymes of *Aspergillus oryzae* and *Aspergillus niger*, used in the baking industry, may lead to occupational asthma in bakers (Dutkiewicz 2004a). Harmful toxic substances produced by moulds include mycotoxins, volatile metabolites, and glucans (Samson et al. 1994; Samson et al. 2002).

Mycotoxins are poisonous, nonvolatile metabolites with a cyclic structure and low molecular mass (average 200–400 daltons). More than 100 types of mycotoxins have been classified; among these, the highest risks are associated with aflatoxins, produced by *Aspergillus flavus* and *Aspergillus parasiticus*; ochratoxin A, produced by *Aspergillus alutaceus* (*A. ochraceus*) and *Penicillium verrucosum*; and trichothecenes and zearalenone, produced by the *Fusarium* species (Mikotoksyny 1984).

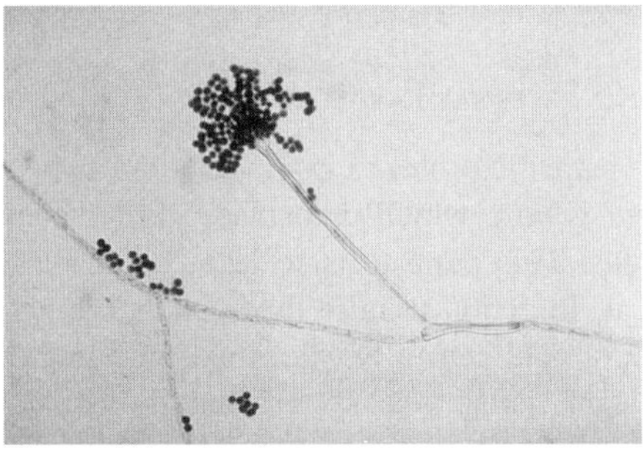

FIGURE 19.5 Light microscope of *Aspergillus fumigatus* (magnification × 1500).

Mycotoxins enter the human body orally and have toxic, carcinogenic, teratogenic, and mutagenic effects. There are reasons to believe that the mycotoxins present in organic dusts may also have toxic and carcinogenic effects on persons exposed to the occupational risk of inhaling these dusts; however, this hypothesis has not yet been fully analysed (Dutkiewicz and Jabłoński 1989; Samson et al. 1994; Samson et al. 2002).

Toxic volatile metabolites, produced by moulds, include low molecular compounds (alcohols, aldehydes, ketones, organic acids, salts), which often have irritative, toxic, and carcinogenic effects. These substances are believed to cause the characteristic syndrome known as 'sick building syndrome' (Samson et al. 1994).

Glucans are polymers of d-glucose molecules and are present in the cell walls of fungi. Inhaled with dust, they may activate macrophages and stimulate the reticuloendothelial system, initiating an inflammatory reaction in the lung tissue.

Geophilic yeast-like fungi, found in soil and on certain plants (*Sporothrix schenckii, Histoplasma capsulatum, Coccidioides immitis, Blastomyces dermatitidis*, and *Paracoccidioides brasiliensis*) may cause mycosis of the internal organs and skin in farmers, gardeners, foresters, miners and workers performing earthworks (Di Salvo 1983; Dutkiewicz 1998). Some of the yeasts commonly found on the mucous membranes of humans and animals, such as *Candida albicans*, are conditionally pathogenic microorganisms. They may cause epidermophytosis in people who work in certain microclimates. Epidermophytosis is characterised by high humidity and temperature levels conducive to high idrosis. The vulnerable occupational groups include miners, bakers, cooks, and workers of flax scutching plants (Di Salvo 1983; Dutkiewicz 1998). Less is known about the allergic and immunotoxic properties of the yeast-like fungi found in organic dusts. Common budding yeast (*Saccharomyces cerevisiae*) recently was identified as the cause of some occupational diseases: asthma in bakers and EAA in farmers (Dutkiewicz 2006).

Some lower fungi, belonging to genus *Trichophyton*, develop on animal skin (*Trichophyton verrucosum, Trichophyton mentagrophytes*) and can cause occupational mycoses in cattle breeders and fur farmers. *Trichophyton rubrum* is a human dermatophyte that poses a threat to the health services workforce, chiropodists, and hairdressers (Di Salvo 1983; Dutkiewicz 1998).

Among the higher fungi (*Basidiomycota*), the spores of some crop parasites and the edible mushroom *Pleurotus ostreatus* exhibit allergenic properties (Lacey and Dutkiewicz 1994).

19.4.5 Internal and External Parasites

The highest risk from parasitic protozoa in moderate-climate zones is posed by the sporozoan *Toxoplasma gondii*, which parasitises many species of vertebrate animals, including humans. This parasite is widespread in the natural environment; its main host is the cat. Infection usually takes place orally, although there could be other routes that have not been fully examined. In humans, the parasite usually attacks the organs of sight, the nervous system, and the lymph nodes; it also poses a significant threat to human foetuses. Occupational toxoplasmosis constitutes about 2%–5% of all cases of the disease (Dutkiewicz 2006).

Among the worms common in Poland's climate zone, *Echinococcus multilocularis* and *Echinococcus granulosus* pose occupational threats to fox and dog breeders. Among the external parasites, ticks (*Ixodidae*), which actively attack foresters, woodcutters, cowherders, and farmers, most frequently cause diseases; the toxic action of the saliva of the ticks causes an inflammatory reaction of the skin when injected into the bite wound, and in cases of greater infestation, they also cause general symptoms. The common tick (*Ixodes ricinus*), often found in broadleaf and mixed forests throughout the territory of Poland, transmits numerous tick-borne diseases (Skotarczak 2006; Cisak 2003; Dutkiewicz 1998).

Some small mites such as the larvae of *Neotrombicula autumnalis*, which are found on grain, *Pyemotes ventricosus* and *Pyemotes tritici*, which parasitise insects feeding on grain, and *Dermanyssus gallinae*, a poultry parasite, may actively attack humans, causing local dermatitis reactions and severe pruritus (Dutkiewicz and Jabłoński 1989; Dutkiewicz 1998).

19.4.6 Plant Toxins and Allergens

Protein and glycoprotein allergens, found in the pollen of grasses, weeds, vegetables, and trees, are well-known causes of pollinosis (also called seasonal allergic rhinitis), asthma, and airborne dermatitis, frequently observed in farmers. Farmers, fruit farmers, gardeners, greenhouse workers, and other such workers may be threatened by pollinosis to a greater extent than the rest of the population (Dutkiewicz and Jabłoński 1989).

Occupational exposure to dusts from various cultivated plants, powdered plant tissue used for production of medications, and plant proteases used in the food industry could cause allergic diseases (asthma, allergic rhinitis, conjunctivitis, and urticaria) among exposed workers (Dutkiewicz and Jabłoński 1989; Frazier 1980). Some disease symptoms, resulting from inhaling crushed plant tissue, are caused by plant toxins (alkaloids, glycosides, toxalbumins, and other substances; Mitchell and Rook 1979; Dutkiewicz 1998).

Direct contact with plants during cultivation and harvest often leads to dermatitis phytogenes in farmers, gardeners, and herbalists. Store workers, salespersons, and cooks exposed to plan products can have similar symptoms. More than 1500 plant species across the world cause dermatitis. The most common in Poland include common rue, celery, garlic, buckwheat, bean, scilla, hyacinth, narcissus, and tulip (Mitchell and Rook 1979; Dutkiewicz 1998; Henneberg and Skrzydlewska 1984).

Wood dust, particularly from broadleaf wood (oak, beech) has carcinogenic effects and may cause occupational nasal adenocarcinoma in carpenters, joiners, cabinet makers, and others in the woodworking industry (Maciejewska et al. 1993). Contact with wood and wood dust, particularly that of exotic trees such as ebony, mahogany, rosewood, and other species, poses a risk of allergic diseases of the respiratory system and skin among woodworkers (Mitchell and Rook 1979; Frazier 1980).

19.4.7 Animal Toxins and Allergens

The excrement and remains of small mites (0.3–0.75 mm, such as *Acaridae* and *Glycyphagidae*) that attack stored agricultural products (cereals, hay, herbs, seeds, dried fruit, cheese) often lead to asthma, allergic rhinitis, allergic conjunctivitis, and dermatitis in farmers and store workers. Similar diseases may be observed among gardeners, fruit farmers, and farmers and workers in plant protection facilities as a result of an allergy to substances produced by spider mites and other mites. Mites are pests that attack crops and orchards (Dutkiewicz and Jabłoński 1989; Dutkiewicz 1998). Another cause of allergies is occupational contact with dusts from captive-bred insects such as bees, as well as dust containing silkworm threads, scales, and excrements (*Bombyx mori*; Dutkiewicz and Jabłoński 1989; Frazier 1980).

Some sea animals (e.g. sponges, jellyfish, bryozoans) have toxic and/or allergenic properties and can cause dermatitis or general symptoms in fishermen and divers (Dutkiewicz and Jabłoński 1989). Occupational asthma is often observed among food industry workers exposed to inhalation of protein allergens from processed sea animals like mussels, squids, shrimps, lobsters, crabs, and fish (Dutkiewicz 2006; Lacey and Dutkiewicz 1994).

Bird breeders and employees of poultry-processing plants are exposed to inhalation of dusts containing allergenic fragments of the feathers, epidermis, excrement, and secretions of birds, which may lead to bird breeder's lung, a specific form of EAA. The estimated prevalence of this disease among poultry breeders in various countries is in the range of 2%–6% (Dutkiewicz and Jabłoński 1989). About 10% of employees of plants producing egg powder are diagnosed with occupational asthma resulting from exposure to high concentrations of the protein antigen found in chicken eggs (Dutkiewicz 1998).

Mammal allergens are found in particles of the epidermis, hair, and excrement, as well as droplets of saliva, milk, and urine released into the air. The most significant are protein allergens produced by laboratory rodents, which cause a specific syndrome known as 'laboratory animal allergy' (LAA; Bush and Stave 2003; Frazier 1980; Dutkiewicz 1998). The syndrome is characterised by asthma and inflammatory reactions of the nose, the conjunctiva, and the skin. It is a particularly frequent consequence of exposure to protein allergens belonging to the group lipocalins, found in the urine of rats and mice (Bush and Stave 2003).

Farmers are often diagnosed with allergies to the substances found in the epidermis and hair of cows and in pig urine (Dutkiewicz 1998). Occupational asthma has also been observed among confectionery industry workers, who inhale powdered milk containing the allergenic α-lactalbumin. Pharmaceutical industry workers also experience occupational asthma as a result of inhaling powdered enzymes (pepsins, trypsins) produced by various mammal organs (Dutkiewicz 2006; Frazier 1980).

19.5 BIOLOGICAL THREATS IN INDIVIDUAL OCCUPATIONAL GROUPS

The following two large occupational groups have the highest level of exposure to harmful biological agents:

1. Health service workers and laboratory employees are exposed mostly to infectious agents, particularly viruses of human origin (Beltrami et al. 2000; *Wirusowe zapalenie*..., 1996) and allergens, for instance, animal allergens (Bush and Stave 2003)

2. Workers of agriculture, forestry, veterinary medicine facilities, and the agricultural, food, and wood industries are exposed mainly to allergenic and immunotoxic agents and to germs causing zoonotic diseases (Cox and Wathes 1995; Gliński and Buczek 1999; Lacey and Dutkiewicz 1994; Rylander and Jacobs 1994; Parnas 1960; Dutkiewicz 1998)

Exposure to biological agents may also occur in many other occupational groups that are not related, for instance, among workers engaged in collecting and processing waste and treating sewage, employees of the biotechnological and pharmaceutical industries, fishermen and employees of seafood-processing plants, workers of the textile industry and employees of machine industry plants, miners, art conservators, librarians and archivists, teachers and tutors, cooks and sellers of food products, and hairdressers and beauticians. In some occupational groups, biological agents have been recognised only recently, for instance, workers of the machine industry, who are exposed to endotoxins and allergens from gram-negative bacteria found in oil mist, or art conservators and librarians, who may be exposed to high concentrations of allergens and toxins produced by moulds developing on the surfaces of damp historic artifacts and books (Dutkiewicz 2004a; Dutkiewicz 2006; Lacey and Dutkiewicz 1994).

19.6 DETECTION AND MEASUREMENT OF BIOLOGICAL AGENTS AT WORK

In most workplaces, harmful biological agents spread through air-dust or air-droplets; therefore, microbiological analyses of workplace air, often referred to as analysis of workplace bioaerosols, is fundamental to the detection of their presence and determination of exposure levels. A bioaerosol is a two-part system, consisting of biological particles (the solid part) suspended in the air (the gas part). According to the Polish standard issued in 2002 (PN-EN 13098. 2002), which resulted from the adaptation of the corresponding European Union standard, occupational exposure to bioaerosols is assessed by measuring the concentration of bacteria, fungi, and bacterial endotoxins in the ambient air at the workplace. In most workplaces, these agents pose the most serious threat; moreover, their measurement methods are generally accessible. The concentrations and species composition of bacteria and fungi are determined using cultivation methods. This involves collecting air samples of a defined volume onto agar culture media, which is selected according to the type of microorganisms to be determined; this determines the number of living microorganisms able to multiply (the colony-forming units [CFU]) in 1 m^3 of air. Limulus tests are commonly used to determine the concentration of bacterial endotoxins in the air. This test is based upon enzymatic reactions that cause the blood of a primitive sea arthropod *Limulus polyphemus* to coagulate in the presence of minimum amounts (within the range of 10^{-12} g) of endotoxin (PN-EN 14031. 2004). The measured concentration is expressed in mass units (ng/m^3) or endotoxin units (EU/m^3).

In assessing the exposure level of employees to biological agents, another factor of possible significance—and this depends upon the type of workplace—is a microbiological analysis of settled dust samples, raw materials (grain, hay), the soil, waste, compost, manure, sewage, water, plants, meat, milk, and animal clinical material (blood, urine, faeces, tissue samples, or skin scrapings), as well as swab tests of walls, floors, and furniture. Analysis of the concentrations of bacteria and fungi should be performed using the dilution plate method, in which dilutions of the sample are evenly distributed on agar plates followed by incubation and colony counting. The measured concentration is expressed as the number of colony-forming units in CFU/g, CFU/ml and CFU/cm^2.

On a global scale, no legislative acts specify the highest allowable concentrations of biological agents in the workplace air and other aspects of the work environment, unlike in the case of most chemical and physical agents. This is due to methodological difficulties resulting from the diversity of the microflora, lack of standardisation of measurement and experimental methods, and difficulties in determining the long-term effects of bioaerosols on the exposed occupational groups. Nevertheless, the need for specification of the threshold values of biological agents has been expressed for a long time by institutions, as well as by individual scientists, particularly with regard to bioaerosols. They have made numerous suggestions concerning the normative and reference values (Brandys and Brandys 2003; Górny 2004). In Poland, the Team of Experts on Biological Agents of the Interministerial Commission recently has formulated such proposals for the Occupational Exposure Limit Values (Dutkiewicz 2006; CIOP-PIB 2004).

19.7 BASIC PREVENTION

The following medical, technological, and organisational solutions can be applied to limit the effects of exposure to harmful biological agents at work (Dutkiewicz et al. 2000; Dutkiewicz 2006):

- Protective vaccinations for particularly vulnerable groups of employees
- Constant medical care and medical checkups for exposed employees
- Special safety measures to ensure sterile conditions for those working with highly infectious and/or genetically modified microorganisms
- Provision of personal protection equipment when work conditions involve exposure to biological agents
- Technologies to prevent the development of microorganisms and mites in stored raw materials
- Efficient ventilation systems
- Air-tight sealing and automated production processes
- Preventive veterinarian measures instituted with regard to animal-borne diseases (zoonoses)
- Health education for exposed workers

19.8 THE LEGISLATIVE SITUATION

The basic legislation concerning occupational exposure to biological agents in Poland is the Ordinance of the Minister of Health of 22 April 2005, on biological health risk agents at work and the protection of employees who are subject to occupational exposure to these risks (DzU no. 81, item 716). It implements the Directive of the European Union 2000/54/EC, which obligates member states to protect workers against risks associated with exposure to biological agents at work.

Both the Directive 2000/54/EC and the Ordinance of the Minister of Health based upon it are legislative acts of great significance. They enable implementation of preventive measures for the large population of workers exposed to biological agents, filling a serious gap in this regard. Both legislative acts precisely specify the obligations of the employer with respect to protection of workers from biological risks. They also lay down exhaustive guidelines on the protection of workers against infectious agents in laboratories, health service facilities, and the biotechnological industry. Their flaws include an outdated definition of biological agents that is limited to microorganisms and internal parasites, an incomplete list of harmful agents, lack of precise specification of the criteria for risk assessment, and a failure to establish exposure limit values (Dutkiewicz 2004b). This has on the whole resulted in the exclusion of such significant biological factors as microorganism-, plant- and animal-borne allergens, and toxins, such as bacterial endotoxins, as well as external parasites, such as ticks, which in many work environments (e.g. the agricultural and food industries) are among the main occupational risk factors. These flaws can be eliminated in the future with appropriate amendments to both of these acts that take into account the broad, modern definition of occupational biological health risk agents presented at the beginning of this chapter. This definition also serves as a basis for the Polish standard (PN-Z-08052. 1980, EN 292), which has been successfully implemented for over 25 years.

REFERENCES

Beltrami, E. M., et al. 2000. Risk and management of blood-borne infections in health care workers. *Clin Microbiol Rev* 13:385–407.

Brandys, R. C., and G. M. Brandys. 2003. *Worldwide Exposure Standards for Mold and Bacteria—Historical and Current Perspectives*. Hinsdale, IL: Occupational & Environmental Health Consulting Services, Inc.

Burrell, R. 1995. Immunotoxic reactions in the agricultural environment. *Ann Agric Environ Med* 2:11–20.

Bush, R. K., and G. M. Stave. 2003. Laboratory animal allergy: An update. *ILAR J* 44:28–51.

CIOP-PIB. 2004. *Biological Agents in the Workplace*. Warsaw: CIOP-PIB.

Cisak, E. 2003. Microorganisms transmitted by ticks as a cause of occupational diseases among forestry and agricultural workers. *Probl Hig* 11:145–158.

Cox, C. S., and C. M. Wathes, eds. 1995. *Bioaerosols Handbook*. Boca Raton, FL: CRC Press.

Di Salvo, A. F., ed. 1983. *Occupational Mycoses*. Philadelphia: Lee and Febiger.

Directive 2000/54/EC of the European Parliament and of the Council of 18 September 2000 on the protection of workers from risks related to exposure to biological agents at work. OJ EC, L 262/21, 21–45.

Donham, K. J., and A. Thelin. 2006. *Agricultural Medicine. Occupational and Environmental Health for the Health Professions*. Ames, IA: Blackwell Publishing Professional.

Dutkiewicz, J., ed. 1998. *Biological Hazards in Agriculture*. Lublin: IMW.

Dutkiewicz, J. 2004a. Biological agents of occupational risk—Current issues. *Occupational Medicine* 55:31–40.

Dutkiewicz, J. 2004b. Directive 2000/54/EC and the strategy of measurements of biological agents in the workplace. *Principles and Methods of Assessing the Working Environment* 3(41):9–16.

Dutkiewicz, J. 2006. *Hazardous Biological Agents in the Working Environment*. 2nd ed. Warsaw: CIOP-PIB.

Dutkiewicz, J., and L. Jabłoński. 1989. *Occupational Biological Hazards*. Warsaw: PZWL.

Dutkiewicz, J., C. Skórska, B. Mackiewicz, and G. Cholewa. 2000. *Prevention of Diseases Caused by Organic Dusts Occurring in Agriculture and Agricultural Industry*. Lublin: IMW.

Dutkiewicz, J., L. Jabłoński, and S. A. Olenchock. 1988. Occupational biohazards: A review. *Am J Ind Med* 14:605–623.

Dutkiewicz, J., R. Śpiewak, and L. Jabłoński. 2002. *Classification of Biological Hazards Occurring in the Work Environment and of Exposed Occupational Groups*. 3rd ed. Lublin: Ad Punctum.

Frazier, C. A., ed. 1980. *Occupational Asthma*. New York: Van Nostrand.

Gliński, Z., and J. Buczek. 1999. *Compendium of Zoonoses*. Lublin: Agricultural Academy in Lublin.

Górny, R. L. 2004. Harmful biological agents: Standards, recommendations and proposals of admissible values. *Principles and Methods of Assessing the Working Environment* 3(41):17–39.

Henneberg, M., and E. Skrzydlewska. 1984. *Intoxications with Higher Plants and Mushrooms*. Warsaw: PZWL.

Lacey, J., and J. Dutkiewicz. 1994. Bioaerosols and occupational lung disease. *J Aerosol Sci* 25:1371–1404.

Maciejewska, A., et al. 1993. Biological effects of wood dust. *Occupational Medicine* (44):277–288.

Mitchell, J., and A. Rook. 1979. *Botanical Dermatology. Plants and Plant Products Injurious to the Skin*. Vancouver: Greengrass.

Ordinance of the Minister of Health of April 22nd, 2005 on biological health risk agents at work and protection of health of employees who are subject to occupational exposure to these risks. DzU no. 81, item 716.

Parnas, J. 1960. *Zoonoses—Diseases Contracted from Animals*. Warsaw: PZWL.

PN-EN 13098. 2002. Workplace atmospheres: Guidelines for measurement of airborne microorganisms and endotoxin.

PN-EN 14031. 2004. Workplace atmospheres—Determination of airborne endotoxins.

PN-Z-08052. 1980. EN 292 Dangerous and harmful biological agents found in the work process.

Rylander, R., and R. R. Jacobs, eds. 1994. *Organic Dusts. Exposure, Effects, and Prevention*. Boca Raton, FL: CRC Press.

Samson, R. A., et al., ed. 1994. *Health Implications of Fungi in Indoor Environments*. Amsterdam: Elsevier.

Samson, R. A., et al., ed. 2002. *Introduction to Food- and Airborne Fungi*. 6th ed. Utrecht: CBS.

Skotarczak, B., ed. 2006. *Molecular Biology of Pathogens Transmitted by Ticks*. Warsaw: PZWL.

Stojek, N. M. 2003. Gram-negative bacteria as a cause of the occupational infectious diseases. *Probl Hig* (11):97–116.

Wilczyńska, U., N. Szeszenia-Dąbrowska, and W. Szymczak. 2006. Occupational diseases registered in Poland in the year 2005. *Occupational Medicine* (57):225–234.

World Health Organization. 1984. *Environmental Health Criteria No. 11: Mycotoxins,* Polish Edition. Warsaw: PZWL.

World Health Organization Collaborating Centre in Occupational Health. 1996. *European Occupational Health Series No. 8: Hepatitis B as an Occupational Hazard,* Polish Edition. Lodz: Institute of Occupational Medicine.

Part IV

The Effects of Hazards on Work Processes

20 Occupational Diseases

Kazimierz Marek and Joanna Bugajska

CONTENTS

20.1 HISTORICAL OUTLINE

Awareness of the influence of work conditions on human health has its beginnings in antiquity. The first mentions of the subject are found in the ancient Egyptian encyclopaedia dating back to 1800 BC. Descriptions of pain and injuries of the spine in pyramid builders are found in Egyptian papyruses. Hippocrates (460–377 BC.) in his treatise *Airs, Waters and Places* described symptoms that occurred in workers extracting nonferrous metal ores. In this treatise he explicitly stressed the need to observe the patient in his or her work environment, including identifying work environment conditions as risk factors of many diseases (Gochfeld 2005).

The first observations of diseases in miners were described by Agricola (Georg Bauer; 1494–1555) and Paracelsus (1493–1541). In his book *On the Nature of Metals*, Agricola described diseases affecting metal ore miners in Bohemia. In his monograph on occupational diseases, Paracelsus presented a similar problem in miners of gold, silver, and other metals in the region of Villach in Austria. In his treatise *Twelve Books on Mining and Smelting* (1557), Agricola described a number of aspects of mining, smelting, and refining of gold and silver. He also advocated the use of ventilation and personal protective equipment such as leather shoes and gloves and loose veils to protect the miners from dust (Gochfeld 2005).

Bernardino Ramazzini (1633–1714), a professor of medicine at Modena and Padua, pioneered scientific research in the areas of occupational hygiene and medicine. He is regarded as the father of occupational medicine because of his masterpiece on occupational diseases entitled *Diseases of Workers*. The quintessence of Ramazzini's school is his advice to doctors, which was 'To the questions recommended by Hippocrates, he should ask one more—What is your occupation?' (Gochfeld 2005). Ramazzini presented a comprehensive approach to workers' health problems, embracing epidemiology, hygiene, and certain aspects of ergonomics. His areas of interest covered many occupational groups, 69 of which he described in his works, including miners, apothecaries, locksmiths, glaziers, painters, mirror manufacturers, tanners, and bakers (Zanchin 2005).

In Poland, Wojciech Oczko, in his treatise entitled *Attribute* (1581), pointed out the occurrence of skin lesions caused by unhygienic work conditions. The treatise *About Sex and Venereal Diseases* by Wojciech Szeliga, published in 1584 and translated into Polish, is considered the first textbook on toxicology. Leopold Lafontaine (1756–1812) promoted Ramazzini's scientific ideas in Poland. His monograph *Studies on the Diseases of Artists and Craftsmen* dealt with the causes of and ways to prevent occupational diseases in various groups of craftsmen (Marek 2006).

The first institutions dealing with the protection of workers' health were established at the turn of twentieth century. In 1898, Great Britain appointed the first medical factory inspector, Thomas Morison Legge (1863–1932; Waldron 2004). In the twentieth century, research progress in the field of occupational pathology, epidemiology and toxicology contributed significantly to the advancement of occupational hygiene and medicine. Achievements in these disciplines led to the implementation of practical solutions for the protection of workers' health.

In the 1920s, the first lists of occupational disease were developed, making it possible for workers to obtain compensation for work-related health impairment. This process was preceded by the establishment of the International Labour Organization (ILO) convention in 1925, which included the first list of occupational diseases and toxic substances and the types of industry and manufacturing processes that could give rise to occupational diseases (Convention 18, 1925).

Although human heath has been threatened by harmful factors since the dawn of mankind, only much later did people become aware of risks related to work conditions. The second half of the twentieth century saw significant progress in occupational medicine, along with the tempestuous development of industry and the introduction of new production technologies, machines and work tools, automation, chemicalisation of agriculture and, most recently, informatics and computerisation. New harmful factors whose health effects were not known before began to emerge, as well as psychosocial factors resulting from work organisation and the mechanisation and automation of production processes. Their effects were connected with the time pressures and increased mental stress. Psychosocial effects are no longer confined to the workplace environment but are carried over to the external environment and probably modify the incidence and course of some chronic diseases.

20.2 IDENTIFICATION OF AN OCCUPATIONAL DISEASE AND A RATIONALE FOR DIAGNOSIS

Occupational disease is a medical-legal term. In order to label a disease occupational, the causal relationship of the disease with work conditions must be established and the disease must be included in the list of occupational diseases. The latter is a prerequisite for obtaining the benefits stipulated in relevant legal regulations.

The probability of various occupational diseases having a causal relationship with work conditions is differentiated. The definition of occupational disease requires this relationship to be indisputable or highly probable. For some diseases, the causal relationship with work conditions may be established with almost absolute certainty, for example, pneumoconiosis and a majority of acute or chronic poisons. Another group lists diseases for which occupational exposure is the most probable causal factor, for example, hearing loss in persons exposed to noise exceeding the permissible level over a long period of time or vibration syndrome in persons exposed to mechanical vibration. In these cases, occupational disease certification requires only high probability and not certainty because the symptoms of the disease are not absolutely specific, that is, similar symptoms sometimes result from causes other than exposure to noise or vibration.

However, there are diseases recognised as occupational for which the degree of probability that occupational exposure causes the disease cannot be defined as high. Chronic bronchitis is a good example; under prescribed conditions in accordance with the Polish list of occupational diseases, it may be recognised as occupational. Medical certification in such cases is extremely difficult because the causes of the disease are complex and tobacco smoking is the dominant factor. For this reason, only about one-fourth of countries regard chronic bronchitis as an occupational disease. The European Union included this disease in their 2003 revised list, but limited it to only miners working in underground coal mines.

According to ILO Convention no. 121 of 1964, each country may employ any of the following procedures to regulate issues connected with occupational diseases:

- A procedure involving a list of diseases, including at least the diseases enumerated in Schedule I of the convention
- A general procedure involving a broad definition of occupational diseases
- A mixed procedure that is a combination of the two above-mentioned methods

Most countries have their own lists of occupational diseases; in Europe, only two countries—Sweden and the Netherlands—have no such lists. In these countries, each case of suspected occupational disease is assessed on an individual basis. Such a solution may create a problem by allowing discretion in decision making by different teams medically certifying similar cases.

When considering occupational diseases contingent upon multiple causes, the inability to exclude an occupational cause cannot be a decisive factor in favour of diagnosing a disease as occupational; the rule of prevailing probability must be followed.

Occupational diseases must be diagnosed based on definite criteria that consider many factors in order to justify the causal relationship between the disease and occupational exposure (Marek 2001). The most important criteria are as follows:

- Symptoms must correspond to the clinical presentation of the disease in question. The extent of diagnostic difficulties varies depending on the specificity of symptoms of the given disease.
- The occupational exposure level must be high enough. This is determined based on the characteristics of harmful factors such as concentration, intensity, and length of exposure. When permissible values for these factors are exceeded, the health risk for the worker increases accordingly.

When assessing occupational exposure to diagnose an occupational disease, a number of aspects should be considered:

- *Chemical and physical factors*: Type of factor, concentration level or intensity (when compared to the maximum admissible concentration or maximum admissible intensity), and the duration of occupational exposure.
- *Biological factors*: Type of and duration of contact with the factor and the mechanism of its effects or dissemination paths. It is not necessary to determine the concentration.
- *Sensitising factors (allergens)*: Type of factor and an ability to determine that the contact took place during work and that the factor was present in the environment, raw materials, or semifinished or finished products. It is not necessary to determine the concentration.
- *Manner in which the work is performed*: Degree and type of physical load (static, dynamic, repetitive) and timing of activities that could impose excessive load on certain organs or systems of the human body.

Regardless of the principal method for assessing occupational exposure, there are situations in which such methods might not be sufficient. The assessment might be incorrect when the worker often moves between workstations with different levels of hazard or different routes of poison absorption, especially through the skin or in which they perform jobs involving physical strain or in hot microclimates. These factors cause increased lung ventilation and dose absorption, and use of personal protective equipment and nonobservance of occupational health principles can cause incorrect assessment. In such instances, the 'biological monitoring' process can be used, which involves measuring the concentration of toxic substances or their metabolites in the blood or urine. For some toxic substances, reference values in nonexposed people and biological exposure indexes that are considered safe are determined.

When diagnosing a disease, the evaluating doctor should always obtain the necessary data on the worker's level of exposure and his or her medical history from the employer and/or the organisation's preventive health care doctor. It should be noted that this information might be different and should be treated with caution.

- Some diseases manifest after a latency period. The length of the latency period is important for correct diagnosis. Due to the fact that the disease manifests after exposure has ceased, long latency periods occur with diseases such as cancers and pneumoconiosis.
- A differential diagnosis should be performed in each case. This is especially important for diseases that can be effectively treated and are diagnosed as occupational.

20.3 EPIDEMIOLOGY OF OCCUPATIONAL DISEASES

Poland has an established system of registering incidences of occupational diseases. Sanitary and epidemiological stations report each new case of a recognised occupational disease to the Central Register of Occupational Diseases at the Nofer Institute of Occupational Medicine in Lodz, where a database on occupational diseases, compiled since 1971, is in use. The register annually publishes a bulletin on the incidence of occupational diseases, which contains the number of new cases of occupational diseases classified according to the number of items assigned to it in a list of occupational diseases, age, sex, national economy sector, and province. In addition to the absolute incidence, the bulletin contains the incidence rate per 100,000 employed. Thus, Poland fulfils the recommendations of the European Commission regarding the maintenance of statistics on occupational disease incidence in member states.

For many years, the annual number of new cases of occupational diseases has stabilised at a level of approximately 10,000–12,000 per 100,000, and a slight upward tendency was observed. In 1999, this trend collapsed and by 2006 the number of new cases plunged more than threefold. This downward trend ceased in 2007 when the incidence rate rose by 156 cases (Table 20.1).

TABLE 20.1
Occupational Diseases in Poland from 1997–2008

Year	Number of Cases	Rate per 100,000 Employed Persons
1997	11,685	116.9
1998	12,017	117.3
1999	9982	98.0
2000	7339	73.9
2001	6007	63.2
2002	4915	53.6
2003	4365	46.6
2004	3790	41.0
2005	3249	34.8
2006	3129	32.8
2007	3285	33.5
2008	3546	34.7

Source: Central Register of Occupational Diseases. *Database of Occupational Diseases in Poland 1980–2008*. http://www.imp.lodz.pl/?p=/home_pl/about_imp/reg_and_databases/work_dissises1/dane_o_zapadalnosci/&lang=PL/ (accessed 24 April 2009).

Such a sudden steep decrease in incidence is not possible without the interference of disturbing factors and therefore should be examined. The Institute of Occupational Medicine and Environmental Health in Sosnowiec drew the following conclusions from a detailed 2004 analysis (Marek and Kłopotowski 2004):

- Statistics on occupational diseases in Poland are subject to errors resulting from underestimation and overestimation of some diseases, caused largely by insufficient or incorrect diagnoses by the primary occupational health services.
- Preventive health care doctors are not inclined to refer affected workers to institutions that can diagnose occupational diseases for fear of incurring additional costs.
- Workers themselves are less motivated to apply for a certification of occupational disease and often preferred to continue in their existing jobs.
- It is unlikely that an improvement in work conditions was significant enough to justify the drastic fall in the incidence of occupational diseases.
- Many chronic occupational diseases like pneumoconioses, hearing loss, diseases of the vocal organs, and vibration syndrome develop after prolonged exposure. They appear after many years of cumulative exposure and are not a result of a current or recent situation.
- A decrease in national employment cannot significantly reduce the number of new cases of recognised occupational diseases, because the number is calculated based on the incidence rate per 100,000 employed.
- The influence of changes to the list of occupational diseases can also be excluded because the changes actually came into force in 2004 while the downward trend in the number of new cases began in 1999.
- The likelihood of errors in the data entered into the system of the Central Register of Occupational Diseases can also be excluded. The system is highly secure, and it can be assumed that all cases are certified by sanitary inspectors and are included in the register.

In conclusion, officials expressed concern that the fall in the incidence of occupational diseases from 1999 to 2000 was not realistic and could not be treated as a positive development. There was a justified concern that a substantial number of occupational diseases were not recognised and/or reported. The rise in the number of cases reported last year might be a sign that this worrying trend will reverse.

There are seven occupational diseases in Poland that constitute over 85% of the general incidence (Szeszenia-Dąbrowska and Wilczyńska 2007). From 1999 to 2000, some of them, such as chronic diseases of the vocal organs, noise-induced hearing loss, and skin diseases, showed a clear downward trend. Other diseases, such as pneumoconioses, maintained similar levels. The number of certified infectious or parasitic diseases increased (Table 20.2). Chronic diseases of the vocal organs mostly affected teachers; the high numbers of incidence in previous years were overestimated.

Pneumoconioses ranked second in the profile mentioned above and in 2007 accounted for 21% of the total number of occupational diseases. Coal miner's pneumoconiosis was dominant, accounting for about 70% of the cases. Moreover,

TABLE 20.2
Most Frequently Diagnosed Occupational Diseases in Poland in 2003–2008

Disease Entity	Number of Cases per Year					
	2003	**2004**	**2005**	**2006**	**2007**	**2008**
Chronic voice disorders	1100	881	681	762	800	809
Pneumoconioses	809	754	672	667	701	697
Infectious and parasitic diseases and their sequels	550	541	615	603	671	956
Hearing loss	738	506	338	295	252	240
Skin diseases	214	181	163	128	147	125
Chronic diseases of the peripheral nervous system	89	84	99	108	158	160
Malignant neoplasms	79	114	100	104	100	85

Source: Central Register of Occupational Diseases. *Database of Occupational Diseases in Poland 1980–2008*. http://www.imp.lodz.pl/?p=/home_pl/about_imp/reg_and_databases/work_dissises1/dane_o_zapadalnosci/&lang=PL/ (accessed 24 April 2009).

105 cases of asbestosis and 90 cases of silicosis were recognised, and evidence exists that the number of pneumoconiosis incidences was underestimated.

Infectious or parasitic diseases rank third in the list with borreliosis accounting for 60% of this type of disease. Viral hepatitis, which was prevalent for many years, now accounts for about one-fourth of this group. Hepatitis B, persisting mainly in health care personnel, was most common for many years. At present, hepatitis C is the most prevalent type and is 2.5 times more common than hepatitis B. This is the result of effective prophylaxis for hepatitis B, that is, widespread use of disposable syringes and needles as well as preventive vaccinations.

A clear, almost threefold decrease in cases of hearing loss in the last five years is probably also the result of better medical prophylaxis. Workers showing audiometric hearing loss close to the criterion for diagnosing an occupational disease must discontinue exposure to noise.

Skin diseases are probably underestimated, as they rank high in many European countries.

An increasing frequency of diagnosis of diseases of the peripheral nervous system is mainly due to carpal tunnel syndrome from the way in which a job is performed.

The rate of diagnosis of work-related cancers is probably greatly underestimated. About 10% of cancers develop as a consequence of occupational exposure, but in Poland as few as 100 cases a year are diagnosed as occupational. The long latency period is the main cause for this underdiagnosis. Due to the latency period, the peak incidence of cancer is found in retirees after their occupational activity has stopped. The incidence of disease in this group is not actively monitored and often remains undiagnosed.

Statistical data on the prevalence of occupational diseases in Poland, although based on a reliable registration of cases reported by sanitary inspectors, are distorted by errors of underestimation or, to a lesser degree, overestimation, and therefore require a critical evaluation.

Certain positive phenomena have been noted over the last 20 years. The number of acute and chronic poisoning cases has decreased considerably. Severe poisoning cases rarely occur. There have been no epidemics of acute poisoning with benzene and carbon disulphide, as had occurred in the past. The frequency of heavy metal poisoning decreased radically and only single cases have been noted recently. Not a single incidence of anaemia and saturnine colic has been observed. Pneumoconioses are now diagnosed in their early phase and cases of tumoural pneumoconiosis among workers occur only as exceptions.

In Poland, as in other countries, statistics on occupational diseases cover only incidence, whereas prevalence, that is, the actual number of sick people, is not known. This number can be estimated based on the assumption that a great majority (about 70%) of occupational diseases diagnosed each year cause irreversible health damage, and the average life expectancy after diagnosis is 15 years. According to this estimate, about 80,000 people in Poland are affected by an occupational disease. This number is as big as the population of a large town, which shows that occupational diseases are a serious problem and have health, social, and economic effects. Occupational disease is especially important because it is due to unsafe work conditions and neglect in prevention.

20.4 WORK-RELATED DISEASES

The adverse effects of work conditions on the health of working people are not limited to disorders relating to classic occupational diseases in the medical and legal sense. Unsafe and harmful work conditions may contribute to the development of some chronic diseases that are very common in the general population. English terminology uses the term 'work-related diseases' and in Poland the term employed is 'paraoccupational diseases'. They are defined as diseases of a multifactorial aetiology in which work conditions are one of several risk factors that affect the manifestation or aggravation of the disease.

In Poland and in many other countries, work-related diseases are not officially registered and are not eligible for compensation. The role of work conditions in the aetiology of these diseases is the subject of extensive research the world over. The importance of work-related diseases was recently highlighted as more significant than classic occupational diseases.

A group of experts from the World Health Organisation (WHO) in Geneva prepared a detailed report in 1985 on the problem of work-related diseases. The recommendations of the report, including proposals for dealing with diseases considered work-related, remain a live issue even today. The experts proposed inclusion of the diseases outlined in Sections 20.4.1 through 20.4.5.

20.4.1 Behavioural Responses and Psychosomatic Illnesses

The risk factors for psychosomatic diseases that are mentioned in the report include work overload, monotonous work, shift work, migration (working abroad), and performing a managerial role in an organisation. Different types of mental disorders may lead to depressive reactions, hypertension, and peptic ulcer disease. Increased job-related tension and anxiety encourage smoking and alcoholism.

20.4.2 Hypertension

The acute effects of stress on increases in blood pressure are well-proven. There is less convincing evidence that repetitive situations, noise, vibration, and adverse microclimates can influence the development of hypertension.

20.4.3 Ischaemic Heart Disease

Studies have shown that acute coronary events are linked to stress, work overload, holding two jobs, and working overtime.

20.4.4 Chronic Nonspecific Respiratory Diseases

This class mainly relates to the effects of dust and aerosol contaminants present in the workplace on the development of chronic bronchitis. A number of studies reveal the effects of exposure to contaminants, mainly organic dusts and sulphur disulphide, on the development and course of this multiple-factor disease. However, there is no doubt that smoking is the main causal factor and therefore only about one-fourth of all countries in the world, including Poland, have added chronic bronchitis to the list of occupational diseases, and only provided that certain conditions are fulfilled. Moreover, the European Union introduced chronic bronchitis into the European Schedule of Occupational Diseases in 2003, but it only applies to miners working in underground coal mines. Therefore, chronic bronchitis in some countries may be considered an occupational disease when certain conditions are fulfilled, but in other countries it remains a work-related disease.

20.4.5 Locomotor Disorders

Locomotor disorders are frequently encountered in people of different age groups. These are multiple-factor pain syndromes and result from such risk factors as degenerative, inflammatory, traumatic, and neoplastic disorders. Evidence shows that some may be work-related. WHO experts selected two syndromes they consider work-related, namely lower back pain and shoulder and neck pain syndrome, for addition to the list.

Lower-back disorders are associated with occupational work involving risk factors such as forced body posture, frequent bending and twisting, lifting heavy objects, and exposure to general vibrations. They are found in jobs such as dock workers, miners, nurses, agricultural machine drivers, and heavy equipment operators.

Shoulder and neck pain syndrome is significantly more frequent in workers who perform work with their hands above shoulder level for prolonged periods. Despite arguments for the work-relatedness of locomotor disorders, a majority of countries, including Poland, do not include them in their lists of occupational diseases.

The European Union also decided not to include them in the new version of European Schedule.

Because of their complex and multifactorial aetiology, occupational diseases are still the subject of intensive epidemiological studies. These studies aim at, among other things, determining the probable degree of the causal relationship between recognised disorders and occupational exposure by using appropriate statistical methods. A demonstration of partly occupational aetiology is more difficult due to the lesser impact of the occupational factor among the possible causes of the disease. This share, called the etiological fraction (EF) is calculated using the following formula:

$$EF = \frac{(RR - 1)}{RR}$$

where EF is the etiological fraction and RR is the relative risk rate.

EXAMPLE 1

Epidemiological studies have found that chronic bronchitis is four times more common in workers exposed to dust at the workplace than in a control group of nonexposed persons. The relative risk rate (RR) = 4. Entering this rate into the formula obtains the following result:

$$EF = \frac{(4 - 1)}{4} = 75\%$$

The EF of the occupational exposure is high, at 75%. In this example, chronic bronchitis fulfils the criteria for an occupational disease.

EXAMPLE 2

Epidemiological studies have found that lower-back disorders are two times more common in miners than in the control group, with an RR of 2. Entering this into the formula obtains the following result:

$$EF = \frac{(2 - 1)}{5} = 50$$

In this example, the share (fraction) of the occupational exposure is lower and does not reach the prevailing probability level. The criterion for an occupational disease is not met, but the condition could be a work-related disease.

The above formula may be applied if the shares of other etiological factors are evenly distributed over both groups under consideration.

Progress in epidemiological studies and the advancement of research methods may lead to the conclusion that some work-related diseases meet the criteria to be occupational diseases and should be introduced into the schedule of occupational diseases. This process will be very difficult, because diseases considered work-related are also very common in the general population. Their aetiology has multiple factors and there are no clinical criteria to distinguish between occupational and nonoccupational natures of these diseases.

20.5 PREVENTION OF OCCUPATIONAL DISEASES

Civilised countries all over the world are engaged in multidirectional activities created to limit the incidence rate of occupational diseases. The WHO, highlighting the need to protect the health of working people, described its aim as 'achieving a state in which the level of general morbidity of different occupational groups will not exceed the level of morbidity of the general population'.

This rule has not yet been fully realised in any country, and the effectiveness of occupational disease prevention varies from country to country. Employers, health care services and, to a great extent, the workers themselves must take the necessary preventive actions. There are three types of prevention:

1. Primary prevention
2. Organisational prevention
3. Medical prevention

Primary prevention, or technical prevention, aims to ensure safe work conditions and is the task of engineers and technicians. Primary prevention starts at the design stage for machines, equipment, and production technologies. Design defects in technologies are very difficult to eliminate once production has started. It is essential to involve a health care physician at the design stage who can assess health risks arising from the introduction of new technologies. In many branches of the national economy, workers are exposed to different harmful or noxious factors. The employer must ensure that these factors are restricted to the limits permitted by hygienic norms. This can be achieved through solutions such as hermetising production processes, using local exhaust ventilation and general ventilation at the workplace, and replacing high-risk technologies with safer ones.

Personal protective equipment such as protective clothing, masks, ear protectors, goggles, and gloves may be placed at the border of technical and medical prevention. However, some protective gear reduces work comfort and can be used for limited durations only. A better solution may be to give up high efficiency in favour of work comfort. Disposable masks used in mining are an example; their efficiency is assessed at 50%, but they are tolerated in underground mining conditions.

Organisational prevention in high-risk conditions and rotating jobs can reduce health risk by shortening working time and lengthening total employment time. The benefits obtained result not only from shortening the time of exposure, but also facilitate the processes of disturbed defence mechanisms, such as eliminating dust from the respiratory tract or detoxification processes. Moreover, organisational prevention tasks involve appropriate arrangements for shift work and work involving great physical effort. These aim to ensure the protection of older employees suffering from a chronic disease. This type of prevention is particularly difficult and requires the cooperation of employers and occupational medicine services.

Medical prevention involves a broadly understood protection of the workers' health at the workplace. The main task of doctors of occupational medicine is to conduct prophylactic examinations—preplacement, periodic checkups, and control. Preplacement examinations cover newly hired employees or those transferred from another post and should detect any health contraindications. Periodic examinations systematically check the employees' health status and assess their fitness for the job. They are particularly helpful in detecting the health effects of exposure to harmful and noxious factors at as early a stage as possible. Periodic examinations also detect diseases not associated with work that appear during employment and which may constitute a contraindication to work at the present job.

All employees on return from a sick leave exceeding 30 days should be subject to control examinations which are intended to obtain a medical opinion on whether the sickness has caused a reduction in their ability to work.

Preventive examinations also provide guidance regarding treatment of occupational or other work-related diseases. A doctor of occupational medicine overseeing the preventive health care of workers should be acquainted with the positions and concomitant health hazards. He or she should prepare a detailed plan for preventive examinations, identify the groups of workers exposed to particular hazards, and classify them according to sex, age, and length of employment. The employer should provide data on exposure levels, results of concentration measurements and the intensities of harmful factors, any record of exceeding exposure limits, and information on daily working hours, shift work, and overtime work.

Doctors of occupational medicine are involved in a number of other tasks apart from preventive examinations, such as propagating health and rehabilitation programmes and medical education, ensuring observance of personal hygiene principles and well-organised rest and leisure time, and providing first aid in cases of emergency.

In order to prevent occupational diseases, it is very important that the employer inform employees about the conditions in the work environment, the potential effects of exposure to these conditions, and the ergonomic principles of safe work. An employee should know what his or her workstation should look like, so that he or she can ask for modifications from the employer. Worker training should be conducted in the form of lectures and practical training at workstations (Dawydzik 1997 and Ordinances of 1996, 2002, 2002).

20.6 FORESEEN DIRECTIONS OF CHANGES IN OCCUPATIONAL DISEASE INCIDENCE IN POLAND

The prevalence of occupational diseases and their clinical forms have been closely associated with the following:

- Development of industry and technologies
- Work conditions
- Work organisation
- Introduction of new substances and theoretical knowledge about their effect on the human body
- Development of health and safety

The development of new technologies and the related changes in work conditions and practices change the workload level and types of occupational hazards. This is caused mostly by changes in the structure of the economy, manifested by marked transfer of workforce to the services sector, more developed technologies that increase the automation of many workstations, and the introduction of computer technologies. The use of computer systems in production processes and services, as well as competition on the free market which forces an increase in productivity, are causes of increased mental stress due to a multitude of incoming information and a need to take responsibility for actions.

Together with improvements in technical and medical prevention, these changes allow hope that the future will see reductions in the incidence rate, structure, and degree of severity of occupational diseases. Progress in technical and medical prevention as well as partial elimination of the most dangerous poisons will bring about further decreases in the number of occupational poisoning incidents, especially severe ones. Improvements in medical prevention will lead to a reduction in the number of occupational hearing loss cases. Progress in the prevention of viral hepatitis, mostly of type B and, to a lesser degree, of type C is another factor that will contribute to the decrease in incidence rate. Assessment of the future incidence rate of borreliosis is difficult, because at present there is no vaccine and it is not certain that it will become available in the future.

Improved diagnostic criteria and progress in prevention involving voice production education will decrease the rate of diagnosis for diseases of voice organs. There are, however, indications that incidence rate of some diseases will grow due to advancement in diagnostic capabilities. Compared to the statistics of other countries, the incidence in Poland of occupational skin diseases and allergic diseases, especially occupational asthma, is underestimated. Poland may also see an increase in the incidence rate of locomotor system disorders, especially in those who work with computers.

Introducing effective medical prevention methods might cause a temporary increase in the number of cases of pneumoconiosis diagnosed. In the long term, the incidence rate will decrease due to advancements in reducing dust levels at workplaces, mainly in mining. The number of diagnosed occupational cancers is certainly underestimated, and not only in Poland. The number of diagnosed cases may be expected to rise, considering the long latency period of cancer.

There are many arguments that occupational work does not always have a negative effect on human health, and is in fact often a factor that promotes health. However, this approach to the effects of occupational work on health has not been the subject of many studies, and should be undertaken in the future.

REFERENCES

Central Register of Occupational Diseases. *Database of Occupational Diseases in Poland 1980–2008*. http://www.imp.lodz.pl/?p=/home_pl/about_imp/reg_and_databases/work_dissises1/dane_o_zapadalnosci/&lang=PL/ (accessed 24 April 2009).

Commission recommendation of 19/09/2003 concerning the European schedule of occupational diseases C (2003) 3297. 2003. Brussels: Commission of the European Communities. http://ec.europa.eu/employment_social/news/2003/sep/occdis_recc_en.pdf.

Convention Concerning Workmen's Compensation for Occupational Diseases, ILO. No. 18, 1925, http://www.oit.org/ilolex/cgi-lex/convde.pl?C018 (accessed 14 April 2009).

Dawydzik, L. 1997. *Health Care According to the Labour Code and Executory Provisions*. Lodz: IMP.

Gochfeld, M. 2005. Chronologic history of occupational medicine. *J Occup Environ Med* 47(2):96–114.

Marek, K. 2006. Occupational diseases. In *Interna*. 2nd ed, eds. W. Januszewicz and F. Kokot, 1372–1398. Warsaw: PZWL.

Marek, K., and J. S. Kłopotowski. 2004. *Assessment of Occupational Medicine Service Effectiveness based on the Statistics of Occupational Diseases Prevalence in Poland*. Sosnowiec: Institute of Occupational Medicine and Environmental Health.

Marek, K., ed. 2001. *Occupational Diseases*. Warsaw: PZWL.

Ordinance of the Council of Ministers of 30 July 2002 concerning occupational diseases. DzU no. 132, item 1115.

Ordinance of the Minister of Heath and Social Welfare of 30 May 1996 concerning medical examinations of employees, scope of preventive health care and medical certificates issued for the purposes stipulated In the Labour Code. DzU no. 69, item 332.

Ordinance of the Minister of Labour and Social Policy of 29 November 2002 concerning the maximum admissible concentrations and intensities of harmful agents; with amendments In the regulation of the Minister of Economy and Labour of 10 October 2002. DzU no. 212, item 1769.

Szeszenia-Dąbrowska, N., and Wilczýnska, U. 2007. *Occupational Diseases in Poland: Statistics and Epidemiology*. Lodz: Institute of Occupational Medicine.

Waldron, T. 2004. Thomas Morison Legge (1863–1932) the first medical factory inspector. *J Med Biogr* 12(4):202–209.

World Health Organization. 1985. *Technical Report Series 714: Identification and Control of Work-Related Diseases*. Geneva: WHO.

Zanchin, G. 2005. Padua, the cradle of modern medicine: Bernardino Ramazzini (1633–1714) on headaches. *J Headache Pain* 6:169–171.

21 Accidents at Work

Ryszard Studenski, Grzegorz Dudka, and Radosław Bojanowski

CONTENTS

21.1 INTRODUCTION

Accidents can happen at home, at school, on sports fields and playgrounds, and while playing, travelling, and working. Injuries suffered in association with work are included as accidents at work. According to statistical information from the International Labour Organisation, the annual average number of accidents at work recorded in the last decade was 270 million, including 350,000 fatal accidents. According to European Union data, each year 4 out of 1000 employees die because of accidents at work. In Poland, an average of 5 fatal accidents are recorded each year per 1000 employees; the highest number of deaths is recorded in the construction trade (17 per 1000) and the mining industry (16 per 1000).

Accidents lead to suffering and losses. Light accidents hinder the performance of current plans and goals, serious accidents lower the quality of life, and the gravest lead to death. Therefore, accident prevention is the subject of many analyses, statistical records, and preventive measures. This chapter presents an overview of accidents, events that are classified as accidents at work and their causes, as well as ways to limit occurrences and consequences.

21.2 CONCEPT OF AN ACCIDENT

The term 'accident' is usually used to describe a sudden event, rarely predictable, that results in injuries or losses suffered by people. Mandatory requirements to record accidents, systems for establishing insurance and compensation claims, and analysis of accidents for purposes of scientific research have created a necessity to strictly define what events are and are not referred to as accidents.

There is no single, generally accepted definition of an accident or of an accident at work, however, accidents are commonly through to be unintentional, adverse, and unplanned events. Other aspects vary greatly, a result of the application of three different approaches to classifying accidents. In the first, an accident must be an injury. In the second, an accident is an event, which together with the situation preceding it, causes an injury. The third assumes that an injury is one of many possible consequences of an accident. In this sense, an accident is a situation or an event that is beyond human control and can precede a bodily injury. In a legal sense, accidents not related to occupational work are distinct from events occurring at work or associated with work.

21.2.1 Accidents at Work

The major feature of an accident at work is that the injury occurs during performance of occupational or work-related tasks. A material loss or bodily injury in itself that is not related to work cannot be considered an accident at work, since such events do not satisfy all the legal criteria associated with accidents at work. Such events are treated as nonoccupational accidents, also known as unfortunate accidents or events attributed to fate.

According to legislation on social insurance (DzU 2002, no. 199, item 1673), an accident at work is a sudden event due to an external cause that results in injury or death and that takes place in association with work under one or more of the following conditions:

- While the employee performs his or her ordinary tasks or the orders of superiors
- While the employee performs activities for the benefit of the employer, even without being ordered to do so
- While the employee remains at the employer's disposal, including travelling between the employer's place of business and a place of performance of duties

This definition of an accident at work correlates the occurrence of an event and the employment of an employee when performing tasks subject to the employment relationship, tasks the superior may order the employee to perform, and other tasks that the employee may perform for the benefit of the establishment or that result from the employee's duty to remain at the employer's disposal. This means that an accident at work may also include an injury suffered by an employee during a lunch break, while changing or taking a shower at work, or en route between the office building and the workstation. Accident assessment takes into account the rules of social life and the practices that are implemented at the establishment. An event does not have to result from inadequacies of work conditions to be considered an accident at work. It is sufficient that the event takes place at work and during working hours.

However, there are occasions when a direct correlation with work may not exist. This can occur if employees perform work during working hours that is not related to their employment or that is contrary to the objectives of employment. An accident that takes place during performance of activities not related to the employer cannot be considered an accident at work.

An accident is the total event and not just a feature of the event. A given event is an accident at work, if, apart from its correlation with work described above:

- It occurs suddenly, that is, its duration does not extend beyond a single working day; injuries do not have to be apparent immediately but may be delayed, revealed after a week, month, year, or later; it is correlated with an event that occurred in the course of employment.
- It is caused by an impulse from a source in the surroundings of the injured person, that is, outside the person's body.
- It causes an injury, defined as damage to bodily tissues or organs, and is caused by an external factor.

An external cause is any external factor, that is, a factor not resulting from a specific illness of the injured person. However, the external cause does not have to be the exclusive reason for the injury. An external factor such as physical overload or stress may lead to a sudden worsening of the condition of an employee suffering from an illness.

Injuries, apart from trauma suffered, can include ordinary cuts, bruises, or needle pricks (frequent occurrences in medical occupations). This definition of an injury means that even the smallest cut qualifies as an accident at work, even if it does not result in a temporary incapacity to work.

21.2.2 Types of Accidents at Work

Accidents are classified by the number of persons subjected to the traumatic consequences of the event, such as individual and collective accidents, as well as—depending on the extent of seriousness of the injuries suffered—fatal accidents, accidents leading to serious bodily injury, and other accidents.

An individual injury is one that results from occurrence of a dangerous event and causes a single person to suffer trauma. A collective accident is an event that causes at least two persons to suffer trauma. A fatal accident is an accident that leads to the death of a person who suffers an injury during or within six months after the accident. Accidents may result in a serious body injury, leading to loss of vision, hearing, speech, limbs, reproductive capacity, or an incurable illness, including life-threatening illnesses. These effects result in an inability to perform occupational work or in permanent serious disfiguration or deformation of the body. Accidents that result in reversible health effects are commonly referred to as 'light' accidents.

21.2.3 Noninjury Incidents

The presence of accident threats in the human environment is signalled by an emergence of dangerous events, which vary in the extent of loss caused. Dangerous events include disasters with many fatalities, collective accidents, individual, fatal, serious, and light accidents, as well as events that do not lead to an unfortunate end, despite a high probability of suffering injury. A noninjury incident is an unsafe occurrence arising in the course of work in which no personal injury is caused (PN-N-18001:2004).

The seriousness of dangerous events and their consequences are related to the frequency of occurrence. Disasters and collective fatal accidents occur less frequently than individual accidents; the most numerous are disturbances and events of noninjury incidents. According to data gathered by Heinrich (1936) at 100 establishments, of 1000 dangerous events, 3 (0.3%) led to serious injuries, 88 (8.8%) to light injuries, and the remaining 909 (90.9%) dangerous events resulted only in the probability of suffering injury. The correlation found between the frequency of dangerous events, which vary in the seriousness of their consequences, is called the *major injury principle*, and the proportion presented as a triangle is referred to as the *Heinrich accident pyramid*, which shows that for every 300 unsafe acts there are 29 minor injuries and one major injury. Noninjury incidents, disturbances, and deviations from the expected, normal course of work, occurring during completion of work tasks are increasingly treated as symptoms of hazards, subject to recording and analysis. They are also used for designing preventive measures. The Central Institute for Labour Protection of the National Research Institute (CIOP-PIB) has designed a system of registration and supervision of noninjury incidents for practical application (Dudka 2003).

Despite research conducted on accident events and the existing systems of registration of noninjury incidents, there is no single definition of these events. The committee responsible for analysis and registration of accidents, which conducted consultations in Geneva in 2002, failed to work out a single unambiguous definition of events referred to as incidents.

One of the few functional definitions of a noninjury incident was presented on 27 April 2001 by the International Labour Organisation in *Guidelines on Occupational Safety and Health Management Systems* (ILO-OSH, 2001): 'Incident—An unsafe occurrence arising out of or in the course of work where no personal injury is caused'. This definition was incorporated into the standard PN-N-18001:2004 *Occupational health and safety management systems and requirements.*

In the company environment, noninjury incidents are also referred to as almost-accidents, incidents leading to material losses, near misses or—generally—incidents.

21.3 ACCIDENT RATE

The accident rate is the total number of accidents occurring within a given time, usually one year, presented using specific indices. In some companies, accidents occur frequently; in others, they are rare. Depending upon the related hazards and the specific nature of occupational risk, some injuries are serious and require prolonged treatment, while others have no significant consequences. The number and seriousness of accidents in a given company depend upon many factors; these include the hazards methods of control present, the duration of exposure to hazards, the number of employees, and so on. Multiple sources of accidents may lead to a situation in which the accident rate for two companies differs significantly, even if the total number of accidents per year is almost identical, because the number of employees is different.

Accident rate indicators illustrate the relationship between the number of accidents and work conditions. Assessments and comparison of accident rates, frequency rates, severity rates, and accident cost rates are applied to compare safety levels in companies.

21.3.1 Accident Rate Indicators

The most widely used accident rate indicators are presented in Table 21.1. Indicators W, W_1 and W_2, called incidence rates, denote the frequency of accidents. W denotes the number of accidents recorded per 1000 employees. W_1 denotes the number of accidents in relation to the duration of exposure to the hazards per 100,000 working hours. W_2 relates the number of accidents to the production size; in the 'mining' version, the number of accidents is determined per million tonnes of coal extracted. It may also relate accidents to other dimensions of production, such as the number of pieces of products manufactured, the number of kilometres of roads constructed, and so on.

W_3 and W_4 relate to the severity of accidents. W_3 shows how many working days are lost due to one accident; W_4 shows the number of employees killed for every thousand who suffer an accident. The last indicator, W_5, is obtained by dividing the total cost of all accidents by their number and shows the average accident-related loss. Other indicators can be constructed by carrying out modifications. For instance, the absence rate, which shows the average absence per 1000 employees or 100,000 work hours, relates the severity of accidents to their frequency.

TABLE 21.1
Accident Rate Indices

Name, Content	Mode of Assessment of Accident Rate
Incidence rate: number of accidents per 1000 employees	$W = \frac{\text{Number of accidents (injuries)}}{\text{The number of employees}} \times 1000$
Incidence rate: number of accidents per 100,000 working hours	$W_1 = \frac{\text{Number of accidents (injuries)}}{\text{Number of work hours}} \times 100{,}000$
Incidence rate: number of accidents per 1 million tonnes of product	$W_2 = \frac{\text{Number of accidents (injuries)}}{\text{Amount of product (tonnes)}} \times 1{,}000{,}000$
Severity rate: number of working days lost due to a single accident	$W_3 = \frac{\text{Total absence caused by accidents}}{\text{Number of injuries}}$
Severity rate: number of fatal accidents per 1000 total accidents	$W_4 = \frac{\text{Number of fatalities}}{\text{Total number of accidents (injuries)}} \times 1000$
Accident cost rate: average loss suffered due to a single accident	$W_5 = \frac{\text{Total cost of accidents}}{\text{Number of accidents}}$

Frequency rates can be calculated for all accidents recorded at work, or they can be determined separately for fatal and serious accidents. Rates are usually calculated annually, although they are also useful when calculated in 3- or 5-year periods. Multiannual rates are useful for comparing fatal accident rates of establishments, trades, or countries.

21.3.2 Applications of Indices

Indicators that use information on accidents can be applied to accident rate characteristics and are used to assess occupational risks. The accident rate is analysed using the number of accidents in relation to employment size, or sometimes to hazard exposure duration (in working hours or working days) or production size, expressed in pieces, kilometres, various weight units, and so on.

Indicators that relate the number of accidents to production volumes are a measure of the 'biological' production cost. They are used mainly for internal sector comparisons. The remaining frequency rates, calculated based on the ratio of the number of accidents to the number of employees or to the hours worked, are indicative of the individual risk of loss of health or life at work at a certain establishment or in the performance of a specific task.

For estimating and assessing risk, incidence rates are applied as a source of information on the number of accidents or losses suffered while manufacturing a specific output quantity. Thus, the number of accidents at work for a given output for two companies or two production lines of similar products that use different technologies or processes can be compared and their safety levels assessed. Indicator variants can be used to compare other types of activities. In service activities, the number of accidents could be related to the number of services performed, or in trade activities to the number of products sold.

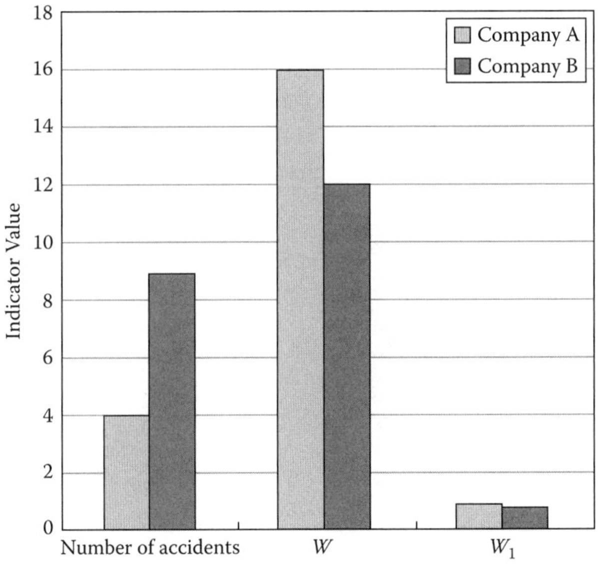

FIGURE 21.1 Comparison of various frequency rates of accidents at work.

The incidence rate measured by the number of accidents per 1000 employees is a universal parameter applicable in many situations. It is most often used to compare accident rates in populations of various sizes. Its limitations become apparent in comparisons of populations that differ with regard to the average working time of employees through the period analysed.

Figure 21.1 presents a comparison of two companies using the number of accidents recorded and two incidence rates W and W_1. Company A employs 250 full-time employees, while company B employs 750 people, of whom 250 are part-time employees (i.e. they work half of the working time). In company A, four accidents at work were recorded during one year, while in company B nine accidents occurred in the same period. Thus, the number of accidents was more than two times greater. However, taking into account the total number of people employed, accidents at work are more frequent in company A than in company B. Considering the part-time employees at company B hires (who are thus exposed to occupational hazards for only half the time), the accident rate in both companies is the same. This example shows that different indicators have different scales and they should not be compared with each other.

The duration of the incapacity to work due to an accident indicates the seriousness of its effects on the health of the injured person. Thus, to describe the accident severity in a company or in a sector of business, the total absences due to accidents at work in a given period (e.g. in one year) should be considered. However, in order to compare the accident severity in various populations (companies, organisational units, business sectors), an average indicator must be utilised. The average accident-related absence (W_3) is one such indicator. The number of fatal accidents can serve as a basis for calculating another severity rate to define the number accidents in 1000 that result in death of the persons injured (W_4). The number of serious accidents can

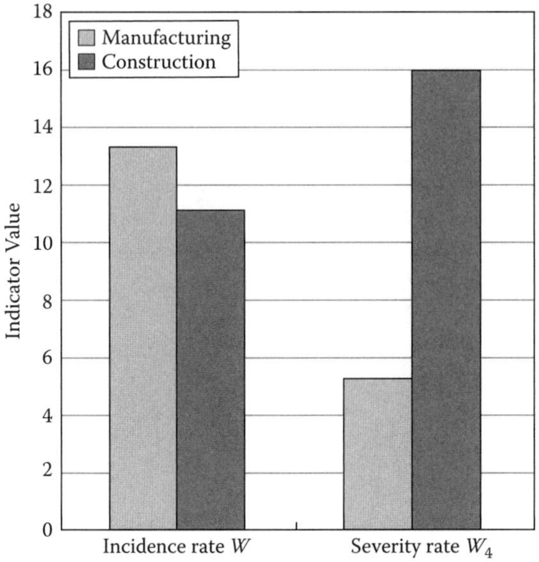

FIGURE 21.2 Frequency and severity of accidents at manufacturing and construction work sites in Poland in 2005.

also be substituted to obtain yet another severity rate. Figure 21.2 presents a comparison of the accident frequency measured by incidence rate W with the severity (W_4) for manufacturing and construction in Poland in 2005. Accident frequency is slightly higher in manufacturing. However, the fatality rate related to accidents at work in the construction industry is five times as high as in manufacturing industry.

21.4 CAUSES OF ACCIDENTS

Identifying the causes of accidents is a prerequisite for formulating preventive action programmes. The legal provisions that require maintenance of accident records define the cause of accident as any flaws and errors which contribute directly or indirectly to the occurrence of the accident. This can be related to material (technical) factors, general work organisation, or organisation of the work station, or can be associated with the employee (DzU 2004, no. 269, item 2672). Descriptions of the causes of accidents may include black ice, old age, or stress. When planning preventive actions, these features of the environment or entity should be considered, as they contribute to a higher probability of occurrence of accidents. However, they are not the causes of accidents.

21.4.1 CONCEPTS OF ACCIDENT CAUSES AND CAUSALITY

In a general sense, the cause of event B is another event A, kinetic or static, that precedes event B and is a sufficient condition and an effective specification condition (Pszczołowski 1978). The latter expresses physical or emotional acts or their causes resulting in a change in event B or leading to occurrence of event B.

The term 'causality' is an expression of the principle that every effect has a cause that precedes it (Reber 2000). *Simple causality* is when events result from a single cause. *Complex causality* is when the observed effect arises from several causes. Accidents are events with many causes. Most correlations between events that lead to an accident have a complex causality. Therefore, causes of accidents must be treated as coexisting events preceding the injury (Kenny 1993).

21.4.2 Types of Causes

Accident causes are classified by their correlation with a dangerous event and an injury as either indirect or direct (Wanat 1964; Zabetakis 1991). *Direct causes* include first-degree causes with the potential to initiate dangerous events and second-degree causes considered to be the source of accidents (Gliszczyńska 1963; Szczurowski 1980; Wanat 1964; Stranks 1990). *Indirect causes* include contact with hazardous energy or materials. Loss of control over energy or over one's behaviour could lead to hazardous contact.

Indirect causes of accident-related injuries and accidents are activities that trigger dangerous events or enhance the probability of their occurrence. This group includes dangerous behaviours and conditions. Dangerous behaviours and conditions are symptoms of flaws in occupational health and safety policy, and in the organisation, supervision, and decision-making processes related to human resources and production, which indicate that occupational health and safety needs are not met. Errors in safety management are classified as those leading to a discrepancy between the expected, normal course of work tasks and their actual course. Some such discrepancies are associated with increased threat levels, which consequently lead to events causing accidents and injuries. The types of causes of accidents are presented in Figure 21.3.

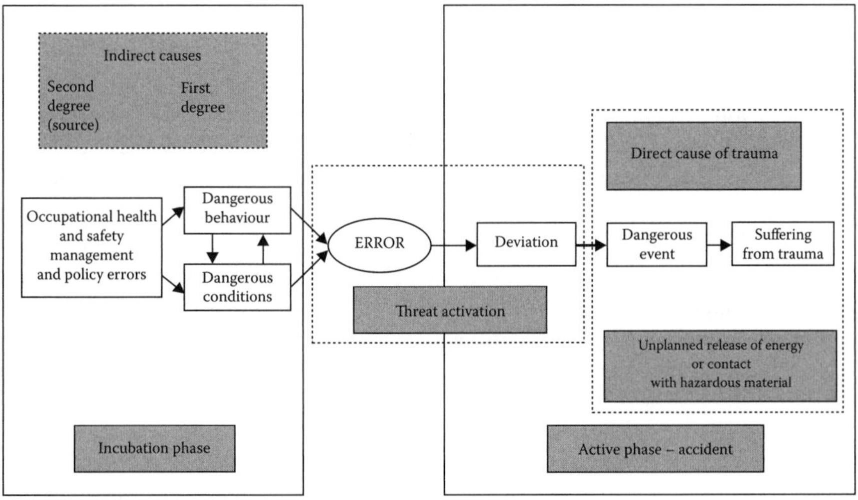

FIGURE 21.3 An accident situation: Accident phases and causes.

Although every accident has individual characteristics, some aspects are common to all accidents. Presented as a model, they illustrate the order of successive cause-and-effect events, which include incubation and active phases.

The *incubation phase* is a set of dangerous activities and conditions that contribute to the triggering of hazards, and 'await' incorporation in a sequence of events as indirect causes of a potential accident. This phase includes, for instance, changes in the order or pace of technological operations, a failure in the use of the means of protection, performance of hazardous tasks by unauthorised persons, equipment shortages, and so on. It originates from source of the cause, that is, errors committed by employers and team managers in the course of their managerial tasks, and—even before that—in devising the plant's safety systems and controls, organising tasks, and preparing people to perform their work while following safety procedures. Examples of such errors include selection of a risky technology, compelling employees to work despite a lack of equipment to monitor the work environment, failure to properly train the employees, failure to provide personal protection equipment, and so on.

The *active phase* is a changed and hazardous course of events that turn into dangerous events and cause injuries. It includes active hazard effects. The active phase is triggered by initiating a hazardous event as a result of natural processes, an unpredicted dysfunction of equipment or the hazardous activities of people. The factor connecting the two phases is the error, which triggers deviations from the expected standards.

21.5 ERRORS

People define their goals with an expectation of achieving positive results. The results attained usually differ from their expectations for various reasons. The undesirable discrepancy between the goal and the result is caused by erroneous actions. Accidents result from the errors committed; however, errors do not cause injuries. The effects of errors in perception, decisions, or action may activate hazards or cause emergence of an active hazard that triggers the occurrence of a hazardous event.

21.5.1 Concept of Error

An error is any effect other than the expected effect and any course of activities different from the programmed activities (Tomaszewski 1976). An error may be defined as a deviation from correctness (Reber 2000). Reason (1997, 71) has described human error as 'the failure of planned actions to achieve their desired ends without the intervention of some unforeseeable event'. People make errors while formulating goals, in planning achievement of these goals, and in implementing programmes to achieve them. They make mistakes in assessing the available information, in planning and defining their goals and the ways to achieve them, and in evaluating the possible consequences of the activities planned.

21.5.2 Types of Errors

Errors are connected with the type of activity being performed. The type of error depends upon the type of activities. Rasmussen (1986) lists three types of

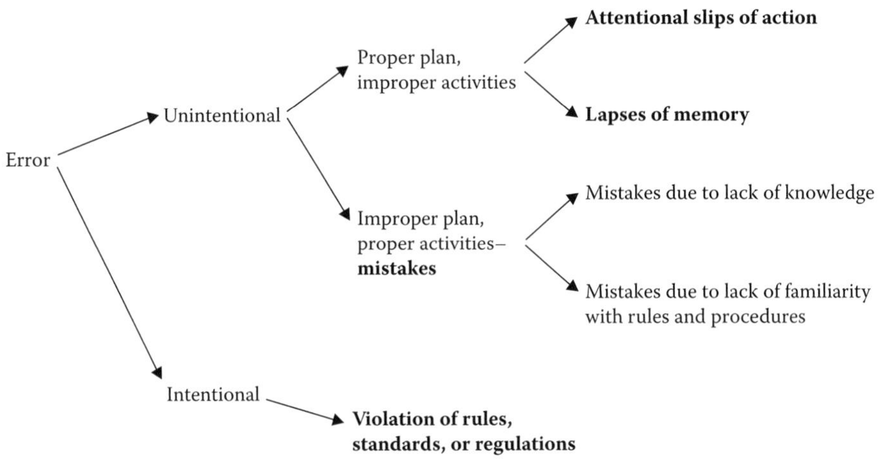

FIGURE 21.4 Classification of errors.

activities: automated, algorithmised, and knowledge-based. Most errors in automated activities result from inappropriate recognition of an initial condition which, despite being dangerous, was assessed as safe. The perception of safety leads to the implementation of automated activities inappropriate for a hazardous situation. An error in algorithmised activities may result from a mistaken recognition of the situation, causing an application of procedures or technologies that are not appropriate for the actual situation, which has been falsely identified.

The third type of erroneous activity is knowledge-based. It occurs when the conditions deviate from the existing standards and classifications, and when the known principles, rules, or algorithms are perceived as risky or insufficient to merit achievement of the desirable result. Erroneous knowledge-based actions are caused by lack of competency in problem solving, reasoning, and use of memory (Chmiel 2003). This classification was used by Reason (1990), who devised a detailed division of human errors (Figure 21.4).

People commit errors that cause accidents unintentionally due to mistakes, omissions, and faults, as well as intentionally, by departing from the canons of knowledge, principles, patterns, and standards of behaviour. Intentionally committing errors is a conscious choice, usually aimed at obtaining additional benefits (Kindler 1990) such as hastening completion of tasks, minimising effort or financial cost, seeking excitement, or improving social image (Studenski 2004).

21.5.3 Causes of Errors

People will choose to act safely if they can, want, and are able to attain their goals safely (Ratajczak 1988; Hale 1995). Fulfilling these requirements means that the individual has the ability to act safely (Sanders and Pay 1988) and is competent, capable, and motivated to perform tasks safely and protect his or her life and health.

The causes of unintentional errors are, among others, carelessness (Arnett 1992; Hallowell and Ratey 2004), tendency to take risks (Studenski 2004), lack of competence, and/or overload of work (Furnham 1992); such errors may also result from intoxication, stress, illnesses, fatigue, and changes caused by ageing. Intentional errors are associated with insufficient motivation to act safely. Research on causes of accidents shows that they occur frequently due to violation of safety rules and regulations; such behaviours are prompted by the particular situation and the personality of the individuals, as well as by specific emotional states and attitudes (Booysen and Erasmus 1989; Pidgeon 1991).

Some behaviours that are excessively risky and dangerous constitute not only intentional violations of safety regulations, but are also ways to achieve self-destructive goals (Suchańska 1998; Franken 2005). Individuals with suicidal tendencies, prior to making a suicide attempt, have been involved in car accidents two times more often than the total population of drivers (Crancer and Quiring 1970). Depression and a risk-taking tendency predict involvement in accidents better than the reasons to take risks; a tendency to take risks due to lack of self-acceptance, a sense of guilt, or low self-esteem are correlated with the frequency of accidents requiring hospitalisation (Studenski 2007). Persons who take risks with a high probability of death feel guilty, hate themselves, and search for opportunities to punish themselves (McMains and Mulins 2001).

21.6 EXPLAINING ACCIDENTS

An appropriate accident report explains the incident and the reasons for its occurrence. Descriptions of the injury suffered, its direct cause, and the error that led to the activation of the potential hazard provide the required information on the cause and consequences of the accident. The order of occurrence of events illustrates the timeline of the accident, and its origin can be explained only by specifying the type and reasons for deviation from the expected standards. An accident's source causes explain the accident itself.

Postaccident activities include investigation of the accident to identify the perpetrators, their responsibility, and the compensation rights of those affected. The causes of an accident are determined based upon the accident causality theories, which explains the mechanisms of accidents and the human contribution to hazardous errors. This is helpful when planning postaccident procedures and identifying the reasons for and describing the results of the analysis of the event.

21.6.1 Accident Causality Theories

In the early twentieth century, railway and ship disasters were caused by errors made due to colour blindness of helmsmen and engine drivers and selecting the wrong people to drive trams caused urban tram accidents (Claparède 1924). The list of accident-prone factors was extended based on research studies by adding work conditions, characteristics of the tasks performed, and occupational and nonoccupational situations that hinder work and increase fatigue or cause distraction.

Theories of accident causality established in the mid-twentieth century explained the emergence of accidents as resulting from the following:

- Lack of professional competence or talent
- Lack of ability to adapt to safety and action procedures
- Lack of acclimatisation to the physical environment
- Management errors leading to diminished work satisfaction and morale of employees

Supporters of the theory of sequence of events proposed another explanation for the causes of accidents. Heinrich (1959) assumed that an accident could be described as an active system of domino components, as illustrated in Figure 21.5. The upper part of the drawing represents the maintenance of safe correlations between the main elements of a work situation. A change in the work environment, such as spillage of oil or unsafe storage of flammable materials, may serve as a trigger for a sequence of events that culminate in the accident. Spilled oil that has not been cleaned up is the first element of the accident sequence. It awaits the second domino component—an employee passing by carrying a load who, due to limited visibility, is unable to bypass the spill. Further elements of the accident scenario include slipping, falling down, and the consequences of the fall. A similar sequence of events can be expected as a result of improper storage of flammable materials, which ignite quickly.

Benner (1975) noticed that the sequence of events in accidents might assume more complex forms than a single chain of dominoes. In the model he designed, he presented accidents as effects of multilinear sequences of events that result in the combination of the dangerous event and the injuries.

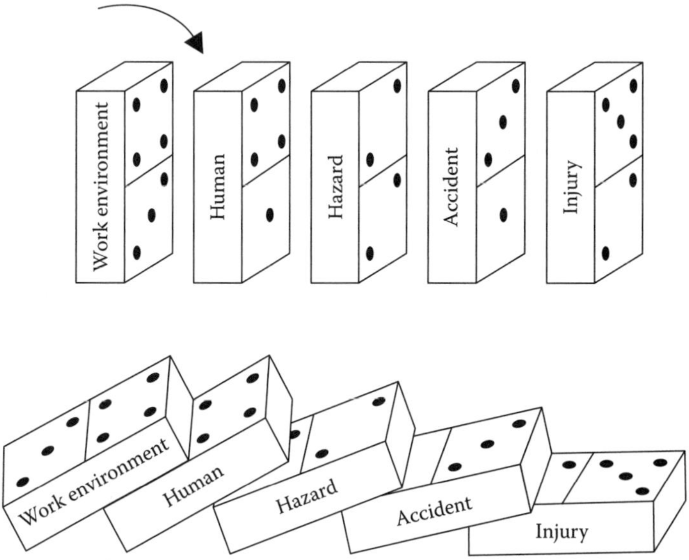

FIGURE 21.5 Classical accident model according to Heinrich.

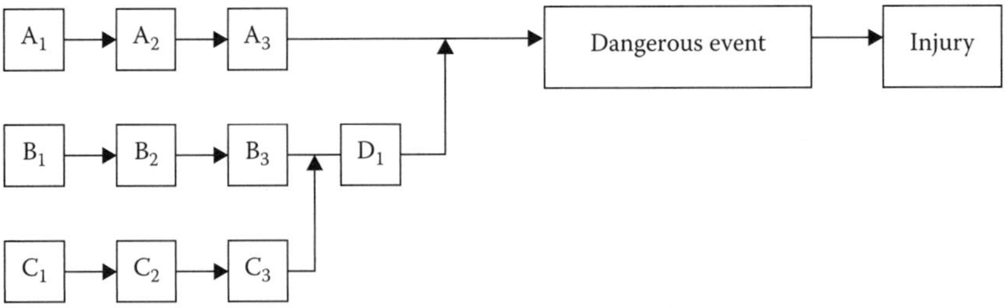

FIGURE 21.6 An accident resulting from a combination of multilinear sequences of events.

The model presented in Figure 21.6 refers to a workshop that employs people (sequence A); flammable materials are also stored there (sequence B) and welding is carried out in the vicinity (sequence C). Due to the welder's error, the flammable materials catch fire (event D1), and the fire (the dangerous event) results in burns suffered by the workshop employees.

These factor-oriented concepts reveal the causes of errors that initiate dangerous events, while event sequence theories enable creation of a chronological sequence of events leading to the errors, and their effects. Attempts to integrate various theoretical approaches (Hale and Hale 1970; Smillie and Ayoub 1976; Corlett and Gilbang 1978) allowed for an analysis of accidents as complex sequences of events, in which three levels of causes can be distinguished (Studenski 1986). The direct cause of the trauma is usually an event arising from dangerous behaviours or use of dangerous materials. On the other hand, the dangerous behaviours of persons responsible for performance of production tasks and the errors committed by them intentionally, as well as work performed under dangerous conditions, could prove to be the effects of errors committed by the employer and the management. These errors take the form of bad safety policy, poor safety management, and erroneous personnel decisions. They are treated as the source causes of accidents (Zabetakis 1991), or their 'hotbeds'.

These accident causes are identified with a poor safety culture. Safety management errors, risk-taking, and committing intentional errors could be eliminated by strengthening the safety culture (Pidgeon 1991; Cooper 1998; Geller 2002).

21.7 POSTACCIDENT PROCESSES

The employer's obligations related to accidents at work are specified mainly in the labour code. The employer is obliged, among other things, to systematically analyse the causes of accidents at work, occupational diseases, and other illnesses associated with the conditions of the labour environment, and—based on the results of these analyses—to institute appropriate preventive measures. For accidents occurring at work, the employer should undertake appropriate actions to eliminate or limit the hazard, provide first aid for the injured persons, and identify and apply appropriate measures to prevent similar accidents in the future in accordance with procedure and the circumstances and causes of the accident. The ordinance of the Council of Ministers of 28 July 1998, in a detailed description of the procedure to be followed in the event of an accident, specified the identification of the circumstances and causes of accidents at work and the ways to document them (DzU no. 115, item 744, as amended). According to the regulations, postaccident processes should include the following:

- Securing the location of the accident
- Appointing a postaccident team
- Investigating the circumstances and causes of the accident
- Specifying preventive measures
- Preparing postaccident documentation

All employees must report accidents. This applies to both the employee who has had an accident (provided their health condition allows it) and any person who witnesses an accident or learns of its occurrence. On obtaining a report of an accident at work, the employer is obliged to secure the accident location until the circumstances and causes of the event are identified and to avoid any changes to the accident location prior to the investigation.

The circumstances and causes of the accident are then examined by a postaccident team appointed by the employer and consisting of an occupational health and safety employee and a social labour inspector. The inclusion of competent specialists in the postaccident team also facilitates an objective analysis of the causes of the event. Plans of preventive measures and accident records serve to document the results of the accident investigation.

21.7.1 Preparing for an Investigation

Occupational health and safety specialists, social inspectors and labour inspectors, as well as the employers may investigate an accident. If the employer is unable to appoint a team due to an insufficient number of employees, a team consisting of the employer and a specialist from an external company should be appointed. To prepare for an investigation of the accident, the inspection area should be specified, the planned scope of investigation defined, and the necessary investigation equipment prepared.

21.7.2 Accident Investigation Methods

An accident is a rare opportunity to learn from mistakes. The plant, its work rules, and the competency and actions of the people should be framed and organised in a manner that ensures safety. An accident is a sign that these conditions were not met. Analysis of the accident is a source of detailed information on the place and time of occurrence of hazards, as well as the type and causes of errors committed. An accident investigation is appropriate if it provides information that enables the design of effective preventive measures. Gathering such information is one of the major objectives of accident investigation. The postaccident team gathers information on the event by:

- Inspecting the place of accident
- Interviewing the injured persons, if their health condition allows it

- Interviewing witnesses
- Familiarising themselves with the doctor's opinion
- Familiarising themselves with experts' opinions, if necessary

The accident investigation programme is usually established from the preliminary data obtained during the inspection. Therefore, the preliminary statements of persons who participated in mitigating the accident situation should be obtained during the preliminary inspection and a layout of the accident location should be prepared, marking the locations of witnesses, machines, hazardous materials, sources of energy, and so on. It is also useful to record information on the conditions prior to the accident, including the processes and control procedures used and any previous reports on difficulties or disturbances encountered in the course of work or deviations from the standards in force.

When interviewing the injured persons or witnesses, questions that might suggest or influence an answer should be avoided. The witness must understand the questions, and if any testimony is inconsistent with a preliminary statement, the reason must be determined. Notes should be taken without distracting the witnesses.

The most probable course of events and probable causes of the accident can then be determined. Identification of the causes of accidents is supported by research procedures, referred to as the 'accident investigation method'.

21.7.3 Accident Investigation Methods

Various methods and procedures are used to investigate accidents. Usually, their theoretical base is the sequence of events. An accident may have sequence of more than ten causal events. Detailed analyses reveal that the direct cause of a traumatic accident is usually one or more dangerous behaviours or the result of working in dangerous conditions; dangerous activities or conditions that are not compliant with the standards should be connected to policy defects and safety management errors as the source causes. To recognise and identify the reasons for occurrence of accidents, the TOL (technical-organisational-human) method (Hansen 1993) can be used, as well as the fault tree analysis (FTA; Delong 1970) or the change analysis procedure and Gantt's sequence sheet (Zabetakis 1991).

21.7.4 TOL Method

This method assumes that the accident is the result of technical (T), organisational (O), and human (L) causes and uses the diagram shown in Figure 21.7. First, the technical causes are analysed, then all of the organisational elements, and finally, the possible human causes are identified. Such an analysis determines the direct and indirect causes of the accident.

21.7.5 Fault Tree Analysis

The fault tree method, also referred to as FTA, is a graphical method that logically presents the combination of events that took place and led to the accident. The

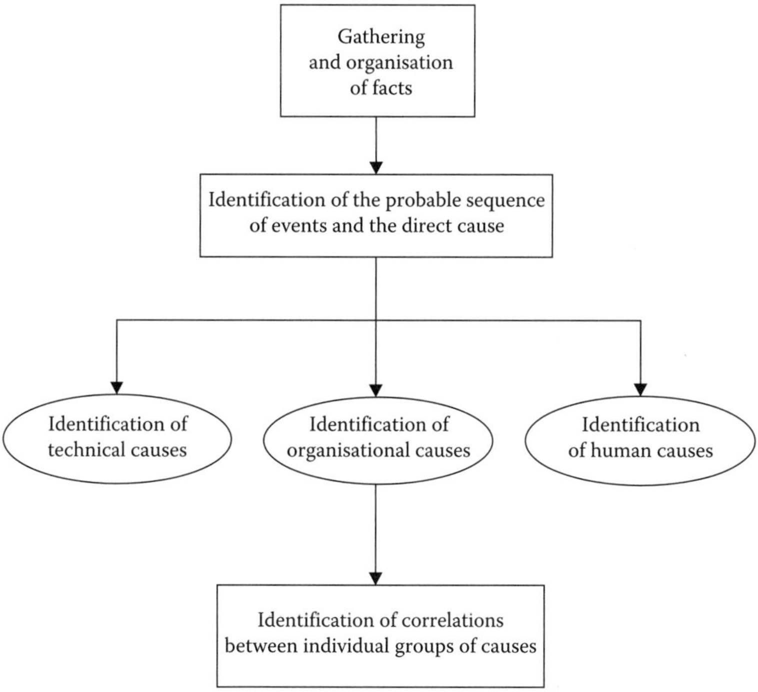

FIGURE 21.7 Diagram of investigation of an accident at work using the TOL method.

construction of the fault tree enables identification of the mechanisms that caused the accident; each of its components may be treated as a separate area of preventive activities.

A fault tree is constructed by gathering the accident data in an organised manner and following all possible paths, which can be identified based on verified evidence. Specific and objective facts, rather than interpretations or judgments, must be gathered for this method.

Construction of a fault tree for the accident at work starts with the last event—most often the injury. Then, systematically going back step-by-step through the sequence of events, the following questions are asked with regard to each of the identified facts:

- What was necessary for this fact to occur?
- Is any other fact (or facts) necessary for the concerned fact to have occurred?

The facts can be connected based on an analysis of the answers to these questions. The types of possible connections between the events are presented in Table 21.2.

When constructing the individual components of the fault tree, it is necessary to go as far back as possible to the final event (injury). This will allow for identification of

TABLE 21.2
Types of Possible Connections between Events Using the Fault Tree Method

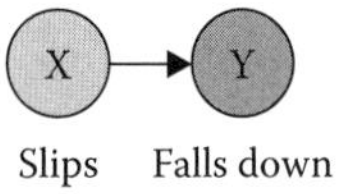	A chain relation (one fact, one preceding event)
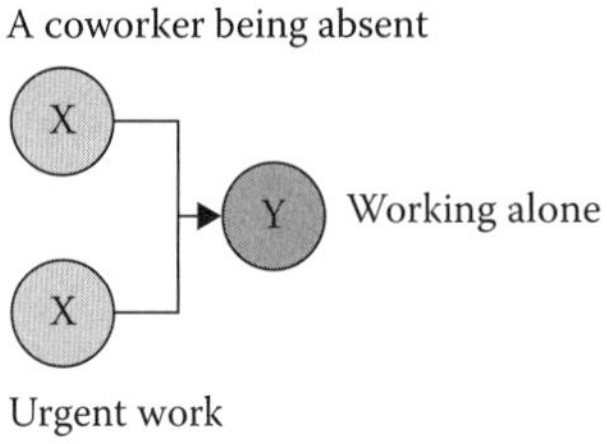	Conjunction (one fact, several preceding events)
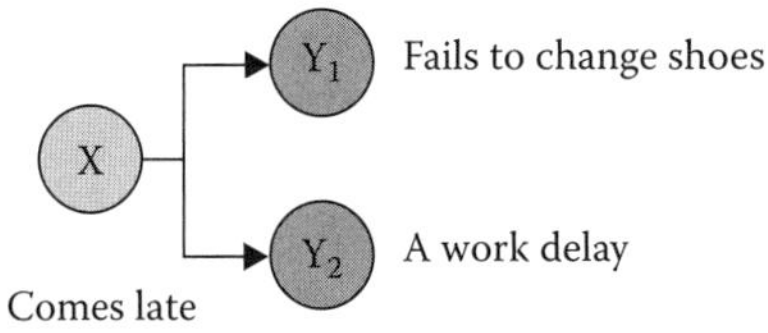	Conjunction negation (several facts, one preceding event)
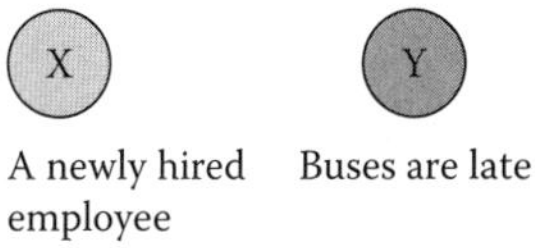	Independent facts

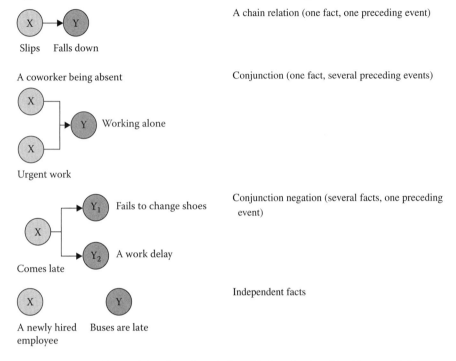

Source: Adapted with permission from Pietrzak, L. 2004. *Investigation of Accidents at Work*. Warsaw: CIOP-PIB.

the indirect causes of the accident, which may be associated with safety management in the workplace.

The construction of a fault tree can help track the course of the accident while taking into account the activities of various correlated factors. However, the most significant goal of an accident investigation is to determine appropriate preventive actions. The fault tree method allows a substantial broadening of the possible preventive methods thanks to a systematised search for activities that might prevent a repeat accident.

To identify these activities, all facts from the fault tree must be examined systematically. Whether one or more ways exists to eliminate the cause, prevent its occurrence, or avoid its adverse effects can then be determined. Methods of introducing preventive actions using the fault tree method are presented in Table 21.3.

Figure 21.8 presents the final part of a sequence that culminated in an accident in which employees suffered burns as a result of a fire caused by flammable materials stored at the entrance of a workshop with an obstructed back exit.

TABLE 21.3
Methods for Introducing Preventive Actions Using the Fault Tree Method

Eliminate directly the accident factor or the factor that is one of the prerequisites for the occurrence of another factor	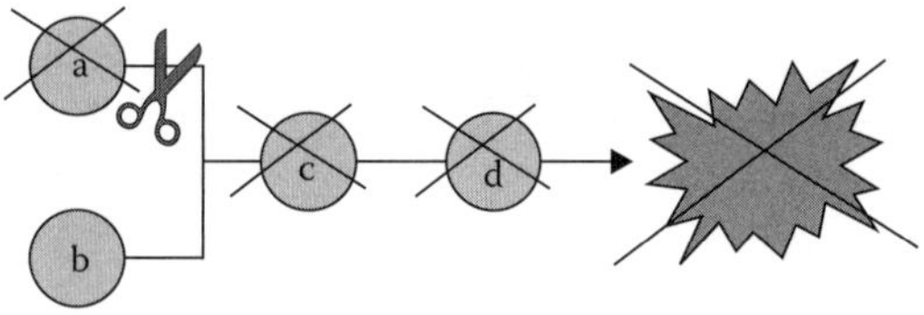
Introduce an element preventing the occurrence of the accident	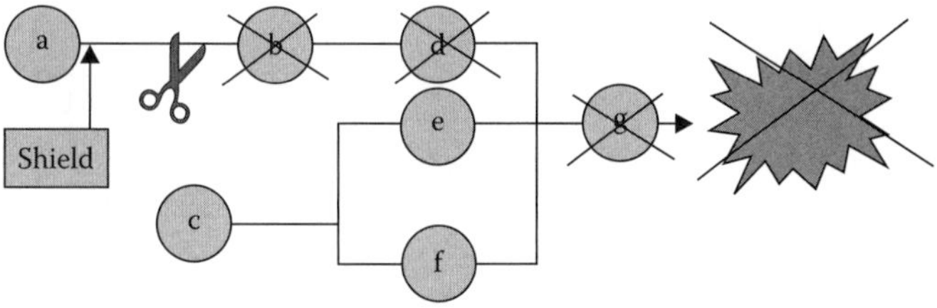
Decrease the probability of one event	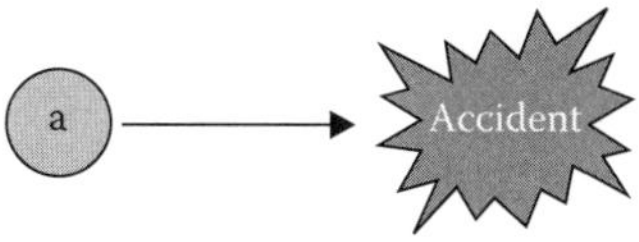
Increase the number of prerequisites by introducing additional events	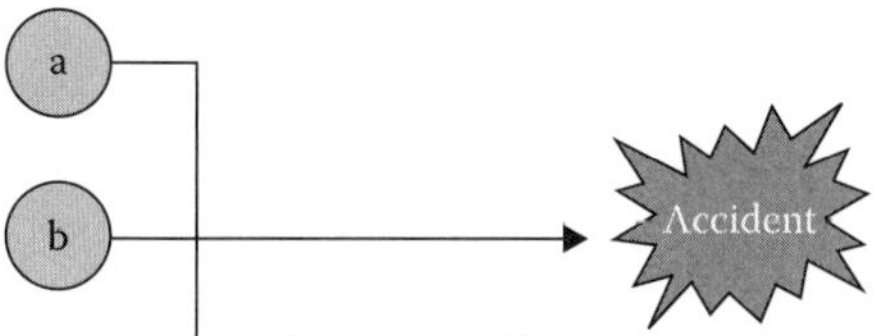 … $P_{acc} = P_{a\ occ} \cdot P_{b\ occ} \cdot ......$

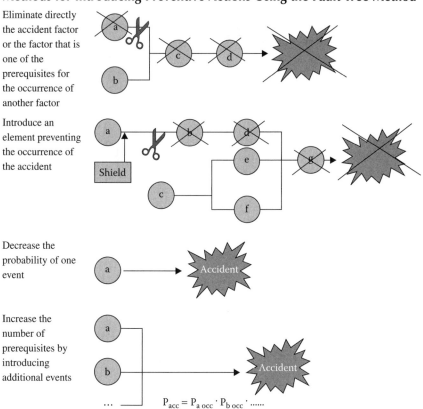

Source: Adapted with permission from Pietrzak, L. 2004. *Investigation of Accidents at Work.* Warsaw: CIOP-PIB.

The accident injuries consisted of burns. The direct cause of the burns was the ignition of flammable materials stored at the workshop entrance. Fire at the entrance and a blocked back exit made a quick escape impossible. Both the fire and the inability to escape quickly were events caused by many indirect first-degree causes and source causes, which are not shown in Figure 21.8. The investigation found that—apart from the fire and the difficult escape—the accident probably would not have occurred if the risk of fire had been identified earlier. The event could have been averted if the risks associated with performing work in a warehouse designed for temporary storage of materials had been assessed and the persons responsible for performing the tasks had been informed of the occupational risk.

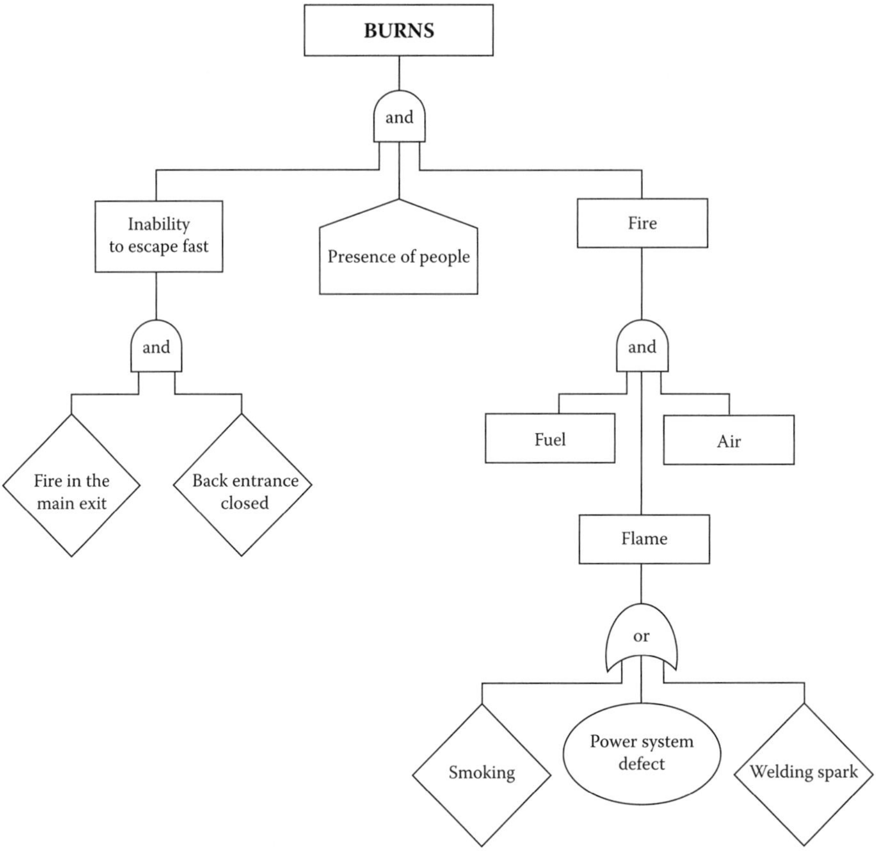

FIGURE 21.8 Investigation and explanation of causes of burns sustained by warehouse employees.

21.7.6 Analysis of Changes

This method is used to identify the causes of accidents by comparing the normal, expected course of work activities and operations with a course of events that was changed as a result of disturbances. This method requires the following for its application:

- Identify the problem—what happened?
- Specify the norm—what was supposed to happen?
- Identify, locate, and describe the changes that occurred.
- Specify what, when, and where things changed, and to what extent, within what limits, size, and degree?
- Specify what was violated, what constituted a deviation from the standard, and what was not caused by the changes?
- Identify the characteristic features of the changes.
- List the possible causes of the changes.
- Decide on the most probable causes of the changes.

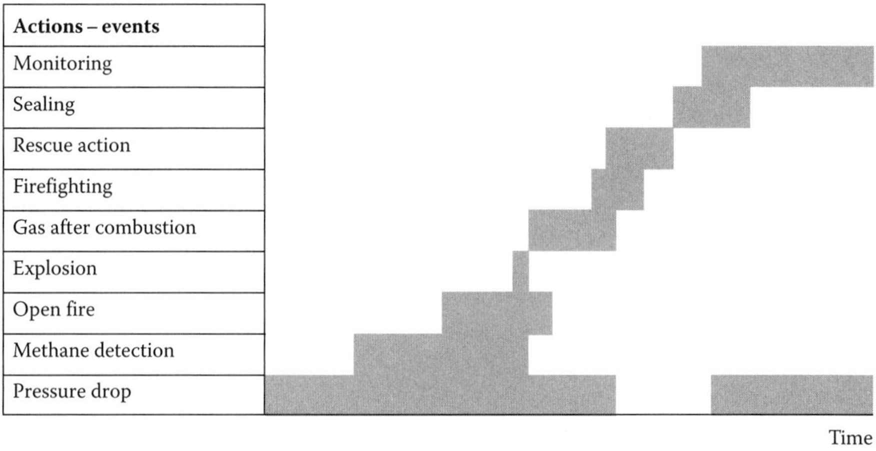

FIGURE 21.9 Gantt's chart illustrating the sequence and duration of activities before and after a methane explosion.

If the accident results due to a deviation from expected norms or standards, the causes of this deviation are the causes of the accident.

21.7.7 Gantt's Sequence Sheet

Gantt's sheet is a diagram representing a given sequence of events. It puts events in their proper order and identifies the sequence of events leading to accidents. It is particularly useful for visualising events that occur simultaneously. Figure 21.9 shows how to document actions that proceed and follow a methane explosion using a Gantt's sheet.

21.8 DOCUMENTING ACCIDENTS AT WORK

The documentation of the accident investigation is the evidence of postaccident proceedings. The documentation involves preparing a protocol for the circumstances and causes of the accident, filling out a statistical card on the accident at work, and recording the accident in the register of the establishment.

21.8.1 Postaccident Protocol

A protocol form for recording the circumstances and causes of accidents at work is attached to the ordinance of the Minister of Economy and Labour (DzU 2004, no. 227, item 2298). The way it should be prepared is specified in the ordinance of the Council of Ministers. In particular, this concerns identification of the circumstances and causes of accidents at work and the manner of their documentation, as well as the scope of information to be included in the register of accidents at work (DzU 1998, no. 115, item 744, as amended).

The protocol for specification of the circumstances and causes of the accident, known as the postaccident protocol, is prepared by the postaccident team no later than 14 days after receiving the accident notification. Attached to the protocol is a record of the explanations of the injured parties and information gathered from witnesses and other documents prepared while examining the circumstances and causes of the accident. This record includes, in particular, the written opinion of a doctor or other experts, layouts or photographs of the accident location, and a separate opinion provided by a postaccident team member with the member's remarks and reservations.

The protocol for specification of the circumstances and causes of the accident includes the following:

- Information on the members of the postaccident team
- Personal information on the injured persons
- Information on who reported the accident and when
- Specification of the type of accident
- A description of the circumstances and causes of the accident
- Specification of the effects of the accident (type and location of trauma)
- Legal qualification of the accident
- Specification of the contribution of the establishment and the injured person to causing the accident
- Conclusions for prevention purposes
- Confirmation that the injured parties or their family members have acknowledged the content of the postaccident protocol
- Signatures of the members of the postaccident team and a signature of the employer confirming the protocol

If the injured party was not an employee, but he or she was covered by accident insurance, an accident sheet is prepared instead of a postaccident protocol. An accident sheet form and the information concerning the mode of its preparation specify the regulations of the Ministry of Labour and Social Policy on the manner of recognition of the event occurring within the period of validity of the accident insurance provisions as an accident at work, the legal qualification of the event, the accident sheet form, and the date of its preparation (DzU 2002, no. 236, item 1992).

21.8.2 Statistical Cards on Accidents at Work

Statistical cards on accidents at work are prepared for each of the injured parties if the records of the postaccident protocol or the accident sheet state that the accident under investigation was an accident at work or an equivalent event. A card form and instructions are provided in the ordinance of the Ministry of Economy and Labour on the statistical sheet for an accident at work (DzU 2004, no. 269, item 2672). The regulation introduced changes to the statistical card for accidents at work. The type and scope of the information on accidents at work that is gathered for statistical purposes have changed because of the necessity to standardise the rules of recording and analysis of this data at the European Union level. The statistical card on accidents at work in its present form is valid from 2005.

Information gathered using the statistical card on accidents at work is divided into four parts, depending on whether it pertains to the employer, the injured party, the accident's effects, or the course of the accident. The information pertinent to the employer includes the address of the employer (the seal), company statistical ID (from REGON, the National Register of Economic Units), the number of persons employed at the company, and the serial number of the statistical accident sheet at the company since the beginning of the year. The injured party data includes the following:

- Gender
- Year of birth
- Nationality
- Status of employment
- Occupation
- Work period at the position occupied
- Number of hours worked since the start of work shift

The circumstances and course of the accident are described using the statistical model of an accident at work, prepared by the Statistical Office of the European Communities–Eurostat (Figure 21.10). The approved model has three phases:

1. Preaccident phase
2. Accident phase
3. Postaccident phase

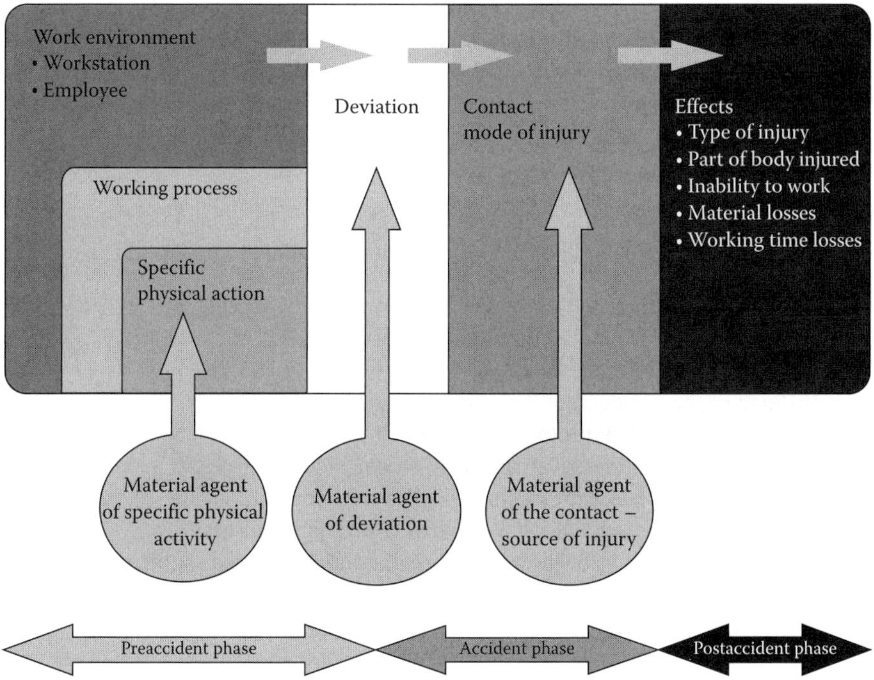

FIGURE 21.10 A statistical accident model according to Eurostat.

The preaccident phase deals with the employee in their work environment, which consists of the following:

- The place the work is performed
- The working process
- The activity being performed by the injured party at the time of the accident and the associated material agents

The work environment is the place the injured party was during the accident. The work process, the activity performed, and the associated materials are the components associated with the work organisation of the injured party and describe the conditions prior to the event that initiated the accident.

The accident phase starts at the moment a deviation from the normal working process occurs—an abnormal event which triggers the accident. The deviation itself is part of the accident phase, although it is the event that separates the preaccident phase from the accident phase, and, as such, does not have to be related to the injury. The materials associated with the deviation are those that caused a disturbance in the work process or that were associated with such a disturbance.

The accident phase also includes the event that caused the injury. This phase describes how the victim was hurt physically or mentally. It is the last event that leads directly to the injury. The description is completed by specifying the material agent that was the source of injury.

The postaccident phase includes the broadly understood consequences of the accident, such as:

- Type of injury and part of body injured
- Duration of inability of the injured person to work
- Material losses

21.8.3 Register of Accidents

The regulations of the Council of Ministers (DzU no. 115, item 1582) require the employer to keep a register of accidents based on postaccident protocols. The register should include:

- First name and surname of the injured person
- Place and date of the accident
- Effects of the accident on the injured party
- Date of postaccident protocol preparation
- Confirmation that the given accident is an accident at work
- Short description of the circumstances of the accident
- Date application was filed at the Social Insurance Institution
- Other circumstances of the accident that are to be entered in the register

21.8.4 Registration and Analysis of Noninjury Incidents

In designing a system for the registration and analysis of incidents at a company, the basic objective of gathering and analysing the data is to be able to plan and implement effective prevention activities. Therefore, the system structure should include three basic components: (1) identification and registration of noninjury incidents, (2) their analysis, and (3) preventive actions.

Data gathered on noninjury incidents and their analysis can be used for planning and implementing effective preventive measures. The usefulness of information gathered for planning of preventive activities is determined by its quality and quantity. This depends upon proper dissemination of the information and the motivational activities accompanying the implementation of the rules of registration and analysis of noninjury incidents at the company (Figure 21.11).

Since the only source of information on noninjury incidents is the employee concerned, he or she must be motivated using appropriate methods to make sure that the information gathered allows implementation of effective preventive measures. Information associated with any dangerous employee behaviour must also be gathered, as such behaviour often leads to accidents that cannot be prevented using traditional methods. Obtaining such information is very difficult; however, planning preventive activities brings measurable benefits, such as reducing the number of injury-related accidents and developing a safety culture in the company.

Efficient prevention of accidents at work through the registration and analysis of noninjury incidents requires the employers to change their approach to work safety management. It is one of the elements of the occupational health and safety

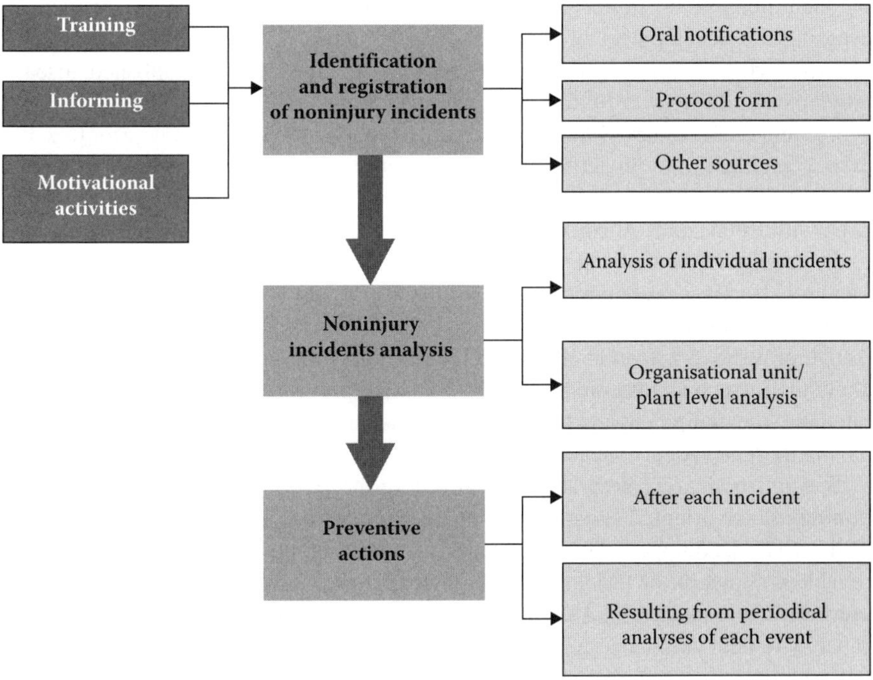

FIGURE 21.11 System of registration and analysis of noninjury incidents at a high-risk establishment. (Adapted with permission from Dudka, G. 2005. *Occupational Safety* 3:12–19.)

management system described in standard PN-N 18001:2004. Moreover, the employer must modify employees' attitudes; this is even more difficult as it cannot be ensured solely by implementation of the system. Devising and implementing procedures is often not sufficient to obtain information on noninjury incidents from the employees; no order can force them to report events that do not result in injuries, and thus—in their opinion—events which are of no consequence. Employees must understand and feel the need to report such events, as this affects the level of safety and the safety culture at the company.

21.9 ACCIDENT PREVENTION

Prevention is a set of procedures aimed at avoiding the occurrence of adverse events. The modern concepts of prevention are based on traditional rules applied in medicine to avoid the spread of diseases, as well as procedures developed to meet the needs of space organisations that detect the potential causes of equipment breakdowns during operations in a space environment that is not sufficiently understood (Johnston et al. 1980).

21.9.1 Prevention Concepts

One of the earliest and the best known medical prevention concepts was based on identification of three prevention levels applied in practice: (1) primary—avoiding the development of a disease; (2) secondary—aimed at early treatment; and (3) tertiary—limiting the negative impact of an already established disease (Anderson and Menckel 1995).

Catalano and Dooley (1980) divided primary prevention (occurring prior to a disease or accident) into proactive and reactive activities. Proactive activities reduce or eliminate the hazard, while reactive activities increase the individual's capacity to cope with the hazard. According to Haddon (1970), accident prevention should aim to reduce the hazardous energy found within the zone occupied by workers or to increase effectiveness of control.

Haddon also introduced the dichotomy of active and passive preventive measures. Active preventive measures are those that require making a conscious decision to apply them, such as fastening a seat belt. Passive prevention measures are incorporated into the equipment and triggered by the hazard itself, without the user acting or having to make a decision, such as an air bag.

Specialists dealing with accident prevention have recognised that human functioning in task-oriented situations is not free from errors; even an individual who is well prepared, equipped, and motivated to act safely cannot always ensure a perfectly reliable completion of tasks (Anderson and Menckel 1995; Hale 1995). Therefore, preventive measures aimed at increasing human efficiency and improving human capacity to deal with hazards are useful but do not guarantee full safety. Such activities are referred to as *relative prevention*, as opposed to *absolute prevention*, which makes it impossible to commit a dangerous mistake or blocks the adverse effects of erroneous actions. In the 1980s, risk management procedures were used to design preventive measures (Roughley et al. 1984). Their application required identification of hazards, assessment of risks and making decisions that would determine the mode of coping with a given risk.

21.9.2 Types of Preventive Actions

Accidents result when people are exposed to dangerous conditions, or when they perform dangerous actions regardless of the conditions. Prevention programmes are therefore designed to change conditions and actions. Programmes focused on conditions recommend the application of technical or organisational safety measures and are referred to as *technical prevention*. Programmes that facilitate actions are referred to as *psychological prevention*.

In practice, various preventive measures and strategies can be applied to eliminate or reduce hazards, to limit human access to hazards, to make organisms resistant to the destructive influence of hazards, and to reduce the effects of hazards that already existed but were not eliminated on time. In devising and selecting preventive measures, a higher priority should be assigned to measures that eliminate or reduce the hazards than to measures that make people resistant to their influence.

Haddon (1980) formulated ten principles for the selection of preventive strategies:

1. Prevent the occurrence of threats at their source.
2. Reduce the quantity of energy generated or consumed.
3. Prevent release of existing hazardous energy.
4. Reduce the speed at which hazardous energy released from its source is dispersed.
5. Separate the threat from the object to be protected in space and in time.
6. Separate the threat from the object to be protected by placing a material barrier between them.
7. Modify the basic characteristics of hazards.
8. Increase protection of the object to make it more resistant to the damage that may be caused by the hazard.
9. Counteract the damage already caused by the active threat.
10. Stabilise, repair, and restore the damaged object.

Rankin and Zabetakis (1977) have proposed two other prevention strategies to reduce human fallibility and raise the individual's resistance to dangerous practices. First, they recommend devising and applying detailed procedures for the performance of particularly hazardous tasks. Second, they propose increasing individuals' reliability by:

- Providing workers with training and knowledge so that they can recognise hazards and take safe actions in a hazardous situation
- Disseminating information on hazards
- Equipping people with appropriate, specialised tools where required
- Allocating tasks suited to the physical and mental abilities of those entrusted with their implementation

- Motivating workers to act safely and avoid risks
- Limiting exposure to physical and emotional stressors

Accident prevention measures should include avoiding situations that have previously caused accidents and thereby reducing the risk present in the processes, tasks, and elements of the workplace. This calls for communicating information on the characteristics of hazards and causes of accidents, promoting life and health, establishing a safety culture of caring for one's life and health, and displaying the desirable personality traits. Well-formulated preventive actions reduce or eliminate the hazards in the human environment and improve the ability to act safely. Selective use of only one type of method, for instance, employing only technical or organisational processes or only activities that improve human performance does not yield satisfactory results. A person who is disinclined to or unaware of acting safely can damage the preventive effects of even the safest technology.

Not all preventive measures proposed can be applied by an organisation. Selecting the most appropriate actions calls for applying the right criteria. Individual preventive actions should be compared with their expected efficiency. The basic criteria for selecting preventive measures are:

- *Compliance with regulations*: The measures proposed should be consistent with the legal provisions and standards in force.
- *Cost for the company*: The financial costs of implementing the proposed measures are often of vital importance in their final selection.
- *Effects on the employee*: Application of the designed preventive actions must not lead to big changes in the work process or increase the employees' workload.
- *Durability of activities*: The effectiveness of the preventive measures selected should not diminish over time.
- *Impact range*: Measures proposed should have wide applicability and be effective in solving safety problems in other situations as well. Preventive measures associated directly with the injury may eliminate the effects of dangerous situations, but will not eliminate the situations themselves. Preventive measures aimed at indirect, distant causes will eliminate these situations.
- *Time of application*: Preventive measures that require more time to implement may have a wider impact than those that can be implemented over a short period. Therefore, it is important not to eliminate preventive measures that are implemented over a longer duration. A specific schedule to introduce such activities can be drawn up (e.g. as part of an annual plan of improvement of work conditions).
- *Possibility of risk relocation*: The measures planned for one location must not have negative effects elsewhere in the organisation or expose the employees to other hazards.

The selection of preventive measures does not guarantee their effectiveness; they must be implemented correctly. Therefore, persons responsible for monitoring the implementation, definition, and control of the completion deadlines must be designated. Moreover, to increase the effectiveness of these activities, employees must understand their usability and significance. This can be achieved by prompt dissemination of the appropriate information on accidents at work and the preventive measures associated with them.

REFERENCES

Act of October 30th, 2002 on social insurance related to accidents at work and occupational diseases. DzU no. 199, item 1673, as amended.

Anderson, R., and E. Menckel. 1995. On the prevention of accidents and injuries. *Accid Anal Prev* 6:757–768.

Arnett, J. 1992. Reckless behavior in adolescence: A developmental perspective. *Dev Rev* 12:339–373.

Benner L., Jr. 1975. Accident investigations: Multilinear events sequencing methods. *Saf Res* 2:67–73.

Booysen, A., and J. Erasmus. 1989. The relationship between some personality factors and accident risk. *Afr J Psychol* 19:144–152.

Catalano, R., and D. Dooley. 1980. Economic change in primary prevention. In *Prevention in Mental Health*, eds. R. H. Price, R. F. Ketterer, B. Bader, and J. Monahan, 21–40. London: Sage Publications.

Chmiel, N. 2003. Safety at work. In *Introduction to Work and Organizational Psychology*, ed. N. Chmiel, 283–303. Gdansk: Gdansk Psychology Publishers.

Claparède, E. 1924. *Occupational Counselling: Tasks and Methods*. Kalisz: The Kalisz District Printing House.

Cooper, D. 1998. *Improving Safety Culture*. Chichester: Wiley.

Corlett, E. N., and G. Gilbank. 1978. A systemic technique for accident analysis. *J Occup Accid* 2:25–38.

Crancer, A., and D. L. Quiring. 1970. Driving records of persons hospitalized with suicidal gestures. *Behav Res Highway Saf* 1:33–42.

Delong, T. 1970. *A fold tree manual.* Research report. Industrial Engineering Department: Texas A&M University.

Dudka, G. 2005. Registering potential incidents. *Occupational Safety* 3:12–19.

Franken, R. E. 2005. *Human Motivation*. Gdansk: Gdansk Psychology Publishers.

Furnham, A. 1992. *Personality at Work*. London: Routledge.

Geller, E. 2002. *The Participation Factor. How to Increase Involvement in Occupational Safety*. Des Plaines: American Society of Safety Engineers.

Gliszczyńska, X. 1963. *Human Participation in Accident Causation*. Warsaw: PWN.

Haddon, W. 1970. On the escape of tigers: An ecologic note. *Tech Rev* 72:44–53.

Haddon, W. 1980. The basic strategies for reducing damage from hazards of all kinds. *Hazard Prev* 16:8-12.

Hale, A. R. 1995. The individual. In *Safety at Work*, ed. L. Bamber, 254–298. Oxford: Butterworth-Heinemann Ltd.

Hale, A. R., and M. Hale. 1970. Accidents in perspective. *Occup Psychol* 44:115–121.

Hallowell, E. M., and J. J. Ratey. 2004. *Driven to Destruction: Recognizing and Coping with ADD from Childhood to Adulthood*. Poznan: Family Media.

Hansen, A. 1993. *Health and Safety at Work*. Warsaw: WSiP.

Heinrich, H. W. 1936. *Industrial Accidents Prevention: A Scientific Approach*. New York: McGraw-Hill.

Heinrich, H. W. 1959. *Industrial Accidents Prevention*. New York: McGraw-Hill.

ILO-OSH. 2001. *Guidelines on Occupational Safety and Health Management Systems.* Warsaw: CIOP.
Johnston, A. G., J. McQuard, and G. A. C. Games. 1980. Systematic safety assessment in the mining industry. *Min Eng* 3:723–735.
Kenny, P. D. 1993. *Correlation and Causality*. New York: Wiley.
Kindler, H. 1990. *Risk Taking*. London: Kogan Page.
McMains, M. J., and W. C. Mullins. 2001. *Crisis Negotiations*. Cincinnati: Anderson Publishing.
Ordinance of the Council of Ministers of July 28th, 1998 on identification of the circumstances and causes of accidents at work and the ways of documenting them. DzU no. 115, item 744, as amended.
Ordinance of the Minister of Economy and Labour of December 8th, 2004, on the statistical sheet for an accident at work. DzU no. 269, item 2672, as amended.
Ordinance of the Minister of Economy and Labour of September 16th, 2004, on the form of a protocol for specification of the circumstances and causes of accidents at work. DzU no. 227, item 2298, as amended.
Ordinance of the Minister of Labour and Social Policy of December 19th, 2002, on the mode of recognition of the event taking place within the period of validity of the accident insurance as an accident at work, the legal qualification of the event, the accident sheet form and the date of its preparation. DzU no. 236, item 1992, as amended.
Pidgeon, N. F. 1991. Safety culture and risk management in organizations. *J Cross Cult Psychol* 22:129–140.
Pietrzak, L. 2004. *Investigation of Accidents at Work.* Warsaw: CIOP-PIB.
PN-N-18001:2004. Occupational health and safety management systems. Requirements.
Pszczołowski, T. 1978. *Little Encyclopaedia of Praxeology and Theory of Organization.* Wroclaw: The Ossolinski National Institute.
Rankin, J. E., and M. G. Zabetakis. 1977. *Job Safety Analysis. MESA Safety Manual No. 5.* Washington: U.S. Government Printing Office.
Rasmussen, J. 1986. *Human Information Processing and Human Machine Interaction.* Amsterdam: North-Holland.
Ratajczak, Z. 1988. *Human Infallibility at Work.* Warsaw: PWN.
Reason, J. 1997. *Managing the Risks of Organizational Accidents*. Aldershot: Ashgate.
Reason, J. T. 1990. *Human Error*. Cambridge, UK: Cambridge University Press.
Reber, A. S. 2000. *The Penguin Dictionary of Psychology.* Warsaw: Scientific Scholar.
Roughley, R., et al. 1984. *Risk Management. Practical Techniques to Minimize Exposure to Accidental Losses*. London: Kogan Page.
Sanders, M. S., and J. M. Pay. 1988. *Human Factors in Mining*. Pittsburgh, PA: U.S. Department of the Interior.
Smillie, R. J., and M. Ayoub. 1976. Accident causation theories: A simulation approach. *J Occup Accid* 1:47–68.
Stranks, J. 1990. *A Manager's Guide to Health and Safety at Work.* London: Kogan Page.
Studenski, R. 1986. *Accident Causation Theories and their Empirical Verification*. Katowice: GIG.
Studenski, R. 2004. *Risk and Risk Taking*. Katowice: Wydawnictwo Uniwersytetu Śąskiego.
Studenski, R. 2007. Self-destructive motivation to risk taking. *Psychological Colloquia* 16: 176–195.
Suchańska, A. 1998. *Psychological Manifestations and Conditions of Indirect Self-destructiveness*. Poznan: Adam Mickiewicz University.
Szczurowski, A. 1980. Determination of accident causes and models of mining accidents. *Research Papers of CIOP* 105:91–97.

Tomaszewski, T. 1976. Basic forms of organizing and regulating behaviour. In *Psychology*, ed. T. Tomaszewski, 491–533. Warsaw: PWN.
Wanat, J. 1964. Taxonomy of factors determining the origin of occupational accidents in coal mines. Studies of the Central Mining Institute. Statement 350.
Zabetakis, M. 1991. *Accident Prevention.* Safety Manual, no. 4.

22 Major Industrial Accidents

Jerzy S. Michalik

CONTENTS

22.1 INTRODUCTION

Major accidents often have catastrophic results and pose serious problems, particularly in industrialised countries. Major accidents are failures of technological and storage systems or transport devices that result not only in explosions or fire but also in the release of large quantities of hazardous chemicals found in such facilities into the environment.

The consequences of industrial and transport accidents for people and the environment, as well as the scope of material losses, mean that many international organisations and states have adopted environmental programmes and incorporated certain tasks in order to protect the life and health of workers, protect the environment, and maintain the safety of the work environment.

The industrial disasters in Flixborough, Seveso, San Juanico near Mexico City, and Bhopal were particularly instrumental to establishing international regulations on control of major accidents, as well as amendments to and improvements in the existing legislation and management systems in this regard. The most significant information on the course and consequences of the gravest industrial disasters in history are presented below.

Flixborough, Great Britain, 1 June 1974. In the chemical plant of Nypro Ltd., manufacturing mainly caprolactam, a raw material used in the production of nylon, about 80 tonnes of hot liquid cyclohexane leaked from a broken pipeline. The mixture of this substance with air caused an explosion equivalent to 30 tonnes of trinitrotoluene (TNT). Twenty-eight plant employees died and another 36 were injured; several hundred people outside the facility suffered various consequences; in 53 cases, the effects were serious. The plant and an area around it of about 5 km were completely destroyed, and substantial damages were recorded further outside the site as well. The material losses were estimated at ECU 110 million.

Seveso, Italy, 10 July 1976. In the ICMESA chemical plant in the suburbs of Seveso, 20 km from Milan, an unexpected exothermic side reaction and a consequent rise in temperature and pressure levels caused the safety valve of a reactor for producing 2,4,5-trichlorophenol (TCP) to open after a process cycle ended. About 2 tonnes of chemical substances, including about 2 kg of TCDD (2,3,7,8-tetrachlorodibenzo-p-dioxin), a highly toxic dioxin (a dose of about 0.1 mg is lethal for humans), were released into the atmosphere. Almost 1500 hectares of a heavily populated area were contaminated; about 700 inhabitants suffered poisoning, many animals were killed, and the about 40 other plants were affected by the contamination. Extensive areas were excluded from agricultural cultivation for almost 10 years. The estimated loss amount was ECU 72 million.

San Juanico, Ixhuatepec, near Mexico City, Mexico, 19 November 1984. A liquefied petroleum gas (LPG; about 80% butane and 20% propane) pipeline burst, releasing a cloud of gas that ignited and damaged the tanks, causing fractures and leaks, and leading to disastrous explosions. Fifteen of the 48 cylindrical tanks, each weighing about 20 tonnes, turned into flaming missiles and travelled a distance of over 100 m (some fragments even to 1000–1200 m), causing catastrophic damage outside the site. Four huge (1500 m^3) spherical tanks were decimated by the boiling liquid expanding vapour explosion (BLEVE). The explosion created fireballs 200–300 m in diameter, accompanied by strong thermal radiation, powerful violent blasts, chips, and flaming gas droplets ('a rain of fire'). About 550 people were killed, more than 4000 were badly burnt or injured and 60,000 residents were evacuated. The material and environmental losses caused by the fire and explosion of about 12,000 m^3 of LPG gas were enormous.

Bhopal, India, 3 December 1984. This was the worst industrial disaster in the world in terms of its human toll and occurred at a plant in Bhopal that belonged to the Union Carbide Corporation that manufactured the insecticide carbaryl and methyl isocyanate (MIC), a raw material used in its production. Due to unexpected exothermic polymerisation and/or hydrolysis of the MIC, the temperature of the system soared to about 200°C, heating the valves and resulting in fractures in the concrete. This released about 30 tonnes of MIC vapour for about an hour, until the situation was brought under control. The disaster resulted in about 16,000 fatalities, while about 100,000 people suffered serious health damage. The timely evacuation of almost 200,000 people avoided even more tragic consequences.

Schweizerhalle, Basel, Switzerland, 1 November 1986. A fire broke out in the warehouses of the well-known Sandoz company, which stored about 680 tonnes of pesticides. The water used during firefighting was contaminated with mercury- and zinc-based pesticides, as well as phosphor-organic insecticides and was then discharged into the sewage system and then into the Rhine river. These substances ranged from about 5 to 20 tonnes. The consequences of the disaster were grave—biological life in the Rhine was destroyed along a section of about 400 km, water intake in Germany and the Netherlands ceased, and on the French shore, economic activity and tourism related to the Rhine vanished.

Baia Mare, Romania, 31 January 2000. Snow melting after a heavy snowfall damaged the banks of settling tanks and postflotation sludge tanks in the gold mining and manufacturing complex of Aurul. Waste containing large quantities of cyanides was discharged into the Tisza river and then into the Danube, causing serious devastation of the aqueous environment.

Enschede, the Netherlands, 13 May 2000. A fire and subsequent explosion of fireworks materials stored in the warehouse of the SE Fireworks Plant caused 20 fatalities, about 1000 injuries, and damage or destruction of about 600 buildings.

Toulouse, France, 21 September 2001. A series of explosions occurred in the ammonium nitrate warehouse of the AZF plant of Grande Paroisse, where about 400 tonnes of this substance were stored. There were 30 fatalities, including eight people outside the plant site, and 2500 people were injured; in 30 cases, the injuries were very serious and one person died. Material losses were substantial (about €1.5 billion). The force of the explosion was equivalent to 20–40 tonnes of TNT.

Czechowice-Dziedzice, Poland, 26 June 1971. This was the worst industrial disaster in Poland. At about 7.50 P.M., lightning struck the dome of refinery tank 1, which caught fire. Crude oil was still being pumped from this tank to the distillation department. Oil being pumped from railroad tank cars to tank 4 continued as well. At 1.30 A.M., burning tank 1 exploded; tank 4 followed immediately. The fire spread to the remaining two tanks as well as to the pumping station, the engine-oil production plant, and other site locations. The evacuation of inhabitants of the surrounding area commenced at dawn. The fire was extinguished after about 60 hours, by which time 37 people were dead and more than 100 suffered severe burns and other injuries. Material losses were extensive and affected areas besides the installations listed and the plant site; 30 fire engines were destroyed.

Issues associated with preventing major industrial accidents and limiting their consequences were regulated in the European Union (EU; EEC at the time) as early as in 1982. Council Directive 82/501/EEC on the major accident hazards of certain industrial activities, also called the Seveso Directive (1982), was the applicable document at the time, along with amendments introduced by Council Directives 87/216/EEC, 88/610/EEC and 91/692/EEC.

The EU's current legal regulations on the control of major accident hazards consist of two fundamental legislative acts:

1. Council Directive 96/82/EC on the control of major accident hazards involving dangerous substances, also called the Seveso II Directive (1996)
2. Directive 2003/105/EC of the European Parliament and of the council amending the Seveso II Directive (2003)

The following decisions govern their detailed provisions:

- Commission Decision 98/433/EC of 26 June 1998 on harmonised criteria for dispensation according to Article 9 of Council Directive 96/82/EC (1998)
- Council Decision 98/685/EC of 23 March 1998 on the conclusions of the Convention on the transboundary effects of industrial accidents (1998)

The provisions of the Seveso II Directive are the basis for Polish legal regulations on the control of major industrial accidents. At present, these issues are governed by the following domestic regulations:

- The Environmental Protection Law (2001)
- The act on implementation of the Environmental Protection Law (2001)
- The act on amendment of the Environmental Protection Law (2006)
- Ordinance of the Ministry of Economy, amending the ordinance on the types and quantities of hazardous substances whose presence in an establishment is conclusive to qualify it either as a lower-tier establishment (LTE) or an upper-tier establishment (UTE; 2006)
- Ordinance of the Ministry of Economy, Labour and Social Policy on the requirements to be met by a safety report (2003)
- Ordinance of the Ministry of Economy and Labour amending the regulation on the requirements to be met by a safety report (2005)
- Ordinance of the Ministry of Economy, Labour and Social Policy on the requirements to be met by emergency plans (2003)
- Ordinance of the Ministry of the Environment on the detailed scope of information that must be made available to the public by the Provincial Commandant of the State Fire Service (2002)
- Ordinance of the Ministry of the Environment that major accidents must be reported to the Chief Inspector of Environmental Protection (CIEP; 2003)

As defined by the Seveso II Directive, a major accident is 'an occurrence such as a major emission, fire, or explosion resulting from uncontrolled developments in the course of the operation of any establishment covered by this directive, and leading to serious danger to human health and/or the environment, immediate or delayed, inside or outside the establishment, and involving one or more dangerous substances.' This definition of a major industrial accident is compliant with the Polish regulations, since the provisions of the Seveso II Directive exclude the transport of hazardous substances from its application. Polish regulations—the Environmental Protection Law—introduced the following definitions:

- A *major accident* is an event, in particular, an emission, fire, or explosion, which occurs during industrial processes, storage or transport, involving one or more dangerous substances and leading to an immediate threat to the lives or health of people or the environment or to a delayed emergence of such threat.

- A *major industrial accident* is a major accident at an industrial establishment.

According to Polish legal provisions, the concept of a major accident pertains to all facilities involving dangerous substances, including mobile transport equipment.

Definitions of the terms are as follows:

- An *establishment* is one or several installations together with the area to which the operator of the installation(s) has legal title, and the equipment located therein.
- An *installation* is a fixed technical unit or a complex of fixed technical units which are technologically interconnected, to which a single entity has legal title, located within the area of a single establishment; and/or building structures which are not technical units or a complex of units, which, when operated, may cause emission.

These definitions indicate that the term 'major industrial accident' refers to stationary facilities.

Major industrial accidents and the threat of potential catastrophes associated with dangerous chemical substances present in production and storage facilities are of particular significance in the countries of Europe because of the large number of establishments that use or manufacture these substances and the high population densities, particularly in industrial agglomerations.

22.2 CONCLUSIONS FROM ANALYSIS OF MAJOR ACCIDENTS IN EUROPE

The Joint Research Centre, within the ambit of activities coordinated by Major Accident Hazards Bureau (MAHB), conducted an analysis of past major accidents as ordered by the European Commission. This was an identification study and intended to gather information on the production of hazardous chemicals and their release into the environment during industrial accidents. Analyses of the results (Cozzani and Zanelli 1997; Michalik and Gajek 2002a) contain a wealth of information on the significant aspects of the emergence, course, and nature of major industrial accidents.

Crucial information on major industrial accidents and on the production of new hazardous substances during such accidents that resulted from these analyses is presented below.

The authors of this study analysed data on 550 accidents, which occurred mostly between the 1950s and 1990s, gathered from the following databases: ARIA (France), FACTS (The Netherlands), MHIDAS, (Great Britain), MARS, and CDCIR (MAHB–UE).

There were 406 accidents involving dangerous chemicals in which additional hazardous chemical substances were or could have been created, and in total, the analysed industrial accidents involved more than 350 chemical substances. Conclusions from

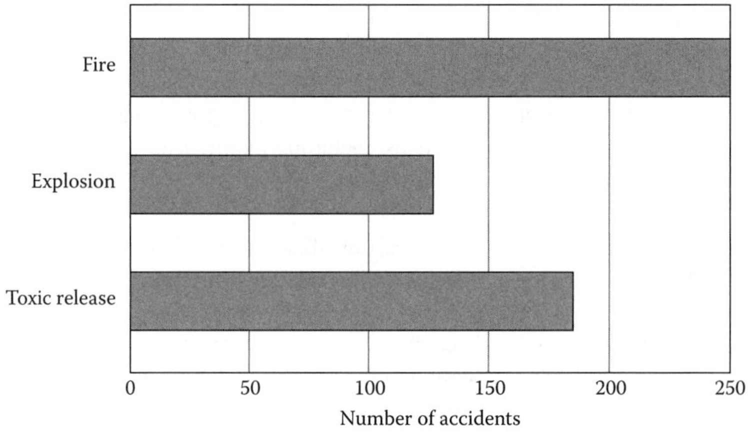

FIGURE 22.1 Numbers and types of industrial accidents. (Adapted from Cozzani, V., and S. Zanelli. 1997. *EUCLID—A Study on Emission of Unwanted Compounds Linked to Industrial Disasters*. Ispra: European Commission.)

an analysis of the data based on the type of accident, that is, fires, explosions, and releases of hazardous substances to the environment, are presented in Figure 22.1. Based on this data, the percentage of each type of major industrial accidents in the total number of events analysed was determined as follows:

- Fires—about 45%
- Explosions—about 22%
- Release of toxic substances—about 33%

The analyses determined the specific causes of these major accidents and their frequency. The two most significant causes were 'operator error' (about 20% of all accidents) and 'component failure' (about 10%). In most cases, accidents caused by operator error were due to insufficient training and lack of clear procedures and instructions.

According to the official records of the EU, the immediate causes of most major accidents were management and/or organisational shortcomings; inefficient management systems contributed to more than 85% of accidents. A concise overview of the most significant conclusions of the analysis of major accidents in Europe is presented in the following paragraphs.

Figure 22.2 illustrates the number of fatalities resulting from various accidents. In 62 events (15% of 406 accidents involving dangerous chemical substances), at least one death was recorded. In 19 accidents, the cause of death was release of toxic substances produced by adverse reactions initiated by the accident.

The consequences of accidents for society and the environment are presented in Figure 22.3. Employees of the establishments involved had to be evacuated in 220 accidents (about 55%), and the population of the surrounding areas in 110 cases (27%); 143 accidents (35%) caused transport bottlenecks. Contamination of soil and water took place in 56 accidents (13.5%).

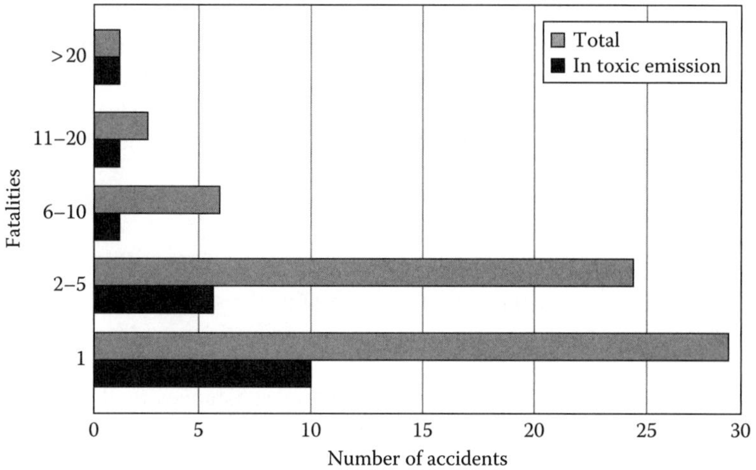

FIGURE 22.2 Number of accidents involving fatalities. (Adapted from Cozzani, V., and S. Zanelli. 1997. *EUCLID—A Study on Emission of Unwanted Compounds Linked to Industrial Disasters.* Ispra: European Commission.)

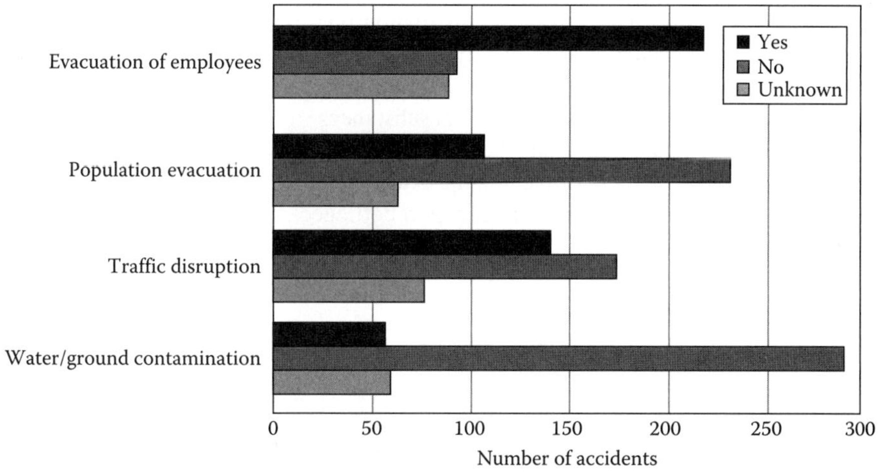

FIGURE 22.3 Consequences of accidents. (Adapted from Cozzani, V., and S. Zanelli. 1997. *EUCLID—A Study on Emission of Unwanted Compounds Linked to Industrial Disasters.* Ispra: European Commission.)

Significant conclusions can be drawn from analysis of the frequency of accidents in which dangerous chemical substances are produced, depending on the type of manufacturing processes and the field of industrial activity. Production of dangerous chemical substances poses an additional threat that is often more significant than the threat associated with the normal use of hazardous substance conditions. Procedures and policies aimed at control of major accidents in accordance with the EU and Polish regulations should take this into account.

Accidents involving the release or production and release of hazardous substances under uncontrolled conditions took place not only in chemical manufacturing processes (reactions) but also during physical operations and simple handling of liquids and solids and—in particular—during storage. Three basic physical and chemical processes that lead to accident-related production of additional chemical substances can be defined:

1. Fires
2. Runaway reactions
3. Unwanted reactions (production of substances resulting from unplanned contact)

Fires played a major role in producing dangerous substances in about 49% of all accidents, runaway reactions in about 32%, and unwanted reactions in about 19% (Figure 22.4). Fires occur mostly in warehouses, especially facilities for fertilisers and pesticides, which are particularly likely to form dangerous combustion products.

Runaway reactions lead to production of dangerous substances that constitute the most serious hazards, which occur in pesticide manufacture, petrochemical processes (mainly periodic polymerisation reactors), and the pharmaceutical industry, mostly in reaction devices during the distillation phase due to heating of chemical systems. Unwanted reactions are mostly the result of errors in handling solids and liquids resulting in accidental contact between inappropriate substances that react with each other to produce new dangerous substances.

Analyses of accidents show that the scenario that leads to production of hazardous chemical substances greatly influences the extent of each type of effect. Most effects, such as evacuation of the population, transport bottlenecks, and some fatalities, result

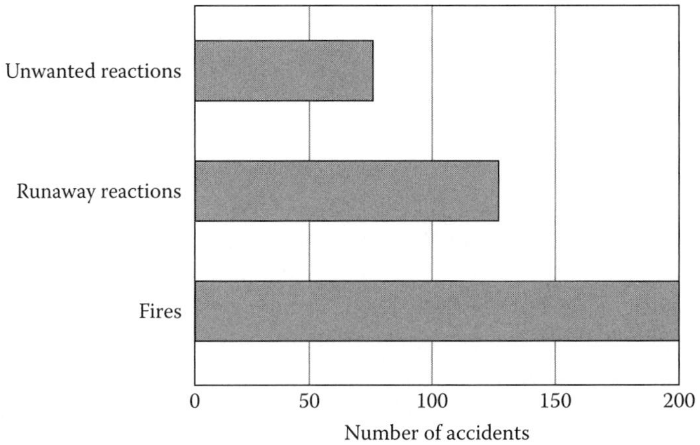

FIGURE 22.4 Accident scenarios that create new and dangerous chemical substances. (Adapted from Cozzani, V., and S. Zanelli. 1997. *EUCLID—A Study on Emission of Unwanted Compounds Linked to Industrial Disasters*. Ispra: European Commission.)

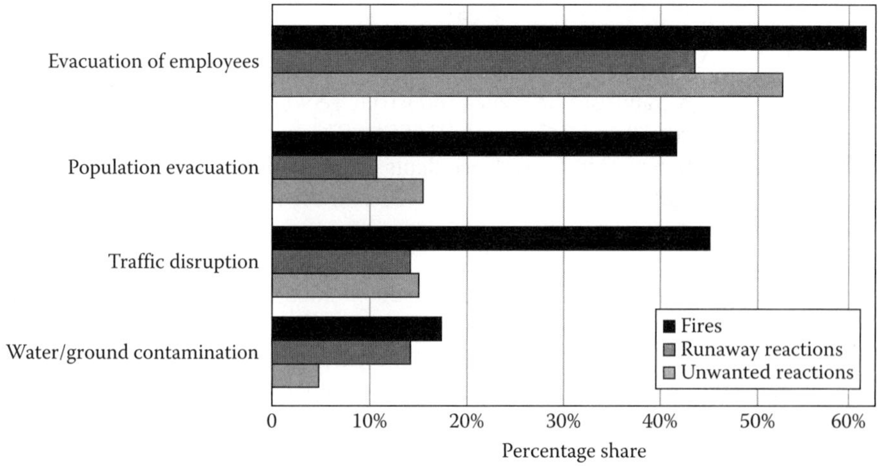

FIGURE 22.5 Percentages of accidents, by scenario, causing various consequences for people, infrastructure and the environment. (Adapted from Cozzani, V., and S. Zanelli. 1997. *EUCLID—A Study on Emission of Unwanted Compounds Linked to Industrial Disasters*. Ispra: European Commission.)

mainly from release of toxic substances produced in adverse reactions during accidents. Therefore, the production and release of toxic substances when loss of control occurs during an accident is the most significant cause of fatalities and environmental damage. Figure 22.5 presents a summary of the consequences of accidents to society and the environment depending upon the accident scenario. As the diagram indicates, there is no single dominant cause for the evacuation of employees. Fires are the most frequent reason for evacuation of the local population and for transport bottlenecks.

22.3 SUBSTANCES PRESENT AND PRODUCED DURING MAJOR ACCIDENTS

Information on substances that are present and produced during accidents is particularly significant for analyses of accident scenarios and assessment of their potential consequences. From the analyses, 352 chemical substances were found to be involved in major accidents:

- 277 substances were present during the processes analysed.
- 75 compounds (chemical substances) were those produced or that could have been produced during the accident.

Most often, industrial accidents involved pesticides, fertilisers, and polymers, as well as—in a substantial number of accidents—groups of chemical substances such as half-finished organic products and solvents and inorganic compounds used widely in various sectors of the chemical industry.

22.4 MAJOR INDUSTRIAL ACCIDENT HAZARDS IN POLAND

Poland has a high incidence of major industrial accidents. Figure 22.6 presents the number of high risk establishments (UTE) and increased risk establishments (LTE) involved in major industrial accidents in individual provinces. As of 31 December 2007, the number of UTEs was 158 and the number of LTEs was 208. The total number of plants subject to regulations on control of major industrial accidents was 366.

According to the CIEP data analysed for 2003–2005, about 29% of UTEs held disposal products of crude oil distillation and flammable substances, about 38% held extremely flammable liquefied gases and natural gas, and about 33% held toxic substances and other dangerous substances. The corresponding figures for the LTEs showed a more even distribution: 33%, 31%, and 36%, respectively.

The level of the hazards of major industrial accidents defined or measured by the number of UTEs allows for comparing the threats in various EU states. The potential of hazard positions Poland among the top 10 member states of the EU, as shown in Figure 22.7. The listing is based on a report of the European Commission adopted on 17 August 2007 assessing the progress made in the implementation of the Seveso II Directive in 2003–2005 (Commission of the European Communities 2007).

This report, prepared based on the reports of each of the member states for 2003–2005 (*Member States country reports...*), puts the number of UTEs at 3939 at the end of 2005, a 7.4% increase over the previous 3 years (up from 3677, i.e.

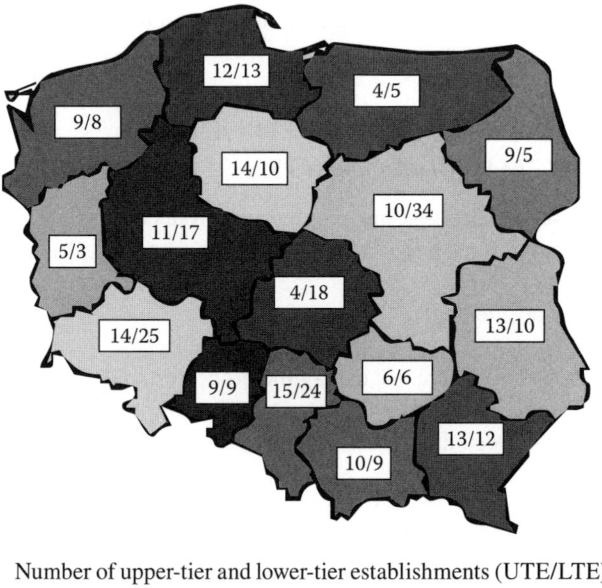

FIGURE 22.6 Number of upper-tier and lower-tier establishments (UTE/LTE) in individual provinces as of 31 December 2007.

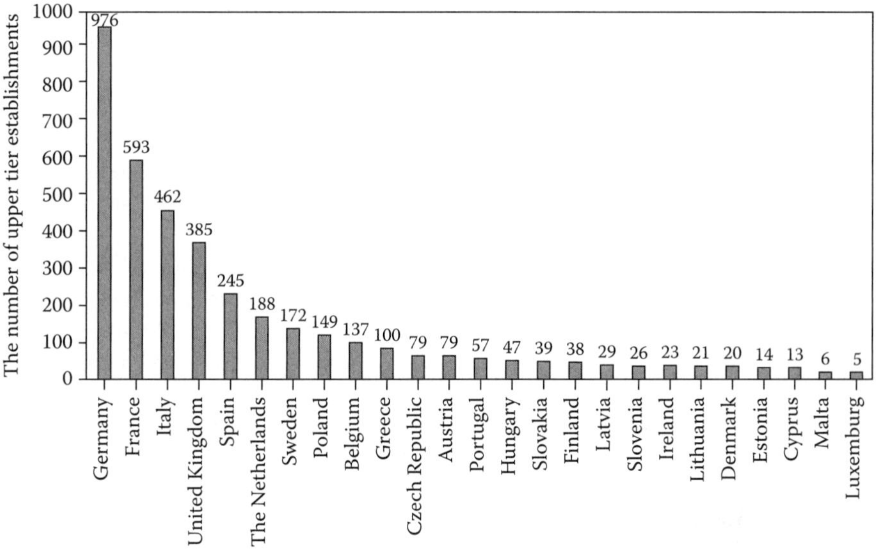

FIGURE 22.7 Comparison of the hazards of major industrial accidents in the European Union member states.

3278 in the 'old 15' member states in 2002 plus 399 in the 'new 10' member states in 2003).

22.5 MAIN ELEMENTS OF THE SYSTEM OF MAJOR ACCIDENT CONTROL

Preventing major industrial accidents and limiting their effects has been regulated in the EU and Poland, presented in Section 22.1. This body of regulations includes the Convention of the United Nations Economic Commission for Europe (ECE) on the Transboundary Effects of Industrial Accidents, ratified by the President of the Republic of Poland in 2003 (DzU 2004, no. 129, item 1352), which came into force in Poland on 7 December 2003 (DzU 2004, no. 129, item 1353).

The main provisions for control of major industrial accidents, that is, those in stationary facilities, are as follows:

- *Qualification criteria* for identification of hazardous facilities
- *Procedure of qualification* of hazardous facilities in the LTE and UTE category conducted by the operator of the establishment
- *Procedure of notification* of the appropriate authorities (for LTE, to the district Commandant of the State Fire Service, and for UTE, to the Provincial Commandant of the State Fire Service and the Provincial Inspector of Environmental Protection [PIEP]) of facilities qualified as hazardous
- Preparation and implementation of an *accident prevention programme* (APP) and its delivery to the appropriate authorities as specified above

- *Preparation of a safety system* which should be fully implemented at UTEs called out in the Seveso II Directive safety management system (SMS)
- Preparation of a safety report (SR) by the manager of a UTE and its delivery to the Provincial Commandant of the State Fire Service and to the PIEP
- Approval of the SR by the Provincial Commandant of the State Fire Service upon issue of an opinion by the PIEP or issue of a different decision (additional information needed, prohibition of operation or launching of the facility)
- Preparation of an *internal emergency plan* by the manager of the UTE and its presentation to the Provincial Commandant of the State Fire Service
- Gathering and delivery by the UTE operator to the Provincial Commandant of the State Fire Service, responsible for preparing the external emergency plan, of the information and data necessary for its preparation
- Development and approval by the Provincial Commandant of the State Fire Service, after consultations with the public, of the *external emergency plan*
- Implementation of these plans in the event of a major accident or in a near-miss situation
- Notifying the authorities of major accidents or near-miss situations, calling an alert, informing the authorities of the situation
- Examining the course of the accident or incidents, emergency activities, assessing the effects, creating a postaccident report, report, and *notification of the accident* to the CIEP
- Providing information to the public and involving the public in some procedures

The Seveso II Directive and Polish legislation specify the duties and procedures of the competent authorities. The most significant are

- *Inspection and supervision* of hazardous facilities
- Implementation of *appropriate land-use policy*, considering the major accident risk
- Implementation of appropriate procedures related to major accidents that may have *transboundary effects*

Full and universal compliance with these regulations in Poland, as well as devising instruments that support adherence to these requirements are among the most urgent tasks for the entities covered by these regulations and for research institutes and technical universities.

The contribution of several scientific and research institutes and universities in the preparation and implementation of tools facilitating conformity to the procedures of the system, as well as to the organisation of postgraduate courses, training, seminars and conferences deserves special mention. These institutions are the Central Institute for Labour Protection—National Research Institute, the Technical University of Lodz (Faculty of Process and Environmental Engineering), the Institute of Atomic Energy (Centre of Excellence MANHAZ), the Industrial Chemistry Research Institute, the Institute of Environmental Protection, the Main School of

Fire Service, Gdansk University of Technology (Faculty of Electrical and Control Engineering), and the Institute of Industrial Organic Chemistry.

The tools devised so far are described in more than 20 books and monographs, and over 100 publications and conference papers. The most significant among them are presented in the references to this chapter. About 2200 participants have benefited from the numerous training courses held.

22.6 MAJOR ACCIDENTS IN TRANSPORT OF HAZARDOUS SUBSTANCES AND MATERIALS

Major accidents often occur during transportation of hazardous substances and materials by cars, railway, and water transport, as well as pipeline transport. Accidents most often happen during motor transport. The effects of most such accidents are not serious and are usually contained. Sometimes, however, they may be very serious, particularly in densely populated areas and when there is a leakage of hazardous substances into waterways. Some examples of transport accidents are presented next.

Los Alfagues, Spain, 1978. A truck transporting 23 tonnes of liquid propane broke into three pieces near a campsite close to the seashore. The permissible load-carrying capacity of the tank was only 19 tonnes; moreover, it was not equipped with a pressure bleed device. 277 people were killed and 67 injured, mainly at the campsite. A discotheque and buildings surrounding the campsite were destroyed. Many houses were damaged, as well as 74 motor vehicles, of which 23 were destroyed. The immediate cause was the explosion of a cloud of vapours of a flammable substance, or BLEVE.

Asha-Ufa, Ural region, former USSR (Russia), 1989. An explosion of natural gas released from a pipeline occurred when two Trans-Siberian Railway trains, carrying about 1200 total passengers, passed each other near a gas pipeline. The pipeline had developed a leak and high concentrations of gas formed in several places. Sparks from under the wheels of the passing trains were the source of ignition. About 640 people were killed and the same number injured. In 1985 the pipeline had been damaged by an excavator, which had led to its unsealing after several years and, finally, to disaster.

The threat posed by transport accidents in Poland is quite serious. According to official CIEP data, among 600 events resembling major accidents in Poland in 2003–2006, as many as 235 (40%) were accidents that happened during transport of hazardous materials, including pipeline transmission. The most serious threat is road transport; here, the number of events that resembled major accidents amounted to 160 (26% of all events) and was even greater than the number of events that resembled major accidents in all industrial establishments (141 events, 23.5%). International regulations on the transport of hazardous substances, materials, and goods by railway (abbreviated RID) and by road transport (abbreviated ADR) are binding in Poland and standardise many of the significant safety issues; however, they contain no provisions on accidents occurring during transport. Considering the data on major accidents during transportation of hazardous materials, particularly motor transport, preventing these events should be considered a vital matter by governmental institutions and scientific and research institutes.

22.7 PROGRAMMES FOR PREVENTING MAJOR ACCIDENTS AND SAFETY MANAGEMENT AT ESTABLISHMENTS

Achieving the major objectives of the system, which for all organisations includes decreasing the risk of a major industrial accident and minimising the effects if one takes place, requires creating employee awareness and encouraging the right attitudes and behaviours at all levels, from top management to the lowest positions related to installations or activities of specific significance for plant safety.

The key components to preventing major industrial accidents at establishments, according to Seveso II Directive, are the *major accident prevention policy* (MAPP) and the *internal emergency plan*. Analysis of the Polish provisions on major APPs, safety systems, and internal operational and rescue plans (i.e. internal emergency plans) reaches the same conclusion. The tasks defined in these plans have two main objectives—preventing major accidents and limiting their effects—and are addressed to all personnel employed at the establishment.

According to the terminology of Article 7 of the Seveso II Directive, MAPP pertains to both categories of plants, that is, LTEs and UTEs. MAPP must be drawn up as an independent document, meeting the requirements specified in Annex III to the directive. The responsibility to ensure implementation of MAPP rests with establishment operator.

Polish regulation Article 251, item 1 of the Environmental Protection Law obligates the operators of establishments belonging to both LTE and UTE categories, to prepare an APP. The programme must include a description of the safety system as a part of the establishment's management structure and guarantee protection of the people and environment. According to these legal provisions, it should specify:

- The probability of occurrence of a major industrial accident
- The procedure to prevent accidents and combat the effects of an accident should it occur
- Ways to limit the effects of major industrial accidents on people and the environment should any occur
- The frequency of APP reviews to assess the programme's validity and effectiveness.

Article 252 of the Environmental Protection Law states that the operator of a UTE must prepare and implement a safety system. The provisions of this article specify the elements of the safety system, which—in short—are as follows (incorporating the amendments introduced):

- Specification of the obligations of employees at all levels of the organisation, that is, the actions they should perform in the event of an emergency
- Training for employees and other persons working at the establishment
- Functioning of mechanisms based on a systematic analysis of a major accident hazard and its probability

- Instructions for the safe use of installations containing hazardous substances, including normal operations, maintenance, and temporary stoppages
- Instructions for proceeding when changes are made to the industrial process
- Systematic analysis of predicted emergency situations for the purpose of devising emergency plans
- Monitoring the installations containing hazardous substances and taking any necessary corrective actions
- Systematic assessment of the APP and safety systems to assess their validity and effectiveness
- Analysis of emergency plans

According to analysis of the Environmental Protection Law, some of the requirements pertaining to MAPP, as stated in the Seveso II Directive, have been separated from the MAPP of major accidents control systems. Requirements have been established that refer to a separate concept the safety system. According to the Environmental Protection Law, APP pertains to both LTE and UTE establishments, so the structuring of the Polish legal provisions would not have to indicate their inconsistency with the EU regulations (i.e., the requirements of the Seveso II Directive). However, the full implementation of the safety system refers only to UTEs in the domestic legal provisions.

Thus, as a result of the separation of some of the requirements for the prevention of major accidents, which are obligatory for UTEs, LTEs in Poland have much more liberal provisions than LTEs in other EU member states. These issues, as well as the recommendations and guidelines pertaining to APP, safety systems, and preventing major accidents in LTEs and UTEs, are subject to separate studies (Michalik et al. 2005; Michalik 2005).

22.8 PROGRAMME FOR PREPARATION OF THE ESTABLISHMENT FOR A MAJOR ACCIDENT AND FOR EMERGENCY ACTION

Article 261 of the Environmental Protection Law obligates the operator of a UTE to prepare an internal emergency plan and implement it immediately in the event of a major industrial accident or a near miss. Operators of UTEs must also (Article 262) ensure that the employees, particularly those who face an immediate threat of effects from a major industrial accident and those who act as social labour inspectors or who are trade union representatives responsible for occupational health and safety, participate in the preparation of the internal emergency plan.

The regulations of the Ministry of Economy (DzU 2003, no. 131, item 1219) specify the requirements of major importance for preparedness of the plant for a major industrial accident and the appropriate reaction in the case it occurs. According to the regulations, the internal operational and rescue plan should, among other things, contain information on the following:

- The method of notifying persons who are to participate in a rescue action
- The method of issuing an emergency alert and the rules and conditions of evacuation

- Plans to be followed in an emergency
- Procedures for the crew, plant rescue service, and plant fire service in the event of an emergency
- The rules for conducting and coordinating the rescue activities of the plant rescue service and the plant fire service
- The rules and plan of action for providing first aid to the injured

A very significant issue should be noted here. One of the objectives, both of MAPP in the EU and of APP in Poland, is to minimise the effects of a major industrial accident, and to specify ways to limit human and environmental damage in the case of an occurrence (Article 252, item 2 of the Environmental Protection Law). These provisions pertain to both UTEs and LTEs. According to the provisions of these articles, response plans for a major industrial accident should also be developed in LTEs, however, the requirements specified in the regulations of the Ministry of Economy (DzU 2003, no. 131, item 1219) that relate to internal emergency plans for UTEs do not apply to LTEs. The content of the LTE emergency plans, as well as the content of the entire APP and SMS, should be adequate to deal with the size of a major industrial accident hazard arising at a given establishment (Michalik et al. 2005).

22.9 SAFETY REPORTS

Preparation of a safety report is a fundamental procedure in the system of preventing major industrial accidents. Article 253 of the Environmental Protection Law introduced the obligation to prepare this document and specified the objectives of the safety report.

The safety report should show that:

- The UTE operator is prepared to implement the APP and combat major industrial accidents.
- The establishment meets the requirements specified for implementation of the safety system.
- The probability of a major industrial accident was analysed and the appropriate actions instituted to prevent it.
- The design solutions for the performance and functioning of installations containing hazardous substances ensure safety.
- Internal emergency plans have been created and information provided for external emergency plans.

In general, the objectives of the safety report in Polish legislation are consistent with the requirements of the Seveso II Directive. There are some differences that are not exclusively of an editorial nature. In fact, according to the requirements of the Seveso II Directive, the safety report confirms that the APP is ready and the appropriate SMS for its implementation has been tested in practice. These requirements are not about the state of readiness the Environmental Protection Law seeks to enforce, but about proving a more advanced state of preparedness by the existence of a safety report, that is, proof that the APP and the safety systems at the UTEs are fully functional. A similar approach has been applied in the Ministry of Economy's regulations of the detailed requirements for safety reports (DzU 2003, no. 104, item 970; DzU 2005, no. 197, item 1632).

The safety report is a type of a report containing, for example, required information about the establishment, its installations, technologies, hazardous substances, and so on. Among other things, it should also contain an assessment of the risk of occurrence of a major industrial accident, identify the sources of threats, and contain a description of emergency event scenarios and an assessment of the potential effects of an event. The Polish legal provisions pertaining to these components of the safety report are similar to the EU's requirements.

The report should also contain information on the safety solutions applied, including a description of the technical, organisational, and procedural ways to prevent major industrial accidents and minimise their effects, as well as information related to the internal operational and rescue plan. It should encompass implementation of APP and safety systems, and it should document the results obtained.

According to item 2 of Article 254 of the Environmental Protection Law, the safety report is subject to approval by the Provincial Commandant of the State Fire Service after receiving the opinion of the PIEP. Any amendments to the report, which are permissible under the legal provisions in force, are subject to an identical approval procedure.

22.10 SHAPING EMPLOYEE ATTITUDES AND BEHAVIOURS IN PLANTS POSING MAJOR INDUSTRIAL ACCIDENT HAZARDS

EU legislation (the Seveso II Directive) and domestic regulations on the programmes for preventing major industrial accidents, safety management systems, and internal emergency action plans serve as the basis for shaping employee attitudes and behaviours in order to prevent major industrial accidents.

Undoubtedly, appropriate training for employees of LTEs, UTEs, and other establishments and organisations that pose a risk of major industrial accidents is imperative. Such training should create awareness of the threats present and convince employees of the compelling necessity to observe the special rules of safety and procedures designed to prevent accidents and limit their consequences. The training programme should list the specific risks, such as the hazardous substances, nature of threats, technologies in use at the establishment, nature of work performed, and safety-oriented tasks and objectives, and should stress the importance of avoiding actions or behaviour that might trigger major accidents or events.

It is important that employees are prepared to perform various tasks in the case of an emergency: chemical rescue, technical rescue, first aid, measuring chemical contamination levels, performing evacuation tasks, firefighting, and many other activities specified in emergency action plans. Training is an essential means of disseminating complete and detailed information on such plans to the employees designated to participate in rescue activities.

The efficiency and level of training should be assessed and documented and the assessments used to upgrade programmes. LTEs and UTEs must update their processes, technologies, safety tools, emergency action plans, and internal requirements while ensuring their training programmes and materials comply with the legislative provisions in force (besides complying with the general rules of occupational safety). When taking advantage of external training services, organisations should search for companies that are certified or renowned for their services.

One vital part of preventing major accidents and limiting their consequences as well as shaping employees' attitudes and behaviours is the involvement and support of the LTE and UTE operators, that is, the top management and their management-level employees (Michalik et al. 2005; Michalik 2003). The establishment operator should provide the necessary means to devise and implement the APP, as well as prepare and execute the internal emergency plans. He or she must define priorities and carry out a policy of implementation, execution, and continuous development of the APP in a manner that is clear and understandable for all members of management and personnel. The operator should also initiate and conduct audits and periodical inspections, initiate training programmes, and encourage employee initiative in safety matters. The operator has authority over the managerial staff and other personnel.

LTE and UTE operators should define the general objectives with regard to preventing major accidents. Lower-level management, in consultation with the establishment's employees, should define detailed objectives for implementation of the APP.

The effective completion of tasks with regard to preventing major industrial accidents and limiting their consequences must be ensured by a clear definition of authority and proper descriptions of responsibility. These must also be presented and communicated to the personnel responsible for implementing the actions and their supervisors. All these matters must be recorded and confirmed and progress monitored periodically (Michalik et al. 2005).

Each employee of an LTE or a UTE is held responsible for safety and should perform the functions assigned to them. Each should be aware of their responsibility for their own safety and that of their coworkers and third persons, and that their actions influence significantly the safety and proper operation of the establishment. Operators of LTEs and UTEs are responsible for establishing the decision-making structure and hierarchy of their organisations and ensuring that the employees concerned possess the knowledge, qualifications, and competence needed to undertake and perform the tasks and duties assigned to them.

To efficiently monitor the implementation of the APP and the internal emergency plans, the following should be ensured:

- Monitoring is conducted by properly qualified and experienced teams
- Monitoring results are made available, whenever needed, to the establishment operator, the managerial staff, authorised employees and their supervisors, and external supervisory authorities

Monitoring should also assess the familiarity of the establishment's personnel with the general and detailed objectives, their commitment to their implementation, and their actual performance (Michalik et al. 2005).

The planning and implementation of tasks to shape employee attitudes and behaviours should ensure that the employees:

- Are appropriately and sufficiently informed of the threats associated with specific hazardous installations and the probable consequences of accidents.
- Are informed of the orders, instructions, and recommendations issued by the appropriate authorities and the establishment's management.
- Participate in consultations to formulate the following documents:
    - A programme for the prevention of major accidents and for the SMS
    - A safety report
    - Emergency action plans and procedures
    - A postaccident report
- Are regularly trained and instructed in the control practices and procedures for accident prevention and in assessment of situations that might cause a major accident, as well the procedure to be followed in dangerous situations resulting from such accidents.
- Act within their scope of activities and without fear of negative consequences, undertake corrective actions, and shut down dangerous systems if they believe there is a risk of a major accident, based on their knowledge and experience, and inform their superiors or initiate an alert prior to doing so.
- Consult the managerial staff about events that they think are a potential cause of a major accident.
- Follow all procedures and practices related to accident prevention and control in dangerous installations and observe all recommended procedures if a major accident occurs.

REFERENCES

Act of April 27th, 2001—Environmental Protection Law. Declaration of the Marshal of the Sejm of the Republic of Poland of January 23rd, 2008 on announcement of the consolidated text of the Environmental Protection Law. DzU no. 25, item 150.

Act of February 24th, 2006 on amending the Environmental Protection Law Act and some other acts. DzU no. 50, item 360.

Act of July 27th, 2001 on introduction of the Environmental Protection Law, the act on waste and amendment of some acts. DzU no. 100, item 1085.

Borysiewicz, M., A. Furtek, and S. Potempski. 2000. *The Handbook on Methods for Assessment of Risk Caused by Hazardous Process Installations.* Otwock-Swierk: IEA.

Borysiewicz, M., and A. S. Markowski. 2002. *Criteria of Risk Acceptability in Relation to Major Industrial Accidents.* Warsaw: CIOP.

Borysiewicz, M., and S. Potempski. 2002. *The Risk of Major Accidents of the Pipelines Carrying Dangerous Substance—Assessment Methods.* Warsaw: CIOP.

Borysiewicz, M., J. Dziembowski, and A. Mielczarek. 2000. *Guidelines for Emergency Plans with Respect to Installations for Storage and Transportation of Dangerous Chemical Substances as well as for Process Installations with a Chemical Reactor.* Warsaw: IChP.

Chief Inspector of Environmental Protection. 2008. *Annual Report of Inspection for Environmental Protection for 2007.* 83–101. Warsaw: GIOS.

Commission of the European Communities. 2007. Report from the Commission. Report on the Application in the Member States of Directive 96/82/EC on the control of major-accident hazards involving dangerous substances for the period 2003–2005, Brussels. http://ec.europa.eu/environment/seveso/pdf/report_2003_2005_en.pdf (accessed 7 August 2008).

Convention on the transboundary effects of industrial accidents, done in Helsinki on March 17th, 1992. DzU 2004, no. 129, item 1352.

Council Directive 96/82/EC of December 9th, 1996 on the control of major-accident hazards involving dangerous substances. OJ L 10, 14.01.1997, 13.

Cozzani, V., and S. Zanelli. 1997. *EUCLID—A Study on Emission of Unwanted Compounds Linked to Industrial Disasters.* Ispra: European Commission.

Directive 2003/105/EC of the European Parliament and of the council of December 16th, 2003 amending Council Directive 96/82/EC on the control of major-accident hazards involving dangerous substances. OJ L 345, 31.12.2003, 97.

Government declaration of December 22nd, 2003 on the setting into force of the Convention on the transboundary effects of industrial accidents, done in Helsinki on March 17th, 1992. DzU 2004, no. 129, item 1353.

Government declaration of July 26th, 2005 on coming into force of annexes A and B to the European Agreement concerning the international carriage of dangerous goods by road (ADR), prepared in Geneva on September 30th, 1957. DzU no. 178, item 1481. Appendix to no. 178, item 1481. Volumes 1 and 2. Warsaw, Chancellery of the Prime Minister 2005.

Koradecka, D., and J. S. Michalik. 2003. Development of tools to meet the EU requirements to support the system of control of environmental hazards connected with industrial major accidents in Poland. In *Technology at the Service of Environment*, Vol. 2. 163–178. I:73–87. Gliwice: Komdruk-Komag.

Markowski, A. 1998. Safety and loss prevention in thermal processing. Part VIII. In *Thermal Processing of Biomaterials*. Vol. 10. ed. T. Kudra and C. Strumiłło. Amsterdam: Gordon & Breach Science Publishers.

Markowski, A. S. 1997. Practical management of safety in industry. In *Risk Analysis in Transportation and Industry*, 59–71. Wroclaw: Technical University of Wroclaw.

Markowski, A. S. 2006. *Layer of Protection Analysis for the Process Industry*. Lodz: Polish Academy of Sciences.

Markowski, A. S. ed. 1999. *Loss Prevention in Industry—Part II: Management of Occupational Safety and Health and Part III: Management of Process Safety.* Lodz: Technical University of Lodz.

Markowski, A. S. ed. 2001. *Management of Risk in Processing and Chemical Industry.* Lodz: Technical University of Lodz.

Markowski, A. S., ed. 2000. Materials for the 9th Scientific Symposium 'Prevention of losses in industry—Computer software for safety reports and operational and rescue plans'. Lodz: Technical University of Lodz.

Member States country reports on the implementation of Seveso II. http://ec.europa.eu/environment/seveso/implementation.htm (accessed August 7, 2008).

Michalik, J. S. 2005. *Prevention of Major Industrial Accidents: Recommendations and Guidelines for Upper-Tier Establishments.* Warsaw: GIP.

Michalik, J. S., and A. Gajek 2002b. *Amended Qualification Criteria and Respective Changes in the 'Seveso Substances' Database.* Warsaw: CIOP.

Michalik, J. S., and A. Gajek. 2002a. *Formation of Dangerous Substances During Major Industrial Accidents.* Warsaw: CIOP.

Michalik, J. S., and D. T. Kijeńska. 2000a. *Identification of Establishments Posing a Major Chemical Accident Hazard. Dangerous Substances and Relevant Procedures. A Guidance.* Warsaw: CIOP.

Michalik, J. S., and D. T. Kijeńska. 2000b. *List of Chemical Substances for Identification of Hazardous Objects: The Seveso II Directive.* Warsaw: GEA.

Michalik, J. S., and W. Domański. 2002. *Major Accidents Prevention Programme and Safety and Management System in Lower-Tier and Upper-Tier Establishments.* Warsaw: CIOP.

Michalik, J. S., D. T. Kijeńska, and A. Gajek. 2001. *Notification Procedure of Major Accident Hazard Lower-Tier and Upper-Tier Establishments. A Guidance.* Warsaw: CIOP.

Michalik, J. S., ed. 2001. The main procedures of prevention of major industrial accidents and limitation of their effects. First National Scientific and Technical Conference, March 13, 2001. Warsaw: CIOP.

Michalik, J. S., ed. *Seveso II Directive: Legal Status as of 2004 (Consolidated Text).* Warsaw: CIOP-PIB.

Michalik, J. S., et al. 2005. *Major Accident Prevention Programme and Safety Management System in Establishments at Major Accident Hazard. A Guidance.* Warsaw: CIOP-PIB.

Michalik, J. S., W. Domański, A. Gajek, E. Łużny, A. Grobecki, J. Lewandowski, and W. Kawa. 2003. Formation of employees' attitudes and behaviours for the control of major industrial accidents. *Occupational Safety* (6):8–11.

Mielczarek, A., J. Dziembowski, and M. Borysiewicz. 2000. *The Package of Supplementary Materials for Training Courses that Refer to Audits of Process Installations Safety.* Warsaw: IChP.

Ordinance of the Minister of Economy and Labour of September 12th, 2005, amending the ordinance on the requirements for safety report of the upper-tier establishment. DzU 2005, no. 197, item 1632.

Ordinance of the Minister of Economy of January 31st, 2006, amending the ordinance on the types and quantities of hazardous substances whose occurrence in an establishment is conclusive of qualifying it as a lower-tier or upper-tier establishment. DzU no. 30, item 208.

Ordinance of the Minister of Economy, Labour and Social Policy of May 29th, 2003 on the requirements for safety report of the upper-tier establishment. DzU 2003, no. 104, item 970.

Ordinance of the Minister of Economy, Labour and Social Policy of July 17th, 2003 on the requirements for emergency plans. DzU no. 131, item 1219.

Ordinance of the Minister of the Environment of December 30th, 2002 on major accidents to be notified to the Chief Inspector of Environmental Protection. DzU 2003, no. 5, item 58.

Ordinance of the Minister of the Environment of June 4th, 2002 on the detailed scope of information required to be delivered to the public by the provincial commandant of the State Fire Service. DzU no. 78, item 712.

Regulations for international railway transport of hazardous materials RID (Journal of Tariffs and Ordinances of the Minister of Communication 1995, no. 7, item 44), constituting Annex I to the Consolidated provisions on the Agreement of international railway transport of goods CIM, constituting annex B to the Convention Concerning International Carriage by Rail COTiF, Berne, May 9th. 1980. DzU 1985, no. 34, item 158, 159; DzU 1997, item 225 and 226; DzU 1998, no. 33, item 177.

Seveso Directive consolidated text with amendments. 1994. In Council Directive on the major-accident hazards of certain industrial activities 82/501/EEC. Office for Official Publication of the European Communities. 1990, CD-NA-12705-EN-C.

Żurek, J., M. Borysiewicz, and E. Lisowska-Mieczkowska. 2001. *Systems of Integrated Health, Environment and Process Safety Management in an Industrial Plant.* Warsaw: CIOP.

Żurek, J., M. Borysiewicz, and W. Kacprzyk. 2001. *A Guide to Integrated Risk Assessments and Hazard Management in Industrial Areas.* Warsaw: CIOP.

Zaleski, B., A. Majka, R. Grosset, W. Kubicki and I. Grunt-Mejer. 2001. *Internal and External Emergency Plans. A Guidance*, ed. J. S. Michalik. Warsaw: CIOP.

Part V

Basic Directions for Shaping Occupational Safety and Ergonomics

23 Occupational Risk Assessment

Zofia Pawłowska

CONTENTS

23.1 INTRODUCTION

At the European level, directive 89/391/EEC—the Framework Directive—is the most important regulation governing occupational safety and health (OSH) management in the EU countries. The directive calls for a systematic, integrated, proactive, and participative approach to OSH management aimed at ensuring continuous improvement of the safety and health of workers. Risk assessment should be the main tool used in OSH management. Employers must evaluate risks to the safety and health of workers and—following the evaluation—take the appropriate (technical and/or organisational) measures to ensure an improvement in the level of workers' health and safety protection. The employer must monitor the effectiveness of the measures and adapt them to changing conditions and technological progress. The introduction of the OSH obligation of risk assessment clearly indicates that OSH management in enterprises should be proactive, that is, all hazards to the safety and health of workers should be identified and the risks arising from them eliminated or controlled to prevent occupational accidents and work-related diseases. Performing and organising risk assessment appropriately, particularly the consultation and participation of workers, ensures its effectiveness.

23.2 OCCUPATIONAL RISK AND ITS ASSESSMENT

Risk is defined as the combination of the probability of an event and the severity of its consequences (ISO/IEC *Guide 73...*). Various types of risks are defined depending on the company's area of activity: strategic, operational, financial, and so on (ISO/IEC Guide 73 2002). Risk management is an organisation's coordinated activities to direct and control risk, and typically includes risk assessment, risk treatment, risk acceptance, and risk communication (ISO/IEC Guide 73 2002).

Workers sustain occupational risks in connection with their work. Definitions of this risk can be found in various standards and guidelines (*Guidance* 1996) and include two elements:

1. The likelihood of injury or disease arising from a hazard in the work environment (even if the word 'loss' is used in the definition, it applies to the worker's health)
2. The severity of the potential injury or deterioration of health

As mentioned above, risk is a combination of the likelihood of an occurrence of a hazardous event and the severity of injury or damage to the health of people caused by that event (*ILO guidelines*..., 2001) Occupational risk assessment is the process by which work-related hazards are identified; the risks resulting from their presence are then estimated and evaluated. This process is often divided into two stages: (1) occupational risk analysis, which involves identifying hazards and estimating the resultant risk; and (2) occupational risk evaluation, which determines if the risk is acceptable or if actions that will eliminate or lower it should be taken.

Risk assessment is the process of evaluating safety and health risks arising from hazards at work (*ILO guidelines*..., 2001). Occupational risk assessment aims to provide the best possible protection to workers' health and safety at work. When assessing occupational risk, all hazards to worker safety and health present in the workplace should be identified. This means that all types of work performed, including those that are not continuous or not directly related to the manufacturing processes, should be considered. A proper understanding of the essence of occupational risk and the principles of its assessment are essential for efficient risk management in an enterprise.

23.3 OCCUPATIONAL RISK ASSESSMENT PROCESS

Occupational risk assessment is a multistage process and is closely related to managing risks. The successive stages of occupational risk assessment are as follows (Figure 23.1):

1. Collect the information required.
2. Identify the hazards.
3. Estimate the occupational risk (i.e. determine the probability and severity of potential harm to workers' health and life).
4. Evaluate the occupational risk (i.e. decide if the occupational risk is acceptable or if actions to eliminate or reduce it are necessary).

After assessment, plans to introduce measures aimed at eliminating or reducing occupational risk should be developed if necessary. The effectiveness of the measures introduced must be monitored.

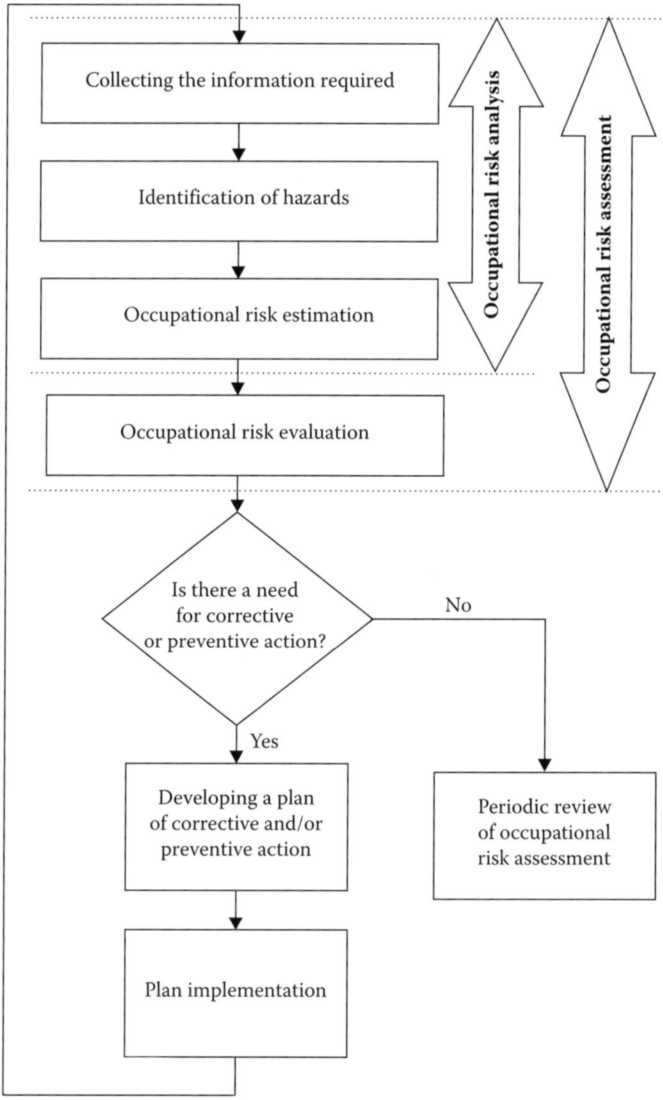

FIGURE 23.1 The process of occupational risk assessment in the context of managing that risk according to PN-N-18002 (Reprinted with permission from PN-N-18002. 2000. Occupational health and safety management systems—Guidelines for occupational risk assessment.)

Occupational risk assessment can be a simple process that does not require specialist knowledge or skills in the workplaces where hazards are well known, easy to identify, and generally do not have severe consequences, and where measures reducing risks associated with those hazards are easily accessible, such as in offices. Complex establishments and processes that pose major industrial hazards require formal methods and expert know-how.

To assess occupational risk at workstations, the following basic information (PN-N-18002) is needed:

- The location of the workplace and the jobs performed there
- The types of people working at the workplace, particularly those upon whom the effects of an occupational hazard may be more severe, such as pregnant women, young workers, or workers with disabilities
- Work equipment, materials, and technological processes involved
- Tasks performed and their duration as well as the methods employed by those working at the workplace
- Legal regulations and standards related to the workplaces in question
- Hazards already identified and their sources
- The potential consequences of existing hazards
- Protective measures already instituted
- Accidents, occupational diseases, and other ailments that are or have been present at the workplace
- Incidents not resulting in injuries, which should be registered and analysed in order to detect a hazard before an accident happens

The training of workers, work instructions, and presence of nonworkers whose activities can be sources of additional hazards should also be reviewed. The basic methods for obtaining such information include analysing existing documents related to OSH, observing the work environment, and interviewing employees (PN-N-18002).

Hazards should be identified using the information gathered. The identification stage can involve various methods, from nonformal analyses simply wondering if anything in the work environment could cause an injury or illness to precisely-defined formal methods that develop models appropriate to the aim of the analysis. When assessing occupational risk at workstations, the methods employed should be as simple as possible. Checklists are often recommended to identify hazards at workstations; the simplest ones list the potential hazards present at a workstation. Checklists help to identify whether these hazards are actually present. Checklists should be open-ended and should enable addition of the hazards workers may refer to in interviews and which might not have appeared on the original list.

The next stage of occupational risk assessment is risk estimation. At this stage, the likelihood of adverse consequences, such as injuries or diseases, and their severity are determined for each identified hazard. The standards recommend using various methods of estimation, usually closely related to risk evaluation.

Standard PN-N-18002 recommends using a three-point risk-level estimator (Table 23.1). A five-point risk-level estimator is optional (Table 23.2). When estimating occupational risk using Tables 23.1 and 23.2, the severity of the harmful consequences of hazards and their likelihood are defined as follows:

- Slightly harmful consequences are traumas and illnesses that do not cause protracted distress or absence at work, for example, temporary deterioration of health, minor bruises and cuts, eye irritations, symptoms of minor poisoning, and headaches.

TABLE 23.1
Three-Point Risk-Level Estimator

Probability	Severity of Consequences		
	Slightly Harmful	Harmful	Extremely Harmful
Highly unlikely	Small 1	Small 1	Medium 2
Unlikely	Small 1	Medium 2	High 3
Likely	Medium 2	High 3	High 3

Source: PN-N-18002. 2000. Occupational health and safety management systems—Guidelines for occupational risk assessment.

TABLE 23.2
Five-Point Risk-Level Estimator

Probability	Severity of Consequences		
	Slightly Harmful	Harmful	Extremely Harmful
Highly unlikely	Very slight 1	Small 2	Medium 3
Unlikely	Moderate 2	Medium 3	High 4
Likely	Medium 3	High 4	Extreme 5

Source: PN-N-18002. 2000. Occupational health and safety management systems—Guidelines for occupational risk assessment.

- Harmful consequences are traumas and illnesses that cause moderate but protracted or periodically recurring distress and lead to short periods of absence at work, for example, cuts, second-degree burns on a limited body surface, dermal allergies, simple fractures, and musculoskeletal disorders.
- Extremely harmful consequences are traumas and illnesses that cause severe and permanent distress and/or death, for example, third-degree burns, second-degree burns on a large body surface, amputations, complex fractures leading to consequential dysfunction, cancer, toxic damage of the internal organs and the nervous system resulting from exposure to chemical agents, asthma, vibration syndrome, occupational hearing damage, and cataracts.
- Highly unlikely hazard consequences are those that should not materialise during the entire occupational career of an employee.
- Unlikely hazard consequences are those that may materialise only a few times during the occupational career of an employee.
- Likely hazard consequences are those that may materialise repeatedly during the occupational career of an employee.

In accordance with the principles noted in Table 23.3 (PN-N-18002), occupational risk can be estimated based on the value of exposure parameters, wherever these

TABLE 23.3
General Guide to Estimation of Occupational Risk on a Three-Point Risk-Level Estimator Based on the Exposure Value

Value of Exposure Parameter	Risk Estimation
$P > P_{max}$	High
$P_{max} \geq P > 0.5\ P_{max}$	Medium
$P \leq 0.5\ P_{max}$	Low

P_{max} is tolerable value of exposure parameter, normally determined based on applicable requirements (this could be the value of maximum allowable concentration [MAC] or maximum allowable intensity [MAI]). When there are no set requirements to determine such a parameter, the opinions of experts and/or the views of employees may be considered.

Source: PN-N-18002. 2000. Occupational health and safety management systems—Guidelines for occupational risk assessment.

values have been established. These principles make risk estimation easy, with expert know-how and values of admissible exposure established in the regulations.

The next stage of assessment is *risk evaluation* (also called *determining the tolerability of occupational risk*), and entails making decisions on the tolerability of the risk or the need to reduce it. The prime criteria for tolerability of occupational risk are the requirements of the applicable legal regulations. If these are not met, the risk is always intolerable. If there are no criteria in the regulations, an expert should decide whether the risk is tolerable. The opinions of the people exposed should also be considered. Table 23.4 illustrates the general recommendations related to determining whether occupational risk is tolerable based on its evaluation and the actions recommended following that assessment (PN-N-18002).

The results of the occupational risk assessment are the basis for planning and carrying out actions to reduce that risk. When planning these actions, it is important to remember the following:

- Immediate action is required to reduce occupational risk related to work in progress assessed as intolerable.
- Planned work cannot begin unless risk is reduced to at least a tolerable level.
- Risk greater than a tolerable level can never be accepted.
- If the risk is estimated to be medium, it should be reduced to the lowest possible level based on a cost-and-benefits analysis.

When planning actions to reduce occupational risk according to legal requirements, the following should be considered:

- Combating the risk at the source
- Using new technological solutions
- Substituting dangerous technological processes, devices, substances, and other materials with those that are safe or less dangerous

TABLE 23.4
General Rules for Occupational Risk Evaluation and Action Recommended Following Assessment of Risk (Risk Estimated on a Three-Point Risk-Level Estimator)

Risk Estimation	Risk Evaluation	Action Required
High	Intolerable	When risk is connected with work currently performed, actions to reduce the risk need to be taken at once (e.g. by changing work organisation or using personal protective equipment); planned work cannot commence until the risk is reduced to a tolerable level
Medium	Tolerable	Planned actions are recommended to reduce the risk level
Low		It is necessary to assure that the risk level will remain the same

Source: PN-N-18002. 2000. Occupational health and safety management systems—Guidelines for occupational risk assessment.

- Prioritising collective protective equipment over personal protective equipment
- Adapting the conditions and work processes to the capabilities of the workers, especially through the following:
    - Appropriate design and organisation of workstations
    - Selection of machines and other technical equipment and work tools
    - Selection of methods of production and work, bearing in mind the need to reduce monotonous work and work at a predetermined pace and reduce the negative effect on workers' health

Records of the results of occupational risk assessment should show:

- That the assessments were carried out and all the risks were assessed
- The manner of assessment and the criteria employed (the requirements, standards, or recommendations used during the assessment)
- The groups of workers particularly at risk that were considered
- Recommendations on measures that should be implemented to reduce risks
- Recommendations made upon review and update of the occupational risk assessment

The results of occupational risk assessment can be registered with the use of computer software such as STER, a computer system for hazard registration, and occupational risk assessment developed at the Central Institute for Labour Protection of the National Research Institute.

23.4 GENERAL PRINCIPLES OF ORGANISING OCCUPATIONAL RISK ASSESSMENT

Occupational risk assessment should help employers and supervisors:

- Identify hazards in the workplace and determine ways to reduce related occupational risks with due consideration of legal requirements.
- Select appropriate workstation equipment, materials, and work organisation.
- Determine priorities for eliminating or reducing occupational risks.
- Demonstrate to all parties (employees and/or their representatives, supervisors, and auditors) that all aspects of work were considered and related risks were appropriately assessed and that proper measures were taken to reduce them.
- Improve workers' safety and health protection.

To achieve these aims, it is very important to establish the following:

- Who will make decisions on occupational risk assessment and the situations in which they will be made
- Who will coordinate and conduct occupational risk assessment
- How will occupational risk assessment be carried out and documented
- How will management and employees be involved in occupational risk assessment

The employer is the person responsible for assessing occupational risk in an enterprise. They should be supported by OSH specialists and occupational physicians. The assessment process can also involve employees, supervisors of employees, and external experts. If possible, teams of workers from the enterprise should perform this assessment. The people delegated to assess occupational risk should have the necessary knowledge and competence; in particular, they should know and understand the general rules for conducting risk assessment and determining the means to reduce that risk. If necessary, they should receive proper training.

Ensuring workers are consulted and cooperate in occupational risk assessment is a basic condition for successful risk management. Workers should take part in occupational risk assessment at their workstations. Workers can provide information on hazards at their workstations and on the ways in which they perceive the hazards. They can point to work-related stress or to painful forced postures that the people carrying out the assessment might not notice. Consulting workers about hazards and actions that result from assessment is necessary. The employer must consult workers or their representatives on all actions related to OSH, especially those related to occupational risk assessment, and inform workers about occupational risks.

Occupational risk assessment is not a one-off activity. It should be done periodically and every time changes are introduced at the workstations—both before and after the changes. In particular, occupational risks should be assessed when the following items change:

- Technological procedures
- Work equipment
- Chemical, biological, carcinogenic, or mutagenic substances
- Work organisation

Assessments should also be conducted after regulations at workstations have changed. This also occurs when the protection measures employed are not adequate to ensure workers' safety and health.

23.5 SUMMARY

Managing occupational risk helps to ensure protection of worker safety and health in the workplace. Occupational risk assessment is a basic element of such management. Its results are significant for

- Establishing OSH goals
- Selecting work equipment, methods, work organisation, and protective measures
- Establishing the worker's competence and training needs, which is especially important for workstations that require special physical or mental abilities or specialist training and certificates

The efficiency of occupational risk management depends on the assessment method used. Workers should be involved for the following reasons:

- To increase their involvement in solving OSH problems.
- To inform them about the occupational risk efficiency.
- To make them aware of the hazards and related risks at their workstations and of the need to use protective measures that reduce such risks.

A well-organised occupational risk assessment improves work conditions, shapes the safety culture of the enterprise, and reduces losses from occupational accidents and diseases, thus increasing tangible benefits.

REFERENCES

Guidance on Risk Assessment at Work. 1996. Luxemburg: European Commission, Directorate General V Employment, Industrial Relations and Social Affairs.

ILO-OHS. 2001. *ILO Guidelines on Occupational Safety and Health Management Systems.* Geneva: ILO-OHS.

ISO/IEC Guide 73. 2002. *Risk Management. vocabulary: Guidelines for use in Standards. A Risk Management Standard.* London: AIRMIC.

PN-N-18001. 1999. Occupational health and safety management systems—Requirements.

PN-N-18002. 2000. Occupational health and safety management systems—Guidelines for occupational risk assessment.

24 Work-Related Activities: Rules and Methods for Assessment

Danuta Roman-Liu

CONTENTS

24.1 INTRODUCTION

Work-related activities should not only be performed in a safe manner but should also follow the rules of ergonomics, which help to prevent musculoskeletal disorders (MSDs).

MSDs are one of the most serious health threats associated with occupational work. In Poland in 2003, 14.4% of total inability to work was caused by MSDs, making them the third most frequent reason for absence from work, after diseases of the circulatory system (21%) and mental disorders (16%). Many studies indicate that biomechanical and psychosocial factors of the work environment, as well as individual factors, cause these disorders. In the European Union (EU), MSDs are among the most frequent health problems and are an important reason for work absences (Parent-Thirion et al. 2007).

Hard physical work, exposure to vibrations and lack of physical activity are typical factors leading to MSDs. Other risk factors associated with the work environment include lifting and moving loads, static loads, repetitive work, and uncomfortable posture. However, activities generally labelled 'light work', such as work at the computer, on an assembly line, and packaging small objects, can also cause MSDs, because such work often involves the upper limbs performing repetitive actions and significant static load to the spine.

MSDs and the resulting sick leaves, disability pensions, and early retirements pose financial burdens for society. Therefore, limiting MSDs will contribute to reducing expenses associated with social insurance and increasing productivity.

24.2 EXPOSURE TO RISK FACTORS RELATED TO WORK ACTIVITIES IN THE EUROPEAN UNION COUNTRIES AND IN POLAND

European research of work conditions (Parent-Thirion et al. 2007) conducted in 2005 shows that in the EU-27 countries, a large percentage (62.3%) of employees perform repetitive movements of hands and arms during at least one-fourth of the total work time (Figure 24.1). About 45.5% work in uncomfortable body postures and 42.9% perform monotonous work. About 24% of employees are exposed to vibrations and performance of repetitive actions shorter than 1 minute, and 39% of employees perform work cycles shorter than 10 minutes. A small percentage of employees (8.1%)

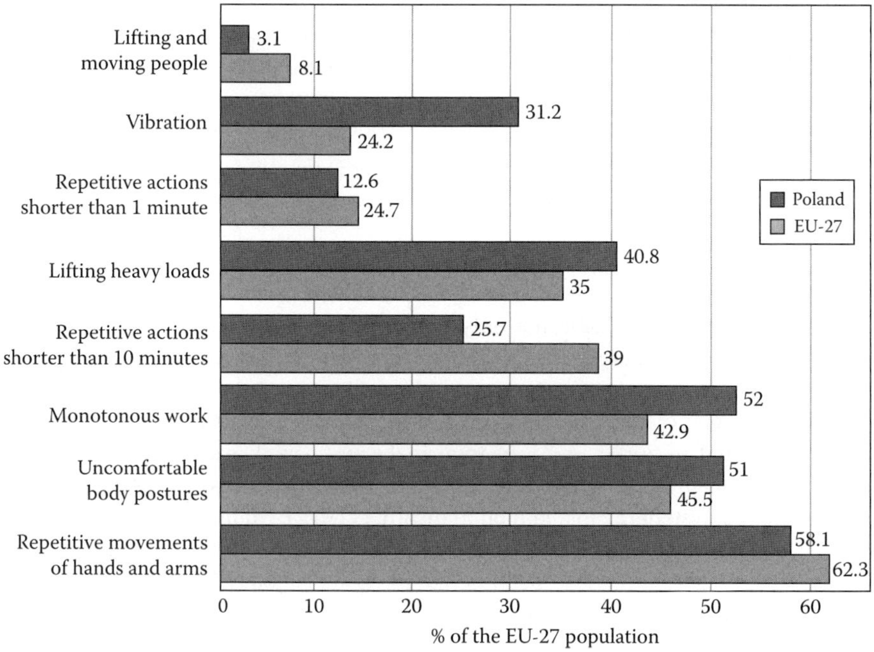

FIGURE 24.1 Exposure to musculoskeletal disorder risk factors in a group of employees in the European Union.

lift and move people, usually while taking care of the sick and disabled. Similar work patterns are found in Polish employees. The exposure of Polish workers to uncomfortable postures, monotonous work, heavy load lifting, and vibration is higher than the average exposure among EU workers. However, in short repetitive movements with cycles shorter than 1 minute, Polish workers' exposure is lower by half. Their exposure to short repetitive movements with cycles longer than 10 minutes is also significantly smaller. These results clearly show that the largest percentage of workers in the EU are exposed to repetitive movements of hands and arms, uncomfortable posture of the body, performance of monotonous work, and carrying heavy loads. In this respect, the exposure of Polish workers is similar to the average exposure in all countries of the EU.

24.3 MUSCULOSKELETAL STRAIN UNDER WORK CONDITIONS

Excessive strain on the body during work causes peripheral fatigue, including in the musculoskeletal system, and changes in the functions of other systems such as the circulatory, respiratory, digestive, and central nervous systems, which increases the subjective sense of fatigue. Physical strain factors leading to fatigue include improper body posture during work, excessive weight of transported materials and tools, and a high frequency of repeated activities, usually caused by bad work organisation and/or workstation design that does not follow the rules of ergonomics.

In addition to factors related to the work environment, certain individual factors, such as gender, age, physical capacity, and lifestyle, are also important, as they can indirectly influence the development of MSDs and cause a gap between the work requirements and the capacity of the worker. Work environment is described in relation to MSDs through psychosocial and physical (biomechanical) risk factors. The largest and most significant increase in the risk of disorders and pains occurs in employees who are exposed to both physical and psychosocial factors of significant intensity.

The development of technology, including mechanisation and automation, has significantly reduced the number of persons performing manual work. However, there still exist certain professions in which many activities cannot be replaced with technical solutions, such as repair work, construction work, or the activities of a fireman or a nurse. Loads on the musculoskeletal system and the risk of developing MSDs also are present while performing work with a low energy expenditure, such as office work, if the technique is improper or if the person performing the work is immobile. This applies especially to jobs that require high intellectual qualifications, and at the same significantly limit muscular activity; this increases the static load on muscles due to long-term immobility in a seated position, as well as mental tension.

The following basic factors should be taken into account during the design of a workstation and the evaluation of related loads:

- Body posture during work
- Type and value of forces during the performance of manual work, performance of work with large external load, transport of objects, operation of steering equipment

- Static load caused by the need to maintain a forced posture at work and to perform monotonous work consisting of repeated similar actions or limitations in actions
- Energy expenditure of the performed work
- Organisation of work, including the rate, time, and number of performed activities and the spatial organisation of the work place

Work-related loads are evaluated based on a properly-performed time study of work—the 'workday photo'—enabling measurement and recording of the duration of individual activities.

The time study should be conducted on days with an average rhythm of work and should cover actions typical for the given workstation. In the time study documentation, all types of activities typical for the position, as well as auxiliary activities and breaks, should be grouped into cycles with the similar workload. The duration of individual activities, expressed in minutes, should be measured several times, for various persons, and with various work intensity, to develop a typical and averaged workday photo for the given position.

The next step in workstation analysis is determining the nature and size of the load for the individual activities. A complete assessment of the load on the musculoskeletal system and the risk of MSDs should cover the worker's whole body, regardless of whether the work is performed in a standing or seated position, and should include an assessment of the load on the upper limbs, lower limbs, and spine. The assessment should also take into account the whole variety of performed tasks, from the standpoint of the body posture, the type and value of force exerted in various phases of the cycle, and the frequency of work sequence repetition. The risk assessment method should be general, applicable to any job and workstation, and, at the same time, should enable accurate risk assessment of each task and activity, taking into account the individual risk factors and the value of force with regard to the force capacity of the employee or employee population (in other words, the maximum force for the given type of force activity). It should also use various assessment methods to get accurate and comprehensive results.

24.4 BIOMECHANICAL CRITERIA AND METHODS FOR EVALUATING WORK-RELATED LOADS

Various disorders of the musculoskeletal system are caused by loads on the system, which depend on the nature of the performed actions. Among all the factors that lead to malfunctions and traumas of the musculoskeletal system, biomechanical factors play a significant role. These factors depend on the construction of the workstation and result from the placement of various parts of the body and the type, direction, and value of external forces. The time for which the posture is maintained or changed, as well as the force, are also significant.

Body posture during work strongly influences the load on the musculoskeletal system. To minimise the risk of MSDs, the person performing the work should be guaranteed appropriate body posture diversity and movement of limbs. The load on the musculoskeletal system increases when the joints are placed at extreme angles and when the body remains in static positions for a long time. Complicated body postures, such as twisted or bent positions, increase the risk.

The sitting position limits movements of the body, especially of the lower limbs and the torso, which may cause an increased load on the back and upper limbs. The standing position often causes discomfort and pain in the lower limbs and the lumbar part of the spine. A load on the upper limbs is present during work performed both seated and standing.

The engagement of muscles necessary to perform an activity or a series of activities is expressed by an external force, exerted on tools, or by an internal force, the tension in the muscles, tendons, and joint ligaments. This force may be associated with a movement, holding a tool or another object immobile, or keeping the body in a specific position. Reducing the load on the musculoskeletal system, and thus reducing the risk of MSDs, is possible by selecting the appropriate parameters that determine the load on the musculoskeletal system (parameters that describe the biomechanical factors).

There are many methods for assessing loads on the musculoskeletal system resulting from the performance of work activities. They include methods for evaluating internal and external loads and methods using observation, questionnaires, or computer modelling. These methods also can be classified based on whether the method assesses absolute load ratios or focuses on assessing risk associated with the development of MSDs by relating the values of load ratios to risk areas. There are also methods to assess risk for the whole musculoskeletal system or for specific areas of the body, for example, the lumbar section of spine or in the wrists.

Questionnaire methods are relatively cheap and frequently used. Subjective assessment questionnaires assess the load on the muscle system, the sense of discomfort (e.g. fatigue, back pain, or pain in other parts of the body), and the scope of the MSDs. Questionnaires and maps for the assessment of the person subjected to the assessment apply to both selected areas and to the whole body of that person.

Values of biomechanical parameters such as force or repetition frequency may be determined with the use of integrated models. In such methods, a simple mathematical relationship is used to determine the value of a specific parameter within the function of other parameters. Methods based on mathematical relationships assess the musculoskeletal load using parameters that describe the position of the individual body parts, the force exerted by the worker, and the time sequences of loads. The load index is calculated as the product of variable coefficients, taking into account the weights attributed to each of these parameters. An increased load indicator value may suggest an increased risk of the MSDs. The integrated models include the National Institute for Occupational Safety and Health (NIOSH) equation (Waters et al. 1993) and the equation used under the occupational repetitive action (OCRA) method (Occhipinti 1998), as well as the strain index (SI; Moore and Garg 1995). The SI was developed to predict the main disorders of muscles and tendons in the wrists, particularly carpal tunnel syndrome, and some disorders of the upper limbs. The SI is calculated using the values of effort intensity, effort duration, effort over 1 minute, position of the wrist, speed of work, and the time spent performing the given actions during one shift.

Theoretical models of the human musculoskeletal system can be effectively applied using computerised methods for analysing and assessing the musculoskeletal load. These models can perform the calculations used to assess the musculoskeletal load. The model's input data are obtained by registering body parts' positions during work. This is achieved with the help of goniometers by registering the angles describing body positions.

The precision of assessment methods varies. Usually, the precision of assessment depends not only on the methodology used, but also on the precision of the input data. The assessment precision closely depends on whether the input values are actually measured or just estimated according to a subjective evaluation by the employee or the person conducting the work analysis.

Risk assessment for MSDs uses methods based on biomechanical criteria, described in the standard PN-EN 1005-1 (2005). The PN-EN 1005 standard consists of five parts describing the dependencies among the individual biomechanical factors and the methods of risk assessment. Each part focuses on one of the three factors listed earlier (body posture, force, duration of load).

24.5 BODY POSTURE AND RHYTHM OF WORK

The optimum body posture that creates the least amount of load is the 'natural posture'—standing, with spine upright and the upper limbs hanging freely along the body. Work performed in uncomfortable body positions—bent or twisted—causes an increase in the value of compressive and transverse forces acting in the spine and in the moment of force in the joints of the upper and lower limbs. The rhythm of work also influences the load on the musculoskeletal system and cause disorders. Even during light work involving small loads on the motor system, the muscular system can bear excessive loads caused by improper technique. Static loads and repetitive loads pose the largest threat to the musculoskeletal system. As illustrated in Figure 24.2, both immobile posture and too-frequent repetition of the same activities

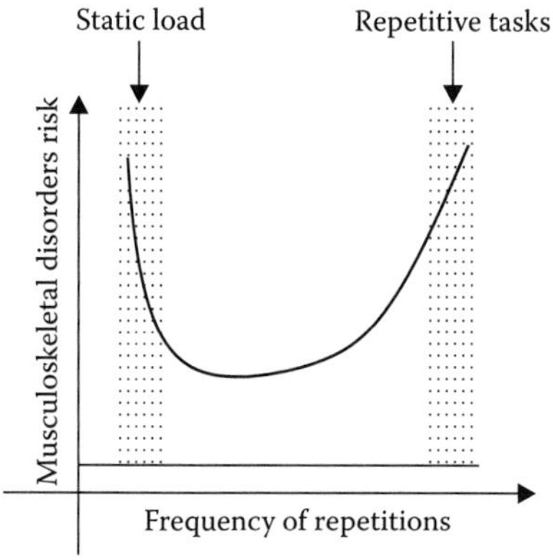

FIGURE 24.2 The risk of musculoskeletal disorders as a function of repetition frequency of work activities. (Reprinted with permission from PN-EN 1005-4. 2005. Safety of machinery—Human physical performance—Evaluation of working postures in relation to machinery.)

increase the risk of MSDs and should be avoided (PN-EN 1005-4). To avoid constant tension of the same muscle groups, the workers should be allowed to perform diverse actions according to the natural rhythm of body movements.

Both the value of force exerted at the workstation and the frequency with which the force or body position changes influence the load on the musculoskeletal system and the development of disorders. The maximum allowable levels of muscle tension during static work have been defined, but an evaluation of total value of external force, as well as of the repetition frequency of the performed activities is required. The effects of static load on body posture and the loads associated with performing repetitive work with the upper limbs are discussed in standards PN-EN 1005-4 and PN-EN 1005-5.

24.5.1 Static Load Resulting from Body Posture

Standard PN-EN 1005-4 (2005) contains guidelines for designers of machines and their components regarding evaluating the risk associated with posture and work activities during the assembly, installation, operation, setting up, cleaning, repair, transport, and dismantling of the machines. The standard contains requirements for body postures during work and the preferred movements, with minimum values of external force. The standard also presents a range of angles for the torso, head, and upper and lower limbs for three zones of loads (acceptable, conditionally acceptable, and unacceptable), taking into account repetitions occurring less than two times per minute (Figure 24.3).

24.5.2 Repetitive Action of the Upper Limbs

The risk of development of disorders of the upper limbs resulting from the performance of repetitive work can be assessed with the OCRA method described in standard PN-EN 1005-5. In practice, this method is applied to define the acceptable number of repetitions of basic actions during one shift and to evaluate the risk associated with work of that type. The recommended number of repetitions is calculated as the basic repetition frequency, which is then reduced depending on various factors that influence the risk of MSDs during repetitive work, such as body posture,

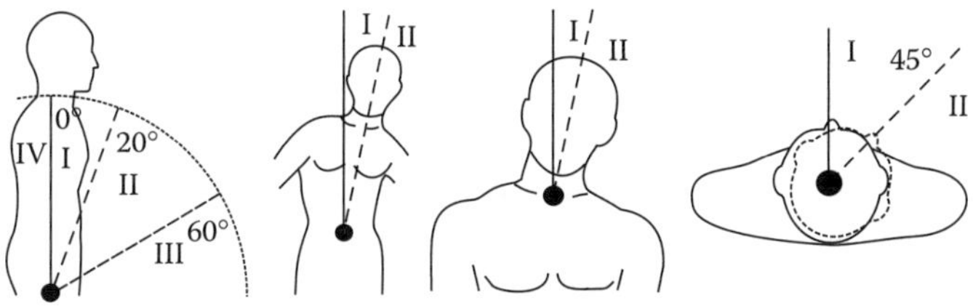

FIGURE 24.3 A body posture during work that exceeds the angle range marked 'I' is cumbersome. A larger angle or twist of the torso, neck or head, and the need to maintain such a position for a long time means it is cumbersome. (Reprinted with permission from PN-EN 1005-4 2005. Safety of machinery—Human physical performance—Evaluation of working postures in relation to machinery.)

repetition frequency, and force. Analysis of the position and position changes (movements) of the upper limbs covers every section of the limb (hand, wrist, elbow, and arm) and aims to define the duration for which the various positions and movements of basic actions are maintained during work. The description of upper limb's position takes into account the pronation or supination of elbow, bending or straightening of elbow, and bending or straightening and adduction or abduction of the wrist. The risk index (the OCRA index) is the product of ratios describing the body posture, repetition frequency of work actions, rest period, and duration of repetitive action. Certain additional factors that may appear during the performance of repetitive work can increase the force requirements (uncomfortable tools or gloves) or cause damage to muscles and tendons (vibration, compression, cold surfaces).

The OCRA method is used only to measure fully repetitive actions performed by the upper limbs, but does not include the shoulders, which significantly reduces the comprehensiveness of the evaluation. Various other activities are treated in the same manner, and the same criteria are applied. The proposed method is relatively complicated.

24.6 ADMISSIBLE FORCE VALUES

24.6.1 Force Capacity

The load caused by maintaining the same position at work increases if the worker has also to exert force. The type of exerted power, for example lifting, and the force values play a significant role. The worker should exert physical force using the muscle groups that are able to overcome the external force. The worker should also maintain an appropriate body posture during the given action. Workstations where actions are performed that require the worker to exert force should be designed in a manner allowing for optimum performance; that is, their design should account for not only the position during work, but also the direction and value of the force and its frequency and duration. The design should also take into account the fact that various muscles and muscle groups should be activated in turns, so as not to cause static overloads, muscular system fatigue, or uneven distribution of loads among various body parts.

The third part of standard PN-EN 1005-3 (2002) specifies methods for determining the muscular efficiency of the adult worker population. It takes into account the muscle forces developed under static and dynamic conditions, as well as the duration and frequency for generating specific values of muscle forces.

Standard PN-EN 1005-3 presents a procedure for assessing the risk of the MSDs caused by the performance of work. It is based on the stated values of maximum isometric forces exerted one time under various conditions. The maximum force values are modified by the appropriate coefficients, depending on the value of the parameters that characterise the speed at which the given move is performed, the frequency of its repetitions, and the duration of work. This enables determination of the maximum admissible force value for specific work conditions. The recommended force is calculated based on a simple formula, using the coefficients of speed, action duration, and the time of work. Risk evaluation, taking into account external force, is performed in three stages. In the first phase, the base force is defined for a specific type of activity, taking into account the force of hand grip, the force of the upper limbs in a seated body posture, the pushing and pulling force exerted standing up, and the force exerted by lower limbs. In the second phase, the value of maximum force is reduced depending on the factors that influence the value of this force—speed of movement, frequency of repetitions, and duration. In the last phase, risk evaluation is performed under a three-level evaluation system. In this system, the load on the musculoskeletal system can be low, medium, or high, depending on the type of force and the value of the individual coefficients. Standard PN-EN 1005-3 also lists additional factors that influence the risk of MSDs, including the body posture during work, acceleration and precision of movement, vibration, man-machine interactions, use of individual protection equipment, and the external environment.

24.6.2 Admissible Force Values during Transportation of Loads by Hand

Next to pulling and pushing, lifting and carrying cause the largest loads on the musculoskeletal system. Lifting heavy loads poses a risk of overload on the musculoskeletal system caused by the force exerted by the mass of the lifted object. This causes degenerative changes to the muscles, ligaments, joints, discs, and vertebrae of the spine.

The minimum health and safety requirements for the manual handling of loads, particularly where there is a risk of back injury to workers, are set forth in the Council Directive dated 29 May 1990 (90/269/EEC). The purpose of this directive was to lay down a legal reason for the introduction of means to reduce the risk of MSDs, especially in the spine, while performing works involving manual handling of heavy objects.

The admissible weight of transported loads is determined based on the admissible maximum value depending on body posture during lifting and the frequency of repetitions according to PN-EN 1005-2 (2003). PN-EN 1005-2 takes into account work tasks that require lifting, lowering, and moving objects; it does not include work tasks that require holding objects without walking, pushing or pulling, or holding objects in a seated position.

The risk evaluation method proposed in this standard is based on the relationship described by the NIOSH equation (Waters et al. 1993). This mathematical relation is used to determine the force—the recommended weight limit—as a function of parameters that describe body posture and repetition frequency. Additional parameters are also taken into account.

The body posture is defined using parameters that directly influence the position, such as the starting height of the load to be lifted, a horizontal location measured from the mid-point of the line joining the inner ankle bones to a point projected on the floor directly below the mid-point of the hand grasps, and the vertical movement of the object, defined by the difference between its vertical location at the start and end of the lift. The body position also depends on the asymmetric angle, which describes the torso twist during the lift.

Other parameters that influence the recommended weight limit and the risk of disorders include grip coefficient, the coefficient determining whether the work is performed by one or two upper limbs, the coefficient specifying the number of persons involved in the lift, and the performance of other actions causing physical load.

24.7 OBSERVATIONAL METHODS FOR EVALUATING WORK-RELATED LOADS

Observational methods are commonly used to evaluate the loads on the musculoskeletal system or to assess the risk of appearance of disorders. The most frequently used observational methods include the rapid entire body assessment (REBA) and the Ovako working posture analysis system (OWAS) methods.

The REBA method enables a rapid assessment of the risk of MSDs (Hignett and McAtamney 2000). The method identifies the physical effort related to the body posture during work, the exertion of forces and the performance of work causing fatigue, taking into account repeated or static loads. Body posture is assessed with appropriate position scores, which match the position of the upper arm, lower arm, and wrist, the flexion or twisting of the neck, and the position of the back and legs. The body position categories are combined with the score determining the exerted power: (1) in the case of the position of back, neck, and legs, or with the grip code and (2) in the case of the positions of the upper arm, lower arm, and wrists. The overall risk assessment takes into account the value and nature of the exerted force. The full assessment is derived by adding an activity score to include an additional load associated with static or repetitive work or a load associated with external force exerted during sudden changes of position. In the REBA method, risk is assessed using a five-degree scale.

The OWAS method for evaluating musculoskeletal system loads allows determination and assessment of the work-related risk as high, medium, or low (Karhu et al. 1977). This method can be used to conduct a quantitative load analysis based on standard body positions during work, taking into account the external force values. This method uses classification of the positions of the back, arms, and legs during subsequent work actions. The external force value is also used to measure load.

Combinations of positions of individual elements (back, arms, legs), taking into account external force, are grouped into four categories in assessing the workstation. Each combination of codes corresponds to one category, which determines the evaluation for the given workstation. As a result, a given action is classified under one of four assessment categories. The duration that each of the body postures is maintained is a very important factor in analysing the load on the musculoskeletal system. This related to the total duration of the work. The combination of the assessment category with the percentage work time provides load qualification for one out of three risk assessment zones.

The REBA or OWAS methods provide guidelines for assessing MSDs risk and prioritising actions and research at the workstations. Load and risk factors vary from one worker to another. The factors that determine a worker's individual reaction can influence the increase of load from an acceptable level to a level that poses a significant problem for the given worker. Therefore, more precise research is needed in most cases where actions are necessary. Because the human body is a very complex organism, simple methods may not be fully effective. Observational methods are of a general and qualitative nature and can be applied to any given workstation. The term 'observational method' means that the method depends on the subjective attitude of the person conducting the work analysis, and hence, the index is also subjective.

24.8 REPETITIVE TASK INDICATOR

Most performed works, especially repetitive ones, involve the upper limbs, often with a static load to the back. The load on upper limbs is therefore of particular interest, and appropriate methods are necessary to assess it during the performance of repetitive tasks. The repetitive task index (RTI) is calculated based on the repetitive work model, which includes all biomechanical factors that influence the load on the upper limbs.

The load on the musculoskeletal system depends both on the time and force characteristics of the work cycle. The time characteristics of repetitive work are defined by the cycle time (CT), number of cycles (k), and the duration of individual phases within a cycle (DP_i, where i is the phase number under a work cycle, $1 \le i \le k$). The force characteristics include both the type of force and the value of each force type. The basic types of force activities are those that reflect the typical, most frequently found work activities. They can involve the hand only or involve the whole upper limb. The hand performs manipulative actions, such as hand grips or finger pinches, including the tip pinch, palmar pinch, and lateral pinch.

Basic types of force activities, given in relation to the performance of work actions that involve the whole upper limb, include lifting a tool held in the hand or moved and keeping the upper limb maintained in a specific position. Lifting force is exerted vertically upward as a reaction to the gravitational force. Other types of force activities often found at the workstation include pushing, that is, the force needed to move an object, exerted outwards along the wrist axis (e.g. a need to exert pressure during work with a drill), the grip force of an object resulting from its weight, and the turning and reversing force (e.g. screwing and unscrewing).

An employee's load, resulting from the exertion of a specific force, depends on his or her force capacities. In the RTI method, the force related to the performance of a specific work is expressed as the percentage of maximum force for the given type of force activity and for the given position of the upper limb. Therefore, the force characteristic of a repetitive work is defined by the relative force (RF), exerted in the individual phases of the work cycle (RF_i) where $1 \le i \le k$. The RF is defined as the value of exerted force, related to the standard value of force of the given type for the same position of the upper limb. The force values are often related to the workstation or work process, which means that they are documented in the work process. In some cases, specific force values should be measured. For a specific force type, both the absolute value of the force exerted to perform the analysed work function and the maximum value for the same position of the upper limb are measured. The method enables the calculation of maximum force value for various types of force as the function of the upper limb position in order to determine the normative force, or the maximum force for the whole worker population. The normative value for the basic types of force activity of the limbs is determined as the function of the upper limb position. The upper limb position is defined using the values of seven angles, which describe the limb's deviation from the neutral position as bending, straightening, abduction, adduction, turning, and reversing (Roman-Liu 2005). The model assumes that bending and straightening are defined using the same angle, with the bending movement positive and the straightening movement negative. The same applies to abduction and adduction and turning and reversing.

All seven angles that define upper limb posture are 0° in the natural position of the body, that is, when standing upright with upper limbs hanging naturally down. In the shoulder joint, q_1 is the angle of horizontal adduction or abduction; movement is in the transverse plane with flexion of 90° from –45° (the arm directed towards the body) to 90° (the arm directed away from the body). The angle of extension or flexion q_2 describes movement in the sagittal plane from –50° (the arm directed towards the back of the body) to 180° (the arm directed upwards). The angle of medial or lateral rotation along the long axis of the arm (q_3) is from 60° (the upper limb flexed at the elbow, the forearm directed towards the body) to 45° (the upper limb flexed at the elbow, the forearm directed from the body). The angle of flexion in the elbow joint (q_4) is between 0° (elbow fully extended) and 135° (maximum flexion in the elbow). The angle of pronation or supination angle q_5 describes movement from −90° (the palm of the upper limb flexed in the elbow directed downwards) to 90° (the palm of the upper limb flexed in the elbow directed upwards). In the wrist joint, the angle q_6 describes adduction or abduction from −45° (the angle between the axis of the forearm and the axis of the hand when the hand is bent in the direction of the fingers) to 30° (the angle between the axis of the forearm and the axis of the hand when the hand is bent in the direction of the thumb). The angle q_7 describes extension or flexion from −80° (the angle between the external plane of the hand and the forearm) to 90° (the angle between the internal plane of the hand and the forearm).

The performance of each task is related to a specific type of force activity. More than one type of such activity is often present, for example, packaging involves maintaining the upper limbs in a specific position and the gravitational force of the packaged object. Therefore, determining the force of a given cycle phase must account for all of the force activities involved in the analysed task. The root of the sum of the square of the RF_i determines the RF of the cycle phase (PRF_i, where i is number of work phase cycle, $1 \leq i \leq k$).

The RTI method uses parameters that describe the performed task to describe the external load caused by that work and quantitatively expresses the load on the upper limbs. This is very important to optimising the work-related load.

The external load that results from performing a single work series of any length is defined by the index of cycle load (ICL). This index is the sum (k) of the products of the relative load of cycle phase (RF), multiplied by the duration of the given phase (DP), divided by CT (Roman-Liu 2005; Roman-Liu and Tokarski 2005).

The RTI defines the external load resulting from the performance of a repetitive work as the function of its components: ICL, CT and the number of cycle phases (*k*; Roman-Liu 2007).

$$\mathrm{RTI} = \mathrm{ICL}\frac{(\mathrm{CT}/k)^2 - 44(\mathrm{ICL} - 0.88)(\mathrm{CT}/k) + 470(\mathrm{ICL} + 0.024)}{0.26(\mathrm{CT}/k)^2 - 16(\mathrm{ICL} - 1.4)(\mathrm{CT}/k) + 152(\mathrm{ICL} - 0.01)}$$

where RTI is the RTI, *k* is the number of phases in a cycle, ICL is the index of the cycle load, and CT is the cycle time.

Because both the static load and an excessive frequency of changes pose a threat of MSDs, the relationship between the CT and the number of its phases is very important. Figure 24.4 illustrates changes in the RTI along with the changes in the V (V = CT/*k*), with the constant value of the ICL. The curves demonstrate the variability of the index values and the minimum functions depending on the value of V and ICL parameters. They also show that the number of cycle phases influences the value of the V parameter for which the load with repetitive work is the smallest. The load index allows the selection of repetitive work parameters in such a way that the load on the upper limb is minimised. The RTI quantitatively describes the load on the upper limbs for different work situations, but it does not allow assessment of the risk of MSDs.

The RTI-based method is comparable with other methods focused on upper limbs load assessment, for example the OCRA method. The OCRA method provides risk

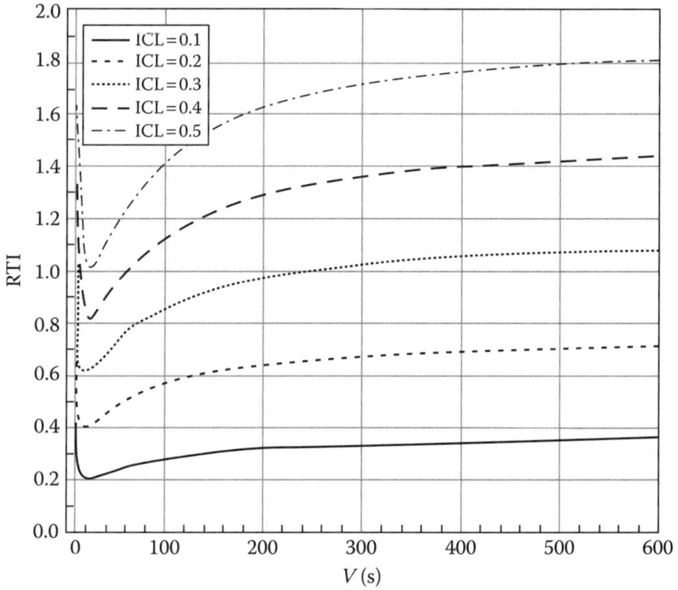

FIGURE 24.4 Values of the repetitive task indicator depending on the ratio of cycle time, number of phases in the cycle, and the index of cycle load for a two-phase cycle.

zones, which RTI does not. However, OCRA is limited to forearm movements only and is less precise when upper limb posture is considered. This method needs stronger verification oriented to development of criteria for risk assessment zones.

REFERENCES

Council Directive 90/269/EEC of 29 May 1990 on the minimum health and safety requirements for the manual handling of loads where there is a risk particularly of back injury to workers. 1992. In *Directives for the European Economic Communities on Labour Protection,* Vol. 1, 63–72. Warsaw: CIOP.

Hignett, S., and L. McAtamney. 2000. Rapid entire body assessment (REBA). *Appl Ergon* 31:201–205.

Karhu, U., P. Kansi, and I. Kuorinka. 1977. Correcting working postures in industry; A practical method for analysis. *Appl Ergon* 8:199–201.

Moore, J. S., and A. Garg. 1995. The strain index: A proposed method to analyze jobs for risk of distal upper extremity disorders. *Am Ind Hyg Assoc* 56(5):443–458.

Occhipinti, E. 1998. OCRA: A concise index for the assessment of exposure to repetitive movements of the upper limbs. *Ergonomics* 41(9):1290–1311.

Parent-Thirion, et al. 2007. *Fourth European Working Conditions Survey. European Foundation for the Improvement of Living and Working Conditions.* Luxembourg: Office for Official Publications of the European Communities.

PN-EN 1005-1. 2005. Safety of machinery—Human physical performance—Terms and definitions.

PN-EN 1005-2. 2003. Safety of machinery—Human physical performance. II. Moving machines and their part by hand.

PN-EN 1005-3. 2002. Safety of machinery—Human physical performance: Recommended force limits for machinery operation.

PN-EN 1005-4. 2005. Safety of machinery—Human physical performance—Evaluation of working postures in relation to machinery.

PN-EN 1005-5. 2007. Safety of machinery—Human physical performance—Risk assessment for repetitive handling at high frequency (February, 2007).

Roman-Liu, D. 2005. Upper limb load as a function of repetitive task parameters, Part 1: A model of the upper limb load. *Int J Occup Saf Ergon* 11:93–102.

Roman-Liu, D. 2007. Repetitive task factor as a tool for assessment of upper limb musculoskeletal load induced by repetitive tasks. *Ergonomics* 50(11):1740–1760.

Roman-Liu, D. 2008. Exposure to risk factors for the development of musculoskeletal disorders in European Union states. *Occupational Safety* 11:16–20.

Roman-Liu, D., and T. Tokarski. 2005. Upper limb strength in relation to upper limb posture. *Int J Ind Ergon* 35:19–31.

Waters, T. R., V. Putz-Anderson, A. Garg, and L. J. Fine. 1993. Revised NIOSH equation for the design and evaluation of manual lifting tasks. *Ergonomics* 36:749–776.

Work-Related Musculoskeletal Disorders—TCRO 2006. 2006. European Agency for Health and Safety at Work (Prevent).

25 Shift Work

Krystyna Zużewicz

CONTENTS

25.1 INTRODUCTION

Shift work has become an integral part of our lives. Society needs 24-hour services such as security, and in many industries, such as power engineering, continuous 24-hour technological processes are necessary, including on Saturdays, Sundays, and holidays. Therefore, the number of employees in this service sector, including services rendered via the Internet, is continuously increasing. White-collar workers, such as bank or stock market employees, also work at night. Receptionists, bakers, daily paper printers, road drivers, physicians, airline pilots, firemen, and policemen perform shift and night work. Sometimes, shift work at industrial plants becomes necessary due to the modernisation of production technologies through the implementation of new instrumentation. In such cases, economic considerations and attempts to achieve a quick return of the costs require a more intensive exploitation of the workforce and equipment.

The number of shift workers, including night workers, is continuously increasing. In the western European countries, one-fifth of employees work in shifts (Harrington 2001).

For effective and safe shift work, the duration and time of the shift and the employee's working capacity should be adequately balanced based on the circadian rhythm of basic life processes. Due to a substantial physiological and social burden, both physical and mental workload should be adjusted to the time of work to limit the risk of errors being committed by shift and night workers, which may result in accidents. In the 1970s, French physician and ergonomist Pierre Cazamian proposed the term 'chronoergonomics' for the branch of ergonomics that applies ergonomics to the management of workers' time (Ogińska 1991).

25.2 CIRCADIAN RHYTHMS OF HUMAN PHYSIOLOGICAL FUNCTIONS AND THEIR SIGNIFICANCE FOR OCCUPATIONAL SAFETY

25.2.1 Circadian Oscillations of Physiological Functions

The human body contains structures that generate internal circadian rhythms, which are dependent on cyclic phenomena observed in the external environment, such as the day–night rhythm. These phenomena optimise body functions by adjusting basic life processes according to environmental time determinants including the time of the day (Figure 25.1).

For example, humans typically sleep during the night and are awake and performing activities during the day, because the ability to perform physical and mental work is maximal during the day. Many rhythmic processes occurring in the human body without our knowledge reach maximal intensity during the day; this is confirmed by scientific findings. Some rhythmic processes reach maximal intensity during the night.

The following processes have a 24-hour rhythmicity:

- Physiological processes (heart rate, arterial blood pressure [ABP], core body temperature)
- Metabolic processes (hormone secretion, changes in blood chemical composition, synthesis of biochemical compounds in tissues and organs)
- Psychophysical processes (ability to perform mental work-related activities, oculomotor coordination, sensitivity to acoustic and visual stimuli)
- Sensitivity to stress of different origins (disease-related stress, sensitivity to pain)

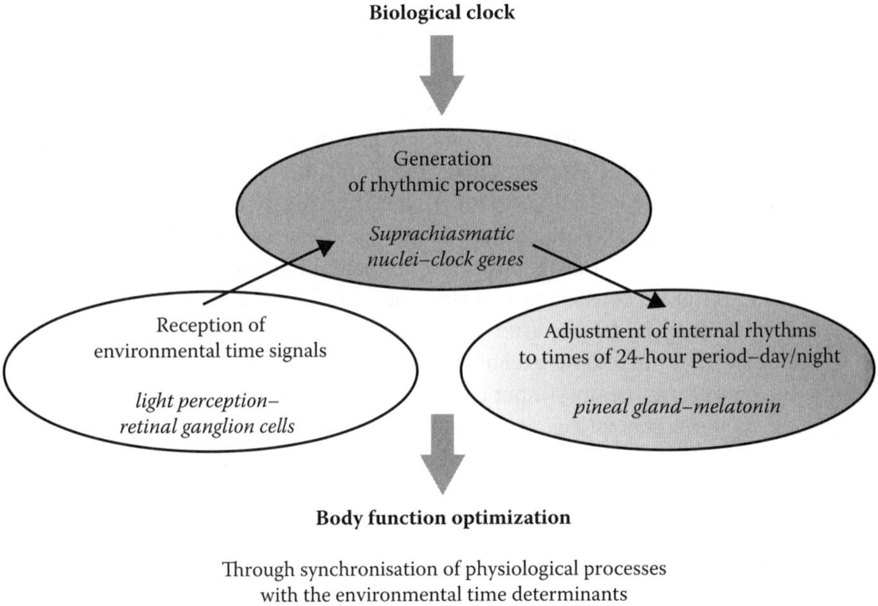

FIGURE 25.1 Basic components of the biological clock, which determines the rhythmicity of basic life processes of the human body and is adjusted to the astronomic day–night cycle.

The rhythmicity of biological processes depends on the human's lifestyle and internal biological clock. These processes are in harmony in individuals who are active during the day and rest during the night. For example, glucose metabolism increases upon awakening and further increases during the day, decreasing in the evening. The metabolism of fatty acids increases in the evening and at night during the rest phase and decreases near the end of sleep. There is a correlation between biological variables and general indicators of feelings, such as alertness and most mental fitness parameters. During the day, body temperature rises and increases the adrenalin serum concentration, which is optimal for well-being and alertness. In the evening, the adrenalin serum concentration and core body temperature decrease, psychophysical fitness deteriorates, and fatigue intensifies (Zużewicz et al. 2001).

During night shift work, a reversed sleep–wake rhythm is forced instead of the natural day–night rhythm, bringing about adverse health and social consequences in shift workers.

25.2.2 Circadian Oscillations of Psychophysical Fitness

The ability to perform physical tasks depends on many biochemical and physiological parameters having circadian rhythms. Higher tolerance to physical tasks during the day compared to night is due to more intensive metabolism, the utilisation of glucose, and increased glycogen decomposition in muscles. Increased blood flow results in improved tissue oxygenation and excretion of metabolic products (CO_2, lactic acid, and so on). Faster breathing delivers a greater amount of oxygen and results in faster CO_2 excretion. An increase in heart rate and ABP delivers a greater amount of oxygen and energy to the working muscles, brain, and other organs. Increased perspiration and dermal blood flow eliminate excess heat produced due to physical effort.

Mental efficiency is also dependent on circadian rhythmicity. Evaluations of psychological tests of mental efficiency (e.g. reaction time, oculomotor coordination, concentration) indicate variations with a characteristic circadian profile. Mental efficiency increases on awakening and decreases in the evening. In many people, particularly the elderly, a transient decrease in mental efficiency occurs in the early afternoon. This is not a consequence of sleepiness after meal consumption, but rather an effect of intensifying fatigue after several hours of activity following the moment of awakening. There are some exceptions to this rule, for example tasks requiring substantial short-term memory input can be performed better at night, but the durability of the acquired knowledge is much shorter. Interestingly, the rhythm of mental efficiency is similar to the rhythm observed in the physiological parameters, such as core body temperature, heart rate, ABP, and adrenalin secretion (Figure 25.2).

25.2.3 Workability Rhythm

In individuals with typical patterns of circadian activity, overall work ability increases from the moment of awakening until the afternoon hours, when there is a slight decrease between 1 and 3, or during the usual lunch time, regardless of meal

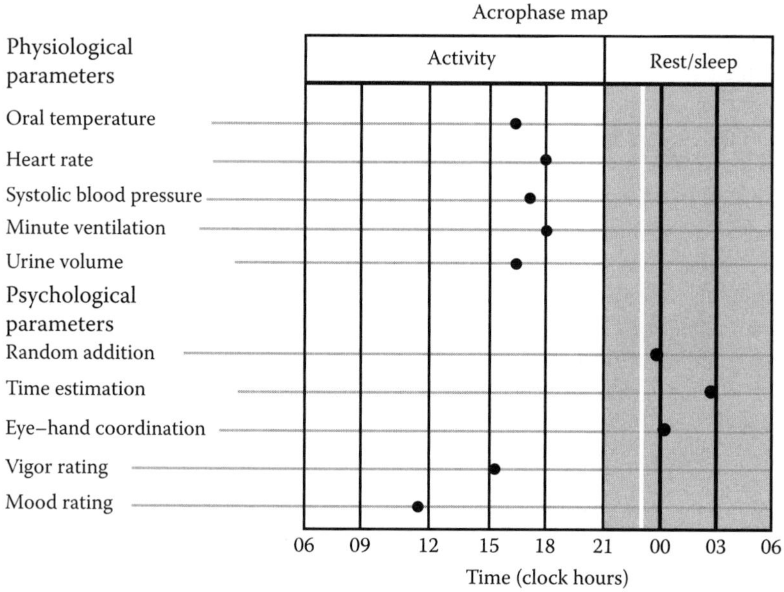

FIGURE 25.2 Time of maximal occurrence (acrophase map) of selected psychophysical parameters. Daytime is the most favourable for physical and mental activities (black dots). *Note:* The black dot indicates the longest time for task performance and the time at which the worst results are obtained.

consumption. In the evening, work ability gradually decreases until the usual time of rest. After prolonged activity, there is minimal circadian rhythm in overall body efficiency between midnight and 4 AM.

The circadian oscillations of work ability, especially physical work, are parallel to the circadian rhythm of the core body temperature, which under the normal conditions of the sleep–wake cycle reaches maximal values in the evening and minimal values in the early morning.

Night shift work disturbs these relationships, and, if combined with sleep deprivation and fatigue, may significantly negatively affect the quality of work and increase the risk of committing an error.

25.3 DEFINITIONS AND TERMS RELATED TO NIGHT WORK*

The term 'shift work' is defined in international laws, particularly in directives 93/104/WE and 2003/88/WE. It refers to any form of shift work scheduling in which employees performing the same duties replace each other according to a given work schedule and that entails a need to work at different times during certain days or weeks. The term 'shift worker' refers to each employee whose work schedule is part of the shift work.

There are different systems for shift work in industry. They involve occupational activities lasting between 6 and 12 hours within the same shift. There may be two, three, or four shifts during a 24-hour period. The most frequently employed is the three-shift system, consisting of shifts starting at 6 AM, 2 PM, and 10 PM, although there may be other systems as well and they may vary in rotation speed, the number of consecutive workdays on the same shift, or direction. Some shift workers work only night shifts.

25.3.1 Night Work

According to the Polish Labour Code (Section VI: *Night work*, Article 151^7, items 1 and 2), the term 'night work' refers to the 10-hour period between 9 PM and 7 AM, and a 'night worker' is defined as an employee whose working schedule has at least three night hours during a 24-hour period or at least one-fourth of the overall working time is at night.

This definition differs slightly from that found in international laws such as convention no. 171 MOP, recommendation no. 178 of June 1990 *Night work*, or UE directive 2003/88/WE, referring to selected aspects of shift and night work. According to these regulations, the term 'nighttime' refers to a period of time no shorter than 7 hours including the hours between midnight and 5 AM.

25.3.2 Work Constantly on the Move

The term 'work constantly on the move' refers to work that cannot be stopped due to the technology being used. This requires hiring three or four crews, working in rotating shifts, and providing uninterrupted 24-hour coverage every day and night.

* According to the Labour Code and the WE Directive.

25.4 HOW DOES SHIFT WORK (INCLUDING NIGHT WORK) DIFFER FROM DAILY WORK?

Table 25.1 compares selected aspects of life activities in day and night shift workers that affect the quality of work performance and the health of the employees.

TABLE 25.1
Differences in Life Activities in Day and Shift Workers

Parameter	Work Schedule	
	Day Work	Shift/Night Work
Work time	Day only	24 hours including night hours
Working system	Inflexible working hours usually between 7 AM and 5 PM	Flexible working hours with the most frequent rotating system: 8-hour shifts in a morning–afternoon–night cycle with different workdays within the same shift
Physical and mental work performance	Always at the same time of day, which is most favourable due to body physiological 'readiness'	Continuous variability of workload and a need for continuous body adjustment to the resultant changes; working also at night when the body's ability to perform physical and mental effort is the lowest
Relations between physiological rhythms and environmental time determinants	Conformity and stability of the relationship between human and environmental rhythms	Lack of conformity between the phases of physiological rhythms in humans and external environment
Sleep	At night, according to physiological readiness for rest	Forced sleep disregarding physiological body readiness under adverse conditions in the external environment (sunlight, noise)
Feeding	Meals consumed at a time of full readiness of the digestive system	Meals consumed at the most favourable time due to physiological conditions of the digestive system and metabolic processes
Response to environmental stressors	Full hormonal and physiological readiness of the body to respond to environmental stressors	Weaker response of the body, especially at night, to adverse and burdensome factors of the work environment (noise, temperature, toxic substances)

25.5 CONSEQUENCES OF SHIFT WORK

Shift work may affect physiological functions, disturbing circadian rhythms and contributing to the development of some pathological disorders. Immediate disturbances of body functions caused by shift work include sleep disorders, chronic fatigue, and digestive system ailments. These symptoms occur because of the body's natural response to atypical work conditions, which are often of a short duration and connected with the work schedules, especially the night shift. They may be relieved by working day shifts or working longer intervals (Knutsson 2003). Shift workers often suffer from chronic ailments resulting from the lack of synchronisation of their internal body clock and a forced, atypical rhythm of activity and sleep. Ailments due to shift work are called 'time debt syndrome' or the syndrome of maladjustment or circadian disruption.

The mechanisms of exactly how shift work affects human health have not yet been fully explored. In healthy individuals who are active during the day, almost all biological functions, even at subcellular levels, have circadian rhythms that are related to a constant sequence of the body's maximal values. Even in isolated parameters, disturbance of these rhythms (such as the values of oscillation amplitude, the time shift of maximal value occurrence, or changes in the average circadian level) may bring about health-related consequences. Research indicates a relationship between night work and functional disorders of the digestive and cardiovascular systems and pregnancy abnormalities (van Mark et al. 2006). Since some diseases develop slowly and only manifest after many years, it is difficult to assess whether they are caused by disturbances of basic internal rhythmic life processes or they result from external reasons such as tobacco smoking, inadequate sleep, or inadequate nutrition.

Studies conducted so far have not determined the relationship between increasing sensitivity to various toxic substances used in production at different times of the day and the health-related problems in shift workers employed in different shift systems. The traditional approach to evaluation of occupational risk assumes that this sensitivity is the same at different times in a 24-hour period. Toxicological studies using animal models indicate that there are different levels of risk depending of the time of the day that the worker is exposed to the adverse factors.

25.5.1 Physiological Consequences of Shift Work—Shift Lag

Basic physiological consequences of night shift work include circadian rhythmicity disorders. Performing occupational activities at night, the usual time for sleep, results in characteristic symptoms with different degrees of intensity, called 'shift lag'. Shift lag results when the body rhythm does not adjust to changes in activity time or the disturbance of the normal relationships between different rhythms as a consequence of shift work. The symptoms most frequently associated with shift lag include:

- Aggravation
- Sleep disorders
- Peristalsis disorders

- Deterioration of oculomotor coordination
- Muscular strength impairment
- Disorders of distance and time perception

The relationship of the severity of circadian rhythmicity disorders to occupational activities at different times of the day depends to a large extent on work scheduling. Shift workers working in rotational shifts with a 'backward rotation' (i.e. the sequence of shifts: morning, night, and then afternoon) and those constantly working night shifts experience the most substantial disturbances. Individuals working in 'forward rotating' shifts with night work for two consecutive days and morning shift work that does not disturb the sleep period experience relatively less substantial disturbances (Pokorski and Costa 1998).

25.5.2 Effect of Shift Work on Sleep

Sleep deprivation is a typical adverse effect of shift work. Overall sleep time during a 24-hour period decreases and the quality of sleep deteriorates. Sleep disorders in shift workers result from the need to rest during the day, when there is an increased noise and light level and when most other people are active. Even experienced shift workers accustomed to resting during the day may experience some problems after reaching middle age because the quality of sleep deteriorates with age. Night workers sleep about 2–4 hours less than they would if they were day workers. Many night workers sleep after returning from work, that is, in the morning, when the majority of people are active after their night rest. A night worker should sleep several hours before the shift in order to feel less sleepy while working a consecutive night shift. This means splitting the usual sleep period into two parts that are shorter than a usual night of sleep. Even a short period of sleep may delay an accumulating feeling of sleepiness during the night shift. Sleeping a lot in the morning, without a nap before the consecutive working shift, may cause a substantial decrease in the worker's psychophysical fitness at the shift end due to a long, over-16-hour period of continuous activity. Sleep problems are noted among shift workers more often than in other occupational groups, leading to more frequent use of hypnotic agents. Workers may excessively consume coffee or other caffeinated beverages and smoke tobacco. Such behaviours increase the risk of gastrointestinal or circulatory system disorders.

25.5.3 Fatigue and Sleepiness

Fatigue is experienced at the end of a night shift by a large population of night shift workers and after many hours of daytime activity. Fatigue adversely affects occupational safety by impairing workers' psychomotor fitness and causing, for example, depression, irritation, and unjustified anger (Gaba and Howard 2002). The effects of sleep deprivation combined with prolonged activity are comparable to the effects of alcohol intoxication. Psychomotor performance decreases to a level corresponding to blood alcohol concentration of 0.5% after 17 hours of wakefulness (from 8 AM to 3 AM), and to a level corresponding to blood alcohol concentration of 1.0% after 24 hours of wakefulness (Dawson and Reid 1997).

25.5.4 Shift Work and Night Work–Related Stress

Night-work-related stress is defined as stress due to disturbances in the natural activity–sleep processes of the natural phases of circadian rhythms. The disturbance of circadian rhythms, sleep deprivation, sleep disorders, and fatigue are typical consequences of night shift work; they cause strain on workers that may result in a decrease in physical capacity, health disorders, worsened general feelings, and decreased performance. These disorders also increase the risk of errors and accidents (Figure 25.3). Stress is modified by many factors that are related not to the work but to the worker, including age, gender, educational background, and lifestyle. The ability to estimate risk level and to work out a problem-solving strategy, for example, by observing proper sleep and nutrition or by napping or other proper relaxation, may also play a role (Smith 1998).

Shift work is a stressor that may occur apart from other stressors associated with work performed at any time within a 24-hour period, such as lack of autonomy, monotonous activities, or environment and time constraints.

25.5.5 Health Consequences of Night Work

The summary health index by Haider et al. (1988) indicates a depleted state of health in shift and night workers when compared to daily workers performing comparable activities.

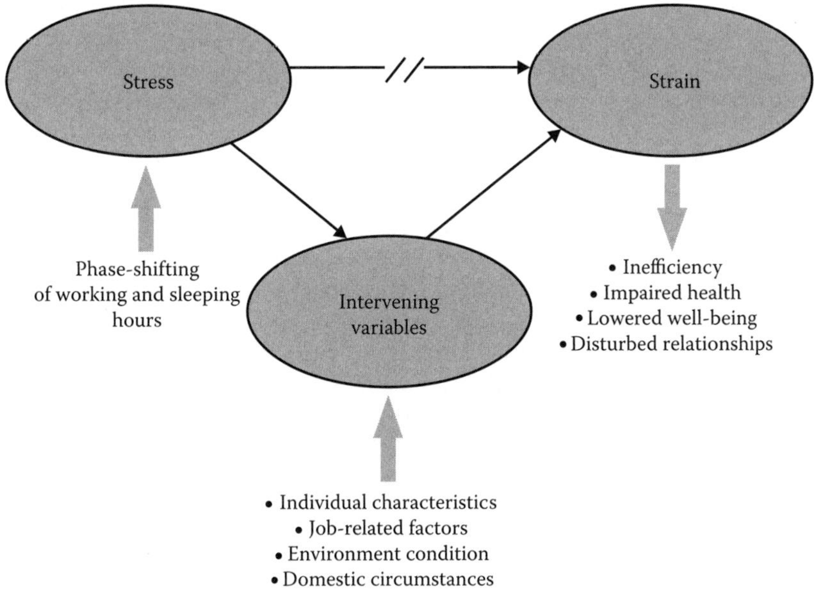

FIGURE 25.3 The 'stress and strain' model of night work load proposed by Colquhoun and Rutenfranz (1980). (Adapted with permission from Monk, T. H., and S. Folkard. 1992. *Making Shiftwork Tolerable*. London: Taylor & Francis.)

25.5.6 Digestive System Disorders

Disorders of normal digestive system functions and metabolism, such as diarrhoea, constipation, or heartburn, may occur more frequently in night shift workers compared to day workers. They result from improper nutrition, in terms of both the food quality and the time of meal consumption. Irregular meal times interfere with the production of the hormones, acids, and enzymes necessary for food digestion, a diurnal function. Like other body rhythms, the digestive process slowly and only partly can become accustomed to an atypical activity–sleep pattern. Karlsson et al. (2001) carried out an analysis of metabolic disorders in a group of 27,000 workers. The study indicated a higher prevalence of obesity, a low high-density lipoprotein cholesterol concentration, and a high triglyceride level in shift workers.

A report by Waterhouse et al. (1990) states that the risk of ulcerous stomach and duodenum disease in night workers is two to five times higher than that of day workers. The development of pathological conditions can be observed 5–6 years after beginning night shift work, which is much earlier than in day workers, who experience such disorders after 12–14 years. Chronic pathological conditions of the digestive system and ulcerous disease have been recently found to be caused by *Helicobacter pylori*. However, shift and night work are still considered risk factors for gastrointestinal tract disorders (Pokorski and Costa 1998).

25.5.7 Circulatory System Disorders

Many studies on estimating the risk of cardiovascular disorders among shift workers suggest that this population has an increased risk of coronary arterial disease, arterial hypertension, and myocardial infarction (Harrington 2001).

Knutsson et al. (1999) studied the relationship between shift work and coronary arterial disease. They compared the percentage of shift workers with coronary arterial disease in two groups of subjects. Each group consisted of 2000 patients with a history of asymptomatic acute myocardial infarction. The study confirmed the relationship between shift work and an elevated risk of myocardial infarction both in women and men, but did not indicate any correlation between the level of work-related fatigue, tobacco smoking, and educational background. The risk of circulatory system disorders in shift or night workers is about 40% higher than that observed in day workers (Knutsson 2003). The risk increases with the duration of shift work and after 15–20 years of shift work. Circulatory system disorders of different degrees are observed in 20% of shift workers, three times more often than in day workers (Zużewicz et al. 2001). The causes of cardiac disorders observed in shift workers include circadian rhythmicity disorders and the resultant changes in social life, lack of social support, stress, tobacco smoking, improper nutrition, excessive coffee consumption, and limited physical activity.

25.5.8 Disorders of Other Systems

Many studies confirm the effect of shift work on pregnancy. An increased risk of miscarriage, low body mass of the neonates, and premature labour are often observed

(Knutsson 2003) in pregnant shift workers. Menstrual cycle irregularities more frequently noted in female shift workers than in day workers may explain problems in conceiving.

More attention has been paid recently to the prevalence of obstructive sleep apnoea (OSA) syndrome, which is a pathological condition characterised by short-lasting periods of breathing obstruction, usually preceded by loud snoring. Although OSA mainly affects male workers who usually drink alcohol before sleep, shift and night work is a risk factor for the development of this condition. Shift and night workers affected by this condition experience excessive sleepiness during the day.

Shift workers frequently report anxiety and depression symptoms. However, it is difficult to assess the range of the problem due to autoselection (the employee's decision to quit the job) within this working population (Harrington 2001). Some reports indicate that disorders defined as 'anxiety or depression, requiring psychotropic agents for over 3 months or hospitalisation' were diagnosed in 4% of daily workers, 22% of shift workers working a three-shift system, and 64% of persons working only at night (Costa et al. 1981).

Shift workers with bronchial conditions may experience exacerbation of symptoms resulting from nighttime bronchial contraction. When night workers are exposed to a high concentration of dust or irritant substances, bronchial failure can develop.

Shift work may prove especially adverse for individuals who take medications regularly. This is because the pharmacokinetic parameters are determined only for day work for most medications. The same doses administered at night may have a weaker or stronger effect than those administered during the day (Zużewicz et al. 2001).

25.5.9 Social Consequences of Shift Work

Some persons, especially young persons, decide to work shifts for family reasons or because they want to earn more, as night work is better paid. If both parents are shift workers, one may look after the children while the second one is at work. Shift work enables some people to look after old or ailing parents. However, workers should consider the accumulating effects of time debt syndrome. Sleep deficiency, a continuous feeling of fatigue and increased irritability, can contribute to the development of family conflicts. Excessive irritability makes interpersonal relations with workmates and closer acquaintances difficult, leading to isolation and feelings of loneliness, especially when the family does not understand the reason for such behaviour. Shift workers are divorced more frequently than their day-working counterparts. Rotating shifts also limit opportunities to participate in extra-occupational activities because socialising usually takes place during afternoon or evening hours. Social life is therefore limited among shift and night workers so that they can only socialise with other shift workers and their workmates. Shift work also limits active participation in political, social, and cultural events, which results in feeling more socially isolated over time.

25.5.10 Shift Work Tolerance

Studies estimate that one in five shift workers quits his or her job, generally due to poor tolerance to ever-changing timing of activities such as work, rest, and sleep.

Reasons for quitting shift work include age, family relations, lack of social life, health problems, and sleep disorders. *Shift and night work intolerance syndrome* is a series of relatively nonspecific ailments or symptoms resulting from a lack of adaptation to shift and night work for a long period of time (Pokorski and Costa 1998). Only about 10% of shift workers believe they tolerate such a working schedule quite well, do not experience the typical health disorders, and have a positive attitude about the system.

The symptoms of shift work intolerance observed in night workers include:

- *Sleep disorders*: Difficulty falling asleep despite fatigue and sleepiness, poor quality of sleep, frequent awakening
- *Persistent fatigue*: Constant sleepiness after awakening or after rest on non-working days, other than a physiological state of fatigue due to physical and mental effort
- *Changes in behaviour*: Irritability for no specific reason, mood changes, feelings of sickness or unsatisfactory performance
- *Gastrointestinal tract disorders*: Dyspepsia, pain in the upper abdomen, exacerbation of ulcerous disease symptoms
- *Use of medication*: Need for regular use of hypnotic agents

Individual tolerance of shift work depends on gender, age, and personality traits. One trait that may affect night workers' shift tolerance selection is their chronotype, or the time of day they are most alert and active. Although the workers with an evening chronotype better tolerate night work, they may experience more serious problems if they must work day shifts, particularly when they have to shorten their usual sleep period if the shift begins in the early morning hours. Better tolerance to shift and night work does not always result from many years of experience. In older workers with many years of service, tolerance to shift work actually deteriorates with time so that such workers can no longer tolerate their working schedule (Baker et al. 2004).

25.6 ACCIDENT RISKS DURING SHIFT WORK

Errors committed at work during night shifts are usually due to workers' inadequate reactions. The main factors that cause accidents include sleep deprivation, a momentary loss of consciousness called 'microsleep', fatigue, and mood deterioration. Most errors committed during night work are harmful not only for the employees themselves, but also for the people they look after, especially in the health service and transport fields. The main causes of errors committed by physicians during their duty hours are a 24-hour variability of fatigue and alertness, circadian oscillations of oculomotor coordination, circadian rhythms of mental performance, a prolonged working period, the effects of fatigue and sleep deprivation, and mental inertia on awakening after a short nap. When occupational activities require the continuous focus of attention in a monotonous environment, such as for road transport drivers, the occurrence of microsleep has a substantial probability of fatal errors.

From an occupational safety viewpoint, fatigue, sleepiness, and monotony increase the risk of overlooking warnings about the defects in technological processes and carry the risk of failure to respond or very slow responses to such events. Examples of such accidents include some widely reported catastrophes. The catastrophe at the atomic power plant on Three Mile Island (USA) occurred when workers did not notice the warning that there had been a radiation release, which occurred at about 4 AM. One of the most tragic accidents was that at a pesticide factory in Bhopal, India, which happened around 12:40 AM. One of the reported reasons of the disaster was insufficient workforce. A cloud of poisonous gas caused the deaths of 2500 people within 8 hours; the estimated number of victims was about 8000, and even 10 years later, about 50,000 people suffer due to exposure to toxic gas emissions. The event with worst consequences for people and the environment was the nuclear reactor catastrophe at Chernobyl, Ukraine, which took place at 1:23 AM (Smolensky and Lamberg 2000).

The probability of accidents and injuries increases even more during consecutive night shifts. The risk is higher during the second night shift by 6%, the third night shift by 17%, and during the fourth night shift by 36% (Folkard and Akerstedt 2004). These data provide further evidence that only up to three consecutive night shifts should be included in a shift work schedule.

25.7 PREVENTION OF ADVERSE EFFECTS OF SHIFT WORK

To prevent negative consequences of shift work, physical and mental workload should be adjusted to the workers' psychophysical capacity, resulting from circadian oscillations of the body's readiness to work. Implementing proper shift work schedules, observing rules that limit the effects of shift work on the physical and mental health, and having concern for the overall wellness of shift workers will also help prevent negative consequences.

The following measures can be taken to limit the adverse effects of shift or night work:

- Analysing and evaluating different shift work systems to select the most suitable system, ensuring the best tolerance to shift or night work, should include the number of working hours, the number of consecutive days working the same shift, the number of nonworking days, shift rotation (backward or forward), multicrew systems, and the time of the shift start
- Identifying individual traits that condition the worker to tolerate the shift or night work and working out the criteria of workforce selection
- Leading a proper lifestyle (proper sleep, naps, meals, active rest) to improve tolerance to shift or night work, as well as propagation of such a lifestyle both among shift or night workers and among their closest family members
- Improving the working schedule by optimising the time and length of breaks and developing strategies to prevent sleepiness, particularly in night workers, including preventing monotony at work
- Monitoring worker health, with particular attention to the increased risk of certain diseases in shift and night workers
- Developing regulations concerning the timing of work and rest, health protection among shift and night workers, and bans on night work for particular

groups at risk, namely pregnant women and juvenile workers, with limits on shift or night work for persons over age 45
- Monitoring core body temperature during a 24-hour period to determine the most favourable shift system that will minimise the adverse effects of shift and night work on health and general feelings of well-being (Knauth et al. 1978)

The most favourable effect of limiting the negative consequences of occupational activity at night is preventing the 'inversion' of body temperature rhythm or phase shifts of other physiological and hormonal rhythms, as this disturbs their reciprocal relations and can cause time debt syndrome.

25.7.1 Work Scheduling

Night work should not be longer than 8 hours daily if the job is particularly dangerous or entails substantial physical or mental exertion (Labour Code, Article 151[7], Sections 3 and 4). The employer and trade unions decide whether a given job meets the above criteria; when there are no trade unions, representatives of the employees make the decision. They are selected according to the procedure given by the employer and following an evaluation by a physician who takes prophylactic care of the employees for the purpose of occupational safety and health protection.

Shift workers and workers constantly on the move have an additional workload apart from night work, namely working Sundays and holidays, which is allowed based on Article 151[10] of the Labour Code. Section 138, Article 1 of the Labour Code provides information on the opportunities for extending working hours when the worker is constantly on the move: 'When the work cannot be stopped due to production technology (work constantly on the move), a system of working time can be applied with acceptable working time prolongation up to 43 hours weekly on average, during the accounting period not exceeding four weeks or one day during some weeks; in these cases the daily number of working hours may be prolonged up to 12 hours.' However, this regulation should be applied with caution, given the previous reports on the adverse effects of shift work.

When the 8-hour workday is exceeded (regardless of the time of work), the risk of accidents rapidly increases during the ninth hour of work. The relative accident risk during the twelfth hour of work is more than twice as high as during the eighth hour (Vogel 2004).

Shift work is connected with stress and disturbance of circadian rhythms; therefore, the optimal shift work schedule provides a compromise between minor and major adverse effects resulting from stress and disturbed biorhythm phases. Considering the possibility of physiological and mental 'adjustment' to night work and the need to minimise health hazards and the effects of decreased physical capacity, solutions for work schedules should be based on the following suggestions:

- The number of consecutive night shifts should not exceed four (Knauth 1993).
- Shift rotation should always be in agreement with the normal course of a day (morning, afternoon, night; Barton and Folkard 1993).

- A system with quick shift rotation is more favourable than a system including several consecutive days of working the same shift.
- A five-crew system is more favourable for shift work and work constantly on the move, compared to a four-crew system (Lillqvist et al. 1997).
- Each shift should begin at a time that does not result in forced sleep shortening (e.g. morning shifts should not begin earlier than 6 AM; Knauth 1993).
- Shift and night workers should work no longer than 8 hours, particularly while working night shifts.
- While planning the workload, the employer should consider the natural decrease in the psychophysical fitness of the worker between midnight and 3 AM.
- A physician should determine if the employees are fit or unfit for shift work based on contraindications to shift and night work and predispositions influencing long-term tolerance to this kind of work (Pokorski 1999).
- In the case of health-related problems or other contraindications, changes to the working schedule should be made and a temporary withdrawal of the employees from night work should be considered.
- Information about the adverse effects of shift work and ways to minimise the risk connected with this kind of work should be well known among managers, medical service, and occupational safety and health staff.

25.7.2 Predispositions and Contraindications for Shift Work

Employees considering shift work or night work should be aware of factors that can help them make an informed decision, for example, job satisfaction and self-esteem. They should ensure that they have the adaptive capabilities required for shift or night work and should also be aware of the effects of this kind of work on family life.

Shift or night work is not recommended for persons with sleep disturbances or circulatory or digestive system disorders because it may exacerbate disease symptoms and make treatment difficult. Contraindications for shift work also include asthma, diabetes mellitus (especially insulin-dependent), and depression. Shift workers should not abuse alcohol, stimulating substances (caffeinated beverages), or hypnotic agents, and they should limit tobacco smoking.

Night work is definitely contraindicated for pregnant women (Article 178 item 1 of the Labour Code) and juvenile employees (Article 203 of the Labour Code). For pregnant women, the employer must change the work schedule to enable the employee to perform work with no night hours, either by giving her another job or by exempting her from night work.

For juvenile employees, the strict ban on night work performance is due to the disturbed relationships between biorhythms, including disruption of the normal rhythm of hormone production that leads to proper development of adolescent individuals, caused by night work. Ageing decreases the ability to adapt to atypical work schedules, so the World Health Organisation recommends using the three-shift system for employees over 45.

Individuals who do not understand the physiological and psychosocial consequences of shift work may not consider the ways to minimise the adverse effects of shift work on occupational health and safety when developing work schedules. Close cooperation between the persons responsible for work scheduling and the employees while developing work schedules, as well as making proper use of the knowledge on health hazards and occupational safety in shift workers (Baker et al. 2004), will improve the work scheduling processes.

REFERENCES

Baker, A., G. Roach, S. Ferguson, and D. Dawson. 2004. Shift experience and the value of time. *Ergonomics* 47(3):307–317.

Barton, J., and S. Folkard. 1993. Advancing versus delaying shift systems. *Ergonomics* 36:59–64.

Colquhoun, W. P., and J. Rutenfranz. 1980. *Studies on Shiftwork—Introduction*. London: Taylor & Francis.

Costa, G., D. Apostoli, F. Andrea, and E. Gaffuri. 1981. Gastrointestinal and neurotic disorders in textile shift workers. In *Night and Shift Work: Biological and Social Aspects,* ed. A. Reinberg, N. Vieux, and P. Andlauer, 215–221. Oxford, UK: Pergamon Press.

Dawson, D., and K. Reid. 1997. Fatigue, alcohol and performance impairment. *Nature* 388:235.

Folkard, S., and T. Akerstedt. 2004. Trends in the risk of accidents and injuries and their implications for models of fatigue and performance. *Aviat Space Environ Med* 75:A161–A167.

Gaba, D. M., and S. K. Howard. 2002. Fatigue among clinicians and the safety of patients. *N Engl J Med* 347(16):1249–1255.

Haider, M., R. Cervinka, M. Koller, and M. Kundi. 1988. A destabilization theory shift work effects. In *Trends in Chronobiology*, ed. W. Th. J. M. Hekkens, A. G. Kerkhof, and W. J. Rietveld, 209–217. Oxford, UK: Pergamon Press.

Harrington, J. M. 2001. Health effects of shift work and extended hours of work. *Occup Environ Med* 58:68–72.

Karlsson, B., A. Knutsson, and B. Lindahl. 2001. Is there an association between shift work and having a metabolic syndrome? Results from a population-based study of 27,485 people. *Occup Environ Med* 58(11):747–752.

Knauth, P. 1993. The design of shift systems. *Ergonomics* 36:15–28.

Knauth, P., J. Rutenfranz, G. Hermann, and S. J. Poppel. 1978. Re-entrainment of body temperature in experimental shift work studies. *Ergonomics* 21:775–783.

Knutsson, A. 2003. Heath disorders of shift workers. *Occup Med (Lond)* 53(2):103–108.

Knutsson, A., J. Hallquist, C. Reuterwall, T. Theorell, and T. Akerstedt. 1999. Shiftwork and myocardial infarction: A case-control study. *Occup Environ Med* 56(1):46–50.

Lillqvist, O., J. Harma, and J. Gartner. 1997. Improving 5-crew shift. *Tyoterveiset: Special Issue* 12–15.

Monk, T. H., and S. Folkard. 1992. *Making Shiftwork Tolerable*. London: Taylor & Francis.

Ogińska, H. 1991. Chronoergonomia. *Ergonomia* 14(1):69–75.

Pokorski, J. 1999. Ergonomics principles in optimization of shift work. In *Occupational Safety and Ergonomics*, ed. D. Koradecka, 2:991–1014. Warsaw: CIOP.

Pokorski, J., and G. Costa. 1998. Effect of shift work on health. In *Shift Work Related Stress, Reasons, Consequences, and Countermeasures*, ed. I. Iskra-Golec, G. Costa, S. Folkard, T. Marek, J. Pokorski, and L. Smith, 75–97. Krakow: Universitas.

Smith, L. 1998. Shift work-related stress models. In *Shift Work Related Stress, Reasons, Consequences, and Countermeasures*, ed. I. Iskta-Golec, G. Costa, S. Folkard, T. Marek, J. Pokorski, and L. Smith, 47–62. Krakow: Universitas.

Smolensky, M., and L. Lamberg. 2000. *The Body Clock Guide to Better Health: How to Use Your Body's Natural Clock to Fight Illness and Achieve Maximum Health*. New York: H. Holt and Co.

van Mark, A., M. Spallek, R. Kessel, and E. Brinkmann. 2006. Shift work and pathological conditions. *J Occup Med Toxicol* 1:25–31.

Vogel, L. 2004. Social/employment gains at risk: The revision of the working time directive. *TUTB Newsl* 26:11–13.

Waterhouse, J. M., S. Folkard, and D. S. Minors. 1990. *Shiftwork, Health and Safety: An Overview of the Scientific Literature 1978–1990. Report to the Health and Safety Executive*. Sudbury, UK: HSE Books.

WHO. 1995. Global strategy on occupational health for all: The way to health at work. Recommendations of the second meeting of the WHO Collaborating Centres in Occupational Health, 11–14 October 1994, Beijing, China.

Zużewicz, K., K. Kwarecki, and J. M. Waterhouse. 2001. *Physiological Effects of Shift and Night Work. Guide for Shift Work Manager.* Warsaw: CIOP.

26 Personal Protective Equipment

Katarzyna Majchrzycka, Grażyna Bartkowiak, Agnieszka Stefko, Wiesława Kamińska, Grzegorz Owczarek, Piotr Pietrowski, and Krzysztof Baszczyński

CONTENTS

26.1 LEGAL STATUS

Personal protective equipment (PPE) belongs to the group of protective devices that directly protect a worker against hazards in the work environment. Before deciding to use PPE, all possible technical and organisational measures to eliminate the risk at the source must be undertaken. When efforts to completely eliminate hazards to life and health or to reduce their admissible values do not succeed, that is, the concentration or intensity values of harmful factors present at workstations are still higher than is permissible, the use of PPE is the final barrier for the worker.

European Union legislation has two areas of regulations. The first is included in directive 89/656/EEC (1989) and determines the obligations of the employer relating to the safe use of PPE. The second area, included in directive 89/686/EEC (1989), covers the rules for placing these products on the internal market, that is, assessment of conformity to basic health and safety requirements.

26.2 RULES FOR SAFE USE OF PERSONAL PROTECTIVE EQUIPMENT

Ensuring the safe use of PPE is the duty of every employer, who should:

- Provide PPE free of charge.
- Select suitable PPE considering the nature and magnitude of hazards.
- Organise training on the appropriate use of PPE.
- Ensure appropriate storage, cleaning, disinfection, maintenance and necessary servicing for PPE.

The appropriate selection of PPE is fundamental. PPE should meet selection criteria that consider the hazards and work environment conditions and at the same time conform to the basic requirements of safety and ergonomics. Conformité Européenne (CE) markings as well as the manufacturer's declaration and the instructions for use confirm the PPE's conformity to regulations.

Before selecting the PPE, all hazards in the work environment should be identified and occupational risk assessment should be carried out. The extent of dangerous and harmful factors should be measured, if necessary, and the findings compared with admissible values such as maximum admissible concentration (MAC) for chemical factors and dusts or maximum admissible intensity for physical factors, noise, vibration, and electromagnetic fields. The number of times an admissible value is exceeded will be the guideline for selecting the protection level. Knowledge about harmful factors activity will indicate the necessary extent of protection.

PPE can be categorised as follows, based on the scope of protection:

- Protective clothing
- Hand and foot protection
- Head protection
- Hearing protection
- Eye and face protection
- Respiratory protection
- Equipment protecting against falls from a height

The first stage of PPE selection is the analysis and assessment of hazards. A technical solution should then be based on additional information related to the following:

- Workstation organisation
- Climatic conditions
- Additional hazards not related to the necessity to use PPE
- Characteristics of the user
- Working time and other conditions that could have adverse effects on a worker's health and well-being

The employer is responsible for determining the terms of use for PPE. While determining the terms of use, the level of hazard, and frequency of exposure to the hazard, the employer should consider workstation characteristics and the efficiency of the PPE. If more than one hazard occurs and the situation calls for use of more than one PPE, the equipment must be designed in such a way that it can be adjusted without reducing its protective properties. The employer is also responsible for training, information and consultations related to the safe use of PPE. Both the employees and the personnel responsible for supervising the use of PPE should be aware of the following:

- The PPE's protective properties
- The consequences of not using the PPE
- How to use the PPE correctly
- How to keep the PPE clean and when it should be discarded

PPE is essentially intended to be used individually. If PPE is worn by more than one person because of special circumstances, all measures should be taken to ensure that such use does not cause any health or hygiene problems.

The need to use PPE in the work environment means the employer must implement an appropriate system for PPE selection with regard to the hazards, the correct use of the PPE and the level of protection it provides. The system should include at least the following areas:

- Risk assessment that will enable selection of the appropriate type of equipment and protection level (i.e. hazards identification, influence on the body, excess of exposure limits)
- Workstation characteristics, including the occupational activity of the worker, microclimate, space limitation, need for movement and communication, evacuation speed from the hazard zone, and additional hazards such as fire or explosion
- Participation of PPE users in the process of selecting technical solutions
- Continuous training for workers, with special attention to increasing awareness of the effects of not using PPE, understanding the instructions for use, practical adjustments in use, time limits, and problems that may occur during use
- Marking areas where PPE must be used
- Ensuring the correct method of storage, maintenance, and necessary servicing
- Constant monitoring by audits of the PPE to ensure correct use, storage, technical conditions, and updating training

Ensuring the correct implementation of the above-mentioned tasks requires thorough knowledge about the chemical and physical properties of harmful substances and the technical aspects of PPE, especially its protective properties dependant on the type of hazards as well as ergonomic requirements, including the psychophysical load related to the use of PPE. Chosen features of individual types of PPE are presented below, including information that may be useful when selecting PPE appropriate to the given hazard and ergonomic requirements.

26.2.1 Protective Clothing

Protective clothing is the most commonly used type of PPE; it protects the worker from dangerous and harmful factors that occur in the work environment. The type of protective clothing depends on the type of harmful or dangerous factors occurring at the workstation. Therefore, clothing should be classified based on its protective properties. Clothing is classified as that providing protection against mechanical, thermal, chemical and biological agents, high visibility warning clothing, and as providing protection from electric shock, electromagnetic radiation, or drowning.

The protective properties of the clothing are ensured by the use of appropriate materials to create a barrier against dangerous factors or to sufficiently reduce their effects so that they are no longer a danger for the user. Depending on the type of harmful factors the worker is exposed to, the clothing can be made of textiles, such as impregnated fabrics, fabrics, nonwovens and knitwear coated on one or both sides with plastic, fabrics and knitwear made of high-efficiency fibres, and systems of materials and leather, plastic or even metal rings. When selecting clothing for protection against hazards, it is important to know the protective properties of the material and choose an appropriate class of parameters that is adequate for the level of risk.

Clothing for protection against mechanical hazards protects the worker from mechanical injuries caused by cuts, punctures, entanglement in the moving parts of devices, and the effects of impact. Clothing made of metallic materials such as wire mesh or appropriately jointed metal elements protects against puncture by hand knives, and multilayer fillers are used to protect against injuries by chain saws. Contact with a chain saw causes the fibres to entangle themselves with the chain in the driving system, block its movement and stop the saw.

High-visibility warning clothing is intended for use when the presence of the wearer must be visually signalled. Suits, jackets, vests, and trousers are examples of typical warning clothing. The clothing should be made of at least 0.5 m^2 background material and at least 0.13 m^2 retroreflective material. The chromaticity coordinates and luminance factors for fluorescent colours provide the protective properties of the background material: yellow, orange–red, and red. For retroreflective materials the coefficient of retroreflection, which depends on the entrance angle of light and the angle of observation, is the protective property.

Clothing protecting against heat and fire protect the worker—depending on its intended use—against the effects of the following agents: flames, infrared radiation, sparks, liquid metals splatter, hot metal splinters, and contact with hot objects and surfaces. The professional groups most exposed are steelworkers, foundrymen, welders, and firefighters. Clothing for protection against heat is obtained by the use of single materials; when the risk increases, multilayer materials or combinations of materials are used. If there is a risk of fire at a low level of infrared radiation and an air temperature below 50°C, materials made of aramid fibres, wool, or chemically modified fabrics made of flame-resistant cotton are used. In a work environment where the risk of a higher level of infrared radiation (up to 20 kW/m^2) is present, clothing made of aluminised material that reflects the radiation is used. The design of this type of clothing is adapted to the hazard level and work conditions. At workstations where the intensity of infrared radiation is greater than 20 kW/m^2, the clothing is made of multilayer material combinations, for example, the outer layer could be an aluminised material made of aramid yarns, glass fabrics, wool, cotton, or nonflammable impregnated viscose; the inner layer could be aramid fabrics, wool, or nonflammable impregnated cotton.

Protective clothing against cold is used for those who work in open space at temperatures lower than standard as well as for those working in closed unheated spaces and in cold storage areas. The thermal insulation of the clothing depends on the thickness of the materials used and the number of layers. Specially prepared nonwovens are often used for the warming layers. If the ambient temperature is very low, the thickness of efficient insulating layers can greatly increase the mass of the clothing. Clothing equipped with heating systems can be used as an alternative. At many workstations in cold microclimates, where the amount of emitted heat and the need for protection against cold changes with the activity of the worker, the desired comfort can be achieved by the use of active clothing whose insulation changes in tandem with the climatic changes of the external environment and the amount of heat emitted by the user (Kurczewska and Leśnikowski 2008). Heating inserts made of steel yarn are the active elements of the clothing. Activation and deactivation of the voltage powering the heating inserts are achieved by a control system that collects and analyses data from temperature microsensors placed on the body under and on the clothing. Figure 26.1 shows heating inserts in active clothing.

Exposure to biological agents occurs at many workstations related to agriculture, sewage treatment and waste disposal and at hospitals and diagnostic and veterinary laboratories, where it is necessary to use appropriate protective clothing. The barrier material of the clothing should be resistant to penetration by infectious agents in fluids, liquid aerosols, and airborne particles. The design of the protective clothing against biological agents depends on the type of infectious agent and resembles the design of protective clothing against chemical agents.

The properties of clothing for protection against chemical agents should be adapted to the aggregation of the chemical substance, its type and the concentrations, and the intensity of its effect on the clothing. Clothing that completely isolates the body from the outside environment—so-called gastight clothing—belongs to this group, as does clothing that protects only against accidental low-volume splashes of chemical substances. Such clothing can be made of textiles, knitwear, or nonwovens coated with plastics or impregnated textiles as well as foil, depending on its intended use. The breakthrough time, that is, the time it takes a chemical to permeate

FIGURE 26.1 Heating insert in active clothing.

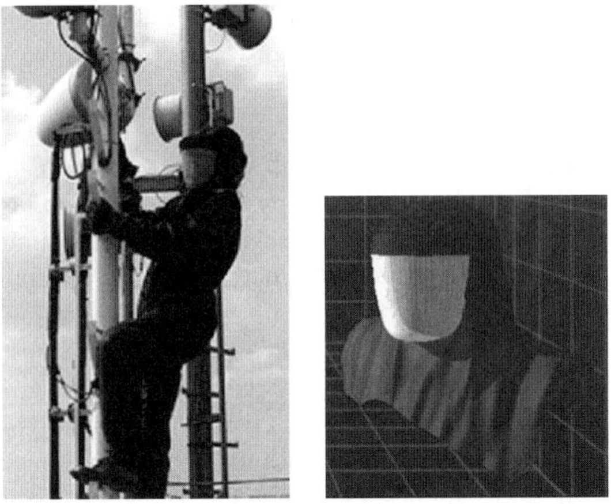

(a) (b)

FIGURE 26.2 (a) Suit protecting against microwave radiation and (b) design of a head protector with a visor, made of a special metal net.

the protective barrier, determines the barrier properties of the materials used. The design of the clothing determines its intended use.

Workers at risk of long exposure to high-intensity, high-frequency electromagnetic fields (in the range of 300 MHz–300 GHz and with a wavelength of 0.001–0.03 m) must use clothing that protects against microwave radiation. Such clothing is intended mainly for operators of equipment that emits microwave radiation and workers who service and maintain transmitters, radio stations, cellular network transmitters, satellite communication systems, radars, and so on (Andrzejewska and Kurczewska 2004). Protective shielding properties are ensured by using electroconductive yarn fabric. An appropriate clothing design is also important (Figure 26.2).

Clothing can also protect against drowning in case of an accidental fall into water. Figure 26.3 shows a safety protection suit (Bartkowiak et al. 2006) worn for work in harbours and industrial buildings on the waterfront or in offshore waters (wharfs, shipyards, docks, river dams, hydroelectric plants, and so on).

Clothing made of rubber filled with lead reduces the exposure to ionising radiation, that is, X-rays. It protects the front and sides of the body from the collarbone to the knees and is fastened at the back with buckles. The protective properties of such clothing are given by the *attenuation equivalent*, that is, the thickness of the lead layer that weakens X-rays of a given energy.

Protective clothing, apart from ensuring a certain level of protection against one or many hazards occurring simultaneously, must not have harmful effects on the user. Harmful substances such as carcinogens or allergens cannot be used for its design; the pH of the materials should be between three and nine. Protective clothing should also meet ergonomic requirements, that is, it must be designed in a way that a user is able to perform tasks without any difficulty or additional load. Anthropometric elements are therefore taken into account in the design of protective clothing in order to adjust the clothing to the body. Thermal properties (enabling the transfer of heat and

FIGURE 26.3 Safety protection suit.

sweat), sensorial aspects (disturbances to senses such as hearing, smell), and biomechanical properties (reducing musculoskeletal system load) are also considered.

The development of plastic and composites has created real possibilities for designing clothing properties according to ergonomic principles. The materials used in protective clothing create a barrier against harmful agents and at the same time are permeable and allow the transfer of excessive heat produced by the body. The microclimate that is created under this type of clothing has a lower humidity and temperature compared to traditional barrier-coated materials. Materials that are permeable to water vapor are made of microfibres, laminates of flat textile products, and permeable membranes and are used mainly to produce clothing for protection against biological and chemical agents. They cannot be used for some types of protective clothing, for example, protective clothing intended for work with aggressive chemicals in liquid, vapour, and gaseous forms. One method used to reduce discomfort in this type of clothing is installing cooling systems in the clothing structure, for example, induced liquid circulation or induced air circulation (Muir 2001; Nag et al. 1998). While it is generally believed that cooling systems lower the thermal load on the user, they have certain shortcomings, such as large mass, reduced mobility of the worker, and water vapour condensation.

Use of the appropriate multilayer underwear with synthetic diffusive layers and hygroscopic sorptive layers improves the microclimate under the clothing, and thereby thermal comfort, when working in barrier impermeable protective clothing (Bartkowiak and Błażejewski 2003). Use of superabsorbent inserts under the clothing reduces discomfort caused by excessive sweating (Bartkowiak 2006).

Use of the new generation of clothing that support the thermoregulation processes of the user's body is increasing. Such clothing is referred to as 'active' or 'intelligent'. Clothing with phase-change materials (PCM; Reinertsen et al. 2008) is an example of such innovative products. PCM (Figure 26.4) have appeared in the new generation of textile fibres and are processed using traditional textile technologies.

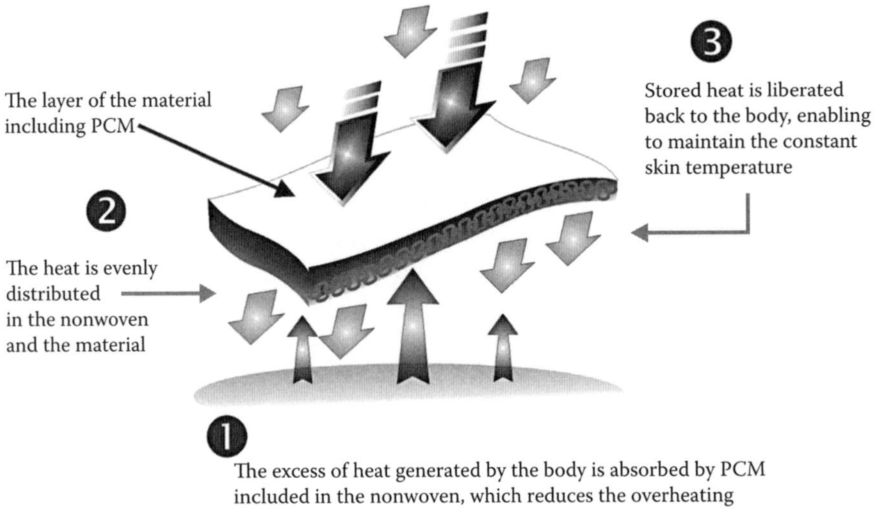

FIGURE 26.4 Diagram of phase-change material performance.

Research was carried out on their use in protective clothing and underwear, and the researchers proved that the products developed with PCM reduce the user's thermal load. This new-generation protective clothing also includes garments for rescue services, for example, firefighters. Electronic microcircuits are implanted into the garment in order to monitor the user's physiological parameters and the level of hazards the user is exposed to. The garments are intended to control the firefighter's physiological condition based on the environmental conditions and the workload (the body energy expenditure and resulting load).

Body systems can be monitored with clothing elements by collecting information on the temporary values measured on the individual user of the garment and transferring them by radio to the monitoring centre, based in a fire car. The centre is equipped with a computer system that analyses the situation unfolding in real time and provides information indispensable for making decisions on the necessity and methods of evacuating firefighters. A diagram of the firefighter's monitoring system during an operation is shown in Figure 26.5.

26.2.2 Hand and Arm Protection

Protective gloves provide basic hand and arm protection. Gloves ensure the protection of the hands, and if their cuffs are long enough they can also protect part of the forearm and the arm. It is particularly important that gloves have the appropriate protective properties but do not restrict the dexterity, accuracy, and firmness of the user's grip when working and thus add to the risk of accidents at work caused by inappropriate or manual work that is too slow. Gloves must ensure protection of the hand against many different harmful and dangerous substances. When seeking to assure protection against several hazards, combinations of different materials are typically used, limiting the comfort. It is essential to strike a balance between protective and

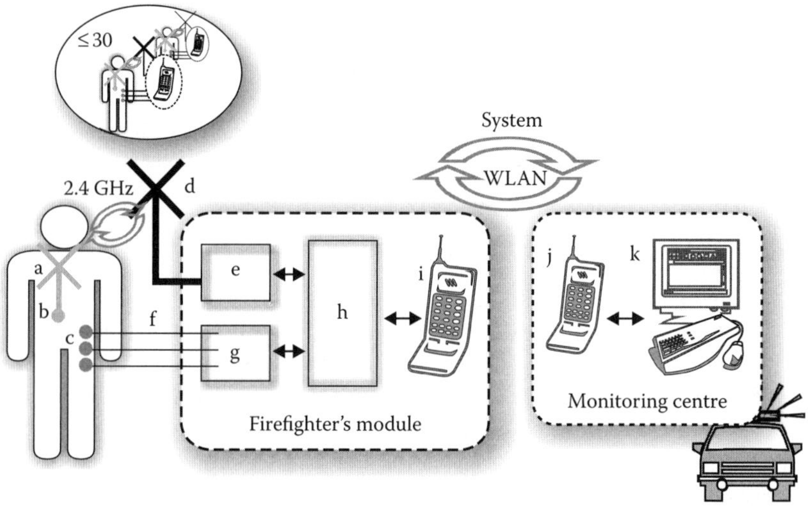

FIGURE 26.5 Diagram of a system monitoring a firefighter's physiological condition, considering environmental conditions and workload: (a) textile antenna for data transmission from the sensor, (b) sensor with radio transmitter, (c) sensors joined by wire, (d) antenna of the firefighter's module for sensor communication, (e) radio transmitter for communication sensor—firefighter's module, (f) wires joining sensors with firefighter's module, (g) interface collecting data from circuit sensors, (h) steering element of firefighter's module, (i) radio transmitter for communication firefighter—centre, (j) centre radio transmitter and (k) centre steering computer. (Adapted with permission from Owczarek, G., K. Łężak, and G. Gralewicz. 2007. Monitoring selected parameters during work in firefighters' clothing. *Occupational Safety* 9:8–10.)

usability properties. Because of the hazardous situations in which gloves are used, the time needed to put on and take off the gloves should be the shortest possible. It is also crucial to fit the size of the glove to the user's hand. Gloves that are too tight not only cause discomfort, but also hasten the loss of protective properties, whereas gloves that are too large do not ensure the necessary protection and additionally make work difficult. Depending on the precision required, gloves with one, three or five fingers can be chosen.

Gloves for protection against minor and average mechanical injuries are often made of combinations of leather and fabric, and fabrics and knitted fabrics partly or wholly coated with plastic or rubber or with polymer dots. Gloves can also be knitted using different types of yarn, for example, polyester, polyamide, aramid, core-spun and polyethylene. These also ensure protection against mechanical risks such as abrasions, cuts or punctures. Knitted gloves, because of their design, are not recommended for protection against punctures, unless the knitted fabric is coated with plastic or rubber. *Double gloves* are also an interesting solution—they are made up of an inner knitted glove with an outer material on the palm side covered with steel plates and an outer glove made of polymer. These gloves provide a high level of protection against cuts and punctures. When a worker needs protection against

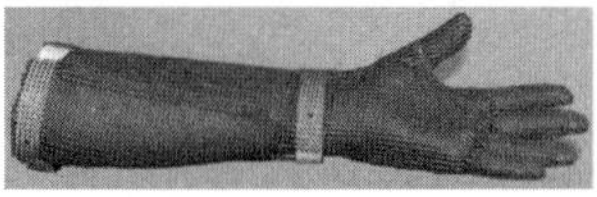

FIGURE 26.6 Example of gloves made of chain mail (Reprinted with permission from Andrzejewska, A., and K. Szczecińska. 2005. Hand protective equipment while handling hand knives. *Occupational Safety* (7–8):38–41.)

serious injuries, he or she should use specialised gloves. Gloves made of chain mail (Figure 26.6) can serve as an example. They are used for protection against cuts and stabs by hand knives. These types of gloves are also recommended for hand protection when working with powered knives. To ensure comfort during use, inner gloves made of cotton yarn are often used. Gloves that protect against cuts by hand-held chain saws are another example of specialist gloves. They protect the hands from the cuts caused by the chain slipping, clogging, or braking.

Gloves protecting against thermal risks are most often made of fabrics or knitted yarn fabrics such as Kevlar, Nomex, Twaron, Preox, PBI, nonflammable impregnated cotton, and wool yarn. Aluminised fabrics and heat-resistant leathers are also used. Gloves ensuring protection against different forms of heat or fire (flame, convective heat, contact heat, radiant heat, small splashes, or large quantities of molten metal) are made by assembling the above-mentioned materials. To ensure thermal insulation, special insulating inserts are used, for example nonwovens. Cattle leather and fabrics are used in the design of gloves protecting against cold. These can also be made of plastic or rubber as well as knitwear or with polymer-coated fabrics. Nonwovens and knitted fabrics, for example, acrylic or woollen artificial fur or polyamide foam, can be used for the thermo-insulating function of the inner layer of the gloves. Water-resistant, permeable membranes are used in certain designs of gloves.

Tight gloves made of different types of rubber (e.g. natural rubber or synthetic rubber—polychloroprene, polyacrylonitrile, butyl, viton) or plastics (polyvinyl chloride, polyvinyl alcohol, polyethylene, hypalon) protect hands from contact with chemical substances. Gloves not wholly made of plastic, rubber, or fabrics and knitwear coated with polymer cannot be used for protection against chemical agents. If all-polymer or all-rubber gloves are used, an inability to evaporate sweat may be a problem and cause discomfort. Moreover, allergies may result from direct or indirect contact of human skin with the components of the rubber mixture. Sensitivity to the latex in natural rubber is particularly common. Additionally, powder or other lubricants used to facilitate putting on and removing gloves can cause skin irritation.

To increase the comfort of all-polymer or all-rubber gloves, some of products are made with knitted supports or are flocked, that is, the inside is covered with a thin layer of cotton dust. Gloves for protection against chemical substances very often protect the skin against micro-organisms as well. Gloves resistant to penetration, that is, the movement of chemical substances through porous materials, seams, pinholes or other imperfections in the material of the glove on nonmolecular level, are an efficient barrier to bacteria and fungi. However, they do not provide protection against viruses.

At many workstations, multifunctional gloves that ensure simultaneous protection against many agents are required. Gloves that offer simultaneous protection against thermal and mechanical risks are currently available. Gloves for protection against chemical, mechanical, and thermal risks (contact with hot objects) that are also suitable for contact with food are also produced. One innovative solution is gloves that protect against mechanical risks (cuts, punctures, abrasions), select chemical agents, micro-organisms, and radioactive contamination. This type of glove is made of many layers of different polymers on a knitted support.

Laboratory tests carried out based on the standards harmonised with directive 89/686/EEC confirm the protective properties of gloves. Gloves are most often rated by level of performance (e.g. 1, 2), which is indicated next to the appropriate pictograms on the glove (Figure 26.7).

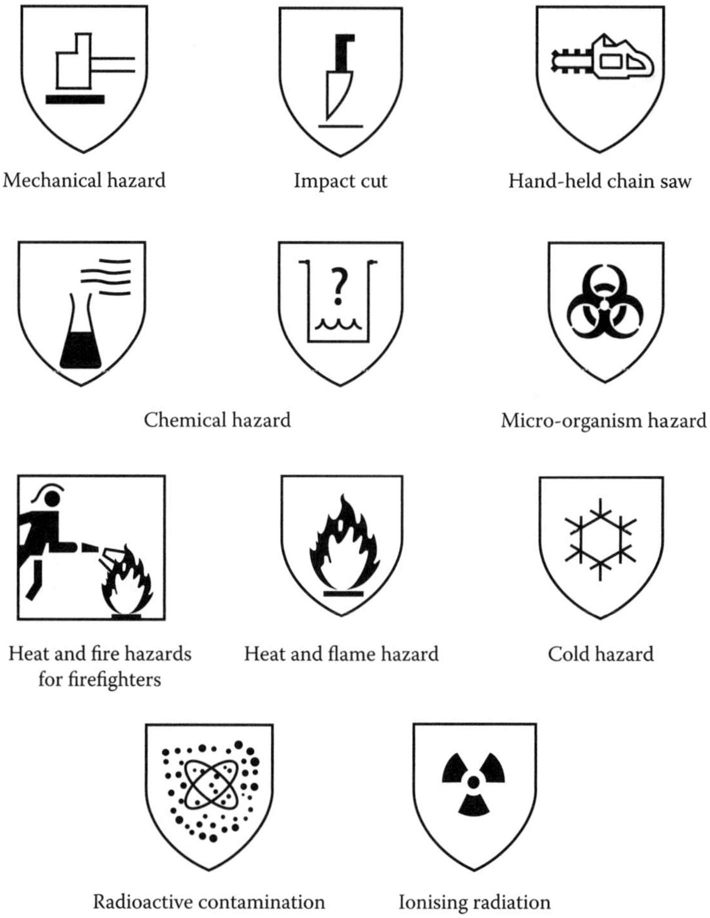

FIGURE 26.7 Examples of pictograms indicating type of protection. (Reprinted with permission from PN-EN 420:2005/AC:2007. Protective gloves: General requirements and test methods.)

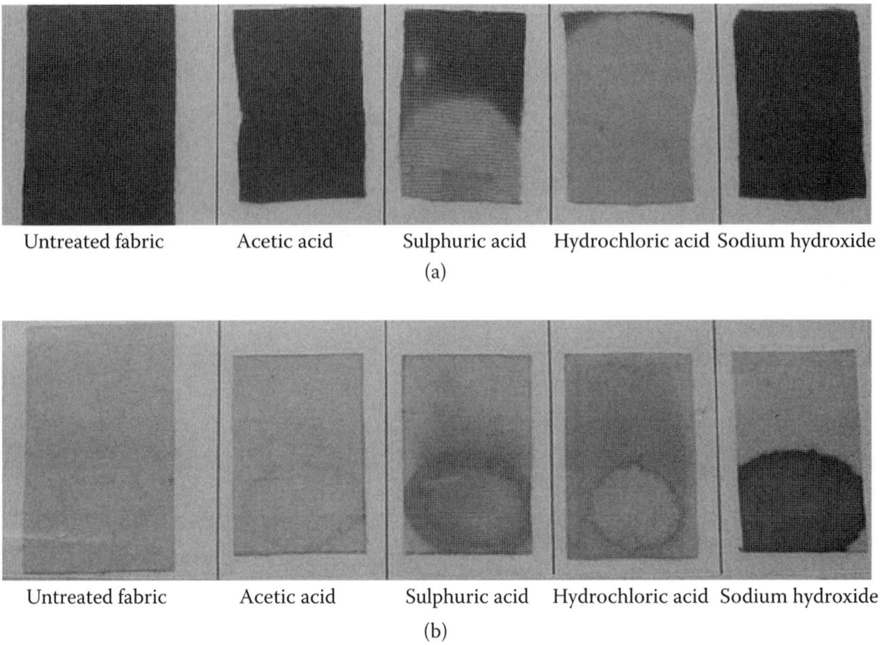

FIGURE 26.8 Example of indicator dyes for signalling the permeation of acids and bases: (a) dyed polyamide fabric and (b) cotton fabric before and after contact with acetic acid, sulphuric acid, hydrochloric acid and sodium hydroxide. (From Szczecińska, K. 2004. Report on the realisation of the task no. I.10 carried out within statutory research activity. [Unpublished work.] With permission.)

Apart from the appropriate selection of gloves to fit the hazards and fulfil ergonomic requirements, users should be able to determine the end of the service life of the glove. Usability, storage, and maintenance conditions influence the protective properties of the gloves. Changes in the material of the gloves observed during use should signal a need for withdrawal from use. This is especially important where protection against chemical risks is required, because the actual level of protection in this category of gloves depends on ambient temperature and humidity, contact with other chemical substances, mechanical wear of the gloves, and the microclimate conditions under the glove. Among the directions for prolonging hand protection is the development of end-of-service-life indicators. These indicators enable the user to decide when the protective properties cease to be effective and discard the gloves. An example of such a solution is presented in Figure 26.8.

26.2.3 Foot Protection

At many workstations, for example, construction sites, metallurgical and machinery industries, and mining, workers are exposed to risks of injury to the lower limbs, especially the feet. The risks are directly related to the work performed as well as to the harmful effects of the floor's surface or the environment. Footwear selected according to the risks provides foot protection. Footwear, depending on the design and materials used, as well as additional protective components, offer protection against various mechanical damages, such as punctures of the sole of the foot, injury or crushing of the toes or metatarsus, thermal injuries caused by excessively high or low temperatures, electric shock, and so on.

According to the European standards (PN-EN ISO 20345 2007; PN-EN ISO 20346:2007; PN-EN ISO 20347:2007), the following categories of footwear are ranked as PPE for the lower limbs:

- *Safety shoes*: Equipped with toecaps to protect toes against impact up to 200 joules (J) and compression up to 15 kilonewtons (kN; safety toecaps).
- *Protective footwear*: Equipped with toecaps to protect toes against impact up to 100 J and compression up to10 kN (protective toecaps).
- *Occupational shoes*: Has protective properties but without toe protection.

As for the materials used, classification I footwear includes footwear made from leather and other materials (excluding all-rubber and all-polymeric footwear), and classification II includes all-rubber entirely vulcanised and all-polymeric entirely moulded footwear.

Depending on the length and design of the upper and particular models of safety, protective and occupational footwear are classified as one of the following models:

- A model—low upper shoes (half boots, sandals, clogs)
- B model—ankle boots
- C model—half-knee boots
- D model—knee-high boots
- E model—thigh-high boots

An open heel is only acceptable in model A of classification I.

Additional protective elements of footwear include:

- Safety and protective toecaps protecting the toes
- Penetration-resistant inserts fixed on the bottom of the shoes protecting the foot against punctures
- Metatarsal protection fixed on the inside or outside of safety or protective footwear
- Ankle protection to absorb impact
- Flaps protecting the metatarsus, preventing sand, stones, sparks, and splashes of melted metal from entering the shoe

For specialist work in some occupations, footwear should fulfil defined requirements—besides general requirements for safety, protective or occupational footwear—arising from the specific character of the work environment conditions, for example, footwear for firefighters, for professional motorcyclists, and footwear protecting against hand-held chain saw cuts and chemicals.

Protecting against many different dangerous and harmful agents in the work environment, especially when occurring simultaneously, often requires the use of footwear that imposes a significant load on the wearer (Mäkinen et al. 2000). Fulfilling the requirements of the protective properties of footwear influences the deterioration of its biomechanical, physiological, and hygienic properties, for example, the use of steel toes for protection and steel inserts to protect the foot against punctures makes the footwear heavier and more rigid. Therefore, the selection of appropriate footwear should take into consideration the requirements and expectations of the user and should not cause significant discomfort.

A key requirement of footwear is its proper adjustment to the shape and size of the foot. The shoes must stabilise the foot, without compressing the dorsum and toes or limiting movement. The seat region should be stiffened. Even small imperfections in the adjustment of the footwear can cause considerable difficulties for the user and increase sensitivity to other factors causing discomfort, such as overheating or excessive humidity. The pressure of the shoe on the foot impedes the dynamic functions and blood flow, which can have health consequences. Uppers made of more flexible material, which stretch more easily, are more advantageous, but cannot stretch too much, as the footwear should not be loose. An important feature of the upper part of the footwear is its ability to maintain its shape. The durability of the shape determines the aesthetics and comfort of the footwear. Natural leather has elastoplastic properties; therefore, some of the deformations occurring from stretching are permanent. The temperature and humidity of the material influence the degree of deformation. The adjustment of the shape of the upper to the shape of the foot differs for natural leather, synthetic leather, and other synthetic materials.

The combination of materials used to form the sole of the footwear in the seat region and its appropriate design are important for reducing limb and spine injuries. Microcell materials of the insole and sole can be used to absorb energy in the heel. The midsole absorbs the forces generated when the foot strikes the ground and prevents injury to the ankles, knees, hips, and lower part of the spine.

The physiological and hygienic properties of the footwear can deteriorate if full protection is desired, in particular against extreme hazards, for example, in metallurgy, during firefighting, or in chemical rescue operations. The footwear used does not assist in the transfer of heat and sweat produced in great quantities during physical effort, so the high temperature and excessive heat in the footwear cause discomfort. If such an unfavourable microclimate is prolonged, the decomposition of organic substances of the sweat occurs, leading to changes in the skin's reaction to alkalinity and the growth of bacteria and fungi. At the same time, the corneum swells because of the high humidity and becomes more susceptible to abrasions and other mechanical injuries as well as more vulnerable to micro-organisms. Therefore, the materials used in footwear design must not only fulfil protective parameters, but also physiology and hygiene requirements, so that they can actively support the thermoregulatory processes of the body. An inappropriate microclimate inside the shoes disturbs the thermoregulatory process of the feet, which in turn disturbs activities in the whole body. Therefore, appropriate regulation is necessary. Modern materials with a significant capacity and sorption dynamism, such as superabsorbents, may be good for use in the inner materials of the footwear (the lining and padding).

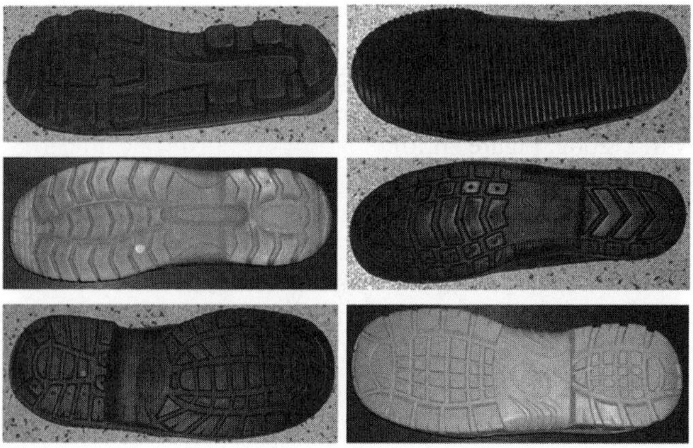

FIGURE 26.9 Examples of patterns of footwear soles with protective properties. (From Kamińska, W. 2006. Report on the realisation of the task no. 03.10 related to services for the State. [Unpublished work.] With permission.)

To counteract the growth of micro-organisms, materials should be disinfected even at the production stage. The materials currently available contain bactericides and fungicides impregnated in the molecular structure of the fibre and even in completely integrated assemblies of materials, combining superior thermal insulation with the capacity to transfer humidity. They also provide active and durable protection against micro-organisms (*Climate control for feet*, 2000). Protection against micro-organisms is a property of the fibre itself, which means that the use of the product does not influence its durability (unlike with the impregnation or surface application of bactericides and fungicides).

Ensuring protection against slipping is a vital factor in the selection of footwear. The majority of the materials used for the production of soles possess good traction on dry and rough surfaces and do not show any tendency to slip. In practice, however, direct friction, the so-called dry friction between two surfaces, is rarely encountered. The surface is usually covered with water, dust, mud, and so on. Smooth surfaces covered with water, oil, grease, or ice are the most dangerous. Suitably selected materials and the design of sole patterns increase the friction coefficient (Figure 26.9). The cleated pattern, open to the sides, enables the diffusion of liquids. The surface of the cleats should be smooth, without roughness, to which the liquid clings. The cleat must be high enough and resistant to wear during the life of the footwear.

26.2.4 Head Protection

In many workstations in industry, for example, mining, power, construction, and forestry, the risk of head injury for workers is constantly present. The most serious are mechanical injuries, which can be the result of the impact of falling objects or collision with fixed objects at the workstation. Because of the nature of these work activities, it is not always possible to eliminate such risks with appropriate organisational solutions or collective protective equipment. Therefore, the only way to ensure the safety of workers is by using safety helmets. The type of helmet depends on the specific nature of mechanical risk.

If the risk is of impact by hard, fixed objects, which can cause superficial head injuries, industrial bump caps can be used, fulfilling the requirements of the standard PN-EN 812 (2002; Figure 26.10).

Workers exposed to a risk of falling objects, which can cause serious injuries to the skull, brain or cervical vertebrae, require industrial safety helmets fulfilling the requirements of standard PN-EN 397 (1997). Examples of designs of this type of helmets are presented in Figure 26.11. Workers exposed to a risk of impact to the head by extremely heavy falling objects from heights and with sharp edges as well as blows to the top of the head by dangerous objects at the workstation (Baszczyński 2002) should be equipped with a helmet fulfilling the requirements of standard PN-EN 14052. 2006(U), Examples of designs of this type of helmets are presented in Figure 26.12.

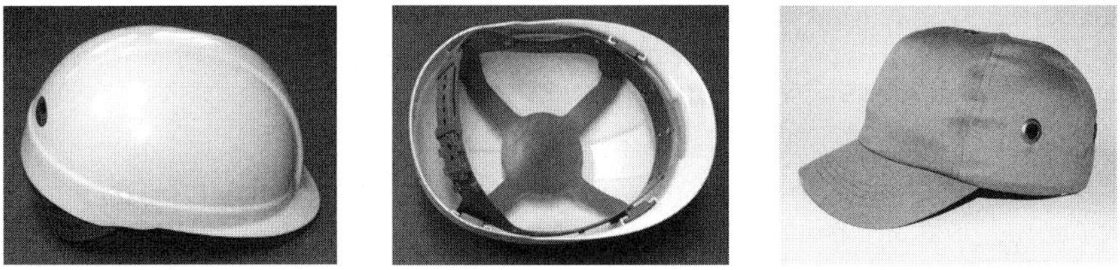

(a) (b) (c)

FIGURE 26.10 Examples of designs of industrial bump caps: (a) Industrial bump cap with a polyethylene shell; (b) construction of the internal part of an industrial bump cap; and (c) industrial bump cap with a shell covered by a textile material.

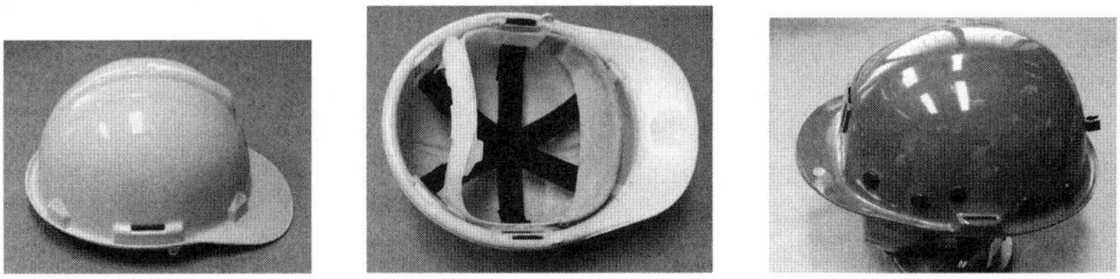

(a) (b) (c)

FIGURE 26.11 Examples of designs of industrial safety helmets: (a) Industrial safety helmet with polyethylene shell; (b) construction of the internal part of an industrial safety helmet; and (c) industrial safety helmet for miners.

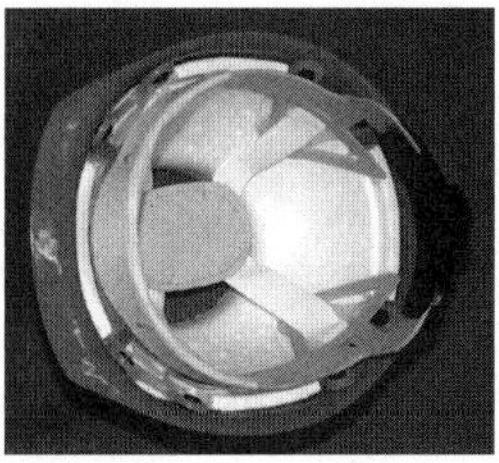

FIGURE 26.12 Example of a design of a highly efficient safety helmet.

Safety helmets, depending on the type, can also protect the head of wearer against dangerous factors such as electric shock, transverse compressive forces, splashes of liquid metal, and high temperatures. Such abilities are given on the helmet as well as in the instructions for use provided by the manufacturer. Helmets can also be a structure for mounting other PPE, for example, eye and face protectors, respiratory protective equipment, nape protection, and supporting equipment such as cap lamps.

The helmet must be carefully selected to ensure the safety of worker's head. Some aspects that must be considered are the risks that arise at the workstation (e.g. falling objects with sharp edges, splashes of liquid metal), the atmospheric conditions in which the helmet will be used (temperature, in particular), the ability to adjust it to fit the head of the individual wearer, and properties indispensable for a helmet used at a given workstation (e.g. enables mounting of hearing protectors).

Helmets lose their protective features during use for many reasons. The most important reasons are solar radiation, especially of the ultraviolet (UV) spectrum, mechanical damages, relaxation of internal stress in plastics, and degradation of materials because of aggressive substances in the work environment. A safety helmet can be used only for the duration indicated by the manufacturer, unless other disqualifying damages occur. A safety helmet that has been hit with great force must be discarded, even if an inspection does not reveal any damage.

26.2.5 Eye and Face Protection

Natural protective mechanisms protect the eye against external factors. For example, a thin layer of slightly oily lachrymal fluid produced by the conjunctiva protects it against less intensive pollution. Eyclids and lashes ensure the protection of the eye against fine foreign bodies. However, this natural protection of the eye against external factors is often insufficient in workplaces as well as in everyday life. In such situations, extra eye protection is required. It should be used in places where the following hazards may occur:

- Impact (e.g. fragments of solids)
- Optical radiation (e.g. radiation related to welding, sun glare, laser radiation)
- Dusts and gases (e.g. coal dust or aerosols of harmful chemical substances)
- Droplets and splashes of fluids (e.g. splatter occurring when pouring fluids)
- Molten metals and hot solid bodies (e.g. sparks of molten metal during metallurgical processes)
- Electric arc (e.g. occurring while working on high-tension equipment)

The eye can be protected against the aforementioned hazards using eye protection equipment falling in the following four basic categories:

1. Spectacles
2. Protective goggles

3. Face shields
4. Welder's face shields (includes hand screens, face screens, goggles, and hoods)

To ensure eye protection and vision, the PPE of the aforementioned categories is equipped with vision systems, oculars, meshes, or filters. Filters include welding filters, UV filters, infrared filters, sun glare filters, and filters protecting against laser radiation. Eye protection equipment may be part of respiratory protective devices (full masks that have vision systems) or head protection equipment (face shields mounted in industrial protective helmets). Eye protection equipment of all categories is composed of a transparent part (vision systems, oculars, meshes, or filters) and a frame (spectacles and goggles) or housing with a harness (shields).

Spectacles are the most widely used eye protection equipment. They should provide forehead protection against dangerous splatters of fluids or fragments of solid bodies. A model of such spectacles is presented in Figure 26.13. If the worker requires a higher degree of eye protection, he or she can use *protective goggles*. Their construction ensures tight adhesion to the user's face, which enables them to protect against biological factors. The ventilation systems of goggles often differ greatly from one from another and are also very important. Figure 26.14 presents goggles with direct and indirect ventilation systems. Goggles with indirect ventilation systems ensure better protection against droplets and splashes of harmful substances. The majority of goggle constructions allow their use with corrective glasses; however, before choosing and purchasing the equipment verification of the quality is recommended. If the hazards require protection of the entire face, face shields should be used.

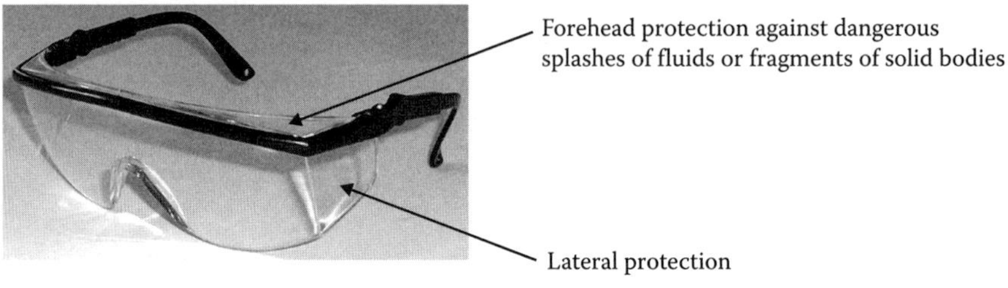

FIGURE 26.13 Spectacles.

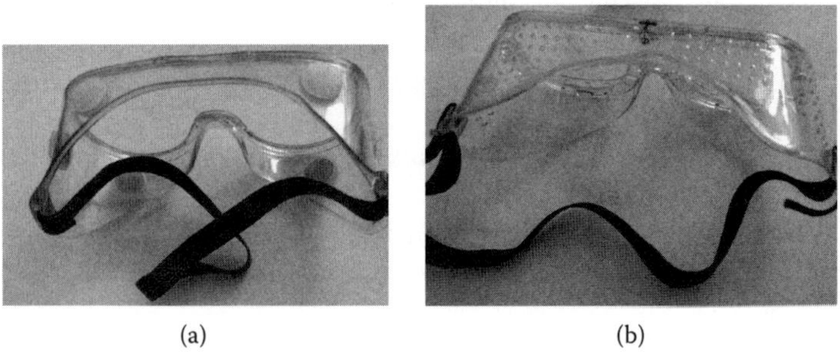

FIGURE 26.14 Protective goggles with (a) direct and (b) indirect ventilation systems.

Face shields protect the entire face; their large protective surface minimises the probability of penetration of dangerous fluid splatter. Face shields may be used with spectacles, corrective glasses, goggles, and some respiratory protection devices.

The last of the basic categories of protective eye equipment are *welder's face shields*, that is, devices protecting the user against harmful optical radiation and other specific hazards arising during welding and/or related techniques. Welder's face shields include face screens, hand screens, goggles, spectacles, and hoods.

The intended use of eye protection equipment and its construction determine its protective properties. If eye protection equipment is intended to protect against impacts of solid bodies, fluid splatters, and drops of molten metal, it must have mechanical resistance (in low and high temperatures) and tightness and resistance to ignition in contact with items of a much higher temperature (up to about 1500°C). To determine if eye PPE protects against the consequences of electric arc flash, the electrical insulation properties of its materials should be verified. Examining spectral characteristics of optical radiation transmittance through transparent elements (spectacles, vision systems, filters) allows determination of the part of the radiation spectrum against which the equipment ensures protection. The requirements for eye and face protective equipment are divided into two categories: (1) nonoptical, all the elements of construction, and (2) optical, only the ocular area.

Examination methods for verifying the fulfilment of requirements are included in the standards PN-EN 167 (2005) and PN-EN 168 (2005). A symbol is assigned to each given purpose of eye protection equipment (digits: 3, 4, 5, 8, or 9); this symbol should appear as part of the equipment markings. The purposes of eye protection systems, the relevant symbols and a short description of the fields of use are presented in Table 26.1.

A common element in the majority of eye protective equipment is ocular. Because its basic function is protection against impact, it is often called the antisplinter ocular, and it is mounted in protective equipment such as antisplinter spectacles, goggles or face shields. If the ocular has filtration properties (e.g. lowers the intensity of UV radiation), it can also act as a filter. Most often, the ocular is made of a polycarbonate

TABLE 26.1
Purpose, Symbols, and Description of Eye Protection Equipment and Fields of Use

Symbol	Designation	Description of the Field of Use
No symbol	Basic application	Unspecified mechanical hazards and hazards arising from ultraviolet, visible, infrared and solar radiation
3	Liquids	Liquids (droplets or splashes)
4	Large dust particles	Dust with a particles >5 μm
5	Gas and fine dust particles	Gases, steam, aerosols, smoke and dust particles <5 μm
8	Short-circuit arc	Electrical arc due to a short circuit in electrical equipment
9	Melted metals and hot solid bodies	Splashes of melted metals and penetration of hot solid bodies

ranging from 0.25 to 3 mm in thickness. This material has a high mechanical resistance. Figure 26.15 presents the effect of the impact of a ball ($\phi = 6$ mm; 120 m/s) on a sample made of 2-mm-thick nonorganic glass and 2-mm-thick polycarbonate.

The high mechanical resistance of polycarbonate is particularly important with regard to the factors responsible for mechanical hazards. Oculars made of this material are an effective barrier against splinters seen, for example, during manual and machine treatment of wood, metals, rocks, concrete, and plastics, as well as splinters that are the result of damaged mobile elements of machines (e.g. drills, grinding discs, and angle cutter discs). Mechanical damages occurring in the polycarbonate structure because of splinter impact can be responsible for simple plastic deformations. The polycarbonate does not crush, as nonorganic glass does; therefore, the fine and sharp parts of the broken material do not cause any hazard.

Another characteristic making polycarbonate a widely used material in the construction of transparent elements of eye protective equipment is its natural capacity to absorb UV radiation and its facility of in-mass tinting of the material to give it appropriate filtration properties. Figure 26.16 shows the spectrum characteristics

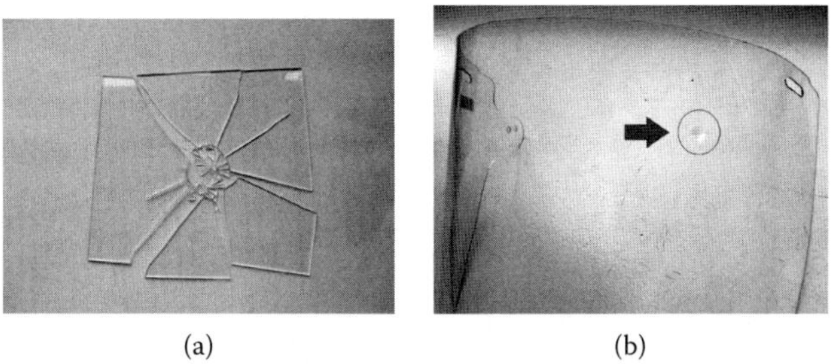

FIGURE 26.15 Results of a resistance examination of an ocular made of (a) nonorganic glass and (b) polycarbonate.

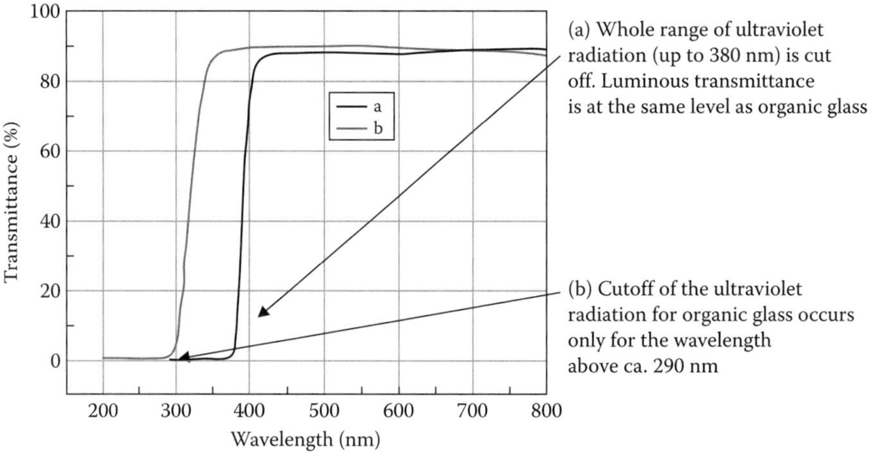

FIGURE 26.16 Spectrum characteristics of optical radiation transmittance (ultraviolet, visible radiation and near infrared) for oculars made of (a) polycarbonate and (b) nonorganic glass.

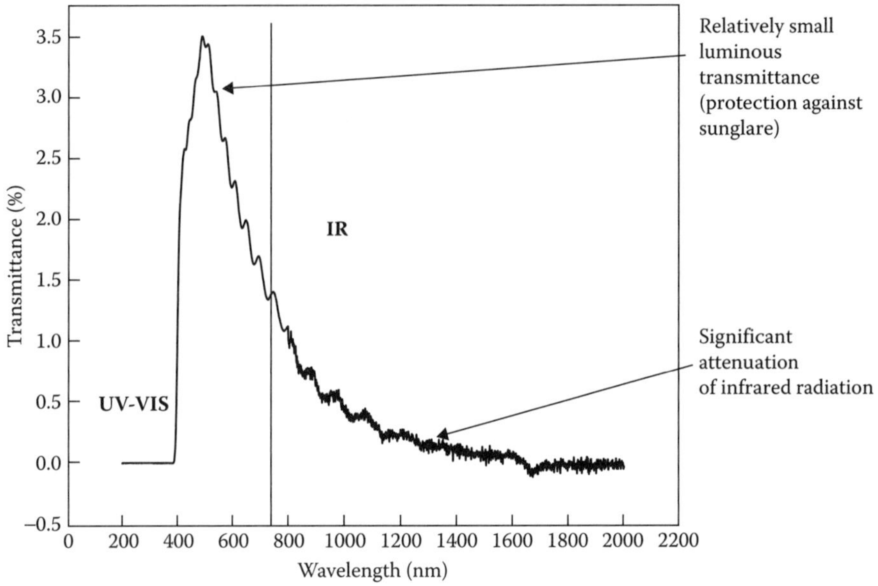

FIGURE 26.17 Spectrum characteristics of optical radiation transmittance (ultraviolet, visible radiation and near infrared) for oculars made of dyed polycarbonate.

of optical radiation transmittance (UV, visible radiation, close infrared) for oculars made of 1-mm-thick nondyed polycarbonate and nonorganic glass.

Figure 26.17 illustrates the possibility of modification of the optical radiation transmittance spectrum characteristics of polycarbonate. With the addition of the appropriate pigment, the polycarbonate may also become an effective barrier against infrared radiation. By correctly choosing the pigments added to the polycarbonate, practically every type of protective filter can be created, against UV, visible (sun glare), infrared, or laser radiation.

A disadvantage of polycarbonate is its relatively low resistance to scratches, compared with typical nonorganic glass. This deficiency is eliminated by coating it with hardening layers. The back surface can be coated with an antifog layer, which is especially popular with goggles.

Regardless of the material used for making filters (e.g. polycarbonate, nonorganic glass, polymetacrylane, cellulose acetate), the basic function is to protect the eyes against dangerous optical radiation. In industrial applications, this is welding, UV and infrared radiation, and the visible radiation responsible for sun glare and laser radiation. The filters are divided into protection classes independently of their purpose, based on the level of protection they provide. This parameter is determined based on the light transmittance factor, with 23 levels of protection defined (1.2; 1.4; 1.7; 2; 2.5; 3; 4; 4a; 5; 5a; 6; 6a; 7; 7a; 8; 9; 10; 11; 12; 13; 14; 15; 16). Code numbers are assigned to filters depending on their use. The complete markings on a filter contain the code number and the protection level. Welding filters are an exception, as they are marked with only one of the aforementioned

TABLE 26.2
List of Markings (Code Numbers and Protection Levels) for Various Filters

Welding Filters	Ultraviolet Filters		Infrared Filters	Sun Glare Filters	
No Code Number	Code Number 2	Code Number 3	Code Number 4	Code Number 5	Code Number 6
1.2	2–1.2	3–1.2	4–1.2	5–1.1	6–1.1
1.4	2–1.4	3–1.4	4–1.4	5–1.4	6–1.4
1.7		3–1.7	4–1.7	5–1.7	6–1.7
2		3–2	4–2	5–2	6–2
2.5		3–2.5	4–2.5	5–2.5	6–2.5
3		3–3	4–3	5–3.1	6–3.1
4		3–4	4–4	5–4.1	6–4.1
4a					
5		3–5	4–5		
5a					
6			4–6		
6a					
7			4–7		
7a					
8			4–8		
9			4–9		
From 10 to 16			4–10		

Code number: 2 = ultraviolet filters (colour perception may be slightly altered); 3 = ultraviolet filters (good colour perception); 4 = infrared filters; 5 = sun glare filters not fulfilling requirements with respect to infrared; 6 = sun glare filters fulfilling requirements with respect to infrared.

protection levels. The list of code numbers and markings for different types of filters is presented in Table 26.2.

Table 26.2 does not include markings of filters used for protection against laser radiation. Because laser radiation has a high degree of cohesion, monochromaticity and orientation, and the angle of beam divergence usually does not exceed several milliradians (mrad), eye protection against this type of radiation is tailor-made for a given type of laser. Laser radiation filters must ensure effective protection against radiation of the wavelength emitted by the given type of laser. Moreover, the housing and filters also must resist the laser radiation, which, in the event of high power or energy density, can damage the protective equipment itself. For individual eye protection against laser radiation ranging from 180 to 1000 μm, spectacles, goggles, and face shields are used. Laser radiation filters are marked with codes from L1 to L10. These markings are determined based on the optical density of the filter (for the wavelength from which the protection must be ensured) and on laser radiation resistance. If dangerous laser radiation remains in the visible spectrum (from 400 to 700 nm), and the eye protection equipment lowers this radiation to the values defined for lasers of class 2 (radiation power $P \leq 1$ milliwatt (mW) for continuous wave lasers—in this case physiological defence reactions, including the blinking reflex, contribute to eye protection), then such a device is called protective equipment for laser adjustment.

Advanced technological solutions are used for the construction of oculars and filters (Owczarek et al. 1997; Kubrak and Owczarek 2001). The use of modification of optical radiation transmittance spectrum characteristics enables the adjustment of the light transmittance factor to values corresponding to given lighting conditions and thus meets the user's requirements in the event of protection against visible radiation (sun glare). The technology of light transmittance characteristics modification offers many possibilities for designing optical filters for radiation ranges that may constitute a real hazard in work conditions (UV, infrared, laser radiation). Modification of the spectrum characteristics is achieved by in-mass tinting, superficial tinting, or by coating the material with reflective or special interference layers. *In-mass tinting* is when pigments are added during the process of production of the material used for making the filter. It is used mainly to produce UV, visible and infrared radiation filters. *Superficial tinting* is tinting by immersing the substrate material for the filter in a dye, and is used mainly for lenses. Lenses fulfilling the requirements specified in the standard PN-EN 166 may be treated as oculars or filters.

Coating the filter with an additional reflective layer (e.g. infrared radiation reflection) reflects relatively high amounts of dangerous radiation from its surface. The protective effect therefore may be achieved by reflecting the radiation, not only by absorbing it. Radiation absorbed by a filter naturally causes heating. When the filters that only absorb radiation (without a reflective coating) are exposed to intensive infrared radiation they may heat to relatively high temperatures. Thus, the filter itself becomes a source of temperature radiation affecting eyes.

It is currently popular to cover the surfaces of lenses and interference filters with an antireflection coating. Such coatings effectively eliminate the reflection of light, for example, that emitted by headlights of a car approaching from the opposite direction, or subdue laser radiation of a given wavelength.

To alter the transmission of optical radiation passing through the filter, welding filter manufacturers also use director orientation modification in the liquid crystal layer occurring under the influence of electrical fields, which may be generated by light impulses (Kubacki et al. 2001a, 2001b) or the photochrome effect. Use of the photochrome effect makes it possible to change the light transmittance factor depending on the external radiation lighting intensity, but is accompanied by the UV radiation responsible for the photochrome effect.

Moreover, the back surfaces of oculars and filters may be covered with antifog coating.

Ageing resistance and the absolute weight of material are as important as fogging resistance, mechanical resistance, and the filtration properties adapted to a given application. These features determine the quality and durability of the product and determine for what contemporary eye protective equipment can be used.

New materials and technologies are also utilised to construct the frames, housings, and harnesses of eye protective equipment (Owczarek et al. 2001; Kubacki et al. 2001a, b). These elements are made mainly of high-quality plastics that do not cause allergic reactions when in direct contact with the user's skin. In manufacture of elements to be mounted on the user's head, a particular importance should be assigned to the comfort of use with regard to adjustment and regulation ability, as well as appropriate ventilation and sweat absorption by materials directly in contact with the forehead.

26.2.6 Respiratory Protective Devices

Respiratory protective devices are divided into two basic categories: (1) filtering devices and (2) breathing apparatus. Filtering devices include filters, gas filters, combined filters, filtering half masks, filtering half masks protecting against gases and particles, powered filtering devices, and power-assisted filtering devices (Figure 26.18). The second category includes self-contained open-circuit compressed-air breathing apparatus and self-contained closed-circuit breathing apparatus, compressed-oxygen or compressed-oxygen-nitrogen types, and continuous-flow compressed air line breathing apparatus.

The different types of filtering devices are classified based on their protection parameters, such as filtration efficiency and sorption capacity, which determine their use. There are three classes of protection for filters, filtration half masks and gas filters: lowest (class 1), medium (class 2), and highest (class 3). Combined filtering elements are classified by their purpose: devices protecting against organic gases and vapours of organic substances of temperatures above 65°C (type A), against organic gases and vapours of organic substances of temperatures below 65°C (type AX), against nonorganic gases (type B; excluding carbon monoxide), against sulphur

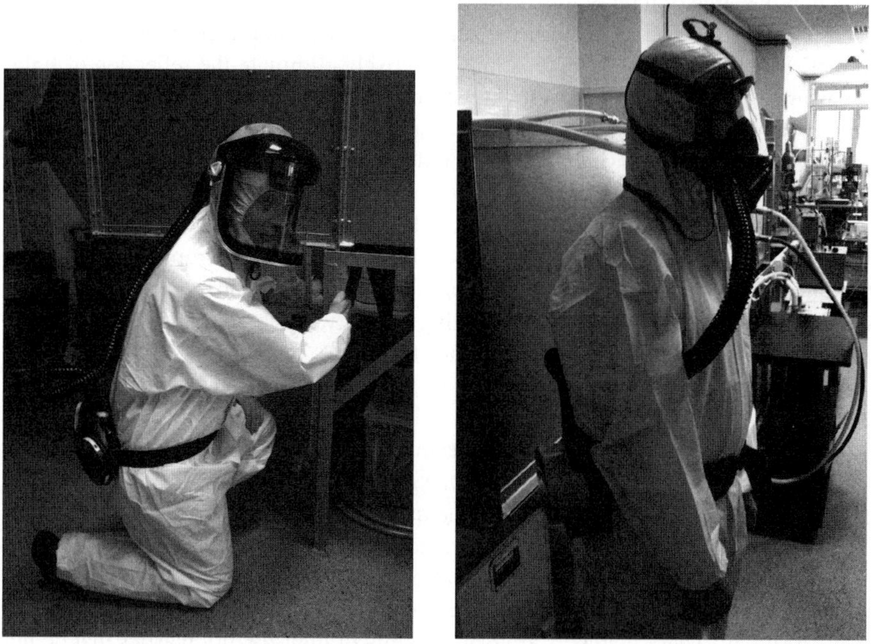

FIGURE 26.18 Powered filtering device incorporating a helmet or a hood and power-assisted filtering device incorporating full face mask, half mask, or quarter mask.

dioxide and other acid gases and vapours (type E), against ammonia and organic ammonia derivatives (type K), and against specific gases and vapours (type SX). Combining different types of gas filters results in multitype gas filters, and combining gas filters with filtering elements results in combined filtering devices.

There is also a separate group of respiratory protective devices intended specially for escape in case of sudden hazard, for example, major accidents, explosions, fires, accidents related to the transport of dangerous substances, and terrorist attacks. These types of devices include lung-governed demand self-contained open-circuit compressed air breathing apparatus with a full face mask or mouthpiece assembly for escape, self-contained closed-circuit breathing apparatus for escape, and filtering elements with a hood. One of the basic requirements of respiratory protective devices intended for evacuation is that they must supply the user with breathable air and that they can be worn easily and quickly.

The escape hood has a head harness made up of an internal half mask, elements allowing a tight fit to the face, visors with appropriate transparency and a fixed combined filtering element, intended to stop contaminants generated during a fire. The proper functioning of hoods with filtering elements is dependant on the ambient atmosphere; therefore they do not provide protection from oxygen deficiency (when the oxygen concentration falls below 17% of the air volume), but they should allow for escape from a contaminated zone.

The remaining respiratory protective devices provide air from a compressed air cylinder or an appropriate closed-circuit system regardless of the ambient atmosphere. The working duration, that is, the duration the device ensures breathable air during evacuation, depends on the construction of the device and ranges from 15 to 60 minutes. These devices have the advantage of small dimensions and mass, guaranteeing a quick and efficient escape from the hazard zone.

The functioning of different elements of respiratory protective devices is based on several physical phenomena. Filtering elements, that is, filters and filtering half masks, are based on the phenomenon of filtration, gas filters on the phenomenon of adsorption, and combined filtering elements on both phenomena. Filtration, based on the type of the filtrated aerosol, includes mechanisms related to the retention of solid and liquid dispersed phases. The base for filtration is the deposition of aerosol particles in porous filtration layers, resulting in their retention within the filtration barrier. The accumulation of a layer of dust around the structural fibres of the filtering material modifies the conditions of the air flowing through the filtering material and separates subsequent inflowing particles. During use of respiratory protective devices, the shapes of the filtering material fibres and the distances between them are modified, resulting in an increase of resistance to air flow. *Dendritic trees* appear and, in the course of the aerosol filtration, a *filtration cake* may appear at the surface of the filtration layer resulting in a rapid increase in the resistance to air flow as well as in higher filtration efficiency.

The filtration mechanism of liquid dispersed phase aerosols differs from the mechanism of solid dispersed phase aerosols. A description of the liquid dispersed phase mechanism is presented by Raynor (2000), who described the process of liquid aerosol filtration as a simultaneous interactions of aerosol components, that is, interaction of volatile phase vapours and filtration elements (fibres). Different components of the aerosol may migrate along the filtration layer, evaporate and condense again. These phenomena occur simultaneously and their mutual proportions change during the filtration process. Raynor also demonstrated that aerosol particles in the filtration material coagulate, forming larger agglomerates that spread out between the available fibres surfaces, forming bridges of liquid, and then coating the entire fibre, creating a kind of 'film' on its surface. The resistance of the air flow is thereby modified, depending on the mass of the deposited aerosol particles. The character of the flow resistance modification during liquid aerosol particles filtration differs from the dynamic changes in resistance achieved by aerosols with a solid dispersed phase. A graph of the filtration fabric's changes in resistance, depending on the mass of the deposited liquid aerosol particles, is presented in Figure 26.19, and of the aerosol with solid dispersed phase in Figure 26.20. The occurrence of deposited aerosol particles delocalisation is attributed to characteristic points of the flow resistance changes curves (a) and (b) in Figure 26.19; because of these, a temporary fall in resistance values appears.

Several materials used as filtration mediums in filtering devices lose protection properties during their utilisation. Prolonged interaction of the liquid dispersed phase aerosol and electret filtration material reduces the device's filtration efficiency, often to below that of its assigned protection class. The effects of aerosols on the protection properties were taken into consideration in the research methodology and evaluation criteria for this type of devices for inclusion in the relevant European standards. This is the basis for correctly classifying a device into one of the protection classes: P1, P2 or P3.

Toxic vapours and gases are removed from the air by a gas filter using the physical adsorption phenomena, which occurs due to the acts of dispersion forces between

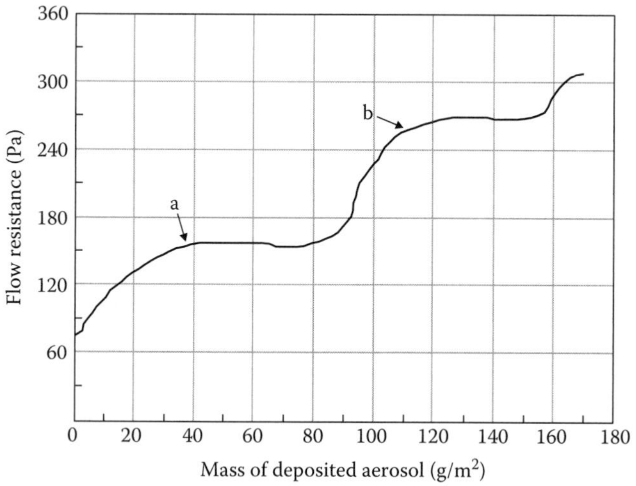

FIGURE 26.19 Change of aerosol flow resistance as the function of mass of the liquid dispersed phase aerosol deposited particles. The mass of accumulated liquid causes the rate of the pressure drop increase to decrease at points (a) between 30 and 70 g/m^2 and (b) between 110 and 150 g/m^2.

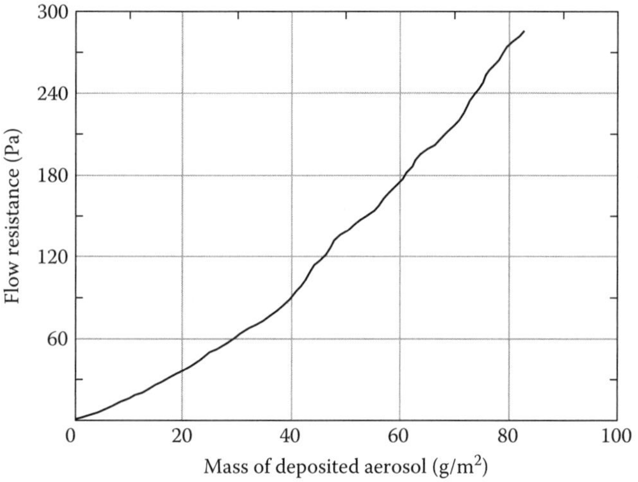

FIGURE 26.20 Change of aerosol flow resistance as the function of mass of the solid dispersed phase aerosol deposited particles.

the adsorbate particles and the surface of active carbon that makes up the gas filter sorption deposit, or due to chemical adsorption and catalysis caused by the interaction of adsorptives and adsorbents resulting from a chemical reaction (Ościk 1976). The course of the adsorption process depends on the physical and chemical properties of the adsorbate, on the characteristics of the porous structure of the adsorbent and the chemical character of its surface, and on the adsorption conditions, that is, pressure, temperature, and humidity.

The influence of changes in the temperature and humidity of the air in which gas filters are used has been documented in numerous researches (Young and Nelson 1990; Pietrowski 2000). For some groups of organic substances, the rise in the environment temperature may positively influence absorption by prolonging the duration of the gas filters' protective action, but for some it can have an inverse effect. Similar correlations were demonstrated for changes in air humidity (Vahdat et al. 1994).

For removing vapour mixtures, for example, organic substances composed of two or more elements, the duration of the gas filters' protective action for organic vapours and gases is closely related to the physical and chemical properties of the different elements of those mixtures. When the air is purified from vapours of substances with different boiling points, the first to appear after the deposit layer is the substance with the lowest boiling point. An interaction of adsorbed mixtures of contaminating vapours usually means a shorter time of the gas filters' protective activity with regard to the duration obtained in standard laboratory conditions during evaluation of their protective parameters.

The utilisation comfort of the respiratory protective devices has also been the subject of many research works (Astrand and Rodhal 1986; Akbar-Khanzadeh and Biesi 1995; Newill et al. 1989). Only 8% of the research participants considered half masks with filtering elements to be comfortable. A lack of comfort during the use of the respiratory protective devices can mean that workers will use them incorrectly or not use them at all (White et al. 1989). The correct fit and comfort of this equipment is therefore very important. Selection should be made not only based on the level of hazard occurring in a given workplace and the environmental conditions (temperature and humidity), but also with regard to the subjective requirements of the user and their characteristic anthropometric dimensions.

Therefore, the United States (ANSI Z88.10.2), Great Britain (BS 4275 1997) and Australia introduced legislation requiring individual fittings of respiratory protective devices to each user. The procedures used are intended to ensure the comfort of the protection devices selected; at the same time they are a form of training (Clayton and Vaughan 2005). The introduction of the requirement for individual fittings gives the user an opportunity to get acquainted with the principles of its operation, as well as check prior use, cleaning and maintenance procedures and methods.

Implementation of individual selection of respiratory protective devices is related to the concept of the experimentally determined protection factor. The protection factor reflects the respiratory protective devices' level of effectiveness, expressed as the ratio of the workplace contamination reduction. The basic principle in selection of respiratory protective devices is that, for a given toxic substance, the equipment effectiveness should guarantee that the concentration of harmful agents under the facepiece is below the MAC or MAC (STEL) value (Zawieska 2007). Therefore, the index value depends on the type of respiratory protective device used, taking into consideration additional conditions such as individual fitting of the device, its correct functioning and the conditions of the environment in which it is used. In Great Britain, BS 4275:1997 sets three types of protection factors, to be differentiated as follows:

1. Assigned protection factor
2. Workplace protection factor
3. Nominal protection factor

They are defined as follows:

Assigned protection factor: The expected level of effectiveness of the respiratory protective device in a workplace, achieved by 95% of users when the following conditions are met: appropriate training is carried out and the proper use, functioning and fitting of the device are monitored regularly.

Workplace protection factor: The protection factor measured in a given workplace for a given respiratory protective device, correctly fitted by the user and that is properly functioning and maintained.

Nominal protection factor: The maximum percentage value of the total inward leakage prescribed for each protection class according to EN standards, determined as:

$$\text{NPF} = \frac{100}{\text{TIL value}}$$

where TIL is the laboratory-determined value of the total inward leakage, in percentage.

The nominal protection factor is determined by the total inward leakage in a laboratory. The protection factor determined in laboratory conditions gives only an approximate indication of the effectiveness of the equipment used. Only by taking into consideration the actual conditions of use or simulation of those conditions can the correct evaluation of the protective properties of a given type of equipment be guaranteed. Correct evaluation is particularly important for devices used in workplaces with a high degree of exposure to harmful agents.

26.2.7 Equipment Protecting against Falls from a Height

Data on work accidents published annually in many European countries indicate that work at a height is still one of the most dangerous types of work. Working at a height in such areas as construction, power engineering, and telecommunications means that often the personal systems protecting against falls are the only method of workers' protection. The systems, depending on the function, can be categorised into three types:

1. Intended for arresting falls
2. Indented for positioning work
3. Intended for restraining falls

Fall arrest systems are intended for use at workstations whose configuration does not allow for elimination of the risks of free fall. The major functions of the system are that it arrest the fall, mitigate its effects by limiting the forces acting on the human body, protect it against collision with dangerous objects on the soil or workstation, and position the body during the fall and after the arrest it in a way that enables the worker to safely wait for help. An example of a personal fall arrest system is presented in Figure 26.21.

A fall arrest system is composed of three basic subsystems, namely anchoring, connecting and shock-absorbing as well as full-body harness. The anchor subsystem (Baszczyn´ski et al. 1999) is the first link of the system and is directly connected to the workstation. It enables linking the connecting and shock-absorption subsystems with the bearing structure of the workstation to arrest the fall. Different types of connectors such as formed wire clamps, lanyards made of wire rope, webbing anchor devices, hook anchor connectors, presented in Figure 26.22, fulfil the role of the universal anchor subsystem. Such subsystems are attached to steel constructions, beams, fentons, and other elements of proper shape and strength. Horizontal flexible lines and anchor rails can be used as anchor subsystems that enable the worker to move horizontally (Baszczyn´ski and Zrobek 1998). For protection of the workers working in hollows, for example, drains and shafts, the best solution for anchoring fall arrest systems is to use stands (tripods) or lateral beams with an anchor point.

The connecting and shock-absorbing subsystems are another part of the fall arrest system, placed between the anchor assembly and full-body harness. The main function of the subsystem is arresting the fall of a person and lessening its effects. The subsystem minimises the distance of a free fall and reduces the risk of striking

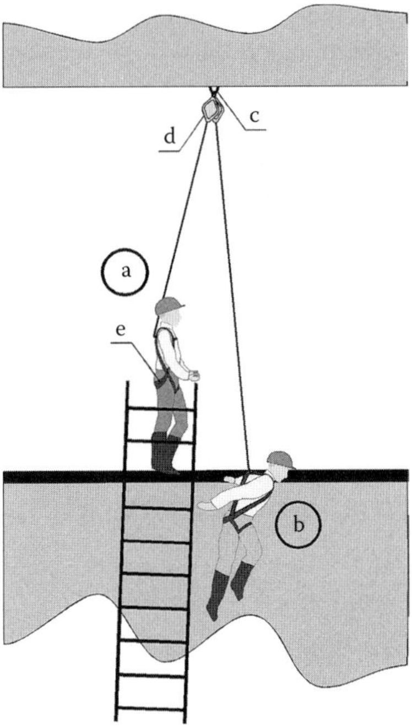

FIGURE 26.21 Fall arrest system: (a) the state before free fall; (b) the state after fall arrest; (c) anchor element on the structure of the workstation; (d) retractable-type fall arrester; (e) full-body harness.

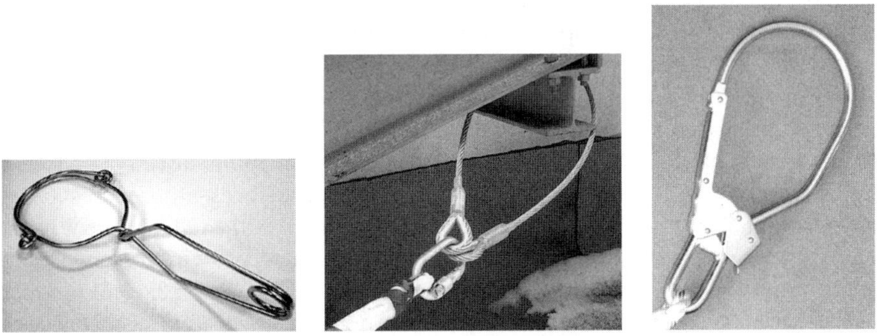

(a) (b) (c)

FIGURE 26.22 Examples of universal anchor subsystem solutions: (a) formed wire clamp, (b) lanyard made of wire rope, (c) connector of large clearance.

elements of workstation construction and the ground, while also reducing the kinetic energy of the fall.

The effect of arresting a fall is that the dynamic force affecting a human body is reduced. This is achieved when the connecting and shock-absorbing subsystem absorbs the kinetic energy of a falling person. The energy absorption is a result of the

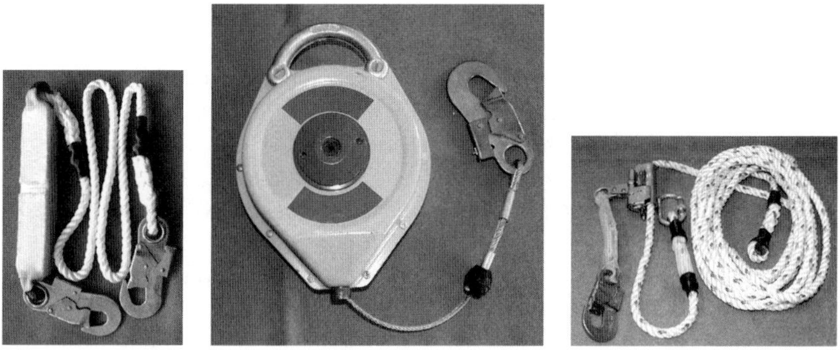

(a) (b) (c)

FIGURE 26.23 Connecting and shock-absorbing subsystems: (a) textile energy absorber with lanyard, (b) retractable-type fall arrester, (c) guided-type fall arrester including a rigid anchor line.

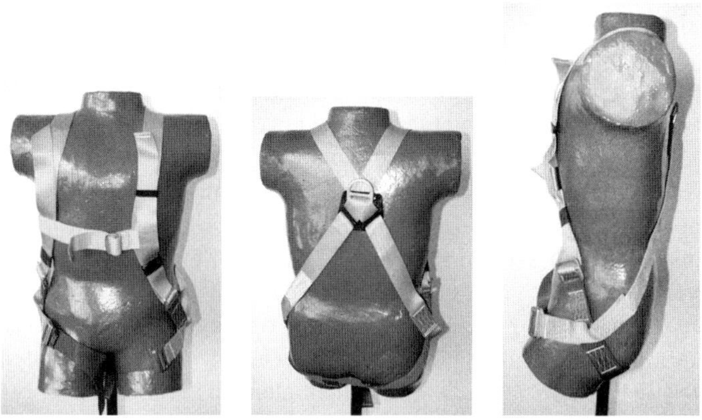

(a) (b) (c)

FIGURE 26.24 Full-body harness with back attachment element: (a) Front view, (b) back view, and (c) side view.

deformation and friction of the subsystem elements. The most popular connecting and shock-absorbing subsystems, presented in Figure 26.23, are as follows:

- *Textile energy absorber*: Kinetic energy is absorbed by tearing two layers of special internal webbing.
- *Retractable-type fall arrester*: Kinetic energy is transformed into the work of friction of brake black plates.
- *Guided-type fall arrester*: Kinetic energy is transformed into the work of friction between the fall arrester and anchor line as well as into the deformation of the anchor line or other intended elements.

The last component of the fall arrest system that remains in direct contact with human body is the full-body harness, examples of which are presented in Figure 26.24.

Because the energy-absorbing properties of the subsystems fulfil the requirement of EN standards, the force acting on the attachment element of the full-body harness does not exceed 6 kN, which is considered safe for a human. The full-body harness is the only type of body harness approved for fall arrest in work environment conditions. The main functions of the full-body harness are to

- Arrest the fall with a distribution of acting forces that will reduce the risk of injuries.
- Properly position the body when arresting the fall to avoid injuries to the inner organs and the spine.
- Properly position the body after arresting the fall to enable a safe and comfortable wait for help.

Work positioning systems are equipment that protect against falls from a height. Their function is to position a worker so that he or she can use both hands while being comfortably supported. The work positioning belt (Figure 26.25a) equipped with lateral attachment elements jointed to the work positioning lanyard with a length adjuster (Figure 26.25b) is an example of this type of equipment. During the use of the equipment, the belt supports the worker's back, while his or her legs lie on the workstation structure. The lanyard, whose ends are connected to lateral attachment elements, is belted around the structural element of the workstation, for example, part of the pillar of lattice construction. Its length is adjusted by the user, which assures a safe and comfortable position. The work-positioning lanyard must be belted around the element that will enable it to slide and, as a result, prevent a free fall. The work positioning system cannot be used as fall arrest system (Baszczyński and Zrobek 2005). If this occurs at a given workstation, the work positioning system must be used along with the fall arrest system and the work positioning belt with lateral attachment elements is then a part of full-body harness.

Restraint systems (Baszczyński and Korycki 2001; Baszczyński 2006) are mainly used to restrict the movement of the user and keep the person away from unsafe areas where there is a risk of falling. The restraint system, an example of which is presented in Figure 26.26, consists of the following:

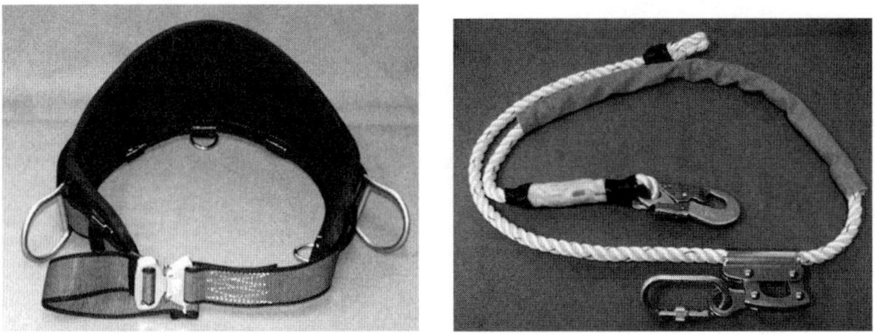

(a) (b)

FIGURE 26.25 Work positioning system: (a) work positioning belt, (b) positioning lanyard.

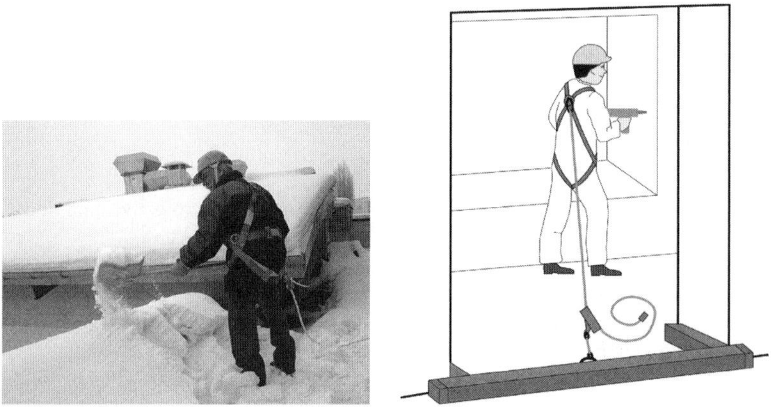

(a) (b)

FIGURE 26.26 Use of restraint systems (a) on a roof and (b) inside a building.

- An anchor subsystem enabling connection with the elements of workstation structure
- A connecting subsystem (e.g., an adjustable lanyard or a guided-type fall arrester equipped with manual locking) that is linked at one end with the anchor subsystem and at the other end with the body harness
- A body harness such as a full-body harness, work positioning belt, and sit harness

Restraint systems are intended mostly for workstations with a large surface area inclined at a small angle. They should be used at workstations where the workers do not have to stay in an area where they are exposed to risk of free fall.

PPE protecting against falls from a height can fulfil its functions properly only when the required protective parameters are maintained. Any of the following factors can cause the loss of these parameters:

- Atmospheric conditions, for example, solar radiation, in particular UV, causing the degradation of plastics, as well as humidity intensifying the corrosion of metal elements
- Mechanical factors such as friction, cuts, and blows by objects that may occur at workstation
- Heat factors such as hot objects, splashes of liquid metal, and open flame that cause damage to plastic elements and especially to textile elements
- Chemical agents causing degradation of plastics and corrosion of metal elements

Hence, thorough checks of equipment that protects against falls from a height are imperative, and should be performed by the user before each use and periodically by another specially trained person, for example the manufacturer's services. The protective equipment should be discarded when any doubts related to the technical condition of the equipment arise, when the period of wear indicated by the manufacturer expires or when the equipment has been used to arrest a fall.

REFERENCES

2000. Climate control for feet. *DuPont Mag* (3):10–11.

Akbar-Khanzadeh, F., and M. S. Biesi. 1995. Comfort of personal protective equipment. *Appl Ergon* 26(3).

Andrzejewska, A., and A. Kurczewska. 2004. Criteria for assessing clothing protecting against microwave radiation in view of the basic requirements of directive 89/686/EEC. Conference materials of the 6th International Symposium EL-TEX—Electrostatic and Electromagnetic Fields—New Materials and Technologies, Lodz.

Andrzejewska, A., and K. Szczecińska. 2005. Hand protective equipment while handling hand knives. *Occupational Safety* (7–8):38–41.

ANSI Z88.10.2. Practices for respiratory protection.

Astrand, P. O., and K. Rodhal. 1986. *Textbook of Work Physiology*. New York: McGraw-Hill.

Bartkowiak, G. 2006. Liquid sorption by nonwovens containing superabsorbent fibres. *Fibres Text East Eur* 55(1):57–61.

Bartkowiak, G., A. Kurczewska, and K. Łężak. 2006. Conference materials of the 3rd Scientific Conference, Lodz.

Bartkowiak, G., and D. Błażejewski. 2003. Reduction of work discomfort in tight protective clothing due to the use of proper barrier clothing. *Occupational Safety* (2):23–24.

Baszczyński, K. 2002. Industrial helmets and head protection against impact from the side. *Occupational Safety* (5):10–13.

Baszczyński, K. 2006. Safety in roof work. *Occupational Safety* (2):2–4.

Baszczyński, K., and R. Korycki. 2001. The use and selection of restraint systems. *Occupational Safety* (11):21–23.

Baszczyński, K., and Z. Zrobek. 1998. Steel horizontal flexible anchor lines. *Occupational Safety* (6):18–21.

Baszczyński, K., and Z. Zrobek. 2005. Protection against falls from poles—efficiency assessment of work positioning equipment. *Occupational Safety* (4):18–21.

Baszczyński, K., M. Karlikowski, and Z. Zrobek. 1999. Anchor devices in fall arresting equipment. *Occupational Safety* (12):6–10.

British Standard BS 4275. 1997. Guide to implementing an effective respiratory protective device programme.

Clayton, M. P., and N. Vaughan. 2005. Fit for purpose? The role of fit testing in respiratory protection. *Ann Occup Hyg* 49(7).

Council Directive 89/686/EEC of 21 December 1989 on the approximation of the laws of the Member States relating to personal protective equipment. OJ EC, L 399, 30.12. 1989.

Kamińska, W. 2006. Development of a method for assessing slip resistance of protective footwear in conditions simulating their use: Report on the realisation of the task no. 03.10 related to services for the State. (Unpublished work).

Kubacki, Z., G. Owczarek, and A. Pościk. 2001a. Composite materials for welder's face shield housing. *Occupational Safety* (5):9–13.

Kubacki, Z., G. Owczarek, and A. Pościk. 2001b. New generation of welding eye and face protections. *Occupational Safety* (4):10–13.

Kubrak, J., and G. Owczarek. 2001. Interference corrector of optical radiation for work safety in textile industry. In *Proceedings of SPIE (Optical Techniques for Sensing, Workplace Safety, and Health Monitoring.)* Vol. 4887.

Kurczewska, A., and J. Leśnikowski. 2008. Variable thermoinsulation garments with a microprocessor temperature controller. *Int J Occup Saf Ergon* 14(1):77–89.

Mäkinen, H., S. Mäki, J. S. Solaz, and D. J. Stewardson. 2000. Fire fighters' views on ergonomic properties of their footwear. In *Proceedings of the 1st European Conference on Protective Clothing*, Stockholm, May 7–10, 2000.

Muir, I. H., P. A. Bishop, and J. Kozusko. 2001. Micro-environment changes inside impermeable protective clothing during a continuous work exposure. *Ergomonics* 44(11):953–61.

Nag, P. K., et al. 1998. Efficacy of water cooled garment for auxiliery body cooling in heat. *Ergonomics* 41(2):179–187.

Newill, C.A., et al. 1989. Utilization of personal protective equipment by laboratory personnel at large medical research institution. *Appl Ind Hyg* 4.

Ościk, J. 1976. *Adsorption.* Warsaw: PWN.

Owczarek, G., et al. 1997. Selected properties of safety oculars. *Occupational Safety* (6):17–20.

Owczarek, G., et al. 2001. Relationship between the difference in prismatic refractive power of an eye-and-face protector and its thickness, radius of curvature, and material. *Int J Occup Saf Ergon* 7(3):277–284.

Owczarek, G., K. Łężak, and G. Gralewicz. 2007. Monitoring selected parameters during work in firefighters' clothing. *Occupational Safety* 9:8–10.

Pietrowski, P. 2000. Adsorbate deposition in organic gas filter beds made of activated carbon. *Environmental Engineering and Protection* 3(3–4):491–500.

PN-EN 14052. 2006(U). High performance industrial helmets.

PN-EN 167. 2005. Personal eye protection. Optical test methods.

PN-EN 168. 2005. Personal eye-protection. Non-optical test methods.

PN-EN 397. 1997. Specification for industrial safety helmets.

PN-EN 420:2005/AC:2007. Protective gloves. General requirements and test methods.

PN-EN 812. 2002. Industrial bump caps.

PN-EN ISO 20345. 2007. Personal protective equipment—Safety footwear.

PN-EN ISO 20346. 2007. Personal protective equipment—Protective footwear.

PN-EN ISO 20347. 2007. Personal protective equipment—Occupational footwear.

Raynor, P. C. and D. Leith. 2000. The influence of accumulated liquid on fibrous filter performance. *J Aerosol Sci* 31(1):19–34.

Reinertsen, R. E., et al. 2008. Optimizing the performance of phase-change materials in personal protective clothing systems. *Int J Occup Saf Ergon* 14(1):43–55.

Szczecińska, K. 2004. Signalling the permeation of dangerous chemicals through protective gloves: Report on the realisation of the task no. I.10 carried out within statutory research activity. (Unpublished work).

Vahdat, N., P. M. Swearengen, and J. S. Johnson. 1994. Adsorption prediction of binary mixtures on adsorbents used in respirator cartridges and air-sampling monitors. *Am Ind Hyg Assoc J* 55(10):909–917.

White, M. K., M. Vercruyssen, and T. K. Hondous. 1989. Work tolerance and subjective responses to wearing protective clothing and respirators during physical work. *Ergonomics* 32:1111–1123.

Young, H. Y., and, J. H. Nelson. 1990. Effects of humidity and contaminant concentration on respirator cartridge breakthrough. *Am Ind Hyg Assoc J* 51(4):202–209.

Zawieska, W. M., ed. 2007. *Occupational Risk. Methodological Principles of Assessment.* Warsaw: CIOP-PIB.

27 Shaping the Safety and Ergonomics of Machinery in the Process of Design and Use

Józef Gierasimiuk and Krystyna Myrcha

CONTENTS

27.1 INTRODUCTION

Man, the creator of machines and other technical devices that serve as workstation equipment, is also the recipient of the positive and negative influences of these devices. Negative influences include accidents, illnesses, and material losses. Positive influences include, for example, a high degree of well-being, job satisfaction, and increased work productivity. Machines can cause significant economic gains or losses, regardless of the human aspect.

Machines, usually produced in batches or sometimes on a mass scale, often function for many years. All of the shortcomings and defects that occur, especially at the design phase and during the use of the machine, significantly influence man–machine interactions and the environment, especially the work environment. According to data from the Polish Central Statistical Office, use of machinery is the direct cause of about 25% of accidents. Reducing these undesirable consequences is necessary for human, social, and economic reasons. This is primarily the duty of the machinery's producers, but also of their users and the employer. Actions to reduce these consequences should be given the same importance as actions to ensure that the machinery properly performs its technological functions.

27.2 TERMS AND DEFINITIONS

Machinery can have any of the following definitions:*

1. An assembly, fitted with or intended to be fitted with a drive system other than directly applied human or animal effort, consisting of linked parts or components, at least one of which moves, and that are joined together for a specific application.
2. An assembly as described in point 1, lacking only the components needed to connect it on-site or to sources of energy and motion.
3. The assembly as described in points 1 and 2, ready to be installed but able to function only if mounted on a means of transport or installed in a building or a structure.
4. Assemblies of machinery as described in points 1–3 or partly completed machinery, which, to achieve the same end, are arranged and controlled so that they function as an integral whole.
5. An assembly of linked parts or components, at least one of which moves, that are joined together, are intended for lifting loads, and whose only power source is directly-applied human effort.

Partly completed machinery:* An assembly that cannot in itself perform a specific application. Partly completed machinery is only intended to be incorporated into or assembled with other machinery or other partly completed machinery or equipment, thus forming complete machinery. A drive system is an example of partly completed machinery.

* According to directive 2006/42/EC (2006).

Interchangeable equipment:* A device that can be assembled with machinery or a tractor by the operator of the machinery or tractor in order to change its function or attribute a new function. According to this definition a tool is not interchangeable equipment.

Safety component:* A component that serves to fulfil a safety function and is independently placed on the market, the failure and/or malfunction of which endangers the safety of workers, and that is not necessary in order for the machinery to function, or for which normal components may be substituted in order for the machinery to function.

Work equipment: Any machine, apparatus, tool, or installation used at work.

Use of work equipment:† Any activity involving work equipment such as starting and stopping the equipment, transporting, repairing, modifying, or servicing, including cleaning.

Operator:† The worker or workers assigned the task of using the work equipment.

Product: An object, irrespective of its degree of processing, intended to be placed on the market or put into service, excepting food and agricultural products and animal feed.

Manufacturer:‡ Any natural or legal person who manufactures a product or has a product designed or manufactured, and markets that product under his or her name or trademark.

Authorised representative:‡ Any natural or legal person established within the community who is duly authorised by the manufacturer to act on his or her behalf to perform specified tasks related to the latter's obligations under community legislation.

Importer:‡ Any natural or legal person established within the community and vested with authority who places on the community market a product obtained from an outside country.

Distributor:‡ Any natural or legal person in the supply chain who makes a product available on the market, other than the manufacturer or importer.

Making available on the market:‡ Any manner of supplying a product for distribution, consumption or use on the community market in the course of a commercial transaction, whether rendered in return for payment or free of charge.

Placing on the market:‡ The first time a product is offered to the community market.

Putting into service:* The first use of machinery for its intended purpose in the community.

CE marking:‡ A marking by which the manufacturer indicates that the product is in conformity with the applicable requirements set out in the relevant community legislation.

Community legislation:‡ Any community legislation harmonising various conditions for the marketing of products.

---

* According to directive 2006/42/EC.

† According to the directives 89/655/EEC and 95/63/EC.

‡ According to regulation (EC) no. 765/2008 of the European Parliament and of the Council of 9 July 2008, setting out the requirements for accreditation and market surveillance relating to the marketing of products and repealing EEC regulation no. 339/93 (text with EEA relevance).

Essential requirements: Requirements relating to a product's features, design, or manufacture, defined in the relevant legislation of the community.

Harmonised standards: European standards developed and adopted by European standardisation bodies* on the basis of a request made by the European Commission (EC), whose numbers and titles are published in the *Official Journal of the European Communities* of the C series.

Conformity assessment:† The process demonstrating whether specified requirements relating to a product, process, service, system, person, or body have been fulfilled.

Certification: The process of conformity assessment carried out by a body independent of the producer, his or her authorised representative, the importer, or the user.

Certificate of conformity: A document issued by the certification body attesting that the product and its manufacturing process comply with the essential requirements.

Declaration of conformity: Declaration of the producer or his or her authorised representative undertaking sole responsibility for compliance of the product with the essential requirements.

Accreditation:† An attestation by a national accreditation body that a body entrusted with conformity assessment meets the requirements set by harmonised standards and, where applicable, any additional requirements including those set out in relevant sectoral schemes, to carry out a specific conformity assessment activity.

Authorisation: Verification by the minister or director of a central state office with regard to the subject of conformity assessment, of the entity or laboratory applying for the notification process.

Notification: Reporting to the EC and member countries of the European Union (EU) of the authorised certification and inspection bodies, as well as the authorised laboratories that are able to perform the actions described in the conformity assessment procedures.

Market surveillance:† An activity carried out and measures taken by public authorities to ensure that products comply with the requirements set out in the relevant community legislation and do not endanger the health, safety, or any other aspect of the general public.

Market surveillance authority:† An authority of a member state responsible for carrying out market surveillance in its territory.

Withdrawal:† Any measure aimed at preventing a product in the supply chain from being made available on the market.

---

* These include the European Committee for Standardization (CEN), the European Committee for Electrotechnical Standardization (CENELEC), and the European Telecommunications Standards Institute (ETSI).

† According to regulation (EC) no. 765/2008 of the European Parliament and of the council of 9 July 2008, setting out the requirements for accreditation and market surveillance relating to the marketing of products and repealing EEC regulation no. 339/93 (text with EEA relevance).

27.3 SOURCES OF DATA FOR DEVELOPMENT OF WORK SAFETY AND ERGONOMICS OF MACHINERY

In order to comply with the concept adopted in the EU to ensure appropriate level of health protection of persons and safety of machinery:

- Producers must construct, produce, and place on the market of the European Economic Area (EEA)—which is comprised of member countries of the EU along with Iceland, Liechtenstein, and Norway, European Free Trade Association (EFTA) members, and parties to the agreement on EEA—or put directly into service machinery that assures the highest possible level of health protection and safety; in other words, producers must comply with the essential health and safety requirements contained in the New Approach Directives.
- Employers—the users of such machinery—must use such machinery for its intended purpose with regard to the processes and environment in which it is used as specified by manufacturer, or, if necessary, must adapt them according to the necessary purpose and existing environmental conditions. They must also maintain the appropriate level of health protection and safety ensured by such machinery throughout its useful life.

Employers in the countries of the EU should also ensure that all machinery and other equipment that was put into service before the country became a member of the EU is brought up to standard within a defined time and fully conforms with at least the minimum requirements for health protection and safety contained in their respective national legislation introducing social directives 89/655/EEC (1989), 95/63/EC (1995) and 2001/45/EC (2001). The current legislation is the basic source of requirements for work safety development and the ergonomics of machinery.

Health protection and safety requirements are formulated in a general manner. In line with the New Approach Directives, the detailed technical requirements relating to conformity with the essential requirements are contained in the European harmonised standards, whose lists are published in the *Official Journal of the European Union*. If a machine conforms with its given standards it is presumed to conform also to essential requirements regarding it. Meeting harmonised standards is not mandatory, but the standards are still often to demonstrate the conformity of machinery with the essential requirements. This does not preclude the possibility of demonstrating conformity by other methods or procedures.

European standards pertaining to the safety and ergonomics of machinery are divided into the following:

- *Standards of A type*: Basic safety standards that contain the definitions of basic terms, principles for design, and general aspects applicable to all types of machinery (e.g. PN-EN ISO 12100-1 (2005); PN-EN ISO 12100-2 (2005); PN-EN 614-1+A1 (2009); PN-EN 614-2+A1 (2009)).
- *Standards of B type*: Generic safety standards that focus on a single aspect of safety, or one type of safeguard that can be utilised in various types of machinery. These standards are divided into

    - *B1 standards*: Covering specific safety issues (e.g. safety distances—PN-EN ISO 13857 (2008); PN-EN 349+A1 (2008); PN-EN 999+A1 (2008))
    - *B2 standards*: Covering protective equipment (e.g. two-hand controls—PN-EN 574+A1 (2008); interlocking devices—PN-EN 1088+A2 (2008); pressure-sensitive protective devices—PN-EN 1760-1+A1 (2009) and 2+A1 (2009); guards—PN-EN 953+A1 (2009))
- *Standards of C type*: Standards relating to machinery and containing detailed safety requirements for specific machinery or small machine tools (e.g. hydraulic press brakes safety standard—PN-EN 12622 (2004); drilling machines—PN-EN 12717+A1 (2009)).

The numerous safety and ergonomics requirements regarding machinery, covered in legislation and standards, are based on the features and properties—anthropometric, biomechanical, physiological and psychological—of their operators. These features and characteristics, many of which are discussed by Gedliczka et al. (2001), can be incorporated directly into the processes of the design, construction and use of the machinery—especially where the data in standards and legislation are insufficient for this purpose.

The experience of users of the machinery—especially the circumstances and reasons for accidents or potentially dangerous events related to the machinery, as well as those related to occupational diseases—are also important in order to incorporate safety and ergonomics (primarily at the design stage of similar machinery), to modernise those that exist and to operate them correctly. Such data is particularly useful to improve protective measures, and to update the information on hazards and protective means provided to employees.

27.4 BIOTECHNICAL SYSTEMS OF MAN–MACHINE–ENVIRONMENT

The regulations described above obligate the designers, producers, and users of machinery to ensure that the machinery performs the intended technological processes and functions properly, with the least possible occupational risk for its operators and effect on the natural environment. This requires the worker's operating conditions to be suitably integrated with the machinery. The design stage is the most appropriate for determining these conditions. At this stage, not only are the conditions for the performance of specific processes and technological functions determined, but also the relationship between the machine and its operator and the conditions under which work occurs.

The design process integrates newly developed technical structures with the worker and the environment to form a single biotechnical system: the 'man–machine–environment'. The design of a machine is in fact similar to the design of a biotechnical system (Figure 27.1).

The fundamental elements of the system are the 'man–machine', where the regulatory processes occur. The work environment is an element that is modified by

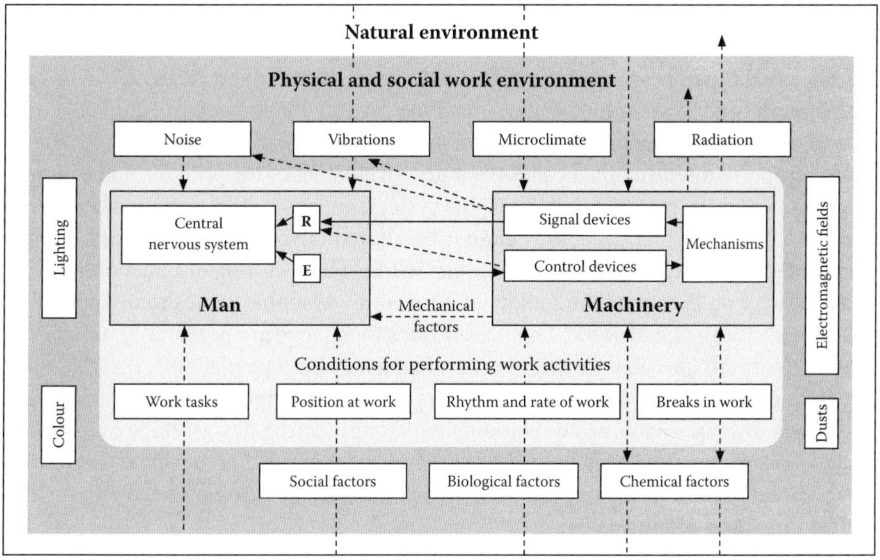

FIGURE 27.1 Diagram of the biotechnical man–machine–environment system; R = receptors; E = effectors.

the influences of the fundamental system, such as emission of chemical substances and noise. This can have either a positive or disturbing effect on the functioning of the biotechnical system.

The man–machine system is a relatively closed one. This means that the processes occurring within it are cyclic and constantly return to the starting point; this does not however exclude entries and exits beyond the system, such as interaction with the environment. In a typical system structure, interactions and feedback between a man and machine are of particular importance. Their functions can be illustrated using an example of the simplified system of driver-car. Information that a car is moving is passed to the speedometer, causing its needle to move; when the driver is speeding, the human photoreceptor transfers this information to the central nervous system where it is processed and passed to the effector, that is, to the lower right limb in the form of an order to reduce pressure on the controlling element (the gas pedal); from there the fuel mechanism reduces fuel flow to the engine. The final effect of this cycle is reduction of the speed of the vehicle.

The process described above applies to all interfaces and regulations of man–machine systems. Of course, a machine can transfer several pieces of information at the same time, which are received by different human senses. After identification and recognition, the information is translated into decisions and passed to effectors for enactment. The direct link from receptor to effector, marked in Figure 27.1, is used in reflexive and automated reactions, which do not require conscious decisions. A direct link from the controlling device to the receptor is used in cases where the behaviour of the controlling device provides information to the operator on the status of the controlled process (e.g. the position, resistance, or vibration of the control device).

To simplify the diagram of the system (Figure 27.1), the man and machine can be considered to be equal elements of a uniform regulatory system. In reality, there are enormous differences both qualitatively and quantitatively between man and machine. In other words, in the man–machine system, man is not an equivalent element, but is actually superior, performs a distinct function and is the entity responsible for properly directing the process, not just an element of the process. The element that determines the reliability of the whole system, and yet is its most vulnerable part, is man. Psychophysical capacities cannot be altered beyond certain specific, rather narrow, boundaries, but condition human work, which is reliable and optimal in terms of load on the organism and its effectiveness. Machine serves man and facilitates the performance of work. The possibilities for change are potentially unlimited, provided they are possible from an economic standpoint. The machine, coupled with the work environment, should be adapted as much as possible to the psychophysical properties of man. Machine adaptation should begin at the design stage and appropriately distribute functions between man and machinery. The designer should be aware of the advantages of one element of the system in question over the other when distributing the functions.

Numerous studies have determined that the advantage of man over machine is primarily due to:

- Ability to think abstractly, generalise, and form ideas
- Ability to combine (integrate) various impressions and concepts into a logical whole
- Ability to better adjust to changing conditions and ability to improve adjustment by training; in machinery this is constrained by its construction
- Creative thinking skills and benefits of experience, which are necessary to generalise conclusions
- Ability to form judgements
- Ability to properly react in emergencies and unexpected situations
- Greater sensitivity to stimulation and disturbances, especially when the signs are weak, due to the ability to maximise the reception of the senses

The advantages of machinery over man include:

- Greater strength, speed, and accuracy, especially in calculations
- Ability to work without fatigue and resist external conditions
- Greater memory capacity and reliability
- Ability to perform numerous activities at the same time (multitask)
- Ability to register signals rapidly

In general, man is the more flexible and universal element of the man–machine system, and machine is the element more effective for specific, narrow functions. These factors must be considered at the early design phase in order to allocate functions between man and machine properly. The next step in the design process is to determine the tasks the operator must perform for his or her allocated functions, as well as the process for performing these tasks under specific environmental conditions.

This allows a definition of the relationships in the man–machine–environment system, and their shaping according to the psychophysical properties of the worker.

An understanding of the range of the tasks performed by the operator, the process needed to achieve this performance and the related interactions in the man–machine system throughout the life of the machinery (production, exploitation, scrapping) allows the designers and producers to identify possible hazards and inconveniences and the situations in which they may arise. This forms a basis to apply the necessary means to eliminate or limit the associated risks, and to inform the users of the remaining risks.

Experience shows that adjusting deficiencies to adapt machinery to man—especially in the area of allocating functions between man and machinery at the operation stage—is usually very costly and does not improve the situation significantly; therefore, this should be done at the design phase.

27.5 STRATEGY FOR REDUCING OCCUPATIONAL RISK IN THE PROCESS OF MACHINERY DESIGN AND OPERATION

Reducing occupational risks associated with machinery involves the whole biotechnical system of man–machine–environment and primarily covers actions tied to the processes of machinery design, manufacture, and operation (Figure 27.2). Hazards associated with the machinery sooner or later cause physical injury or damage to health. Therefore, the aim should be to eliminate hazards or implement protective measures, especially safeguarding, reducing the occupational risk associated with hazards that cannot be eliminated.

Protective measures are a combination of measures taken by the designer and the user. Measures applied at the design phase are usually more effective than measures implemented by the user. For this reason, the designer of the machinery, using the experiences of other users of similar machinery, should:

- Determine the intended use of the machinery and its limits.
- Identify hazards and associated hazardous situations, and estimate the risk of each.
- Evaluate the risk and decide whether it is necessary to reduce it.
- Eliminate the hazards or reduce the risks associated with the remaining hazards by developing protective measures.

These actions are intended to minimise risks, in other words, to achieve the highest possible level of safety and ergonomics of the machinery at all stages of its working life, from construction, installation, and exploitation until decommissioning and disposal. They should also ensure that the machinery performs its functions optimally within the scope defined by the designer. This means it performs its functions according to the machine's intended purpose and considering the different operating modes, phases of use, and intervention procedures for the operators. However, the reasonably foreseeable misuse of the machine also has to be taken into account.

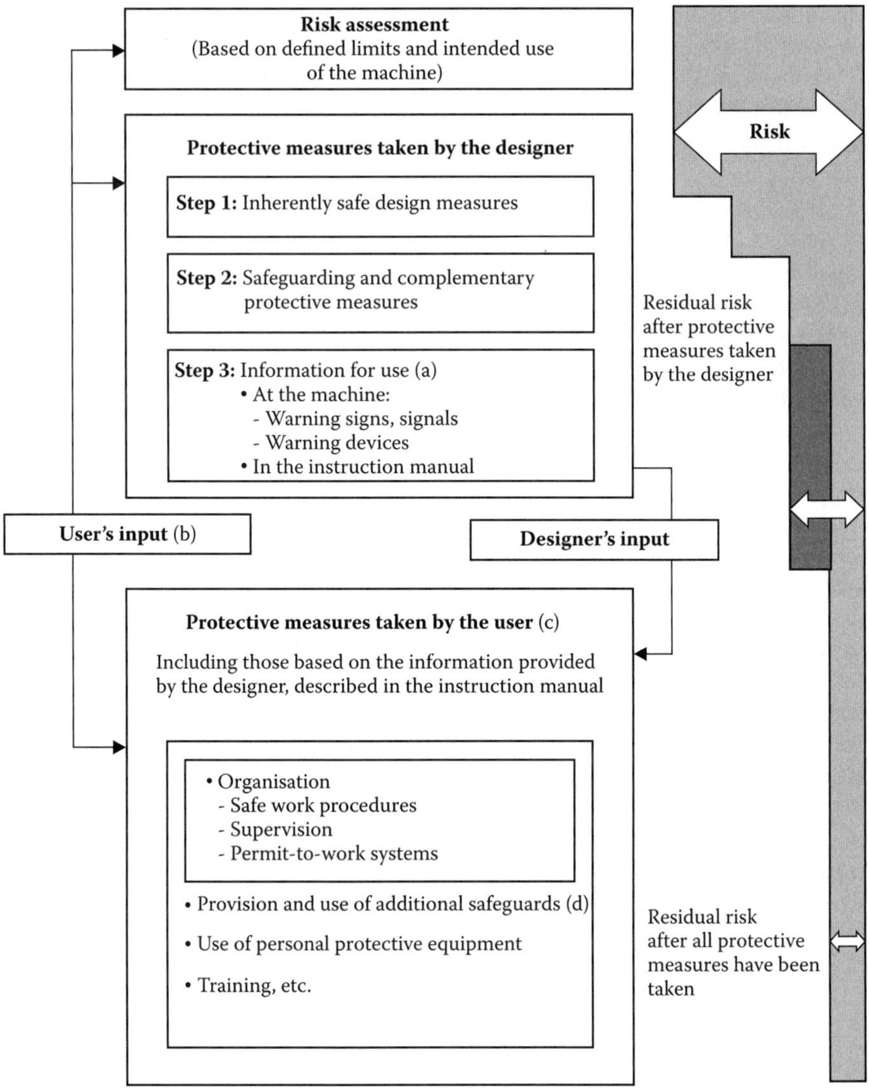

FIGURE 27.2 Process of reducing risk associated with the machinery: (a) the preparation of proper information on exploitation is the designer's input into reduction of risk, but the suggested protective measures will be effective only if the user implements them; (b) the user's input is in the form of information regarding the intended use, passed to the designer by groups of users or by a specific user; (c) there is no hierarchy as to protective measures implemented by the user; and (d) the protective measures required due to specific process (processes), not envisaged in the machine's intended use, or due to specific installation conditions on which the designer has no influence. (Reprinted with permission from PN-EN ISO 12100-1. 2005. Safety of machinery—Basic concepts, general principles for design. Part 1: Basic terminology, methodology.)

Assumed limits also have to be taken into account:

- *Spatial limits*: The scope of the machinery's movements, space requirements for its installation, operation, and interactions in the man–machine–environment system, including the characteristics of foreseen operator population
- *Temporal limits*: The expected life of the machinery and/or its elements that undergo wear and tear in their intended use

The reasonably foreseeable misuse of the machinery may result from events such as:

- The operator losing control over the machine—this applies specifically to hand-held or mobile machines
- Reflex actions of the operator if disruptions occur in the machinery's normal operation
- Behaviours arising from inattention or lack of concentration, from choosing the easiest path when performing a task, or from pressure to keep the machine working regardless of circumstances
- Behaviours of immature persons

The design and manufacture of the machinery should attempt to prevent nonstandard use if such use would likely cause significant risk. At a minimum, the user's manual should inform the user of prohibited uses.

The following items must be considered in the design and manufacture of machinery:

- Limitations resulting from necessary or foreseen use of personal protective equipment, such as boots or gloves
- Principles of ergonomics, in order to minimise discomfort as well as physical and mental work load on the operator associated with the use of the machinery

Effective elimination of hazards and reduction of occupational risk is ensured by the application—especially at the design phase—of the following actions, known as the *triad of safety*:

1. Avoiding hazards by applying inherently safe design measures, for example, selecting appropriate construction features for the machine (Chapter 18, *Mechanical Hazards*), and allocating tasks properly between the man and machine.
2. Safeguarding to protect persons from hazards that cannot be eliminated (e.g. with guards, electrosensitive protective equipment, local exhaust ventilation with filtration) and instituting complementary protective measures (e.g. emergency stops, measures for isolation, and energy dissipation).

3. Providing information on hazards that cannot be eliminated and the associated risks:
   a. Cautionary information on the machine (e.g. pictograms, safety colour codes and safety signs, written warnings)
   b. In the machinery's instruction handbook prepared by the manufacturer

These actions should be performed in the order in which they are listed above.

Information for the user should not be considered a measure that replaces the inherently safe design of machinery or safeguarding. The user should take the protective measures recommended by the machinery's designer, including from the information contained in the instruction manual. This information is the basis for the user's organisational activities, which include:

- Providing safe working procedures in workstation instructions; permit-to-work systems; supervision of adherence to safety rules; inspection of used equipment according to directive 89/655/EEC (1989).
- Use of personal protective equipment.
- Training.

If the designer and producer have not considered the actual conditions under which the machinery is installed and used, the user should take adequate additional precautions.

27.6 ESSENTIAL HEALTH AND SAFETY REQUIREMENTS REGARDING MACHINERY

The essential requirements for designers and manufacturers of machinery and safety components placed on the market or put directly into service within the EEA, including machinery manufactured in countries of the EU for their own use, are set forth in directive 2006/42/EC (2006), referred to as the machinery directive (MD). Provisions of directive 2006/42/EC cover:

- Essential health and safety requirements pertaining to the design and manufacture of machinery and partly completed machinery, safety components, interchangeable equipment, lifting accessories, chains, ropes and webbing, and removable mechanical transmission devices
- Procedures for assessing a machine's conformity with these requirements
- The manner for marking and form of the CE marking
- Minimal requirements regarding notified bodies

Aside from the provisions of the MD, the machinery, depending on its construction and intended use, can be also regulated by provisions of the other directives, for example:

- 2004/108/EC (2004), on electromagnetic compatibility
- 94/9/EC (1994), on equipment intended for use in potentially explosive atmospheres (ATEX)

- 2000/14/EC (2000), on noise emission in the environment by equipment for use outdoors (NOISA)

27.6.1 Structure of Essential Health and Safety Requirements

The essential requirements contained in the MD include the following (Figure 27.3):

- Essential requirements for all machinery and safety components (ES)
- Supplementary essential requirements for the following machinery:
  - Machinery for foodstuffs and cosmetics or pharmaceutical products (F)
  - Portable, hand-held and/or hand-guided machinery (H)
  - Portable fixing and other impact machinery (I)
  - Machinery for woodworking and working materials with similar physical characteristics, such as cork, bone, hardened rubber, hardened plastics, and similar rigid materials (W)
- Supplementary essential health and safety requirements for machinery presenting specific hazards
  - Machinery presenting hazards due to its mobility
  - Machinery for lifting of unit loads
  - Machinery intended for underground work
  - Machinery for lifting of persons

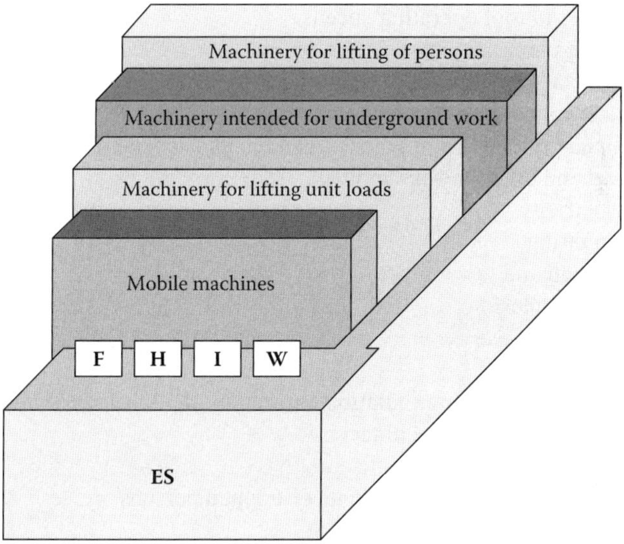

FIGURE 27.3 Structure of the essential health and safety requirements according to directive 2006/42/EC. ES = essential requirements for all machinery. Supplementary requirements are required for machinery for: F = foodstuffs; H = portable hand-held and/or hand-guided machinery; I = portable fixing and other impact machinery; and W = woodworking machinery. (Adapted with permission from Gierasimiuk, J. 2004. *Occupational Safety* 3:7–12.)

The essential health and safety requirements for all machinery according to directive 2006/42/EC (2006) cover the following matters:

- Principles of safety integration as discussed in Section 27.5
- Materials used for constructing machinery and the products used and created in the course of their usage
- Local lighting
- Design of machinery to facilitate its handling
- Ergonomics—discomfort, fatigue, physical, and psychological stress
- Operating position
- Seat for the operator
- Control systems and control actuators
    - Safety, reliability, construction, placement, and operation of the control systems and control actuators
    - Starting and stopping (normal or operational and emergency)
    - Selection of operating mode
    - Failures of power supply and control systems
- Protection against mechanical hazards
    - Loss of stability
    - Breaking up (e.g. of cables, pipes, linkages, moving elements)
    - Falling and ejected objects
    - Surfaces, edges, and angles (rough, sharp, etc.)
    - Multitasking of the machinery (i.e. combined machinery)
    - Changes in tool operating speeds
    - Moving parts
    - Improper selection of protective equipment
    - Uncontrolled movements
- Guards and protective equipment—general and special requirements
- Electrical and other power supply
- Static electricity
- Errors in fitting
- Extreme high and low temperatures
- Fire and/or explosion
- Noise
- Vibrations
- Ionising and nonionising radiation emitted by the machinery and external radiation affecting the machinery
- Emission of dusts and gases
- Measures for the escape and rescue of trapped persons
- Lightning
- Maintenance
    - General requirements
    - Access to operating points and servicing points
    - Isolation of energy sources
    - Other types of operator intervention
    - Cleaning of internal parts and interior of the machinery

- Means and elements of information
    - Signals and warning devices
    - Signs (pictograms), written warnings
    - Markings
    - Information for use contained in the instruction manual provided with the machinery

27.6.2 Assessment of Machinery Conformity with Essential Health and Safety Requirements

The manufacturer or his or her authorised representative who places machinery on the market of the EEA or puts it directly into service must ensure and document conformity with essential health and safety requirements per the MD and all other New Approach Directives applicable to the machinery. The manufacturer or his or her authorised representative may independently assess the conformity of machinery not referred to in Annex IV to directive 2006/42/EC (2006). This does not preclude potential consultations and cooperation with external bodies, especially if they are notified or accredited.

Assessment of conformity according to the MD requires:

1. For machinery referred to in Annex IV to the MD
    a. That is not designed and manufactured in accordance with harmonised standards, or only partly in accordance with such standards, or if harmonised standards do not cover all relevant essential health and safety requirements or if no harmonised standards exist for the machinery in question, the manufacturer or his or her authorised representative shall apply one of the following procedures:
        i. The EC-type examination procedure provided for in Annex IX and the internal checks on the manufacture of the machinery provided for in Annex VIII; if the type satisfies the provisions of the MD, the notified body shall issue an EC-type examination certificate.
        ii. The full quality assurance procedure provided for in Annex X; the manufacturer must have a full quality assurance system that is assessed and approved by a notified body
    b. That is designed and manufactured in accordance with the harmonised standards, provided that those standards cover all relevant essential health and safety requirements, the manufacturer or his or her authorised representative may apply:
        i. One of the above-mentioned procedures
        ii. The procedure for assessment of conformity involving internal checks on the manufacture of machinery, provided for in Annex VIII
2. For machinery not referred to in Annex IV, the procedure for assessment of conformity with internal checks on the manufacture of machinery, provided for in Annex VIII, is applied. However, the alternate procedure may be applied at the request of the manufacturer or his or her authorised representative.

Before placing machinery on the market and/or putting it into service, the manufacturer or his or her authorised representative will ensure that the technical file referred to in Annex VII is available. The technical file shall contain the following:

- A general description of the machinery
- An overall drawing of the machinery and diagrams of the control circuits, as well as pertinent descriptions and explanations necessary to understand the operation of the machinery
- Full and detailed drawings, accompanied by calculation notes, test results, certificates, and other information necessary to check the machine's conformity with essential health and safety requirements
- Documentation of risk assessment along with a description of the essential requirements that apply to the machinery, a description of protective measures employed to eliminate identified hazards or reduce the risks associated with the machinery, and an identification of residual risks
- A list of applied standards or other technical specifications indicating the essential health and safety requirements covered by those standards
- Any technical reports containing results of tests conducted by the manufacturer or by a body selected by the manufacturer or his or her authorised representative
- A copy of the instructions for the machinery
- Declaration of incorporation of partly completed machinery and appropriate instructions for the assembly of such machinery, if applicable
- Copies of the EC declarations of conformity for the machinery and other products incorporated into the machinery, if applicable

For series-produced machinery, steps must be taken to ensure the machinery's conformity with essential requirements. The scope of the required technical documentation for partly completed machinery is set forth in Annex VII, item B, of directive 2006/42/EC (2006).

Before placing the machinery on the market and/or putting it into service, the manufacturer or his or her authorised representative must draw up an EC declaration of conformity with the essential health and safety requirements for the manufactured machinery, if it satisfies the requirements. Next, he or she affixes the CE marking according to Annex III.

27.7 MINIMUM HEALTH AND SAFETY REQUIREMENTS FOR THE USE OF WORK EQUIPMENT AT THE WORKPLACE

27.7.1 Scope and Structure

The minimum health and safety requirements for the use of work equipment, and equipment for temporary work at heights are set forth in the provisions of directives 89/655/EEC (1989), 95/63/EC (1995) and 2001/45/EC (2001). These directives define the minimum technical and organisational requirements for all used work equipment and requirements related to the specific features of the following:

- Mobile work equipment
- Fork-lift trucks
- Work equipment for lifting loads and workers
- Ladders, scaffolding, and ropes for temporary work at heights

The minimum technical requirements for all types of work equipment apply to the following:

- Control actuators and systems, for example, for starting and stopping both normal and emergency isolation from all energy sources
- Protection against hazards such as emission or ejection of substances, materials or objects, emission of gases, liquids and dusts, moving parts, fire and explosion, and electrical current
- Guards and protective devices
- Access to all the areas necessary for production, adjustment, and maintenance operations
- Markings, information and signalling—explicit, visible, and clear

These requirements state, among others, that to protect against hazards the user must:

- Prevent or limit access to moving parts in the work equipment area.
- Prevent access to moving transmission parts.
- Use appropriate safeguards if the work equipment generates a hazard of falling or ejecting objects.
- Provide the work equipment with appropriate containment and/or extraction devices located near the sources of the hazard of emission of gases, fumes, liquid, or dust.
- Assure the stability of the work equipment, if necessary by use of appropriate clamping or some other means.
- Apply appropriate safeguards if there is a risk of rupture or disintegration of a part of the work equipment.
- Provide appropriate lighting for the workstations and maintenance points.
- Protect work equipment parts at high or very low temperatures to avoid the risk of workers coming into contact with these parts or coming to close to them.
- Prevent the risk of fire and explosion.
- Protect against electric shocks.

The minimum organisational requirements are:

- Ensure appropriate dimensions of passages between the pieces of work equipment and between the work equipment and fixed room elements; ensure means of access to maintenance or service points based on the

anthropometric body measurements of the operators and the type of operations performed and tools used, in order to reduce the risk to an acceptable level.
- Set rules and criteria for selecting accessories, for example, for lifting loads.
- Set conditions for performing operations in the process of machinery usage.

The provisions of these directives, while identifying appropriate actions, allow also for replacement actions and define the conditions that must be observed in such cases, such as:

- Loads may not be moved above unprotected workplaces usually occupied by workers. If, however, work cannot be carried out properly any other way, the employer should ensure that appropriate procedures are applied.
- Workers may be lifted only by means of work equipment and accessories provided for this particular purpose. In exceptional cases, work equipment that is not specifically designed for the purpose of lifting persons may be used for this, provided the employer provides safe conditions for the operation and supervision of this equipment.

27.7.2 Obligations of Employers

The employer should primarily ensure that the work equipment used by the workers

- Is appropriate for the given work or is properly adapted.
- Is adapted to the environment in which it shall be used.
- Assures the lowest acceptable risk level.

In order to maintain the work equipment's conformity with health and safety requirements, the current legislation obliges the employer to:

- Take the measures necessary to ensure that, throughout its working life, work equipment is kept, by means of adequate maintenance, at a level that complies with the provisions of the relevant regulations.
- Ensure the inspection of work equipment:
    - Initially—after installation, but before being put into service
    - After assembly at a new site or in a new location, to ensure that it is installed correctly and works properly, where the safety of work equipment depends on the installation conditions
    - Periodically, with testing when appropriate
    - Special inspections each time there are exceptional circumstances that may jeopardise the safety of the work equipment, such as modification work, an accident, natural phenomena or prolonged periods of inactivity; this is to ensure that health and safety conditions are maintained and that deterioration can be detected and remedied

Inspections and/or tests should be carried out by bodies acting based on the regulations, or by persons authorised by the employer and possessing the relevant qualifications. Persons with appropriate qualifications, who the employer authorises to carry out the inspections and/or tests, should be legal or natural persons competent to perform the tasks required for the specific work equipment. These persons should, above all, have knowledge of the structure of the work equipment to be inspected and/or tested, the scope of the inspection and/or test, the manner in which it is performed, and the criteria for evaluating its results. These persons can be hired by the employer—in which case verifying their qualifications to competently perform the inspection and/or test is easiest. The inspection and/or test results should be recorded and kept for a 5-year period. They must be kept at the disposal of the concerned authorities. The results of the most recent inspection or test should physically accompany the machinery if it is used outside the enterprise.

The employer must also:

- Cooperate with the workers in order to ensure safety when using the work equipment ensure that the workers understand the safety rules during all actions involving the use of the work equipment. It is vital that the employer take the measures necessary to ensure that workers have adequate information at their disposal, and when necessary, issue written instructions on the use of the work equipment. The minimum requirements state that the instructions must contain at least adequate safety and health information concerning:
  - Conditions of use of the work equipment
  - Foreseeable abnormal situations
  - Good practices in using work equipment
- Inform the workers of hazards generated by the work equipment located at adjacent workstations, and of any changes that can affect safety.
- Ensure that workers receive training on any risks that the use of such equipment may entail, including abnormal situations; workers who perform repairs, modification, maintenance, or service should also receive adequate training.
- Consult workers or their representatives on all matters relating to health protection and safety, especially matters involving introduction of new technologies and selection of work equipment.

27.8 ADAPTING WORK EQUIPMENT TO MINIMUM HEALTH AND SAFETY REQUIREMENTS

27.8.1 Description of the Most Frequently Found Deficiencies

Polish experiences adapting work equipment to the minimum health and safety requirements led to the identification of some frequent deficiencies. Employers evaluating their work equipment should pay special attention to these deficiencies;

the most frequent are related to hazards due to moving elements and are caused by the following:

- Lack of guards and protective devices (Figure 27.4)
- Only partial restriction of access to the danger zone (e.g. lack of side guards when light curtains are used)
- Nonfulfilment of protective functions by guards and protective devices (e.g. placement of guards, especially open-work ones, two-hand control devices, and other protective devices at distances that do not guarantee safety, that is, are not in accordance with standards PN-EN ISO 13857 (2008) and PN-EN 999+A1 (2008); Figure 27.5)
- Interlocking devices that can be easily defeated
- Ineffective inspection of protective devices monitoring access to hazard zones, such as light curtains

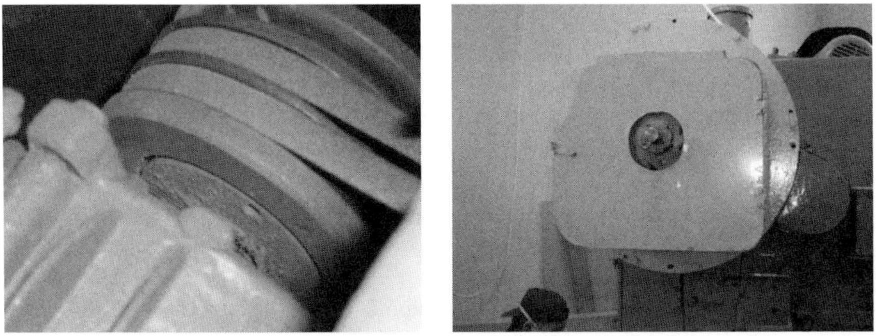

FIGURE 27.4 Lack of protection against rotating parts.

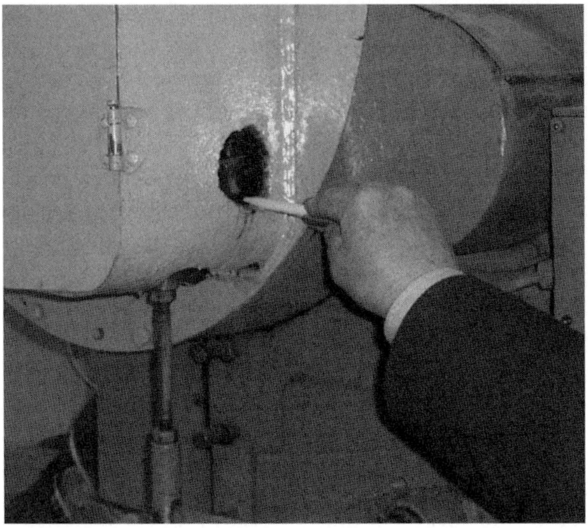

FIGURE 27.5 Requirements for safety distances are not satisfied.

Deficiencies involving identification of control actuators refer primarily to the following:

- Illegible markings
- Unclear information or information in foreign languages
- Lack of markings
- Incorrect colours or placement of the control actuators

Typical deficiencies in control systems include the following:

- Lack of transformers
- Lack of emergency stop devices
- Insufficient reliability of parts

Deficiencies relating to isolation from energy sources refer primarily to the following:

- Lack of ability to lock the isolating device in the isolating position
- Lack of compliance with the requirement for an isolation feature in the device itself
- Lack of an isolator for isolating the equipment from pneumatic and hydraulic energy sources

The most frequent deficiencies in terms of protection against electric shock include (Figure 27.6) the following:

- Insufficient protection provided by housing
- Insufficient identification of electrical equipment
- Lack of measures to prevent electrical equipment from being accessed by unauthorised persons

27.8.2 Procedure for Adapting Used Work Equipment to Minimum Health and Safety Requirements

Experience shows that employers usually follow the procedure presented in Figure 27.7 when adapting used work equipment to the minimum health and safety requirements. The review is intended to collect overall information on the kinds and types of work equipment, its current technical condition and its technological application. This information can be the basis to draw up a list of obsolete work equipment designated for decommissioning.

The general assessment identifies deviations from the minimum health and safety requirements. The user can then determine the necessary changes and the estimated costs and time necessary to introduce these changes. *Work equipment* with the CE marking or the national safety markings (in Poland, the B marking) and that has the

FIGURE 27.6 Lack of protection against electrocution.

required documentation attesting to its conformity should comply with the minimum requirements if no changes were made and if no hazards were found during the inspection.

The results of the review and general assessment are the basis to draw up lists of equipment:

- That complies with the minimum requirements
- That is designated for decommissioning for technological reasons and/or due to a lack of profitability in adapting it to the requirements
- That will be adapted to these requirements

Work equipment placed on the last of these lists must be analysed in detail using checklists specifying the minimum requirements and the requirements that take into account specific features of the machinery contained in the detailed regulations and related standards. The actual condition and the deficiencies and their locations in the machinery are assessed based on these checklists. The assessment results are documented with photographs. The actions and measures necessary to assure the machinery's conformity with the minimum requirements, especially of a technical nature, are determined with respect to the assessed risk associated with the hazards that could result from the revealed deficiencies. These actions and measures are included in the adaptation plan.

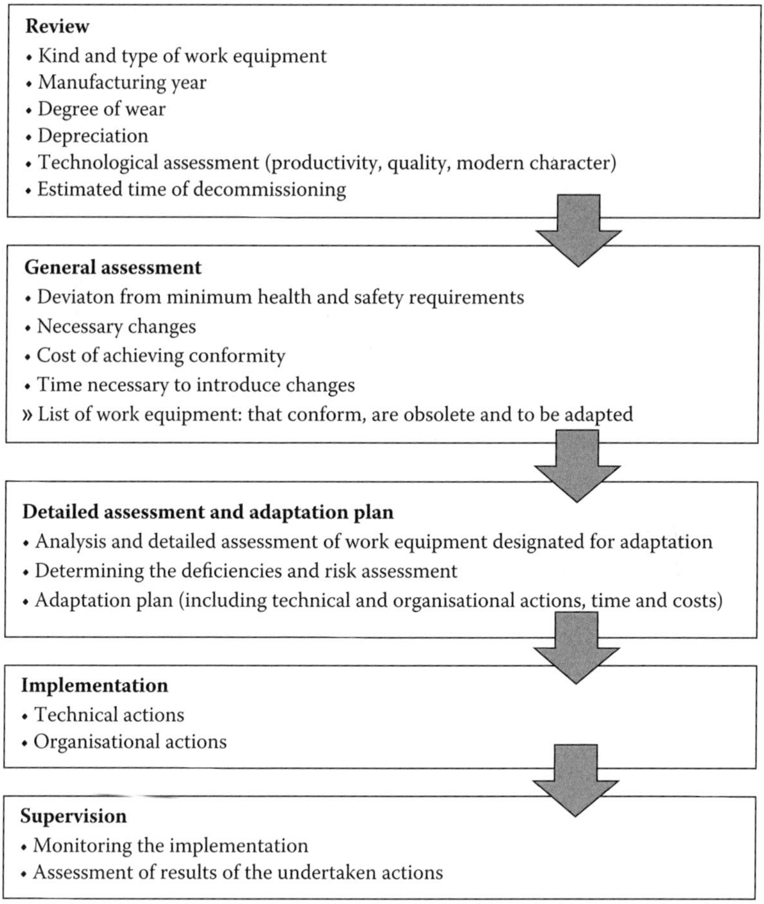
Review
• Kind and type of work equipment
• Manufacturing year
• Degree of wear
• Depreciation
• Technological assessment (productivity, quality, modern character)
• Estimated time of decommissioning
General assessment
• Deviaton from minimum health and safety requirements
• Necessary changes
• Cost of achieving conformity
• Time necessary to introduce changes
» List of work equipment: that conform, are obsolete and to be adapted
Detailed assessment and adaptation plan
• Analysis and detailed assessment of work equipment designated for adaptation
• Determining the deficiencies and risk assessment
• Adaptation plan (including technical and organisational actions, time and costs)
Implementation
• Technical actions
• Organisational actions
Supervision
• Monitoring the implementation
• Assessment of results of the undertaken actions

FIGURE 27.7 Diagram of the procedure for adapting used work equipment to minimum health and safety requirements. (Adapted with permission from Dąbrowski, M., and J. Gierasimiuk. 2002. *Occupational Safety* 7–8:30–32.)

Harmonised standards, especially of the C type, can be also used for assessment and planning actions. These standards specify the hazards generated by given, small groups of machinery and their locations, as well as requirements and solutions. Solutions proposed under the harmonised standards assure the highest possible level of safety complying with the essential requirements. When applying them to modify a control system or protective device, conformity with the harmonised standards must be ensured. The last task is to implement the plan successfully and assess the results from the standpoint of fulfilment of the minimum requirements.

The rules do not specify the document to be used to confirm that the adaptation of the used work equipment conforms to the minimum health and safety requirements, leaving this decision to the employer. Polish employers usually use documents developed in the course of the adaptation process, such as checklists and adaptation plans confirming positive assessment of the actions undertaken, or introduce an internal document of conformity.

27.9 REQUIREMENTS FOR MANUFACTURERS AND OTHER SUPPLIERS OF MACHINERY

Manufacturers and other suppliers of any machinery (including used machinery) from countries that do not belong to the EEA, as well as of new and modified machinery in the EEA, must ensure the machinery complies with the provisions of the MD and other related New Approach Directives (legislation implementing these directives into the national legal system) and that the machinery is delivered along with the following:

- Full markings (including CE markings)
- Machinery instruction manuals
- Required special equipment and accessories that enable the adjustment, maintenance and use of the machinery without creating hazards
- The EC declaration of conformity

Markings on the machinery, according to directive 2006/42/EC (2006), should be visible, legible and durable and should contain at least the name and address of the manufacturer or his or her authorised representative, identification of machinery by name or symbol, the CE markings, identification of the machinery's series or type, a serial number, the manufacturing year, special markings regarding conditions for use (e.g. explosion-proof), and information necessary for safe use of machinery (e.g. the maximum speed of rotating parts, largest diameter of tools, weight, rated load).

The instruction manuals that should accompany the machinery should contain at least:

- The name and full address of the manufacturer or his or her authorised representative
- The designation of the machinery as marked on the machinery itself, with the exception of the serial number
- The EC declaration of conformity or a document presenting the contents of the EC declaration of conformity, identifying detailed data on the machinery, with serial number and signature optional
- A general description of the machinery
- Drawings, diagrams, descriptions, and explanations necessary for the use, maintenance, and repair of the machine and for checking its correct functioning
- A description of workstation or stations likely to be occupied by operators of the machinery
- A description of the intended use of the machinery
- Warnings about ways in which the machinery must not be used that experience has shown might occur

- Instructions for assembly, installation, and connection, including drawings, diagrams, and the means of attachment of the chassis or installation on which the machinery is intended to be mounted
- Instructions relating to installation and assembly for reducing noise or vibration
- Instructions for the service and use of machinery and, if necessary, instructions for operator training
- Information on the residual risks that remain despite inherently safe design measures, safeguarding, and complementary protective measures
- Information on the protective measures to be taken by the user, including, where appropriate, personal protective equipment to be provided
- The essential characteristics of tools that may be fitted to the machinery
- The conditions in which the machinery meets the requirements for stability during use, transportation, assembly, dismantling, testing, and/or foreseeable breakdowns
- Information on safe conditions for transport, moving, and storage, specifying the weight of the machinery and of its parts (if there is a need for transport in parts)
- The procedure to be followed in the event of accident or failure; if a blockage is likely to occur, the operating method to be followed in such an event that enables the machinery to be safely unblocked
- A description of the proper adjustment, preventive, and maintenance operations
- Instructions for safe adjustment and maintenance, including protective measures that should be taken during these operations
- The specifications of the spare parts to be used when this affects the health and safety of the operator
- The following information on airborne noise emissions:
  - The A-weighted emission sound pressure level at workstations where this exceeds 70 dB(A); where this level does not exceed 70 dB(A), this fact must be indicated
  - The peak C-weighted instantaneous sound pressure value at workstations where this exceeds 63 Pa (130 dB in relation to 20 μPa)
  - The A-weighted sound power level emitted by the machinery where the A-weighted emission sound pressure level exceeds 80 dB(A)
- Information on the radiation emitted by the machinery, especially nonionising radiation that may cause harm to the operator and other exposed persons, specifically if they have implanted active or nonactive medical devices

The above information should be supplemented with typical data for the various groups of machinery, for example, vibration parameters for portable, hand-held, and hand-guided machinery, or the maximum working load, for example, by means of load tables for hoisting equipment. Instructions should be provided in at least one official language of the EU. Such language version or versions, verified by the manufacturer or his or her authorised representative, should be marked with the inscription 'original instructions'. If the original instruction does not exist in the official language or languages of the country where the machinery will be used, the manufacturer, his or her authorised representative, or the person placing the machinery into the market of the given area should provide a translation. The translation should be marked with the inscription 'translation of original instructions'. The machine should be provided with the original instructions and, if applicable, the translation of the original instructions.

In justified cases, the machinery's maintenance instruction, to be used by specialised personnel employed by the manufacturer or his or her authorised representative, may be provided only in the language that this personnel uses. The instructions should be written in a clear, transparent manner that is easy for all users to read.

The EC declaration of conformity for machinery should contain:

- The name and full address of the manufacturer and/or his or her authorised representative, if applicable
- The name and address of the person residing within a member country of the EU, who is authorised to compile technical documentation
- A description and identification of the machinery, including a general description, function, model, type, serial number, and commercial name
- A statement that the machinery complies with all relevant provisions of the MD (2006/42/EC) and the provisions of all other directives, if applicable; all must reference regulations published in the *Official Journal of the European Union*
- The name, address, and ID number of the notified body that conducted the EC-type examination and the certificate number of the EC-type examination if applicable
- The name, address, and ID number of the notified body that approved the full quality assurance system, if applicable
- References to the relevant harmonised standards, if applicable
- References to other applied standards and technical specifications, if applicable
- Location and date the declaration was drafted
- The first and last names and signature of the person authorised to draft the declaration on behalf of the manufacturer or his or her authorised representative

The EC declaration of conformity refers to the machinery solely in the condition in which it was placed on the market or put into service. It does not cover parts added by a user or actions undertaken by him or her later. The EC declaration of conformity for machinery should be drafted in the same language as the machinery instruction manual. The EC declaration of conformity and its translations should be typewritten or handwritten in block letters.

Second-hand machinery coming from EU countries that will be placed on the market or put into service should comply with the minimum technical requirements stemming from national legislation implementing directives 89/655/EEC (1989), 95/63/EC (1995) and 2001/45/EC (2001).

The obligations of the machinery user are to:

- Adapt the machinery used before it can adhere to the minimum technical requirements stemming from the above-named legislation.
- Maintain the health and safety levels ensured by the machinery at the time it was put into service by complying with the remaining minimum requirements contained in the legislation.

REFERENCES

Council Directive 2001/45/EC of 27 June 2001 (amending Council Directive 89/655/EEC) concerning the minimum safety and health requirements for the use of work equipment by workers at work. OJ L 195, 19.7.2001.

Council Directive 2004/108/EC of 15 December 2004 relating to electromagnetic compatibility and repealing directive 89/336/EEC. OJ L 162, 3.7.2000.

Council Directive 89/655/EEC of 30 November 1989 concerning the minimum safety and health requirements for the use of work equipment by workers at work. OJ L 393, 13.12.1989.

Council Directive 94/9/EC of 23 March 1994 concerning equipment and protective systems intended for use in potentially explosive atmospheres. OJ L 100, 19.4.1994.

Council Directive 95/63/EC of 5 December 1995 amending Directive 89/655/EEC concerning the minimum safety and health requirements for the use of work equipment by workers at work. OJ L 335, 30.12.1995.

Dąbrowski, M., and J. Gierasimiuk. 2002. Adaptation of machinery and other work equipment used by workers at work to conform with safety requirements of directives. *Occupational Safety* 7–8:30–32.

Directive 2000/14/EC of the European Parliament and of the Council of 8 May 2000, on the approximation of the laws of the Member States relating to the noise emission in the environment by equipment for use outdoors. OJ L 162, 3.7.2000. (Amended by Directive 2005/88/EC—OJ L 344, 27.12.2005).

Directive 2006/42/EC of the European Parliament and of the Council of 17 May 2006 on machinery, and amending Directive 95/16/EC. OJ L 157, 26, 9.06.2006.

Gedliczka, A., et al. 2001. *Atlas of Human Body Measurements: Data for Design and Evaluation in Terms of Ergonomics.* Warsaw: CIOP.

Gierasimiuk, J. 2004. Essential requirements with the mode and procedures for conformity assessment of machines and safety components. *Occupational Safety* 3:7–12.

PN-EN 1088+A2. 2008. Safety of machinery: Interlocking devices associated with guards—principles for design and selection.

PN-EN 12622. 2004. Hydraulic press brakes—safety.

PN-EN 12717+A1. 2009. Safety of machine tools—Drilling machines.

PN-EN 1760-1+A1. 2009. Safety of machinery, pressure sensitive protective devices, Part 1: General principles for design and testing of pressure sensitive mats and pressure sensitive floors.

PN-EN 1760-2+A1. 2009. Safety of machinery, pressure sensitive protective devices, Part 2: General principles for the design and testing of pressure sensitive edges and pressure sensitive bars.

PN-EN 349+A1. 2008. Safety of machinery—Minimum gaps to avoid crushing of parts of the human body.

PN-EN 574+A1. 2008. Safety of machinery—Two-hand control devices—Functional aspects—Principles for design.

PN-EN 614-1+A1. 2009. Safety of machinery—Ergonomic design principles, Part 1: Terminology and general principles.

PN-EN 614-2+A1. 2009. Safety of machinery—Ergonomic design principles, Part 2: Interactions between the design of machinery and work tasks.
PN-EN 953+A1. 2009. Safety of machinery—Guards—General requirements for the design and construction of fixed and movable guards.
PN-EN 999+A1. 2008. Safety of machinery—The positioning of protective equipment in respect of approach speeds of parts of the human body.
PN-EN ISO 12100-1. 2005. Safety of machinery—Basic concepts, general principles for design, Part 1: Basic terminology, methodology.
PN-EN ISO 12100-2. 2005. Safety of machinery—Basic concepts, general principles for design, Part 2: Technical principles.
PN-EN ISO 13857. 2008. Safety of machinery—Safety distances to prevent hazard zones being reached by upper and lower limbs.

28 Basic Principles for Protective Equipment Application

Marek Dźwiarek

CONTENTS

28.1 INTRODUCTION

The Industrial Revolution, which took place in the nineteenth century, depended on the development of more effective and all-purpose machines. New sources of power such as steam and electric engines paved the way for more advanced technological procedures. Higher productivity and better work quality were accompanied by new, unknown hazards. Bigger and faster machine elements posed greater hazards. The effective protection of machine operators without affecting their productivity therefore became crucial.

Initially, machine operators were protected by preventing access to dangerous mobile elements, especially in the case of moving power transmission parts. However, protection from working elements has become more complex. For technological reasons, operators often penetrate the hazard zone, at least for a short time, for example, to provide materials or collect the manufactured parts. Therefore, solutions that allow for automatic control of the operator's position relative to hazard zones are necessary. Historically, the first protective equipment consisted of two-hand controls used in presses. In the course of technological development, especially due to computerisation, protective devices have become more effective, provided that they are applied properly.

Due to growing technological needs and the need for higher productivity, the operation of machines involves more monotonous and repetitive tasks. The operator becomes fatigued and is no longer able to concentrate, increasing the probability of errors that can result in a dangerous accident (Dźwiarek 2004, 2007).

28.2 GENERAL CHARACTERISTICS OF INDUSTRIAL PROTECTIVE EQUIPMENT

Protective devices generally perform their functions by generating a signal that stops a dangerous motion of a machine element after an operator or part of his body has been detected to approach a danger zone, ensuring that the dangerous motion (e.g. that of the press slide) is stopped before the operator penetrates the danger zone. These protective devices do not present any physical barrier. Therefore, their use requires the machine design to allow for an automatic stop of a dangerous element within a short time; for example, in hydraulic presses. They are not applicable to presses with a rotating key design, which, despite the stop function, can be stopped only after executing the whole operational cycle or a given part of it. Therefore, the protective devices perform their functions by using the machine control system.

Applying protective devices is a phase involved in the use of a machine that should be considered by both the designer and the user (Dźwiarek 2006). In the scheme presented in Figure 28.1, the chance to use protective equipment occurs at two stages: (1) during the design phase and (2) during its use (since one cannot foresee all the aspects of machine operation at the design phase). When a risk assessment performed by the user after the machine is installed shows a need for additional safety measures, protective devices should be considered. In either case the procedures of installation and periodic checking should be followed.

28.3 DEFINITIONS

Electrosensitive protective equipment (ESPE): An assembly of devices or components working together for protective tripping or presence-sensing purposes and comprised of the following at a minimum:

- Sensing device
- Controlling or monitoring devices
- Output signal switching device

Safety function: The function initiated by the input signal and processed by the safety-related parts of the control system to ensure that the machine (as a system) achieves a safe state or that a machine whose failure can result in an immediate increase of risk(s) continues to function.

Safety-related part of a control system: The part of a control system that responds to safety-related input signals and generates safety-related output signals.

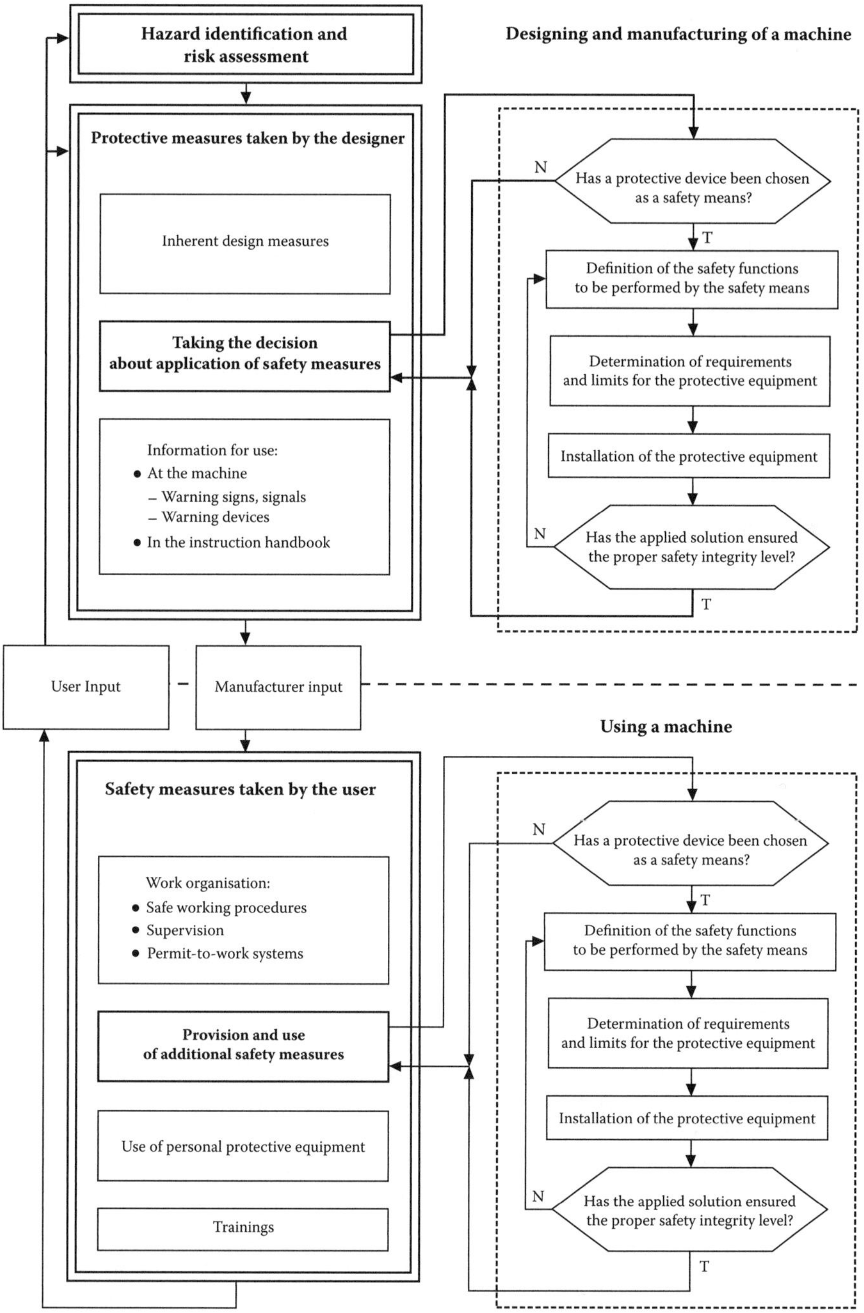

FIGURE 28.1 Scheme of the procedure for reducing the risk involved in using a machine, including the application of protective equipment.

Fault: The inability to perform a required function, excluding inabilities during preventive maintenance or other planned actions or due to lack of external resources. A fault is often the result of a failure of the item itself, but may result even without any prior failure.

Failure: Termination of the ability of an item to perform a required function. 'Failure' is an event, whereas 'fault' is a state.

28.4 CHARACTERISTICS OF PROTECTIVE EQUIPMENT

Protective devices continuously monitor a detection zone, that is, the area where the test piece is detected. Therefore, as each detection zone is penetrated, the output signal of the device changes. Protective devices can de divided into the following three types:

1. Electrosensitive protective equipment
2. Pressure-sensitive protective devices
3. Two-hand control devices

28.4.1 Electrosensitive Protective Equipment

Electrosensitive protective devices are a group of commonly used advanced protective equipment (Dźwiarek 1998). Electrosensitive protective equipment includes systems of devices or components working together for protective tripping or presence-sensing purposes and comprised of, at a minimum, a sensing device, controlling or monitoring devices, output signal switching devices, and secondary switching devices (optional). Their functions are based on a variety of physical phenomena (e.g. electromagnetic radiation within the ranges of microwaves, infrared and visible light, respectively, acoustic waves including ultrasounds, or changes in capacitance and inductance) that indicate the presence of a human or a part of his or her body. Active optoelectronic protective devices are now used and are based on the application of infrared radiation. Figure 28.2 shows most common types of electrosensitive protective devices.

Figure 28.3 presents the principle behind a light curtain operation, which also applies to a majority of electrosensitive protective devices. Both the transmitter and receiver in the device comprise a series of optical channels, the number of which depends on the height of the detection zone. The detection zone is divided into parts corresponding to the respective channels. The height of each part determines the detection capability, which is characteristic for a given device.

In vision systems, the supervised zone is monitored by a CCD camera with a light-sensitive surface divided into pixels. Each pixel monitors its respective part of the detection zone. Image reading consists of reading subsequent pixels (Boemer 2003; Gardeux et al. 2003).

Laser scanners are the most advanced type of contactless protective devices (CPDs). A laser beam scans a detection zone continuously and then measures the reflected beam level. The application of lasers allows not only for the detection of each zone penetration, but also for determination of the place being penetrated. The

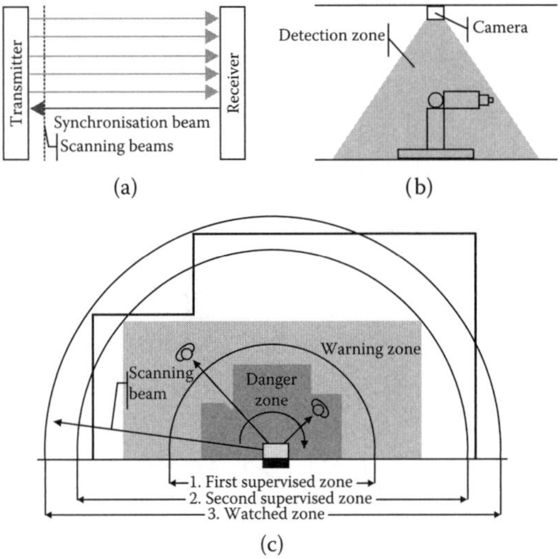

FIGURE 28.2 Sample designs of electrosensitive protective devices: (a) light curtain, (b) vision system, and (c) laser scanner.

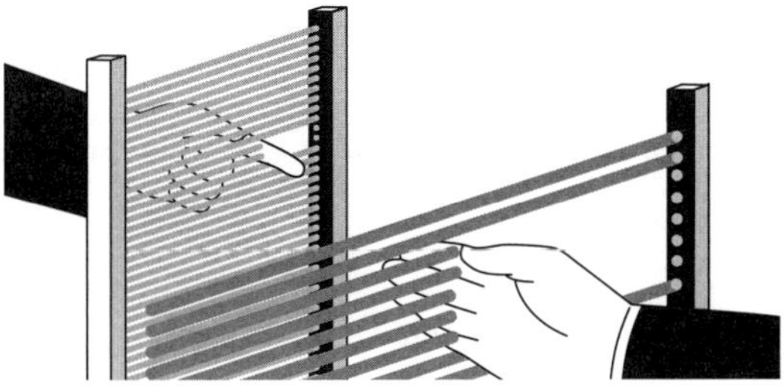

FIGURE 28.3 Principles of light curtain operation.

user can therefore distinguish zones of different hazard levels (usually three zones) within the watch zone, depending on the current workstation configuration.

An automated workstation in which various protective devices have been installed to prevent accidents is shown in Figure 28.4. Some of them are electrosensitive protective devices. Light beam device *A* is situated at the side that has free access to the dangerous zone. It detects the crossing of a dangerous zone border. The light beam device and its response time should be designed in such a way that objects similar to human parts moving at speeds characteristic of humans will be detected. Laser scanner *F* detects the presence of a human within a dangerous zone. It responds when all other protective devices fail. In contrast, light curtain *C* prevents access to the box containing the manufactured elements. In this case, the dangerous zone can be reached by hand, and the curtain detects objects of the diameter and speed similar to a human hand. A single light beam *D* detects the entry of a truck into the work stand to collect the manufactured elements.

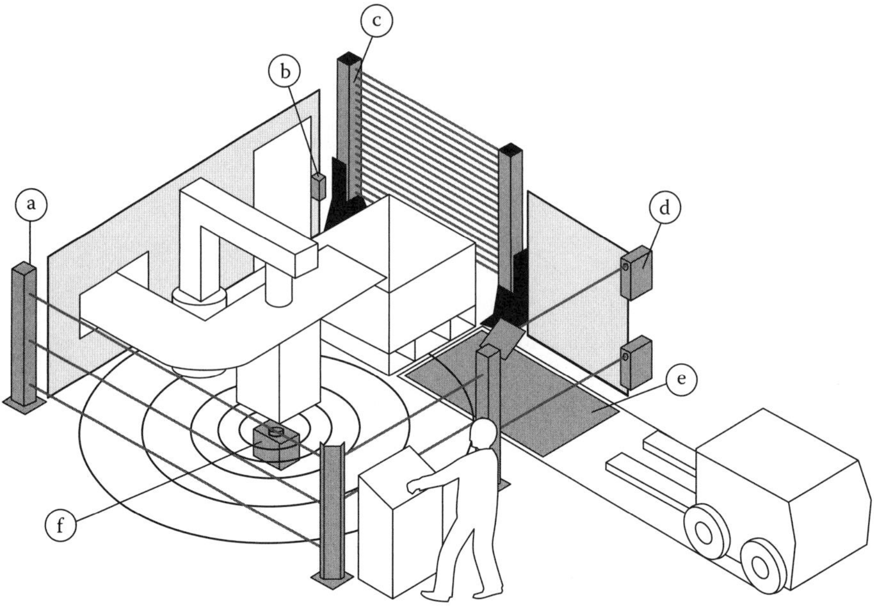

FIGURE 28.4 Automated workstation with protective devices: (a) light beam device that detects crossing of the dangerous zone; (b) interlocking device with key and locking that detects opening of the entrance door; (c) light curtain that detects reaching into the dangerous zone; (d) single beam that detects driving into the dangerous zone; (e) pressure-sensitive mat that detects a presence in the dangerous zone; and (f) laser scanner that detects a presence in the dangerous zone.

28.4.2 Pressure-Sensitive Protective Devices

Pressure-sensitive protective devices detect pressure (force) that can mean the presence of a human or a part of his or her body within the supervised zone. These devices have the following two functional elements (which can form one part or comprise many different parts):

1. *Sensor*: The part of the pressure-sensitive mat or pressure-sensitive floor that contains an effective sensing area on which the application of an actuating force causes the signal from the sensor to the control unit to change state.
2. *Control unit*: The device that responds to the condition of the sensor(s) and controls the state of output switching device (OSSD).

Many types of pressure-sensitive devices are available, for example, mats, floors, edges, bars, bumpers, plates, and wires. Pressure-sensitive protective devices have the following parameters:

- Actuating force
- Effective sensing area

- Direction (angle) of effective actuation
- Pretravel (distance to activation)
- Response time
- Dead zone

28.4.3 Two-Hand Control Devices

Two-hand control devices allow for dangerous motion of a machine only when both hands handle the control elements. Like CPDs, they are usually applied in high-risk machines such as presses, hammers, and bending machines. A two-hand control device is usually made up of two control actuating devices, signal converters, and a signal processor or processors (Figure 28.5).

The basic function of a two-hand control device that determines both its safety and control functionality is simultaneous actuation, that is, generation of an output signal without any breaks (the delay between starting the first input signal and the second one is negligible) when both control elements are actuated. This allows for dangerous motion only when the operator uses both hands to move the control elements. The operator's hands are therefore protected by checking their positions during a dangerous motion.

The effectiveness of two-hand control devices depends how easily they can be defeated (Figure 28.6). The design and installation of a two-hand control device

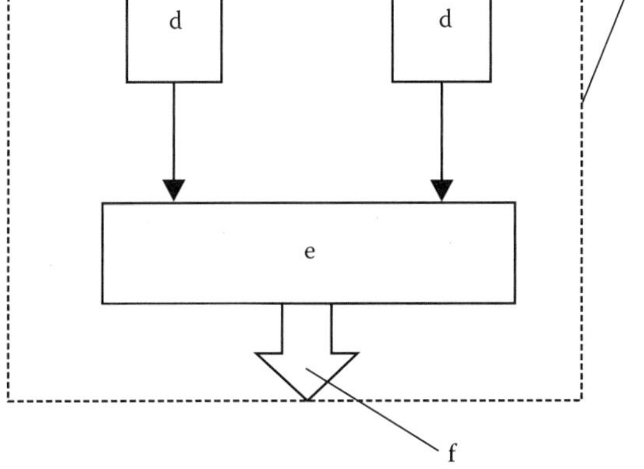

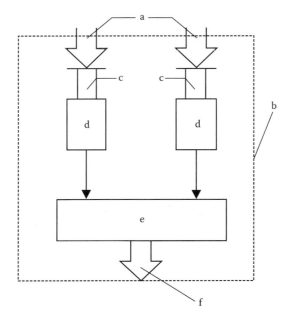

FIGURE 28.5 Functional schematic representation of a two-hand control device: (a) input signals, (b) two-hand control device, (c) control actuating device, (d) signal converter, (e) signal processor, and (f) output signal. (Reprinted with permission from PN-EN 574+A1. 2008. Safety of machinery—Two-hand control devices—Functional aspects—Principles for design.)

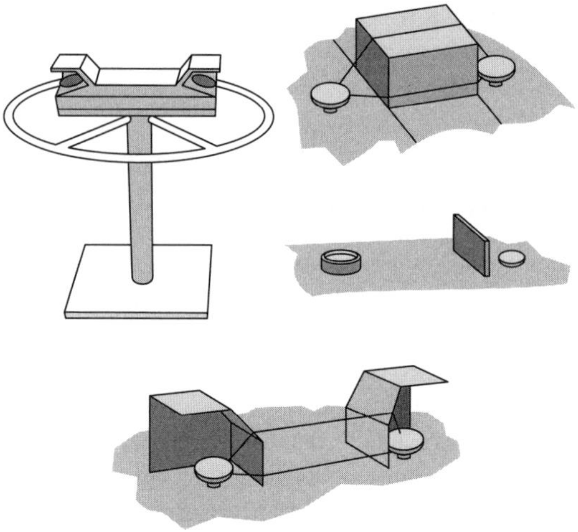

FIGURE 28.6 Sample defeat prevention measures, using barriers making access more difficult and a ring that separates the control buttons. (Adapted with permission from PN-EN 574+A1. 2008. Safety of machinery—Two-hand control devices—Functional aspects—Principles for design.)

should prevent the operator from defeating it by locking one of the control elements or by preventing the entry of the following:

- Fingers of one hand
- Hand and elbow of the same arm
- Forearm(s) or elbow(s)
- One hand and any other part of the body

Two-hand control devices are divided into four types depending on their design and their resistance to faults. The minimum safety requirements and fault resistance categories for different types of two-hand control devices are presented in Table 28.1.

28.5 INSTALLATION OF PROTECTIVE DEVICES

Protective devices belong to the group of tripping devices for technical safety since they supervise the distance between the human and the dangerous zone. For the task to be executed effectively, each penetration of the detection zone should be detected early (Dźwiarek 1996) and the response should be prompt. The selection of safety distance using a light curtain with a press as a test case is shown in Figure 28.7 (Dźwiarek 1998).

The safety distance d can be determined using the following formula:

$$d = v(t_1 + t_2) + e_3$$

where t_1 is the response time of the CPD, t_2 is the time to stop the moving element, e_3 is a constant, which depends on the detection capability of the device, and

TABLE 28.1
Types of Two-Hand Control Devices by Functionality

Minimum Safety Requirements	Type of Two-Hand Control Devices				
	I	II	IIIA	IIIB	IIIC
Use of both hands (simultaneous actuation)	×	×	×	×	×
Relationship between input signals and output signal	×	×	×	×	×
Cessation of the output signal	×	×	×	×	×
Prevention of accidental operation	×	×	×	×	×
Prevention of defeat	×	×	×	×	×
Reinitiation of the output signal	×	×	×	×	×
Synchronous actuation	[a]	×	×	×	
Use of category 1 (EN ISO 13849-1:2006)	×		×		
Use of category 3 (EN ISO 13849-1:2006)		×		×	
Use of category 4 (EN ISO 13849-1:2006)					×

[a] According to the risk assessment results.

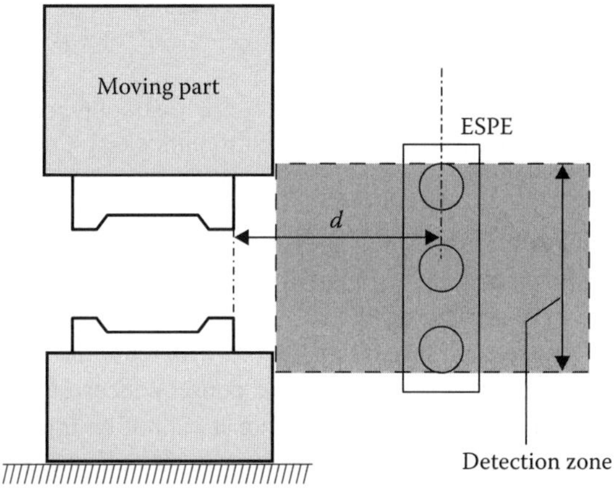

FIGURE 28.7 Choosing a safe distance. (Reprinted with permission from PN-EN ISO 13849-1. 2008. Safety of machinery—Safety-related part control systems—Part 1: General principles for design.)

v is the speed at which a part of the human body moves; for a hand one assumes $v < 2000$ mm/s.

Light curtains with a detection capability P_w less than 16 mm are used to detect a finger penetrating a dangerous zone. The detection capability should not exceed 40 mm to detect a hand. In such a case, one should assume

$$v = 2000 \text{ mm/s}, \quad e_3 = 8(P_w - 14)$$

A light curtain with a detection capability of 40–70 mm detects the arm or trunk of the operator. In such a case, one should assume

$$v = 1600 \text{ mm/s}, \quad e_3 = 850 \text{ mm}$$

28.6 RESISTANCE TO FAULT OF PROTECTIVE DEVICES

The safety level of a system depends on its performance in case of a fault. Disruptions in the safety function, such as the dysfunction of moving elements, may cause an accident. Fault resistance is thus crucial for ensuring safety. The design of a manufacturing system should consider safety issues, and should ensure safety even if a fault has appeared in the control system (Dźwiarek 2000; BGIA 2008). The aforementioned problems arise both during the choice of accident prevention strategy and when defining the ways to maintain assumed safety levels in the workstation. Consider the following examples.

EXAMPLE 1

An interlocking device with locking is mounted in the access door to the working area of an industrial robot. Its basic task is to generate an EMERGENCY STOP signal when the device is unlocked.

EXAMPLE 2

A light curtain has been installed on the border of a working zone. The light curtain stops a dangerous motion of a press slide or interlocks a start signal in case of penetration of the detection zone.

In Example 1, the safety function is used not more than a few times a day. As the interlocking device is usually used along with other protective devices, its fault does not always pose a hazard to the operator. High reliability is therefore not required of the interlocking device, as it might considerably increase the cost. However, even though the probability of the event is low, the consequences of a possible accident might be serious and irreversible, and therefore it cannot be ignored. This problem can be solved by introducing an additional function, the self-checking device, which detects a fault early enough. The term 'early enough' means that the interlocking device operation should be checked by a self-checking device before calling the safety function, that is, before unlocking.

Example 2 presents a completely different situation. In this case, the safety function is used a few times a minute. Any improper operation of the device may immediately cause an accident leading to serious consequences. Therefore, the fault should be detected as soon as possible—a shorter amount of time than the device response time. Not only single faults should be detected, but also their accumulations, which might result in the loss of safety function. As the safety function is required frequently, the reliability of the device is important. Faults appearing too frequently make the work more difficult, resulting in improper actions by the operator and finally in an accident.

Standard ISO 13849-1:2008 is a basic document devoted to the issue of control system resistance to fault. According to the standard, the safety-related elements of machine control systems are classified into five categories (B, 1, 2, 3, and 4), depending on the device's resistance to faults and its performance under fault conditions, determined by the device structure and its reliability regardless of the technology applied.

B is the basic category, which includes devices constructed according to product standard recommendations and a predicted work environment. A fault appearing in devices of this category may lead to a loss of safety function. Category 1 devices have a higher resistance to faults mainly due to selection and application of components. Category 2, 3, and 4 devices achieve higher fault resistance by extension of the device structure. Category 2 does this by periodical safety function checking, and categories 3 and 4 use continuous assurance to ensure that a single fault will not cause safety function loss. In category 3 devices, faults should be detected when it is reasonable, while in category 4, when it is possible. In category 4, resistance to fault accumulation should also be determined.

The capability of a control system to perform a safety function depends on its basic parameters:

- Architecture, defined by the category
- Reliability of the applied subsystems characterised by the middle time to failure (MTTF)
- Diagnostic coverage (DC)

The safety function performance is indicated by the performance level Pl. In standard ISO 13849-1:2008, five performance levels have been introduced from Pl = *a* to Pl = *e*. Figure 28.8 presents the relationship between the performance level and the parameters of the control system.

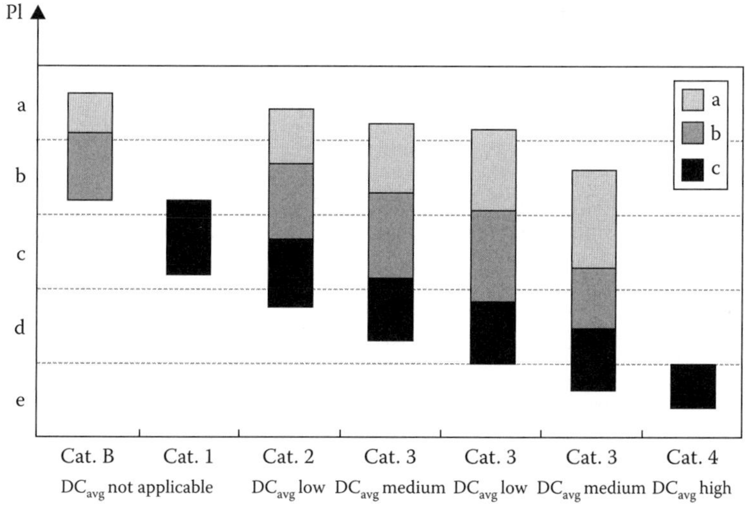

FIGURE 28.8 Relationship between the performance level (Pl) and parameters of the control system: (a) $MTTF_d$ of each channel (low), (b) $MTTF_d$ of each channel (medium), and (c) $MTTF_d$ of each channel (high). (Reprinted with permission from PN-EN ISO 13849-1. 2008. Safety of machinery—Safety-related part control systems—Part 1: General principles for design.)

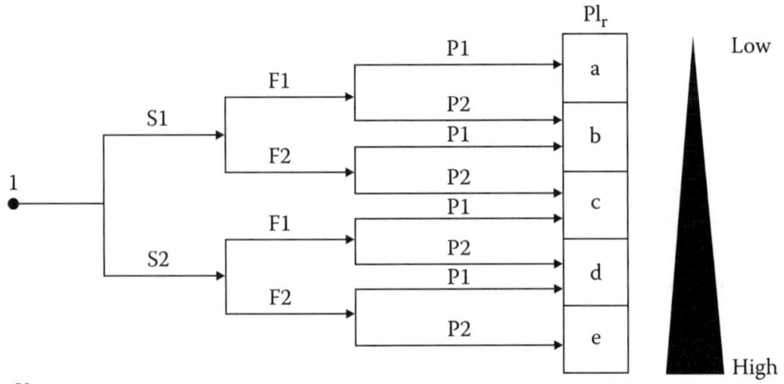

Key:

1 – Starting point for evaluation of safety function's contribution to risk reduction

L – Low contribution to risk reduction

H – High contribution to risk reduction

Pl_r – Required performance level

Risk parameters:

S – Severity of injury

S1 – Slight (normally reversible injury)

S2 – Serious (normally irreversible injury or death)

F – Frequency and/or exposure to hazard

F1 – Seldom-to-less-often and/or exposure time is short

F2 – Frequent-to-continuous and/or exposure time is long

P – Possibility of avoiding hazard or limiting harm

P1 – Possible under specific conditions

P2 – Scarcely possible

FIGURE 28.9 Risk graph for determining the required Pl_r for safety functions. (From PN-EN ISO 13849-1. 2008. Safety of machinery—Safety-related part control systems—Part 1: General principles for design. With permission.)

When choosing a protective device, the required performance level Pl should be determined based on the risk analysis results using the graphs shown in Figure 28.9.

REFERENCES

Boemer, T. 2003. Vision based protective devices (VBPD)—A vision becomes reality. In *International Conference Safety of Industrial Automated Systems*, Nancy, 7–11, 7–15.

Central Division of BGIA. 2008. Functional safety of machine controls.

Dźwiarek, M. 1996. A method for response time measurement of electrosensitive protective devices. *Int J Occup Saf Ergon* 2(3):234–242.

Dźwiarek, M. 1998. Electrosensitive protective equipment. *Measurements Automisation Robotics* 6:47–51.

Dźwiarek, M. 2000. Application of complex electronics in machinery control systems. In *Manufacturing Agility and Hybrid Automation*, eds. T. Marek and W. Karwowski. 341–344. Krakow: Institute of Management, Jagiellonian University.

Dźwiarek, M. 2004. An analysis of accident caused by improper functioning of machine control systems. *Int J Occup Saf Ergon* 10(2):129–136.

Dźwiarek, M. 2006. Documenting functional safety as a part of conformity documentation according to machinery directive. In *Production Engineering Knowledge—Vision—Framework Programmes*, ed. E. Chlebus. 141–147. Wroclaw: Technical University of Wroclaw.

Dźwiarek, M. 2007. Functional safety of machinery control systems—General consideration. In *Functional Safety Management in Critical Systems*, ed. K.T. Kosmowski. 101–114. Gdansk: Foundation for the Development of Gdansk University.

Gardeux, F., J. Marsot, and J. P. Buchweiller. 2003. Detection of persons in danger zones. Contributions and difficulties of artificial vision. In *International Conference Safety of Industrial Automated Systems*, Nancy, 7–21, 7–26.

PN-EN 574 +A1. 2008. Safety of machinery—Two-hand control devices—Functional aspects—Principles for design.

PN-EN ISO 13849-1. 2008. Safety of machinery—Safety-related part control systems—Part 1: General principles for design.

29 Methods, Standards, and Models of Occupational Safety and Health Management Systems

Daniel Podgórski

CONTENTS

29.1 DEFINITION OF AN OCCUPATIONAL SAFETY AND HEALTH MANAGEMENT SYSTEM

According to the International Labour Organization (ILO), an *occupational safety and health management system* is a set of interrelated or interacting elements used to establish occupational safety and health (OSH) policies and objectives and to achieve those objectives (ILO 2001).

However, standard PN-N-18001 (2004) defines the OSH management system as part of the overall organisational management system, including an organisational structure, planning, responsibilities, rules of conduct, procedures, processes, and the resources necessary to develop, implement, enforce, review, and maintain an OSH policy.

29.2 DEVELOPMENT OF AN OCCUPATIONAL SAFETY AND HEALTH MANAGEMENT STRATEGY

The development of OSH management strategies started as early as the nineteenth century and occurred in three stages (Frick and Wren 2000). In the first stage, which ended in the 1960s, industrialised countries developed the first OSH-related laws. In the second stage, which lasted until the 1980s, these countries made intensive reforms and improvements to work safety and health protection systems in Europe and the United States. In the third stage, which started in the late 1980s and continues today, a systematic approach to OSH management methods was standardised and popularised globally. In this stage, there is a change in the approach of concerned parties to the creating and observing detailed technical regulations concerning voluntary implementation of operational procedures that ensure continuous improvement of OSH. The concept of occupational risk has been included in OSH management, and the evaluation and minimisation of this risk is now the core element of OSH management systems. Since the objectives and subjects of OSH management have changed so significantly between stages 2 and 3, OSH measures implemented by companies during stage 2 are often considered *traditional* or *nonsystematic*, as opposed to the *system management* measures practiced in stage 3.

According to Weinstein (1997), there are four main levels of enterprise OSH management strategies—from measures aiming to avoid penalties for noncompliance to those focused on the continuous improvement of all processes affecting OSH.

The evolution of these strategies, with regard to both the respective companies and their entire population, is classified from level I through level IV. The hierarchy of these levels is presented in Table 29.1.

The OSH management approach corresponding to levels I and II is referred to as *traditional* or *reactive*, whereas the approach corresponding to levels III and IV is called *systematic* or *proactive*, since it involves the implementation of a system for managing resources, measures, and processes that aim to improve the OSH standards in an enterprise.

A similar approach to the typology of OSH management was adopted by Bottomley (1999) and then by Coelho and Oliveira Matias (2006). They isolated the effects of traditional management, which is based on reacting to existing circumstances or incidents, and compared them to the effects of systematic management, which ensures continuous improvement in OSH (Table 29.2).

TABLE 29.1
Occupational Safety and Health Management Levels

Safety Level	Motivation	Description	Typical Assessment Method	Typical Learning Method	Typical Safety Goal	Typical Safety Result
I	Fear	Inactive	OSHA inspection only	Basic training only	No fines, penalties	Less than full compliance, worse than average record
II	External punishment	Reactive	Paperwork audit, inspection	Classroom instructions, testing	No non-compliances, citations	Rote compliance, no improvement, average record
III	External reward	Active understanding and belief	Work observation	In-depth, instruction, coaching	All jobs done correctly	Appropriate behaviours, better than average record
IV	Self and internal	Proactive passion and commitment	Peer and subordinate, interviews, work results	By example, self-learning	No accidents, best methods	Continuous improvement and leadership, excellent record

Source: Weinstein, M. B. 1997. *Total Quality Safety Management and Auditing.* Boca Raton, FL: CRC Press.

TABLE 29.2
Differences between the Traditional and Systematic Approach to Occupational Safety and Health Management

In a Reactive Approach to the Osh Management in Organisations	In a Systematic Approach to the Osh Management in Organisations
Hazards are dealt with reactively	Hazards are identified preventively
Risk controls are dependent on individuals	Risk controls are described in procedures
Risk controls are not linked to each other	Risk controls are linked by a common method
OSH activity happens but is not planned	OSH activity is planned
Controls are reviewed after an incident	Controls are monitored and reviewed regularly
Responsibilities are not defined	Responsibilities are defined for everyone
Only the risk within the organisation's own 'backyard' is focused	Public and supplier risks are managed in a planned way
There is no company policy on OSH to communicate	Company policy on OSH is communicated

Source: Coelho, D. A., and J. C. Oliveira Matias. 2006. The benefits of occupational health and safety standards. In *Handbook of Standards and Guidelines in Ergonomics and Human Factors*, ed. W. Karwowski, 413–40. Mahwah, NJ: Lawrence Erlbaum.

29.3 STANDARDISATION OF OCCUPATIONAL SAFETY AND HEALTH MANAGEMENT SYSTEMS

29.3.1 First National-Level Standards

British standard (BS) 8800 (BSI, 1996) was the world's first OSH management system standard established within a national standardisation system. This standard was based on the systematic approach to enterprise management, which was gaining popularity globally and particularly in the United Kingdom, and included quality management systems (ISO 9000) and environmental management systems (ISO 14000). BS 8800 (1996) laid down guidelines for designing and implementing OSH management systems. These guidelines enable the integration of such systems with the overall enterprise management system. The standard offered two alternative and equivalent management system models: (1) a model based on the HS (G) 65 document developed by the UK Health and Safety Executive, which contains guidelines for efficient OSH management (Health and Safety Executive 1993) and (2) another model based on a continuous improvement cycle, in accordance with the model adopted in the ISO 14001 (1996) standard.

Similar normative documents for voluntary implementation were subsequently developed and applied in other countries who lead in the systematic OSH management approach, which primarily includes the Dutch standard NPR 5001 (Nederlands Normalisatie-instituut [NNI] 1996) and the Australian SAA HB53 handbook (Standards Australia 1994) to be used in the construction sector and a joint standard

AS/NZS 4804 (Standards Australia and Standards New Zealand [SA and SNZ] 1997/2001) developed by the technical committees of Australia and New Zealand. Those documents were considered to be guidelines and were not used for certifying the management systems as was the case with ISO 9001 and ISO 14001 (1996). However, the AS/NZS 4804 (1997/2001) standard became the basis for the development and establishment of AS/NZS 4801 (SA and SNZ 2001), which contains specifications for third-party assessment of the compliance of such systems. Standards and draft standards governing systematic OSH management were also developed and published in Spain as a series of six documents (Abad et al. 2002).

In several other countries, such as United States, the Nordic countries and Japan, the systematic approach to OSH management was promoted using mandatory national laws rather than voluntary standards. The United States had voluntary protection programs (VPPs) starting in 1982 that were supervised by the US Occupational Safety and Health Administration (OSHA; 1982). As part of this program, enterprises implemented the OSH management programs based on principles published and updated by OSHA (2000). Enterprises participating in VPPs are exempted from routine and planned OSHA inspections, which are replaced by periodical management system audits (Dyjack and Levine 1996). Experiences with VPPs in the first years were used to develop guidelines for the implementation of systematic OSH management systems in enterprises (OSHA 1989).

In addition to the activities of OSHA, the American Industrial Hygiene Association (AIHA) also promoted the OSH management systems and published system management guidelines (AIHA 1996), which were strictly based on the concept of the quality management system and the structure of the ISO 9001 (1994). Recognising the need to provide more practical guidelines for the implementation of OSH management system guidelines based on ISO 9001, Kozak and Krafcisin (1997) published relevant guidelines to that effect in the United States.

The Nordic countries initiated a systematic approach to OSH management in the 1990s by enacting mandatory laws on internal control. In Norway, guidelines for the implementation of such systems were established and published in 1991 (Kommunaldepartementet 1991), and in Sweden in 1992 (Swedish National Board of Occupational Safety and Health [SNBOSH] 1992). After several years of experience in both countries, the laws on internal control of the work environment were revised and updated in 1996 (Kommunaldepartementet 1996; SNBOSH 1997). These guidelines ensure practical implementation of the European Framework Directive on OSH (European Communities 1989) in enterprises and define elements of the OSH management system reflecting the systematic management approach based on standards ISO 9001 and ISO 14001.

In Japan, guidelines for an OSH management system were put in place according to an ordinance of the Minister of Labour of 30 April 1999 (Japanese Ministry of Labour 1999). Both the structure and the content of those provisions correspond to the British standard BS 8800 (1996) and the Australian standard AS/NZS 4804 (1997/2001) as well as other OSH normative documents. Despite their governmental origin, the Japanese guidelines are voluntary for employers striving to enhance their performance by improving work conditions in their enterprises.

29.3.2 Development of International Standards for Occupational Safety and Health Management Systems

The above-mentioned initiatives to develop and implement national standards for systematic OSH management resulted in an increased interest in the development of such standards at an international level. In 1996, the International Organization for Standardization (ISO) considered the need for international standards governing OSH management systems (Podgórski 1997; Zwetsloot 2000). After debating this during international workshops conducted by the ILO in Geneva in September 1996 and after analysing the results of a vote taken among national organisations for standardisation, the ISO Technical Management Board decided to discontinue any further efforts to standardise specifications for OSH management systems (ISO 1997) in February 1997. This was due to significant differences in OSH management methods and cultures of developed and developing countries, which are reflected in their different legal frameworks.

The ISO's decision to discontinue the development of OSH management system standards triggered other international efforts in this area. The lack of standards that could be used to certify OSH management systems prompted a group of international private certification and consulting bodies who specialised in the certification of quality management and environmental management systems to act; thus documents OHSAS 18001 (1999) and OHSAS 18002 (2000) were established.

29.3.3 Creation and Establishment of the Polish PN-N-18000 Standards

The need to standardise the OSH management principles emerged in Poland in the mid-1990s. A systematic approach to OSH management, based on quality and environmental management system principles, was then being applied by several leading companies and promoted by the National Labour Inspectorate and several scientific institutions, mainly the Central Institute for Labour Protection (CIOP) and the Central Mining Institute. Because the Polish companies and organisations were interested in establishing OSH management system standards and since the ISO had decided to discontinue the standardisation of requirements for such systems, in 1997 the CIOP requested that the president of the Polish Standardisation Committee (PKN) establish a standardisation commission (NKP) for OSH management. The commission was set up on 26 February 1998 and assigned no. 276. NKP no. 276 in was largely funded by CIOP in 1998–2001 as part of the Strategic Government Program (SPR-1) on Occupational Safety and Health Protection. After the amendment of the standardisation law in 2002, NKP no. 276 transformed into a technical committee (KT no. 276) for OSH management; however, its essential composition, programs, and operating principles remained unchanged.

As a result of the efforts of NKP no. 276 and KT no. 276, Polish standard PN-N-18000 was established. The first standard of the series, PN-N-18001 (1999), contained specifications for the establishment of OSH management systems to be evaluated by independent certification bodies. The standard was amended in 2004 to comply with ILO-OSH 2001 guidelines. The subsequent standards PN-N-18002

(2000) and PN-N-18004 (2001) were considered practical guidelines supporting the implementation of the OSH management system in enterprises. In 2006, KT no. 276 completed work on the last standard of the series PN-N-18011 (2006), which pertains to audits of OSH management systems and is based, to a very large extent, on the ISO 19011 standard implemented in Poland as PN-EN ISO 19011 (2003).

29.3.4 Development and Implementation of International Labour Organization's Guidelines for Occupational Safety and Health Management Systems

The ISO's decision to discontinue the standardisation of requirements concerning OSH management systems triggered initiatives from various bodies to establish documentation to standardise the principles of systematic OSH management at an international level. Due to some developed countries' lack of support for OSH management system certification and the growing conviction among safety and health professionals that ISO was not the right organisation to set requirements governing relationships between employers and employees, development of long-awaited international guidelines was picked up in 1998 by the ILO. In 2000, the ILO's draft guidelines for OSH management systems were reviewed by all ILO member states and then submitted to the International Experts Forum, representing employers, employees, and government bodies, to discuss, improve, and finalise the draft. The final text of the ILO's guidelines was adopted on 27 April 2001 as the ILO-OSH 2001 document (ILO 2001). The ILO's guidelines were initially published in English, French and Spanish only, but the guidelines were translated into other languages and published in various local versions in a number of ILO member states.

The introduction to ILO-OSH 2001 addresses the guidelines to all persons responsible for OSH management. The direct application of such guidelines is not compulsory for member countries of ILO, enterprises, or any other bodies operating in such countries. These guidelines do not aim to supersede any laws or voluntary standards existing in such countries. However, provisions adopted in the document reflect the requirements of the ILO's most important conventions, including but not limited to ILO Convention no. 155 on OSH (ILO 1981) and ILO Convention no. 161 on occupational health services (ILO 1985).

29.3.5 The Role of ILO-OSH 2001 Guidelines in the Promotion and Standardisation of Occupational Safety and Health Management Systems at a National Level

The ILO's guidelines are meant to be applied at two levels: (1) national and (2) organisational (enterprise-wide). This distinguishes the document from and gives it a significant advantage over other OSH management system standards, which only apply to the organisational level. This is reflected in the three major sections of the document: the first section outlines the general objectives of the guidelines; the second contains provisions to be applied at the national level; the third contains guidelines for the OSH management system to be used at the enterprise level.

National-level ILO guidelines are for the establishment and promotion of a systematic approach to OSH management. These measures should be supported by relevant national laws if possible, and include the following elements:

- A *competent body* for developing and implementing national policy concerning the implementation and promotion of OSH management systems
- A consistent *national policy* concerning OSH management systems
- *National* and *tailored guidelines* concerning voluntary implementation and maintenance of OSH management systems in organisations

In addition, the ILO's guidelines specify the scope of the required *national policy* on OSH management systems. The most important aspects to be dealt with in such a policy include the following:

- Promote OSH management systems as an integral part of overall enterprise management
- Avoid unnecessary bureaucracy, administration, and costs related to systematic OSH management
- Arrange for the promotion of OSH management systems by labour inspectorates and OSH services

The ILO's guidelines recommend that *national guidelines* be developed and implemented in different countries, and that those national guidelines should conform to the given country's laws and practices and also to the relevant *tailored guidelines.* Tailored guidelines should be given for voluntary application in enterprises and should be based on the OSH management system model defined in the ILO's guidelines. The tailored guidelines should incorporate the most important elements of the national guidelines and the general objectives of the ILO's guidelines and also primarily reflect the specific circumstances and needs of the enterprise or group of enterprises, including, but not limited to, the following:

- Size (small, medium, or large) and infrastructure
- Types of hazards encountered and the degree of occupational risk

The provisions of the national-level ILO-OSH 2001 guidelines offer a number of different solutions for the transposition, application, and promotion of a systematic approach to OSH management in different countries. These provisions essentially suggest three principal ways to use the guidelines at an enterprise level, as illustrated in Figure 29.1 (ILO 2001).

The ILO's guidelines provide a unique approach to the standardisation of systematic OSH management principles at an international level. Due to a built-in mechanism ensuring the transposition of its provisions at the national and enterprise level, many countries initiated the creation of new normative documents that were adapted to those countries' needs and to the different sizes of the enterprises. In Germany, for example, the document was applied at the national level through guidelines developed by the Federal Institute of Occupational Safety and Health (Bundesanstalt für

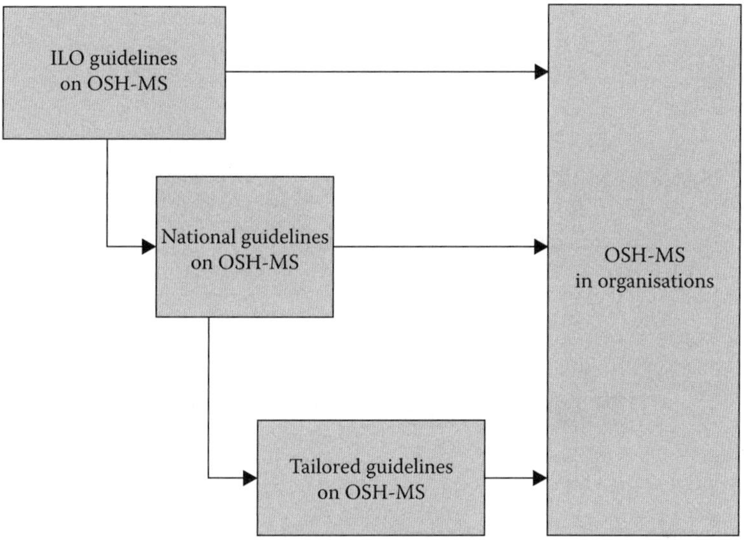

FIGURE 29.1 Rules of application of ILO-OSH 2001 guidelines at an enterprise level. (Reprinted with permission from ILO. 2001. *Guidelines on Occupational Safety and Health Management Systems*. Geneva, Switzerland: International Labour Office.)

Arbeitsschutz und Arbeitsmedizin 2002). In Poland, on the other hand, the ILO's guidelines were implemented at the national level through the revision and amendment of standard PN-N-18000. The most important step in that approach was the establishment of standard PN-N-18001 (2004), which fully reflects the ILO's guidelines that are applicable to enterprises.

29.4 PRINCIPAL OCCUPATIONAL SAFETY AND HEALTH MANAGEMENT MODELS

29.4.1 The Cycle of Continuous Improvement as the Basis for Occupational Safety and Health Management System Models

According to standard ISO 9000 (ISO 2000), *continuous improvement*, which means repetitive action aiming to increase compliance ability, involves an ongoing search for opportunities to improve the efficiency of all members and processes of an organisation. To demonstrate and promote continuous improvement and its application in all processes and organisations, the *cycle of continuous improvement*, also known as the *Deming cycle*, the *Shewart cycle*, or the *PDCA cycle*, was adapted. The acronym PDCA comes from the first letters of the names of subsequent phases of the cycle used by Deming: Plan, Do, Check, and Act. The different phases of the cycle are given below:

P: Plan changes that aim to improve a process
D: Do (test) the process after such changes are implemented

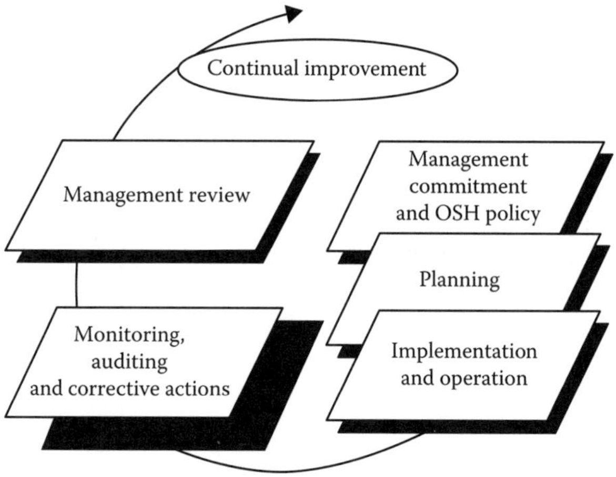

FIGURE 29.2 Graphical representation of the occupational safety and health management system model according to PN-N-18001 (2004). (Reprinted with permission from PN-N-18001. 2004. Occupational health and safety management systems—Requirements.)

C: Check (analyse) the results of such changes
A: Act—take one of the following actions:
  Adopt the changes
  Abandon the changes
  Repeat the cycle with other parameters

Because the PDCA cycle presents a concept that can be applied regardless of the type of action, it is used as a basis to design models for quality management, environmental management, and OSH management, systems. In particular, the PDCA cycle was used in the management system models defined in standards ISO 14001 (1996), BS 8800 (1996) and in the Polish standard PN-N-18001 (Figure 29.2).

29.4.2 Occupational Safety and Health Management System Models According to the ILO-OSH 2001 Guidelines

The ILO's guidelines are meant to be applied at the national level and organisational (enterprise) level. National-level guidelines refer to the creation and functioning of national structures responsible for promoting a systematic approach to OSH management. On the other hand, enterprise-level guidelines apply to the employer, who should demonstrate commitment and leadership towards improving OSH and should introduce organisational arrangements to implement the OSH management system according to a model based on the continuous improvement cycle. Figure 29.3 is a graphical presentation of the management system model in its original version in the ILO-OSH 2001 document.

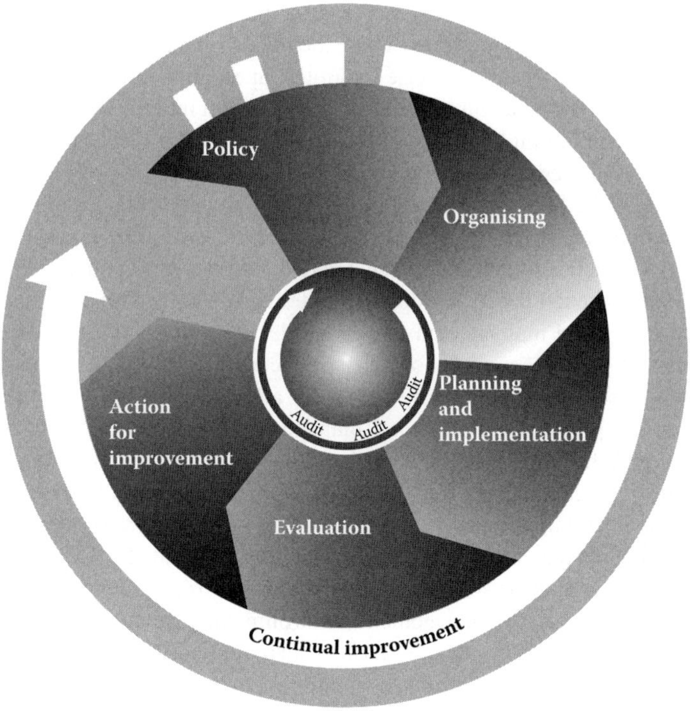

FIGURE 29.3 Occupational safety and health system model adopted in the ILO-OSH 2001 guidelines. (Reprinted with permission from ILO. 2001. *Guidelines on Occupational Safety and Health Management Systems*. Geneva, Switzerland: International Labour Office.)

29.5 INTEGRATED MANAGEMENT SYSTEMS INCLUDING OCCUPATIONAL SAFETY AND HEALTH

29.5.1 THE CONCEPT OF MANAGEMENT SYSTEM INTEGRATION

The coincidence of the establishment of international environmental management standards (ISO 14001 in 1996) and the implementation of the first national OSH management system standards (e.g. BS 8800 in 1996) and the compatibility of the management system models adopted in those documents triggered initiatives to establish integrated management principles in both areas. Those initiatives at first referred to only very specific and narrow sectors of the economy that posed significant OSH-related and environmental hazards. At the same time, suggestions concerning the integration of the OSH management systems with quality management systems emerged, for example, the US AIHA guidelines (1996) concerning OSH management system based on ISO 9000 requirements or the Norwegian guidelines for an integrated management system covering quality, natural environment, and OSH (Norges Standardiserings-Forbund 1996).

The popularisation of integrated management systems in these three areas occurred mostly through marketing activities carried out by certification bodies and consulting firms and has led to an increased number of enterprises implementing and maintaining integrated systems. Thus, the hands-on experience in management system integration necessary to revise the initial integration concepts was obtained. This revision led to more advanced management system integration guidelines, such as the joint Australian and New Zealand standard AS/NZS 4581 (1999) and a BSI handbook (2000) based on the standard.

According to the BSI handbook, an integrated management system is defined as the combination of processes, procedures, and practices used by an organisation to implement its policies. Integrated systems may be more effective in achieving policy goals than an approach based on separate systems. AS/NZS 4581 (1999) guidelines provide an integrated management system model including quality, environmental protection and OSH. The model shows that only certain components of an integrated management system may be equally useful for all areas of activity covered by that system, that is, quality, environmental management, and OSH. Due to the differences in the specific natures of these areas, other components of the system must not be forced into integration, but may be applied on an individual basis in respective areas so as to ensure the highest possible efficiency of the entire system.

Management system integration must take into account the fact that the term 'integrated management system' can also be used to mean something other than a system including quality management, environmental management, and OSH or other areas of business activity. The term is also applied to the computerised systems, such as ERP/ERP II systems, that support management, which includes all areas of an enterprise. Sometimes, the term 'integrated system' can also be used to refer to different activities conducted within a single, specific management system, such as a quality management system.

29.5.2 Management System Integration Strategies

Quality management, environmental management and OSH management systems can be integrated in order to achieve *synergy*. Synergy ensures that different components of a management system interact in such a way that the results are better, more cost-efficient, and less labour-intensive when compared with that achieved through independent operation of those components. A successful achievement of synergy effects in enterprise management can be supported or hindered by different organisational and human factors (Zwetsloot 1994).

Integrating an OSH management system with a quality management system or environmental management system is not the only way to achieve synergy. Another strategy is *analogy*, which involves implementing different components of subsequent management systems in a way analogous to the methods adopted and tested during the implementation of the first system. Research conducted by Zwetsloot (1994) on strategies for synergy in enterprises implementing quality, environmental, and OSH management systems indicated that in the 1990s the *analogy* strategy was considered more flexible and was preferred by enterprises with such systems. Advanced enterprises, in which the management systems were already mature and could be further developed through the optimisation of all activities through an integrated management system, preferred the *integration* strategy.

Kirkby (2002) described another strategy leading to the synergy of management systems, called an *aligned systematic approach*. This is similar to the previously mentioned *analogy* strategy, but involves a different approach on two different levels of enterprise management. On the higher level, requirements shared by management systems referring to different areas are identified, and then identical or shared procedures are implemented, such as policy announcements, training, internal audits, management reviews, and corrective and preventive measures. On a lower process management level, however, three separate subsystems are in place, each of them covering a different area.

To summarise the synergy strategies for management systems and the models used to integrate those systems: the implementation of an integrated management system should enable *equally* efficient management of areas that the enterprise wants to manage. Thus, an integrated system should not prioritise any single area. Since enterprises implemented quality management systems first, and certification bodies usually focused mostly on marketing their services in that area, the integrated systems that came later are often based on quality management systems that include elements of environmental management and OSH management. As a result, quality-related goals and activities are usually the priorities in such systems.

Despite the adoption and popularisation of integrated management systems in many countries, they are still controversial, and there is no consensus about their effectiveness and usefulness. Research conducted in Singapore on 96 building companies (Low and Chin 2003) indicates that a large number of managers consider the implementation of an integrated system including quality management (ISO 9001 2000) and OSH management based on OHSAS 18001 (BSI 1999) to be relatively easy and that it benefits the enterprise. This research also demonstrates that the integration of management systems requires greater management skills and training expenditure and generates increased costs because of the need for enhanced competency and experience among the management personnel. Similar research conducted later by Low (2005) on the managers of 30 building companies in Singapore, which used integrated systems including the three areas of management, showed that the integration of those systems was beneficial for the enterprise due to the improvement of internal management procedures and of the company's market position. As observed in previous research, the integration of management systems required greater costs and more time for training personnel and management staff.

Questionnaire-based research conducted among 151 managers in the construction industry in the United States confirmed that the main characteristics of quality management systems and OSH management systems are highly correlated and equally important (Loushine et al. 2004). In addition, according to the majority of these polled managers, OSH can be managed using quality management techniques, which proves that there is a great potential for the integration of management systems.

29.6 OCCUPATIONAL SAFETY AND HEALTH MANAGEMENT BASED ON THE BEHAVIOURAL SAFETY CONCEPT

Behavioural safety is a proactive safety and health management method that involves preventing accidents or potential accidents at work by observing the behaviour and reactions of personnel. The method is based on an assumption that people tend to act dangerously, not as a result of an incorrect attitude toward safety but as a result of their experiences. Research conducted by Behavioural Science Technology Inc., a consulting firm, indicates that behaviours at work can be classified into three types: easy, difficult, and impossible (Polok and Koźlik 2006). Easy behaviours are those that are easy for employees to perform; difficult behaviours require that employees deviate from their usual course of action or make an extra effort; impossible behaviours involve employees being physically unable to act differently or those that will not be welcome in an organisation.

Programs for implementing the behavioural safety method are offered and promoted globally by Dupont, a leading company in the chemical industry and in efficient OSH management (Wokutch and VanSandt 2000). According to Krause (1997), one of the main authors and promoters of the behavioural safety concept, such a program should include identification and definitions of employee behaviours that are critical to their safety, observation of such behaviours, and the collection of data on the frequency of such behaviours, followed by implementation of corrective and preventive measures to ensure OSH improvement.

Examples of good organisational solutions supporting the implementation of behavioural safety methods and stimulating employee involvement in OSH improvements include the Tuttava program established in Finland (Saari 1998). This program is based on the efforts of working teams, which include representatives of the management, direct supervisors and employees, established provisionally to observe dangerous behaviours and implement relevant improvements based on such observations. The application of the program in many companies was not only highly effective in reducing the number of accidents at work by up to 80% and in reducing postaccident absences from work, but also brought significant benefits through the reduction of financial losses.

The implementation of the behavioural safety method in an enterprise where an OSH management system compliant with one of the abovementioned system models is already in place does not require significant changes to that system or any organisational changes in the enterprise. The behavioural safety management processes can be considered additional components supporting the existing OSH management system, introduced to ensure the application of the continuous improvement cycle (Fleming and Lardner 2002).

29.7 EFFECTIVENESS OF OCCUPATIONAL SAFETY AND HEALTH MANAGEMENT SYSTEMS

Opinions about the effectiveness of formal OSH management systems are often ambiguous (Pawłowska 2004). Depending on the design and implementation of a system within an enterprise and the OSH objectives set within that system, OSH management systems can be an efficient enterprise management tool or a complicated bunch of documents that do nothing to facilitate practical activities. Test results and indicate that systematic OSH management systems, if correctly understood and implemented, can have considerable benefits for the enterprise.

According to data published by OSHA (2006), many American companies have significantly improved their safety by applying systemic OSH management guidelines. In one of the refineries of the Mobil Oil-Joliet Refining Corporation in Illinois, the implementation of systematic OSH management methods brought an 89% reduction over 5 years in the cost of compensation paid to employees for accidents at work. Ford New Holland plant recorded a 13% increase in work efficiency and a 16% decrease in the number of absences within 3 years of implementing the system. Another example is Thrall Car Manufacturing Company, which started the implementation of an OSH management program according to OSHA's guidelines in 1989; by 1992, the number of days missed due to accidents at work decreased more than threefold, and by 1994, it had decreased 30 times. The amount of compensation paid for accidents at work also decreased nearly seven times (from USD 1,376,000 in 1989 to USD 204,000 in 1992). At the Occidental Chemical Company, the number of accidents at work and occupational diseases declined nearly four times between 1987 and 1993, after implementing OSHA's program.

Norwegian aluminium plants implemented an internal control system (Kjellén et al. 1997), which, according to Norwegian law, should be used in all companies. The system greatly reduced costs related to accidents and absence due to sickness, and after several years, system maintenance costs and other running costs also decreased significantly.

The effectiveness of OSH management systems in reducing the number of accidents, occupational diseases and related economic losses and their positive influence on the quality and productivity of enterprises were confirmed by a report by Robson et al. (2007), which was based on an analysis performed on nine different types of OSH management interventions, as described in scientific publications. The results of the analysis clearly indicated that OSH management systems, if properly implemented, bring a number of benefits to enterprises. The report described many advantages; the following are particularly significant:

- Reduced numbers of accidents at work and sickness absence days
- Reduced accident insurance premiums
- Better understanding of occupational risks among managers and employees
- Better safety climate and employee involvement in activities towards improvement of OSH
- Improved productivity in the entire company

Research conducted at Polish companies by Pawłowska et al. (2001) also indicated a correlation between the level of OSH management and the number of accidents. The research included 71 Polish companies of different sizes in the industrial processing sector. It involved a questionnaire containing questions on the structure of the OSH management system, its different components, and how they function, which are related to a given company's adopted occupational safety culture. The questionnaire provided a quantitative evaluation of systems—systems or their components that were fully compliant with the standard and operating in a favourable safety culture were rated 100%; if a given component of the system was absent, a 0% rating was awarded for that component.

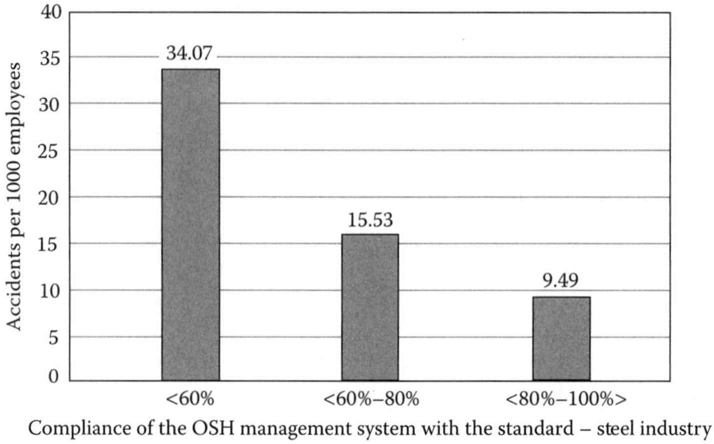

FIGURE 29.4 Accident figures for steel industry companies depending on the level of occupational safety and health management system compliance with standard PN-N-18001. (Adapted with permission from Pawłowska, Z., M. Pęciłło, and G. Dudka. 2001. *Occupational Safety* 1:20–22.)

In the steel industry, in OSH management systems that are largely compliant with standard PN-N-18001, the number of accidents was approximately twice as low as enterprises focusing exclusively on compliance with applicable laws (Figure 29.4). In service and sales companies, where OSH management systems are also largely compliant with the requirements of PN-N-18001, accidents were approximately 30% lower than in other companies in which compliance with the abovementioned standard was low.

The abovementioned research also stated that the implementation of management systems in other areas of the company's operations (e.g. quality management systems or environmental management systems) improved the overall OSH level. However, in several of the analysed enterprises, accident figures were still high despite the high rating on the compliance of the OSH management system (in some cases it was even declared 100% compliant with standard PN-N-18001). More detailed investigations conducted in those areas showed that the OSH management systems in those enterprises were primarily formal structures, and the number of employees actually involved in their implementation was very small.

29.8 CERTIFICATION OF OCCUPATIONAL SAFETY AND HEALTH MANAGEMENT SYSTEMS

29.8.1 Definition and Procedures of Management System Certification

According to the definition quoted in the ISO/IEC Handbook no. 2 (ISO/IEC 1996), *certification* is a procedure involving a *third party* providing written assurance that a product, process, or service complies with the specifications. A third party in this context is a person or organisation considered independent of any parties involved in the issue in question. According to the explanatory note of the definition of a third party, the other parties involved are usually the supplier (first party) and the buyer (second party). Instead of 'third party', the term 'certification body' is often used, meaning a certifying organisation. In the certification of OSH management systems, one must establish the parties who are considered second parties. Only the relationship between the supplier (first party) and the certification body (third party) is visible. Zwetsloot (2000) analysed this issue in the context of the new market for certification services in OSH management systems that emerged in the 1990s. According to Zwetsloot, the market grew in response to the expectations of large companies who felt an urgent need to show the government, current and prospective employees, and the general public that they had proper management systems in place. The term 'second party' is very broad and encompasses several different categories of parties interested in certification of OSH management systems.

Management system certification procedures used by certification bodies differ depending on reference documents (criteria) used to asses the compliance of a given system. However, most of the procedures include four main stages:

1. Reception and registration of an application for certification
2. Audit of the enterprise's management system
3. Analysis of audit results and delivery of the certification decision
4. Supervision of the issued certificate in the form of periodical audits

The most important stage of the certification procedure is the audit. According to the definition in standard PN-EN ISO 19011 (2003), an *audit* is a systematic, independent and documented process for obtaining and objectively assessing audit evidence in order to establish a level of compliance with the audit criteria. The methodology for auditing quality management and environmental management systems is described in detail in PN-EN ISO 19011. This standard, along with PN-N-18011, should be applied when auditing OSH management systems in Poland.

Management systems are audited not only for certification purposes, called *external audits.* According to quality management (ISO 9000) and environmental management system models (ISO 14000) and the majority of OSH management system models, audit methodology is used mainly for self-evaluation of management systems to ensure the continuous improvement of such systems. These are called *internal audits.*

Figure 29.5 shows a diagram of the procedure for certification an OSH management system's compliance with standard PN-N-18001 (2004), from the Management System Certification Centre of CIOP-PIB. A certificate is issued for an OSH management system if no significant irregularities are detected during the certification audit and after the enterprise implements any corrective measures resulting from that audit. If any significant irregularities are identified during the certification audit, the certification bodies will have to repeat the audit after the irregularities are corrected. The certification body also supervises the certified management system

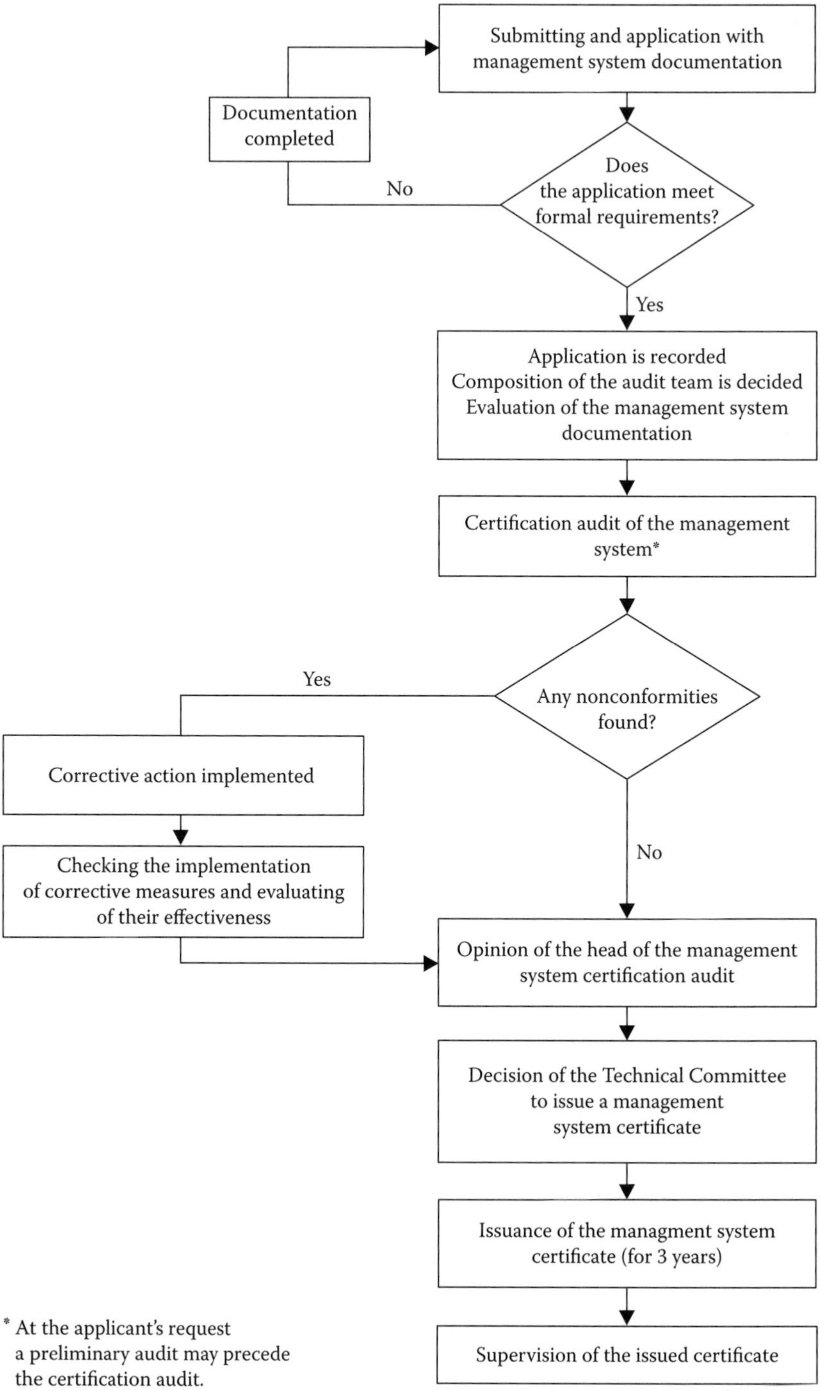

FIGURE 29.5 Example of an OSH management system certification procedure used by CIOP-PIB.

after the certificate is issued. In most cases, this involves annual supervision audits covering selected areas of the management system or selected departments of the organisation. Some bodies perform more frequent supervision audits (e.g. once every 6 months) covering fewer system components. Management system certificates are usually issued for 3 years, after which another full audit of the management system should be performed and another certificate must be issued.

29.8.2 Advantages and Disadvantages of Occupational Safety and Health Management System Certification

Certification bodies often claim that companies with a certified OSH management system enjoy a number of benefits, such as:

- Fewer accidents, potential accidents, and occupational diseases at work
- Compliance with applicable OSH laws
- Increased competitiveness due to higher standards and reduced OSH losses

Still, these benefits result from the implementation of an *efficient* OSH management system in an organisation; mere certification does not offer any significant advantages. These benefits are thus not benefits of system certification but rather are benefits of system implementation. Such advertising by certification bodies is misleading to their potential clients. The actual potential benefits from the certification of OSH management systems include:

- Implementing a program of periodical internal audits along with periodical external audits, which ensures continuous system improvement through consistent identification of nonconformities and initiation of corrective and preventive measures.
- Objectively demonstrating to government administration and labour regulation bodies that the enterprise maintains an OSH management system and that the system ensures the planning, organisation, control, monitoring and review of protective and preventive measures aiming to effectively eliminate or reduce occupational risks.
- Demonstrating to accident insurance companies that the enterprise conducts activities in compliance with applicable laws and in a way that reduces the number of accidents, in order to obtain additional discounts on accident insurance.
- Demonstrating to shareholders and potential investors that the company operates in compliance with applicable laws and the continuous improvement principle and that it uses 'best management practices', which promise long-term growth and high competitiveness.
- Enhancing the enterprise's goodwill and improving its recognition among clients and other concerned parties, both domestically and globally.
- Creating an opportunity to conquer international markets, where a certified OSH management system is expected, and an opportunity to create good business partnerships, which require efficient OSH management.

- Improving personnel involvement in grassroots activities toward continuous improvement, which boosts morale and reduces staff rotation.

All management system certifications, including OSH management, have a number of disadvantages as well. Disadvantages of certification of such systems most commonly described in publications include the following:

- The difficulty of designing and implementing an OSH management system compliant with the specified certification requirements
- The need to develop and update extensive management system documentation, commonly perceived as a sign of excessive bureaucracy
- The long amount of time required to implement and certify a management system, often required due to excessive documentation of all activities in the system and implementation of all procedures from scratch without building on existing solutions
- The implementation and certification of an OSH management system as part of an integrated system in which the quality management system is considered the most essential component, which may diminish the importance of OSH as perceived by employees and the management
- The need to perform time-consuming internal and external management system audits, which leads to a significant increase in the stress levels of managers and employees
- The significant financial costs of obtaining a certificate and maintaining a certified OSH management system
- The need to overcome resistance against changes in the enterprise, both at the organisational level and among staff members; this may cause a deterioration of performance in the field of OSH even in well-managed enterprises after the implementation and certification of a formal system
- Increased focus on compliance with formal management system procedures and decreased focus on human factors in OSH management, such as the need to monitor employees' safety-related behaviours and to involve them in activities to improve OSH
- Difficulty in keeping staff committed on a long-term basis to activities geared towards maintaining and improving the implemented OSH management system

Potential disadvantages of OSH management system certification were discussed and publicised during an international debate initiated in 1996 and devoted to the need to establish an ISO standard in this area. In particular, representatives of employers and industrial organisations did not expect a significant improvement in OSH at the enterprise level through the implementation of an ISO standard and were afraid that the standard would be misused to increase the scope of operations of certification bodies (Lambert 2000; Zwetsloot 2000).

Considering the advantages and disadvantages of OSH management system certification, standardisation processes initiated around 1995 and the resulting development of certification services are relatively recent and still unfinished. The adverse consequences of OSH management system certification can be alleviated by taking note of the following items when implementing these systems:

- Enterprises' demand for external certification of OSH management systems should not hinder activities geared toward the development of internal continuous improvement processes in those management systems and the promotion of internal audits.
- External certification audits should not be viewed merely as examinations to pass or fail, but rather as a stage in the process of OSH improvement.
- If there is no demand for external certification of OSH management systems, internal audit processes should be developed and disseminated primarily in that area.
- Where possible, external audits should be based on the results of internal audits; this includes considering the possibility of linking external auditors to teams of internal auditors instead of increasing the total number of audits.
- The competency of external auditors and internal auditors selected to audit OSH management systems should be equal.

With reference to the last recommendation, if the auditors (both internal and external) have no hands-on experience in hazard identification, occupational risk assessment and the implementation and evaluation of protective and preventive measures typical to the organisation they are auditing, their audit reports will not be useful in practice (IOSH 2003), despite their knowledge of the model's scope and theory.

29.9 NEED FOR FURTHER DEVELOPMENT OF OCCUPATIONAL SAFETY AND HEALTH MANAGEMENT SYSTEM STANDARDS

Due to the extent of potential applications, international guidelines on OSH management systems in small- and medium-sized enterprises are urgently needed. Such guidelines can be created and established in countries that have the competent bodies and long-term experience in the implementation of management systems. In other countries, the establishment of such a document can be difficult and requires support in the form of an internationally accepted management system model.

Such a model does not yet exist, and further research and development is needed to create it, particularly in terms of customising the model to the needs of enterprises of different sizes (from very small- to medium-sized), different economy sectors, and organisations and institutions with different interests (e.g. employer organisations, trade unions, insurance institutions, health inspectors, and other government bodies). To ensure that the model will be useful and commonly used in the future, it should be verified by pilot implementations in a large number of enterprises representing different sectors and countries, taking into account different legal environments, management practices, and safety cultures.

REFERENCES

Abad, J., P. R. Mondelo, and J. Llimona. 2002. Towards an international standard on occupational health and safety management. *Int J Occup Saf Ergon* 3:309–319.

AIHA. 1996. *Occupational Health and Safety Management System: An AIHA Guidance Document.* Fairfax, VA: American Industrial Hygiene Association.

AS/NZS 4581. 1999. *Management System Integration—Guidance to Business, Government and Community Organizations*. Homebush, NSW: Standards Australia; and Wellington, NZ: Standards New Zealand.

AS/NZS 4801. 2001. *Occupational Health and Safety Management Systems—Specification with Guidance for Use*. Homebush, NSW: Standards Australia; and Wellington, NZ: Standards New Zealand.

AS/NZS 4804. 1997. *Occupational Health and Safety Management Systems—General Guidelines on Principles, Systems and Supporting Techniques*. Homebush, NSW: Standards Australia; and Wellington, NZ: Standards New Zealand.

AS/NZS 4804. 2001. *Occupational Health and Safety Management Systems—General Guidelines on Principles, Systems and Supporting Techniques*. Homebush, NSW: Standards Australia; and Wellington, NZ: Standards New Zealand.

Bottomley, B. 1999. *Occupational Health and Safety Management Systems: Strategic Issues Report*. Canberra, Australia: National Occupational Health and Safety Commission.

BS 8800. 1996. *Guide to Occupational Health and Safety Management Systems*. London: British Standards Institution.

BSI. 2000. *Management System Integration—A Guide*. London: British Standards Institution.

Coelho, D. A., and J. C. Oliveira Matias. 2006. The benefits of occupational health and safety standards. In *Handbook of Standards and Guidelines in Ergonomics and Human Factors*, ed. W. Karwowski, 413–440. Mahwah, NJ: Lawrence Erlbaum.

Dyjack, D. T., and S. P. Levine. 1996. Development of an ISO 9000-compatible occupational health and safety standard: Defining the issues. In *New Frontiers in Occupational Health and Safety: A Management System Approach and the ISO Model*, eds. C. F. Redinger and S. P. Levine. Fairfax, VA: American Industrial Hygiene Association.

European Communities. 1989. Council Directive of 12 June 1989 on the introduction of measures to encourage improvements in the safety and health of workers at work (89/391/EEC). OJ EC No. L 183, 29 June 1989, 1–8.

Federal Institute for Occupational Safety and Health. 2002. *Guide to OSH Management Systems.* Dortmund, Germany: The Federal German Institute for Safety and Health at Work.

Fleming, M., and R. Lardner. 2002. Strategies to promote safe behavior as part of a health and safety management system. Contract Research Report 430/2002. Sudbury, UK: Health and Safety Executive.

Frick, K., and J. Wren. 2000. Reviewing occupational health and safety management—Multiple roots, diverse perspectives and ambiguous outcomes. In *Systematic Occupational Health and Safety Management—Perspectives on an International Development*, eds. K. Frick, P. L. Jensen, M. Quinlan, and T. Wilthagen, 17–42. Oxford, UK: Elsevier Science.

Health and Safety Executive. 1993. *Successful Health and Safety Management.* (HS(G) 65). London, UK: Health and Safety Executive.

ILO. 1981. *Occupational Safety and Health Convention No. 155*. Geneva, Switzerland: International Labour Office.

ILO. 1985. *Occupational Health Services Convention No. 161*. Geneva, Switzerland: International Labour Office.

ILO. 2001. *Guidelines on Occupational Safety and Health Management Systems*. Geneva, Switzerland: International Labour Office.

IOSH. 2003. *Systems in Focus—Guidance on Occupational Safety and Health Management Systems*. The Grange, UK: Institution of Occupational Safety and Health.

ISO. 1997. Confirmed report of the international workshop on occupational health and safety management system standardization in Geneva, 5–6 September 1996. Geneva, Switzerland: ISO.

ISO. 2000. *Quality Management Systems—Fundamentals and Vocabulary.* Geneva, Switzerland: International Organization for Standardization.

ISO/IEC. 1996. *Standardization and Related Activities—General Vocabulary* (ISO/IEC Guide 2: 1996). Geneva, Switzerland: International Organization for Standardization and International Electrotechnical Commission.

Japanese Ministry of Labor. 1999. *Guideline for occupational safety and health management systems.* Ministry of Labour Notification No. 53, 30 April 1999. Tokyo: Ministry of Labour.

Kirkby, A. 2002. *The One-Stop Shop. Quality World.* London: Institute of Quality Assurance.

Kjellén, U., K. Bae, and H. L. Hagen. 1997. Economic effects of implementing internal control of health, safety and environment—A retrospective case study of an aluminium plant. *Saf Sci* 21(2/3):99–114.

Kommunaldepartementet. 1991. *Internal Control. Ordinance with Guidelines.* Oslo, Norway: Ministry of Local Government and Labour.

Kommunaldepartementet. 1996. *Regulation on Systematic Health, Environment and Safety Activities in Enterprises et al. The Internal Control Regulation.* Oslo, Norway: Ministry of Local Government and Labour.

Kozak, R. J., and G. Krafcisin. 1997. *Safety Management and ISO 9000/QS-9000: A Guide to Alignment and Integration.* New York: Quality Resources—A Division of the Kraus Organization Limited.

Krause, T. R. 1997. *The Behavior-Based Safety Process. Managing Involvement for an Injury-Free Culture.* New York: Van Nostrand Reinhold—A Division of International Thomson Publishing Company.

Lambert, J. 2000. The German position with respect to standardisation of OH&S management systems. In *Ergonomics and Safety for Global Business Quality and Productivity. Proceedings of the Second International Conference ERGON-AXIA 2000*, 19–21 May 2000, ed. D. Podgórski and W. Karwowski, 315–318. Warsaw: CIOP.

Loushine, T. W., P. Hoonakker, P. Carayon, and M. J. Smith. 2004. The relationship between safety and quality management in construction. In *Proceedings of the Human Factors and Ergonomics Society 48th Annual Meeting*, September 20–24. New Orleans, Louisiana, 2060–4.

Low, S. P. 2005. ISO 9001, ISO 14001 and OHSAS 18001 management systems: Integration, costs and benefits for construction companies. *Archit Sci Rev* 43(4):523–541.

Low, S. P., and Y. P. Chin. 2003. Integrating ISO 9001 and OHSAS 18001 for construction. *J Const Eng Manage* 129(3):338–347.

NNI. 1996. *Technical Report NPR 5001: Guide to a management system.* (International version, 1997). Delft, the Netherlands: NEN.

Norges Standardiserings-Forbund. 1996. *Management Principles for Enhancing Quality of Products and Services, Occupational Health and Safety, and the Environment.* Oslo, Norway: Norwegian Standards Association.

OHSAS 18001. 1999. *Occupational Health and Safety Management Systems—Specification. Occupational Health and Safety Assessment Series.* London: British Standards Institution.

OHSAS 18002. 2000. *Occupational Health and Safety Management Systems—Guidelines for the Implementation of OHSAS 18001.* London: British Standards Institution.

OSHA. 1982. *Voluntary Protection Programs.* Federal Register, 47, 29025. Washington, DC: Occupational Safety and Health Administration.

OSHA. 1989. *Safety and Health Program Management Guidelines; Issuance of Voluntary Guidelines.* Federal Register, 54, 3904–16. Washington, DC: Occupational Safety and Health Administration.

OSHA. 2000. *Revision to the Voluntary Protection Programs to Provide Safe and Healthful Working Conditions.* Federal Register, 65, 45649–63. Washington, DC: Occupational Safety and Health Administration.

Pawłowska, Z. 2004. Effectiveness of systematic management of occupational safety and health. In *Fundamentals of Systematic Management of Occupational Safety and Health*, ed. D. Podgórski and Z. Pawłowska, 137–144. Warsaw: CIOP-PIB.

Pawłowska, Z., M. Pęciłło, and G. Dudka. 2001. *Occupational Safety* 1:20–22.

PN-EN ISO 19011. 2003. Guidelines for auditing quality management and/or environmental management systems.

PN-N-18001. 1999. Occupational health and safety management systems—Requirements.

PN-N-18001. 2004. Occupational health and safety management systems—Requirements.

PN-N-18002. 2000. Occupational health and safety management systems—Guidelines for assessing occupational risk.

PN-N-18004. 2001. Occupational health and safety management systems—Guidelines.

PN-N-18011. 2006. Occupational health and safety management systems—Audit guidelines.

Podgórski, D. 1997. Occupational safety and health management systems in enterprises—Perspective of national and international standardization. *Standardisation* 4:4–11.

Polok, A., and M. Koźlik. 2006. Technology of behavioural accident prevention process of BST company. In *Proceedings of the 5th National Conference on Occupational Safety and Health Management in Enterprise*. Toruń, 2–3 October 2006, Warsaw: CIOP-PIB, 163–8.

Robson, L. S., J. A. Clarke, K. Cullen, et al. 2007. The effectiveness of occupational health and safety management systems interventions: A systematic review. *Saf Sci* 45(3):329–353.

SAA HB53. 1994. *A Management System for Occupational Health, Safety and Rehabilitation in the Construction Industry (Handbook).* Homebush, Australia: Standards Australia.

Saari, J. 1998. Participatory workplace improvement process. In *Encyclopaedia of Occupational Health and Safety*, 4th ed., ed. J. M. Stellman, 59.11–59.16. Geneva, Switzerland: International Labour Office.

SNBOSH. 1992. Ordinance (AFS 1992:6) internal control of the working environment. *Statute Book of the Swedish National Board of Occupational Safety and Health.* Solna, Sweden: Swedish National Board of Occupational Safety and Health, Publishing Services.

SNBOSH. 1997. Ordinance (AFS 1996:6) internal control of the working environment. *Statute Book of the Swedish National Board of Occupational Safety and Health.* Solna, Sweden: Swedish National Board of Occupational Safety and Health, Publishing Services.

Weinstein, M. B. 1997. *Total Quality Safety Management and Auditing.* Boca Raton, FL: CRC Press.

Wokutch, R. E., and C. V. VanSandt. 2000. OHS management in the Unites States and Japan: The DuPont and the Toyota models. In *Systematic Occupational Health and Safety Management—Perspectives on an International Development.* eds. K. Frick, P. L. Jensen, M. Quinlan and T. Wilthagen, 367–389. Oxford, UK: Elsevier Science.

Zwetsloot, G. 1994. *Joint Management of Working Conditions, Environment and Quality. In Search of Synergy and Organizational Learning.* Amsterdam: Dutch Institute for the Working Environment.

Zwetsloot, G. 2000. Developments and debates on OHSM system standardization and certification. In *Systematic* Occupational Health and Safety Management—Perspectives on an International Development, eds. K. Frick, P. L. Jensen, M. Quinlan and T. Wilthagen, 391–412. Oxford, UK: Elsevier Science.

30 Education in Occupational Safety and Ergonomics

Stefan M. Kwiatkowski and Krystyna Świder

CONTENTS

30.1 INTRODUCTION

All European Union (EU) member states consider education in occupational safety and health (OSH) and ergonomics an important prevention tool, and one that should be implemented as soon as possible. According to the EU, safety and health education and training are key factors that strengthen the culture of hazard prevention. Safety and health education should not begin when a worker enters the labour market, but rather should begin at the earliest stages of education, as part of the curriculum.

The regulation of Council Directive 89/391/EEC of 12 June 1989 on the introduction of measures to encourage improvements in the safety and health of workers states that 'The employer shall ensure that each worker receives adequate safety and health training, in particular in the form of information and instructions specific to his work stand or job'. Regulations in this field have been incorporated into the labour code within the framework of transposing EU directives into Polish law; according to these, OSH education should occur at various levels of school and adult education.

Much effort has been made in Poland in recent years to lay the foundations for OSH and ergonomics education. As a result, many national programmes, system solutions, modern educational materials, and teaching techniques have been created, making it possible to create an integrated system for shaping safe behaviours in both the work and living environments.

30.2 SPECIFIC EDUCATION IN OCCUPATIONAL SAFETY AND ERGONOMICS

Modern education is the sum of teaching and learning processes, and lasts for a lifetime. This applies to OSH and ergonomics education in a special way. The goals and content of this kind of education include everyone, from the very young to the very old, independent of the degree of their awareness of this field during education, work, or leisure.

Education in OSH and ergonomics has specific aims of teaching and training. Teaching takes place in the academic system, whereas training occurs outside that system. The goals of such education are generally to gain knowledge, skills and habits. The first two components (knowledge and skills) are universal, but the last one (habits) is especially characteristic of OSH and ergonomics. The knowledge and skills of specialists in a field become apparent only though their habits. Habits are not only shaped by information on a specific subject, but are also related to belief in the validity of the ideas that one is learning about (Kupisiewicz 2005). When discussing ideas and convictions, remember that these goals take years to shape, and thus, it is important to know the basic psychological characteristics of various phases in human development. The content should be relevant to the goals, which vary depending on age and occupational and life experience.

The academic system has many subjects and subject groups, and various aims are also understood outside the academic system. However, the specificity of aims leads to the specificity of contents. Because of the lack of a separate subject on OSH and ergonomics, these topics should be integrated with subjects covered by the curricula. The teacher or lecturer is responsible for this integration. The ability to teach OSH and ergonomics along with another subject depends on how well the teacher is prepared. The views and convictions of teachers and lecturers, as well as those of academic decision makers (directors of schools, heads of university-level schools, and so on) on OSH and ergonomics should be shaped early.

Cooperation between schools and parents at the earlier stages of education (primary and lower secondary school) is especially significant. Parents' views and convictions are responsible to a large extent for the prestige of the subjects and the 'borderline' contents (parents have an influence on the teaching and learning process). The variety of goals outside the academic system requires the development of content to match these goals. Modular curricula—flexible contents focused on specific aims of education or training—can help. This kind of curricula can also be used successfully for self-education.

In addition to the aims and contents, principles of teaching are also important for education in OSH and ergonomics. Teaching principles should consider the students' age, knowledge, skills and habits they have already acquired. The *principle of using visual methods* is very important. It stresses hands-on experience with elements of observation and doing practical things. The *principle of accessibility* also corresponds with that goal; it helps not only children and young people, but also adults without appropriate experience carry out their goals. This principle states that easier content precedes more difficult content and the known precedes the unknown.

The *principle of awareness and active participation in the process of education or training* is important when teaching students from upper secondary and university-level schools as well as workers. Awareness applies to the aims, and active participation helps to meet them. Students form important views and convictions when they solve occupational problems and face real tasks under real conditions. The *combining theory and practice principle*, which concerns the correlation between the lectures and training, applies to attempts to complete those tasks. The lifelong education or training process, additional training, and in-service training in OSH and ergonomics also require the *principle of being systematic*—that constant learning leads to durability of knowledge, skills, and habits. The *principle of balance between an individual and a team* is of fundamental importance for students of any age. It considers the individual qualities of every student and ensures the cooperation and collaboration of all participants in the education or training process.

Methods and techniques for education or training in OSH and ergonomics are less specific. However, there are conditions related to practical tasks in some elements of the teaching process. Problem-focused and practical methods should be highlighted, and methods focusing only on knowledge should be minimised. The effectiveness of problem-focused methods (e.g. situational, case study, drama, or decision-making games) and practical methods (e.g. exercises or workshops) depends on the appropriate selection of teaching aids, which include real machines, equipment, and tools as well as audiovisual materials such as those presenting work stands and technological process or simulating specific situations or workers' behaviours.

Education that starts with a formulation of its goals should end with an assessment of the achievement of those goals. Education in OSH and ergonomics is related more to skills and habit than simply to knowledge. Knowledge is passive; it can rarely be directly transformed into skills and habits, but the degree to which it has been mastered can be assessed easily. Skills and habits are more difficult to assess. Skills can be assessed by assigning students relevant practical tasks and habits can be assessed through observation of behaviours in various situations (Kwiatkowski 2001).

30.3 SCHOOL EDUCATION

The recently updated Polish laws on the education system include a new task for the school system: 'disseminating among children and young people education for safety, and shaping correct attitudes towards hazards and extraordinary situations'. The problems of OSH, safety culture, and ergonomics should now be included in the foundation curriculum, which defines a range of educational tasks for individual schools. Students should learn about the principles and methods of safe work during workshops and laboratories. This should cover the use of dangerous substances, preparations, machines, and technical equipment.

In primary schools, the curriculum covers the following:

- *Grades 1–3 (the first stage of education)*: Recognising danger signals; behaving safely near public roads; using common technical equipment safely; taking care of one's health and cleanliness; making sure the environment is clean; eating healthy; playing games safely; cultivating safe habits for handling dangerous tools such as knives, scissors, and matches.

- *Grades 4–6 (the second stage of education)*: Learning about civilisation hazards; learning skills for safe behaviour during contact with dangerous, toxic, flammable, and explosive objects; being safe on the road and while playing games; using first aid for some injuries; handling difficult situations; looking for help; avoiding behaviours hazardous to one's health; learning about the safety of machines and installations; following safe and responsible procedures in a technical environment, including using roads, safely using tools, and household equipment; reading and understanding technical instructions.
- *Lower secondary schools (the third stage of education)*: Civil defence; pro-health education; education in ecology. The school's task is to shape students' awareness of and responsibility for their own health and life and that of others and to shape their convictions to respect laws and regulations that are important to human life and health; shape safe behaviours at school and at home, on the road, when playing games and during study and leisure; correct behaviours when life, health, or property is at risk; instil the ability to recognise risk factors and hazards to the environment that result from manufacturing and transporting energy or wrong waste disposal and administer first aid in emergencies.
- *Upper secondary schools (the fourth stage of education)*: A range of tasks related to broadening knowledge and improving skills related to safety, depending on the type of school (general secondary schools, secondary schools that specialise in some subjects, vocational schools, secondary technical schools, and postsecondary schools). Foundation curricula for vocational subjects include OSH problems, whereas general secondary and secondary technical schools have included 'foundations of business'.

Polish vocational schools recognise 195 occupations, including OSH technicians. For each occupation, contents related to OSH regulations and ergonomics are included in the description of the qualifications of a graduate and in the foundation curricula for each occupation.

The European Agency for Safety and Health at Work project 'Mainstreaming Occupational Safety and Health into Education' has supported Polish schools in their task of shaping safe habits. The project's findings resulted in, among other things, supplementary materials for teachers teaching safety culture, developed at the Central Institute for Labour Protection–National Research Institute. It led to the development of educational materials for teaching safety culture at all levels of school education and for teachers from all types of schools that complement one another and consider the differences among different age groups. The modular structure permits flexible use of the materials, and the teacher is equipped with complete education tools and methodological guidelines—including those for the student—as well as computer presentations and films.

30.4 UNIVERSITY-LEVEL EDUCATION

Changes in Polish university-level schools are subject to the developments of private university-level schools, who have some independence in creating teaching standards and defining requirements related to education. Such independence means that OSH education is covered in a variety of ways, even within the same type of school. Education is usually covered by the following:

- Obligatory training for first-year students
- Obligatory lectures and classes within individual subjects, amongst others, OSH and ergonomics, assessment of hazards in the work environment, and legal labour protection
- An introduction of the elements of this knowledge into specialised subjects in individual courses

University-level school management, faculty and students must also take part in training courses on safety in the workplace and at school. Several dozen university-level schools in Poland run postgraduate OSH courses (Marcinkowski 2006).

30.5 ADULT EDUCATION

Because of the complexity of the modern world and the speed of changes in technology, the knowledge and skills acquired at school must be updated throughout the working life. The learning process has changed from a one-off limited by time and the curriculum cycle to a lifelong modification of occupational qualifications. The need for OSH education always comes up when introducing a new technology and a new work organisation. However, training courses are necessary even in stable situations. An increase in accidents at work and excessive turnover at work indicate a need for further training. The employers' expectations are the most significant element for identifying training requirements. These expectations are usually related to the operation of the enterprise as a whole and to the workers' needs at a given work stand.

Lifelong education in OSH is a significant part of adult education. By law, all employers, supervisors, OSH staff, engineers and technicians, and office workers in Poland participate in various kinds of OSH education at least every 5–6 years. The adult education market is highly diversified in the number of training centres, the forms of education and the level of the services. In Poland, training centres run by state institutions, community organisations and private owners provide OSH adult education—usually commissioned by employers—both for large companies, in which OSH education is just one concern, and small companies with only a few employees. Recent years have brought changes in this market. The number of consulting companies offering OSH services has significantly increased. Those services include education services, especially for small- and medium-sized enterprises. The level of education has improved, and more companies have up-to-date classrooms and professional staff with European-level competency certificates.

30.5.1 Education for Occupational Safety and Health Services

In 2004, new educational requirements for people seeking employment in OSH services were prepared in Poland to ensure professionalism. To be employed in OSH services, an OSH technician or inspector should have a university-level education in OSH or have taken a postgraduate course in OSH. According to the labour code, employers can also perform the tasks of OSH services if they have relevant training.

30.5.2 Organising and Running Courses

Training in OSH takes the form of preservice and periodic in-service training. *Preservice training* is carried out according to curricula developed for groups at similar work stands:

- New employees, university-level students during internship, and vocational school students employed to learn their occupation must undergo general training before they are allowed to work. General training ensures they learn about the basic OSH regulations in the labour code, collective work agreements or work rules, the principles of first aid, and the regulations and principles in OSH that are in force in the given enterprises.
- Workers must undergo training at the work stand before they are allowed to work, which ensures that they learn about harmful and strenuous factors, occupational risks, ways to protect themselves against hazards and ways to work safely at the work stand.

Periodic training aims to update and consolidate knowledge and skills and teach workers about new technical and organisational OSH solutions. The following people take part in periodic in-service training:

- Employers and supervisors, in particular managers and foremen
- Blue-collar workers
- Engineers and technicians, including designers, constructors of machines and other technical equipment, technologists, and organisers of the production process
- Staff of OSH services and other people responsible for those services
- Administrative or office workers and others who are exposed to harmful and strenuous factors of the work environment or who have OSH responsibilities

Periodic training for blue-collar workers is conducted at least once every 3 years; those whose safety or health are especially exposed to hazards are trained at least once a year. Periodic training of administrative and office workers should take place at least every 6 years.

OSH courses can be organised and run by employers or by upper secondary, vocational or university-level schools, research and R&D institutes, associations, and corporations, if commissioned by them. The organisers do not have to be licensed or accredited, except when training welders or forklift operators.

Increasing competition in the education market has caused some training centres to voluntarily implement a quality-control system according to standard ISO 9001:2000 *Quality Management Systems and Requirements.* Moreover, various institutions have begun to acknowledge the competence of educational bodies by certifying, accrediting or registering them. All establishments and centres that run lifelong educational programs outside the school system can undergo voluntary accreditation. An establishment can be accredited if it

- Provides classrooms equipped with teaching aids.
- Employs staff with qualifications and professional experience relevant to the training provided, whose work is regularly assessed and who have a constant opportunity to improve.
- Develops and makes available new teaching materials, if new techniques and technologies are included in its curricula and if the education evolves.

To improve the quality of training conducted by training bodies, the Centre for Certification of Personnel's Competence was set up in 2000 at the Central Institute for Labour Protection–National Research Institute. The centre, which is accredited by the Polish Centre for Accreditation, provides voluntary certification of the competence of people working in OSH services and of training centres. The criteria for certification applies to the conditions in which training is done, the competence of the lecturers and organisers of training courses and the level of education materials, teaching methods and effects. A certificate is awarded for 3 years and can be extended for another 3 years. The network of training centres with certificates from the Central Institute for Labour Protection–National Research Institute is supported in educating its staff and preparing new educational aids and in consultations and exchange of information. For training organisers, the institute issues a lecturer's certificate, which guarantees the highest standard of services.

REFERENCES

Kupisiewicz, Cz. 2005. *The Principles of Didactics*. Warsaw: WSiP.

Kwiatkowski, S. M. 2001. *Occupational Education. The Dilemmas of Theory and Practice.* Warsaw: IBE.

Marcinkowski, J. 2006. *A Diagnosis of Education in Occupational Safety and Health and Ergonomics at Polish University-Level Schools*. Warsaw: ROP.

Index

E

M

T

U

V

W

Z